The Complete Guide to Business Risk Management

To my partner Jayne, with love.

The Complete Guide to Business Risk Management

Third Edition

Third Edition

KIT SADGROVE

Routledge
Taylor & Francis Group

LONDON AND NEW YORK

First published in paperback 2024

First published 2015 by Gower Publishing

Published 2016 by Routledge
4 Park Square, Milton Park, Abingdon, Oxon OX14 4RN

and by Routledge
605 Third Avenue, New York, NY 10158

Routledge is an imprint of the Taylor & Francis Group, an informa business

Publisher's Note
The publisher has gone to great lengths to ensure the quality of this reprint but points out that some imperfections in the original copies may be apparent.

British Library Cataloguing in Publication Data
A catalogue record for this book is available from the British Library.

The Library of Congress has cataloged the printed edition as follows:
Sadgrove, Kit.
  The complete guide to business risk management / by Kit Sadgrove. -- Third Edition.
     pages cm
  Includes bibliographical references and index.
   ISBN 978-1-4724-4219-2 (hardback) -- ISBN 978-1-4724-4221-5 (ebook) -- ISBN (invalid) 978-1-4724-
4220-8 (epub) 1. Risk management. I. Title.
  HD61.S2 2014
  658.15'5--dc23

                                                                                    2014029452

ISBN: 978-1-4724-4219-2 (hbk)
ISBN: 978-1-03-283855-7 (pbk)
ISBN: 978-1-00-307507-3 (ebk)
ISBN: 978-1-315-61491-5 (eBook+)

Contents

List of Figures

List of Tables

Acknowledgements

I have drawn on information from many sources. They include The Association for Financial Professionals, Conference Board of Canada, FERMA, Amcor, the *Financial Times*, Association of Assurance and Risk Managers, IDW.de, Ifac.org, Electrical Safety Council, Hackett Group, Fair Labor Association, Achilles, Aravo, Transport Asset Protection Association, International Energy Agency, The Fire Protection Association, *CIO Journal*, The Loss Prevention Council, Control Risks Group, the Institute of Risk Management, *Science Insider* magazine, MIT Center for Transportation Studies, Association of Certified Fraud Examiners, RiskWorld, RiskInfo, *Risk Management* magazine, Business Continuity Online, AIRMIC, NHS, Institute of Internal Auditors, HM Treasury, ALARM, Procurement Strategy Council, *Purchasing* magazine, *Guardian* newspaper, *Observer* newspaper, McKinsey, Kenexa, *Wall Street Journal*, ACAS, CCH, Unison, BP Law, Office of Fair Trading, Data Monitor, Arlington Institute, Carnegie Mellon University Robotics Institute, International Chamber of Commerce, IFPI, Just Food, Sunday Times, StoraEnso, Sustainability, Cow & Gate, UK Foreign Office, DEFRA, UK Home Office, Office of Government Commerce (Efficiency and Reform Group), Green Party, Business Report South Africa, Talk Left, Reebok, Rits, Dejan Kosutic, Bankruptcy Action, Dun and Bradstreet, CIMA, Foresight Institute, Mongabay.com, Ars Technica, Imperva, Internet Identity, *Journal of Global Information Management*, Information Commissioner's Office, Staff Monitoring Solutions, *The Scotsman*, *Sunday Business Post*, Amnesty International, Institute for the Future, Human Capital Management Institute, Proudfoot Consulting, Burnt Oak Partners, Foresight Technology, Whitakers Almanac, Ash, Wikipedia, Arson Prevention Bureau, Coso, *Inc* magazine, *Real Business* magazine, *Entrepreneur* magazine, Institute for Crisis Management, ICAEW, Law Society, *The Times*, Standish, Credit Services Association, AT Kearney, Project Management Institute, Financial Executives International, *HR* magazine, *The Economist*, Clearly Business, Standards Australia, AON, *Computer Weekly*, Seth Godin, *Chicago Tribune*, *Computer World*, Ziff Davis, Eusprig, Lacie, *Computer Business Review*, Project on Government Oversight, Worker Rights Association, Commodity Futures and Trading Commission, Kable, Logitech, Breaking News, Financial Services/Conduct Authority, European Union, Council of European Municipalities and Regions, International Organization of Securities Commissions, Scottish Environment Protection Agency, Rolls Royce Motors, Agency for Toxic Substances and Disease Registry, Fairtrade Foundation, USDA, Natural Resources Defense Council, Said Business School, Hull University and the BBC.

Statistics and survey data was also provided by The National Fraud Authority, International Labor Organization, Industry Market Trends, Ernst and Young, The Audit Commission, FERMA, Cass Business School, Institute of Crisis Management, FM Global, Price Waterhouse Coopers, Business Software Alliance, IBM Business Recovery Services, Gartner, Packaged Facts, Millward Brown, PA Consulting, Carlsberg Group, Infoplan, Symantec, Health and Safety Executive, Pew Research Center, Chubb Fire, Transparency International, KPMG, International Monetary Fund, Sedgwick, AT Kearney, XpertHR, Aveco, *PC Magazine*, Powerchex, Society for Human Resource Management, Department for Transport, Kroll Ontrack, DataFort, Axa, UK Intellectual Property Office, US Bureau of Labor Statistics, Marsh Inc., International Coffee Organization, British Retail Consortium and the Royal Academy of Engineering.

I am grateful to people who have commented on selected chapters including Jan Cottrell, Erik Engstrand of Ikea, Ciaran Delaney, Harvey Betan and Betty Kildow. I'm also grateful to my researchers Magdalena Kurkowska, Stephen Murage and Tom Gibbs, as well as all the individuals and organizations who have helped to create this book; but any errors are my responsibility alone.

Kit Sadgrove

Glossary and Abbreviations

Some of the vocabulary of risk management differs from other specialisms. For example, the word 'control' is different from what scientists mean by it. Even the varying risk standards don't always agree on the definition of the words.

These definitions have been taken from various sources, including COSO and the Institute of Internal Auditors, and have in some cases been modified to provide more clarity. For more detail, the ISO/IEC Guide 73-2009 provides a standardized vocabulary.

ABAC	Anti-bribery and corruption
ACA	Associate Chartered Accountant
ACM	Asbestos containing materials
AFD	Automatic fire detection
AIRMIC	Association of Insurance and Risk Managers
ALARP	As low as reasonably possible. Managing a risk without imposing undue costs or limitations on the organization
Appetite	See 'Risk appetite'
Assessment	See 'Risk assessment'
ATSDR	Agency for Toxic Substances and Drug Registry
Audit	In risk management, an inspection of an organization's procedures, to identify whether they are complying with the organization's policy
AUP	Acceptable use policy
B2B	Business to business
Basel	Basel II and III are global, voluntary standards for banking regulators, designed to ensure that banks have enough capital to avoid insolvency
BCM	Business continuity management
BCMS	Business Continuity Management System
BHO	Browser helper objects
BIS	The Department for Business, Innovation & Skills
BOD	Biochemical oxygen demand
BPO	Business process outsourcing
BPR	Business process re-engineering
BRIC	Brazil, Russia, India and China
Business Continuity	A strategy for managing major business risks so that the organization can continue to operate after a crisis
Business Impact Analysis (BIA)	An analysis of how much disruption would be caused by a hazard on the organization's output
Business risk	The organization's operational risks. This contrasts with credit risk.
BYOD	Bring your own device. The risks and problems caused to an organization by users own smartphones, tablets and laptops
BYOS	Bring your own software

CAT	Cable avoidance technology
CCO	Chief Compliance Officer
CCP	Critical control point
CDO	Collateralized debt obligation
CEO	Chief Executive Officer
CFO	Chief Financial Officer
CFTC	Commodity Futures and Trading Commission
Chief Risk Officer	A senior manager responsible for enterprise risk management
CIA	Certified Internal Auditor
CIO	Chief Information Officer
CIP	Certified/Chartered Insurance Professional
CIPD	The Chartered Institute of Personnel and Development
CMO	Chief Marketing Officer
CMS	Compliance Management System
Compliance	Conforming to laws and regulations, a standard or a management system
Conformance, conformity	Complying with legislation; meeting specified standards
Consequence	Outcome on an event
Control	A process or practice that minimizes or prevents negative consequences
COO	Chief Operating Officer
Corporate Governance Code	A code of practice issued by the Financial Reporting Council for UK publicly quoted companies. It sets out standards of good practice for Boards
Corrective action	A measure taken to eliminate a non-conformity in order to prevent recurrence
COSO	Committee of Sponsoring Organizations of the Treadway Commission. An influential US private-sector organization that produced a framework for risk management
CPA	Certified Public Accountant
CPM	Critical path method
Credit risk	For a financial services organization, the risk that a borrower or bond issuer may default and be unable to pay their debt or the interest
CRM	Crew resource management
CRO	Chief Risk Officer
CRSA	Control and Risk Self Assessment
CSA	Control self assessment
CSR	Corporate social responsibility. Relates to a company's economic, social and environmental impacts
CTO	Chief Technology Officer
D&O	Directors and officers; relates to liability and insurance
DDOS	Distributed denial of service attack. Malicious attempt to render a server unable to cope by flooding it with requests
DEFRA	Department of Environment, Food and Rural Affairs
DfT	Department for Transport
Documented	Information that must be controlled and maintained to avoid waste and errors (for example price lists or engineering plans)

EA	Enterprise agreement
EAS	Electronic article surveillance
EBITDA	Earnings before Interest, taxes, depreciation and amortization. A way of measuring profitability
ECGD	Export Credits Guarantee Department
EDI	Electronic data interchange. A standard that lets different organizations exchange data
EIA	Environmental impact assessment
ELD	European Liability Directive
EMEA	Europe, the Middle East and Africa
EMS	Environmental management system
EPC	Energy Performance Certificate
EPOS	Electronic point of sale
ERM	Enterprise risk management
ERP	Enterprise resources planning. Software that integrates the data from the different departments of an organization
ESA	Enterprise Software Advisors
ETF	Exchange traded fund
ETTO	Efficiency thoroughness trade-off
EU	European Union
Event	A set of circumstances leading to a consequence
External context	Anything outside the organization that contributes to its risks. This includes competitors, the political and social environment, and technology
FCC	Federal Communications Commission
FCPA	Foreign Corrupt Practices Act
FEI	Financial Executive International
FERMA	Federation of Risk Management Association
FMCG	Fast moving consumer goods
FMEA	Failure mode effect analysis. An assessment that examines a process to check for risks and possible failure
FSA	Financial Services Authority
FSR	Freight Security Requirements
GDP	Gross domestic product
GIGO	Garbage in, garbage out
GMO	Genetically modified organism
GRC	Governance, risk and compliance
GRI	Global reporting initiative
HACCP	Hazard analysis and critical control points. A systematic approach to managing the safety of food and other products mainly in production plants
Harm	Loss or damage to people, assets, the environment or the organization
Hazard	Source of potential harm
HAZOP	Hazard and operability study. A systematic examination of a process in chemical and nuclear plants to check for risks

HRO	Highly reliable organization
HSE	Health and Safety Executive
HVTT	Heavy vehicle transport technology
ICM	Institute of Crisis Management
ICT	Information and communications technology. Technology used in computers and electronic communications
IEA	International Energy Agency
IIA	Institute of Internal Auditors
IM	Instant messenger
Impact	Result or effect of an event. This could be positive or negative.
Inherent Risk	The level of risk before any action is taken to reduce it. See also Residual risk
Interested party	Any individual or group that might be affected by the organization and its processes. Similar to Stakeholder
Internal audit	An audit controlled by the organization itself to assess whether processes are conforming to requirements
Internal context	The people, processes, systems and culture within the organization that contribute to its risks. The converse of External Environment
Internal control	A process designed to assure managers that policies are being met, such as compliance with laws or regulations
IOSCO	International Organization for Security Commissions
IP	Intellectual property
IRM	Institute of Risk Management
ISMS	Information Security Management System
ISO	The International Organization for Standardization, a body that creates global standards
ISO 14001	The environmental management standard, issued by ISO
ISO 26000	The corporate social responsibility standard issued by ISO
ISO 31000	The risk management standard issued by ISO
ISO 45001	The health and safety standard issued by ISO. Successor to BS OHSAS 18001
ISO 9001	The quality management standard, issued by ISO
ISO/IEC 27001	The information security management standard issued by ISO
ISO22301	The business continuity standard issued by ISO
ISP	Internet service provider
IT	Information technology
JIT	Just in time
JV	Joint venture
KISS	Keep it simple, stupid
KSF	Key success factors
Likelihood	The possibility that an event will occur
M&A	Mergers and acquisitions
Management system	A set of elements (policy, procedures, audits and review), that an organization uses to manage its processes and meet its goals

Market risk	For financial companies, the risk that price of stocks or other investments will fall, usually caused by outside forces such as recession or inflation. Also known as Systemic Risk
Maximum Acceptable Outage (MAO)	The maximum period of time before an impact becomes unacceptable. Usually measured in hours or days
MCP	Management control plan/programme
MIS	Management information system
Mitigation	Actions to reduce the probability or severity of a risk
Moncza and MSU model	A business model used to identify and manage suppliers
MRO	Maintenance, repair and operations
MTAS	Medical Training Application Service
MTPD	Maximum tolerable period of disruption
MVP	minimum viable project
NATS	National Air Traffic Service
NBIC	Nanoscience, biotechnology, information technology and cognitive science
NDA	Non-disclosure agreement
NEST	National Employment Savings Trust
NFC	Near field communication
NGO	Non-government organization
Non conformity, non conformance, non compliance	Failure to comply with requirements. Used by auditors to refer to processes than do not meet the standard. Can also refer to products or services that are outside acceptable tolerances
OECD	Organization for Economic Co-operation and Development
OFR	Operating and Financial Review
Opportunity Risk	The risk that a better opportunity may arrive after an irreversible decision has been made.
PBS	Public Broadcast Service
PERT	Program (or Project) evaluation and review technique
PESTLE	Political, economic, social, technological, legal, environmental and ethical. Headings for a review of external risk factors
PM	particulate matter
PMBoK Guide	Project management body of knowledge guide. Divides a project into its five stages: initiating, planning, executing, monitoring and controlling, and closing
PMI	Primary mortgage insurance
Poka yoke	Japanese for 'mistake proofing. Strategies adopted by organizations to prevent or warn of error
PPI	Payment protection insurance
PR	Public Relations
PRINCE2	Projects in controlled environments, version 2. A project management methodology
Probability	The likelihood that an event will occur. A number between 1 and 0, which represents how likely some event is to occur. A probability of 0 means the event will never occur, while a probability of 1 means that the event will always take place

Procedure	Written instructions, in a series of steps, for carrying out a process.
Process	An activity that is repeatedly carried out within the enterprise.
PRS	Performing Rights Society
PWC	PricewaterhouseCoopers
R&D	Research and Development
Recovery Time Objective (RTO)	The maximum amount of time allowed to for an activity to resume after a disruption
Repetitive Strain Injury (RSI)	Pain from muscles, nerves and tendons caused by repetitive movement and overuse, especially in the arm, hand or wrist
Residual risk	Risk remaining after risk treatment. See also inherent risk
Resilience	The ability to recover
Revenue drivers	Activities that are essential to an organization's income, such as the IT system, supply chain or corporate reputation
RFID	Radio frequency ID tags
RIDDOR	UK Reporting of Injuries, Diseases and Dangerous Occurrences Regulations. Requires organizations to report certain kinds of injuries
Risk	1. The impact of a currently unknown event on the business 2. A potential problem 3. The chance of something happening, usually with negative consequences 4. The opportunities, uncertainties and threats faced by the organization
Risk analysis	Systematic process to understand the nature and scale of risk
Risk appetite	How much risk the organization is willing to bear
Risk assessment	1. The process of identifying, analysing and evaluating risks 2. Defining the likely outcome of uncertain events
Risk Criteria	Measures or standards set to decide whether a risk is acceptable
Risk Management Plan	A document that sets out the organization's risk management
Risk Management Policy	An organization's written statement that sets out its approach to risk management
Risk management	The process of identifying, controlling, and minimizing the impact of uncertain events
Risk owner	The person responsible for managing a risk
Risk register	A list of the organization's risks
Risk reporting	Publishing information to stakeholders about the organization's risks, often in an annual report
Risk retention	Accepting the burden of loss, or benefit of gain, from a specific risk
Risk sharing	Sharing the burden of risk with another party, such as an insurance company or joint venture partner. Also known as 'transferring risk'
Risk tolerance	The amount of risk an organization is prepared to accept
Risk treatment	Measures that modify risk
RMP	Relationship management plan
RMS	Risk management system
ROHS	Restriction of Hazardous Substances Directive
RRP	Registered risk practitioner
RSI	Repetitive strain injury

RTO	Recovery time objective
SaaS	Software as a service. Software that is hosted and owned by a third party on the web, rather than in the business
Sarbanes-Oxley (SOX)	US Act that places more controls on annual reports and directors' behaviour
SBT	Scan-based trading
SEC	Securities and Exchange Commission
SEPA	Scottish Environment Protection Agency
SFAIRP	So far as is reasonably practicable
SHE	Safety, health and environment
Six sigma	A movement designed to improve business processes. Six sigma relies on statistical techniques to measure quality
SME	Small and medium-sized enterprises
Solvency II	A set of regulatory requirements for insurance firms in the European Union. It is mainly concerned with setting a minimum amount of capital held by each firm, to reduce the risk of insolvency
SOP	Standard operating procedures
SRM	Supplier relationship management
Stakeholders	People interested in or affected by the organization's risks
Supply Chain Management (SCM)	Managing the flow of goods from initial suppliers to the end user.
SWOT	Strengths, weaknesses, opportunities and threats. Headings for assessing risk
TAPA	Transport Asset Protection Association
TBL	Triple bottom line. A method of accounting that includes social, environmental and financial costs.
Threat	A risk. An indication of impending danger or harm
TI	Transparency International
TOM	Target operating model
TPM	Total productive management
TQM	Total quality management
Treat	Methods of managing a risk. These are Accept, Avoid, Share or Control. See also Risk Treatment
TSR	Trucking security requirements
UAV	Unmanned aerial vehicle
UNESCO	United Nations Educational, Scientific and Cultural Organization
UPS	Uninterruptable power supply
VaR	Value at risk
Verification	Establishing that specified requirements have been met
VOCs	Volatile organic compounds
VOIP	Voice over internet protocol
VPN	Virtual private network
VSM	Virtual stakeholder meeting
Vulnerability	The degree to which an organization is susceptible to threats. A weakness that could lead to a problem

WBS   Work breakdown structure

ZHC   Zero hours contracts

Preface

If you want to be in business in ten years time you have to ask three questions:

1. What are the worst things that could happen to us?
2. How likely are they to happen?
3. Are we taking the right steps to prevent them?

This is the art of risk management, and the subject of this book.

The Complete Guide to Business Risk Management helps you identify risks and prevent them from taking place. It then shows what steps you should take in the event of a crisis.

Business is inherently risky. At any moment, a competitor may launch a superior product, a fire may engulf your premises, or a government decision may tie the company up in red tape.

We have to live with those uncertainties, because without them, the business can't serve its customers or make profits.

There is no point in trying to avoid risk completely. 'Hide from risk and you hide from its rewards' said a sign in the New York dealing room of Bankers Trust.

Inaction merely creates new risks. If you don't innovate you stagnate, and that leads to failure. So this book doesn't warn you off risk taking. Rather, it helps you understand your risks and manage them professionally.

It emphasizes the really big risks – being caught napping by a change of technology, a change in the business cycle, or being left behind by competitors' activity.

We start by asking the questions, 'What is risk?' 'How do we assess it?' and 'How can it be managed?' Each subsequent chapter covers a major area of risk, such as finance or the environment.

The book shows how to audit the main areas of risk. It outlines the questions that need to be asked, and then helps you develop a risk management plan for each area.

Changes to the Third Edition

You might think that risks remain fairly constant over time. But the past few years have brought many changes, and they are reflected in this new edition.

Some earlier concerns have faded. They include repetitive strain injury, pollution and the product liability directive, matters that greatly exercised managers in past years. Even a relatively recent issue as diversity is fading from view as a major risk.

Corporate interest in governance and regulation has also declined, but the scale of regulation and growing penalties have led to its inclusion as a new chapter. Ethics now have their own chapter, too, bearing testimony to the continued problems of cartels, bribery and corruption, the latter being a special problem for businesses that operate in developing countries.

As the evidence of climate change becomes clearer, I have added a discussion on the hazards of adverse weather. Having distant suppliers is a related issue, with some businesses repatriating their offshored departments. The exhaustion of natural resources and competition for raw materials are two other contemporary risks we look at.

While terrorism doesn't affect many businesses directly, home-grown extremism and violent pressure groups are always a possibility, especially for organizations in the public eye or in controversial markets. The threat of business interruption, for whatever reason, is something that businesses are taking more seriously, and this has been given extra space in the book.

Technology continues to create surprises for any business that hasn't scanned the horizon. Customers' behaviour continues to be affected by smartphones, tablets and the internet. Managers using their own devices at work and reputation damage caused by social media sites are just two examples of technology risks.

With the need to manage risks in a systemic way, I have also included an assessment of different brands of risk software.

The book's emphasis was originally more towards the manufacturing industry. While this is still important in the West, more people are now employed in service businesses. Hence this edition includes the risks relating to staff behaviour, the need for creativity and the management of the workforce as a whole rather than just top management.

Finally, international standards have grown in number and maturity, not least with ISO 31000, the standard for risk management, which has been included in the new edition.

1
A Powerful Tool for Protecting the Business

It's 1,700BC, and you're a merchant in the busy city of Babylon. You have several rolls of cloth you want to ship 200 miles (320km) down river to the distant town of Ur where there's a demand for your merchandise.

But what happens if the boat sinks, or is set upon by pirates? Until recently you suffered the loss. But now you can manage your risk. Thanks to the new code of Hammurabi, King of Babylon, there is a new service called 'insurance'. You can borrow money to buy your cargo. And the lender will cancel your loan if your ship is lost at sea. The Code of Hammurabi is the earliest known form of risk management.

History of Risk Management

Cargo insurance was introduced nearly 4,000 years ago. And until recently insurance was still the main way that companies managed risk. In turn, insurance companies sought to reduce their potential losses by encouraging businesses to make their premises safer.

This was the *first age of risk management*, as shown in Figure 1.1. Businesses considered only hazard risk (such as fire or IT failure). They also used risk reactively, to see how much insurance they should buy.

In the 1970s and 1980s, businesses started to introduce quality assurance, to ensure that products conformed to their specifications. This was epitomized by ISO 9000, successor to the British Standards Institution BS 5750, the quality standard, itself the successor to US military standard MIL-Q-9858 which had been launched in 1959.

In this, the *second age of risk management*, companies treated risk in a more proactive or preventative way.

Risk awareness was fostered by government legislation that aimed to make businesses think about the risks they posed to workers and customers. New concerns also emerged in the 1980s about environmental risks. And risks to shareholders caused by bad governance became an issue in the 1990s. In 1993, James Lam became the world's first chief risk officer (CRO), at the US financial services firm GE Capital.

Finally, the *third age of risk management* arrived in 1995 with the publishing by Standards Australia of the world's first risk management standard, AS/NZS 4360.

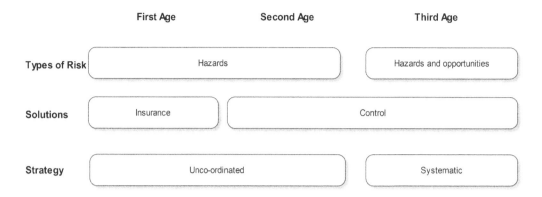

Figure 1.1   The three ages of risk management

RISK MANAGEMENT STANDARDS

After Australia's risk management standard, two others followed in quick succession. In 2001 Japan launched a risk management system (RMS) called JSI Q 2001, which introduced continuous improvement. And in 2002 the UK's Institute of Risk Management (IRM) introduced its own risk management standard.

Finally the International Organization for Standardization (ISO) launched its ISO 31000 in 2009, based largely on the Australian standard.

Meanwhile, the New York Twin Towers disaster had taken place in 2001, and companies came to think more about business continuity. In 2003 the British Standards Institution launched PAS 56, a specification for business continuity, which ultimately emerged as ISO 22301, the international continuity standard.

PUBLIC COMPANIES AND FINANCIAL REPORTING

From the 1990s onwards, successive failures in public companies led to demands for greater accountability and more visibility of the companies' risks.

The scandals at Polly Peck, BCCI and Robert Maxwell led to the UK Government's Cadbury Committee report in 1992. It recommended measures for better governance, such as the separation of the roles of Chairperson and CEO.

Following public concern about directors' rising pay, the Greenbury Report advocated controls on boardroom pay through the creation of remuneration committees.

In 1998 the UK's Department for Trade and Industry launched a review of company law aimed at developing a more modern framework for doing business in twenty-first century. A year later the Institute of Chartered Accountants in England and Wales published the Turnbull Report. This called for stronger internal financial controls and better monitoring of risk.

The European Union (EU) was equally concerned. In 1999 it decided to harmonize accounts across Europe, so that investors in one country could understand and trust annual reports from a company based in another country. The EU Accounts Modernization Directive required, among other things, a report on 'environmental and employee matters'. From then on, company reports were to be broader in scope.

But the scandals continued to erupt. In 2001 the $101bn energy business Enron was found to have committed massive accounting fraud. Its auditor, Arthur Andersen, was found guilty of criminal charges and collapsed. The scandal led to the US's Sarbanes–Oxley Act of 2002 which demanded more risk management and better annual reporting. To meet this requirement, the Committee of Sponsoring Organizations of the Treadway Commission (COSO) organization (a respected US-based private sector grouping that sets out best practice in enterprise risk management) launched a document called 'Enterprise Risk Management – Integrated Framework'. It outlined how public companies should implement risk management and report on it.

In Europe, meanwhile, Italian shareholders discovered that nearly €4 billion of funds purportedly owned by dairy company Parmalat, and supposedly held in a Bank of America account, did not actually exist.

Partly in response to the EU Accounts Modernization Directive, the UK published its Operating and Financial Review (OFR) in 2004. This required companies to publish information in their annual report about their principal risks, as well as non-financial information about environmental and employee matters.

In 2006 the EU passed the 8th Directive which formally embedded risk management into public companies and 'public interest entities' such as banks. These businesses were to have an audit committee to whom the external auditors would report findings about weaknesses in internal controls. This directive ensured that what was good practice in many countries was applied all across Europe.

Today all organizations have to comply with a raft of legislation and are watched by regulators. There is no going back to the buccaneering days when companies could do as they liked. These measures ensure that large companies are better managed and that they have systems for identifying risks. This means that organizations are less likely to fail. However critics point out that such controls failed to prevent Western banks from precipitating the 2008 global recession.

THE EMERGENCE OF ENTERPRISE RISK MANAGEMENT (ERM)

The phrase Enterprise Risk Management (ERM) has come to the fore. It means managing risk systematically. A RMS ensures that the company manages its threats in a proactive, co-ordinated, cost-effective and prioritized way.

There is a certain inevitability about all this. ERM sits neatly alongside company-wide audits and enterprise resource planning (ERP) software that links all departments.

Nevertheless the power of risk management is limited. Every time a company scandal erupts, it becomes apparent that risk management is only as good as the integrity and commitment of the players. If regulators turn a blind eye or are captured by the industries they are supposed to manage, if rogue traders manage to hide big losses, or a powerful CEO browbeats the Board, all the risk systems will be of no avail.

Getting Corporate Strategy Right

As we will see in Chapter 20, the organization has to be in the right markets and have the right products. This is the most basic of all risks: finding that customers no longer want the service that the company offers.

In 1999, at the height of the dot com boom, Marconi (formerly GEC) got rid of its dull retail, defence and food businesses, and bought exciting telecomms companies instead. Two years later, the telecomms market collapsed, and Marconi's share price fell 54 per cent in one week. In the six months to September 2001, it lost £5.1bn. But it's easy to be wise after the event.

Conglomerates were once seen as a way of reducing risk. If one market was doing badly, another would be performing well. Many companies diversified, only to find that they owned too many loss-making businesses that they were unable to turn around.

Since then, companies have tended to return to their roots. However, some conglomerates do well. In the past these have included GE, Virgin and Mitsubishi.

Diversified companies can use their core skills in marketing, management, strategy and raising capital to direct a range of businesses. However they tend to be short lived and depend on one individual's management or entrepreneurial skills.

What are Business Risks?

As we have seen, there are two types of business risk. The first and more traditional type is *hazard* risk. It is found in fire, pollution or fraud. Companies used to protect themselves by buying insurance but, as we shall see, insurance is only one way to protect the company: there are many others.

The second type is entrepreneurial or *opportunity* risk. This happens when a company builds a new plant, launches a new product or buys a company. If the company gets its forecasts wrong, it loses money. There are ways of reducing entrepreneurial risk, as we shall see. In this book we don't seek to eliminate risk. It's a necessary part of the enterprise. It's a precondition for innovation; and without innovation the business will fail. An organization that tries to obliterate all possible dangers can't create value.

Risk applies to any management decision that could have a good or bad outcome. It follows that most management decisions and projects contain risk. Most risks are not catastrophic, but as Table 1.1 opposite shows, the major ones cause loss of life and great damage. Better risk management could have forestalled some of them.

In other cases, organizations have been overwhelmed by the forces of nature, whether tornados, earthquakes or war. At that point, the business needs a continuity plan, something we examine in Chapter 22.

Risk is also a future event that results from actions taken now. That is why managers should consider different options for any problem, and evaluate the consequences.

It is easy to focus on obvious risks, such as workplace accidents. Important though they are, the company must be alert to the big or unexpected risks. The company that is not expecting change is especially prone to suffer.

Risks often defy conventional thinking. For example, what is the most likely cause of death for a New York police officer? It is not being killed by a drug dealer's bullet, but his own poor driving ability. In one year alone, 1,230 officers of the NYPD (known as Not Yet Proficient Drivers) were hurt in car crashes, compared with 20 who were wounded in shootings. The force now requires its officers to wear seat belts and take driving lessons, especially as the accidents cost the NYPD $3 million in sick pay.

What Kind of Organizations are Affected?

All organizations face risk. Even those whose future is assured, such as government bodies, can be harmed by fraud, staff negligence or failure to adapt.

Small companies are often more vulnerable, having fewer resources and therefore less resilience. A disaster at a single-site company could leave the business with no production facilities. On the other hand, scale also brings its own problems. A multinational business has more complex financial arrangements and more processes, making it impossible for any individual to foresee every risk.

Table 1.1   Some modern day catastrophes

Date	Event	Cause	Outcome
1980	The North Sea oil platform Alexander Keilland collapses	One leg of the rig snapped in heavy seas	123 crew members died
1983	South Korean airplane shot down over Kamchatka peninsula	Soviet Union fighter pilots shoot the plane down	269 passengers die, including 61 Americans
1984	Toxic gas released at the Union Carbide plant in Bhopal, India		3,000–10,000 people die
1985	Fire sweeps Bradford football stadium	Match or cigarette dropped on to rubbish underneath the stadium	56 people die
1986	*Challenger* space shuttle exploded on take-off	Malfunctioning rocket seal	The three crew members die. US space programme jeopardized
1986	Nuclear reactor at Chernobyl explodes	Power surge causes the nuclear rods to disintegrate, causing overheating and explosion	250,000 people may have died, plus damage to agriculture and the environment
1986	Series of explosions on the Piper Alpha oil rig	Excessive flare from gas safety release led to fire	167 crew members die
1987	*Herald of Free Enterprise* ferry sinks in Zeebrugge harbour	Bow doors left open	193 people die
1987	*Doña Paz* passenger ferry sinks after colliding with the *MT Vector*, leading to the deadliest peacetime maritime disaster in history	Collision at sea	4,375 deaths (only 26 survivors)
1988	Pan Am flight explodes over Lockerbie	Terrorist bomb	259 passengers die
1988	Iranian airplane shot down during Iran–Iraq conflict	US destroyer *Vincennes* mistook the airliner for a fighter plane	299 passengers die

Table 1.1   Some modern day catastrophes (continued)

Date	Event	Cause	Outcome
1989	Hillsborough stadium disaster, Sheffield	Overcrowding. Fans pushed to get into the stadium, where the match had started. This was made worse by police mistakes	95 people die
1990	*Exxon Valdez* shipwrecked off the Alaskan coast	Navigation error	Two million gallons of oil covered the coast. 2,000 sea birds and 300 otters die
1991	Iraqis set fire to Kuwait's oil wells	Economic sabotage at the end of the Gulf war	Environmental catastrophe
1994	Estonian ferry sinks in the Baltic	Faulty bow doors	900 people die
1995	Kobe earthquake	Earth movement	5,500 people die, 300,000 made homeless, 180,000 buildings destroyed or badly damaged
2001	Foot and mouth disease on UK farms	Viral disease	£8 billion losses
2001	Two aircraft fly into the twin towers, New York	Terrorism	3,000 die, $20 billion losses
2001	AZF fertilizer factory explosion, Toulouse, France	Improper handling of ammonium nitrate	29 deaths, 2,500 seriously wounded, 8,000 light casualties
2003	Nissan recalls 2.55 million cars	Faulty engine sensors may prevent the vehicles from starting	Costs of $140 million
2003	Space Shuttle *Columbia*	Pieces broke off the craft on take-off	All crew die
2003	Power blackouts in Eastern USA and Canada	Power fluctuation and overload	80 million people affected, losses of $6 billion
2004	Madrid train bombing	Terrorism	191 people killed
2005	Buncefield explosion	An oil tank overflowed at night, and then ignited possibly due to a spark from an electrical generator	Total UK fined £3.6 million, plus £2.6 million costs. £700 million civil claims

Table 1.1   Some modern day catastrophes (concluded)

Date	Event	Cause	Outcome
2008	Global financial near meltdown	Irrational exuberance among banks; lack of regulation	Recession cost the USA alone $12 trillion
2009	Toyota recalls 5.2 million vehicles due to pedals being trapped in the floor mat, and 2.3 million more vehicles recalled due to a sticking accelerator pedal	The floor mat problem partly related to manufacturing or design faults. The accelerator pedal problem is thought to be due to driver error. Driver fraud may also have been involved	Toyota paid $1,000 million in legal payouts. $2,000 million in lost sales
2010	Deepwater Horizon oil spill	Defective cement, cost-cutting, and inadequate safety systems	11 deaths, $60– $100 billion losses
2011	Explosion at Fukushima nuclear power plant, Japan	An earthquake causes a tsunami which floods the emergency generators and causes them to fail, leading to overheating	480,000 people evaluated. Unknown other effects
2011	Earthquake, Christchurch, New Zealand	Earthquake	$13 billion losses, 185 deaths
2011	Civil war in Syria	War	More than 21,000 deaths
2012	*Costa Concordia* cruise ship sinks on the coast of Italy	Inattention by the captain	32 deaths
2013	Clothing factory collapses, Dhaka, Bangladesh	Faulty construction	Over 1,000 people die. The worst industrial accident since Bhopal in 1984
2014	Malaysian Airlines Boeing 777 lost over the South China Sea	Cause unknown	All passengers and crew presumed dead
	A second Malaysia Airlines jet was later shot down over Ukraine	Civil war	All passengers and crew dead
	Ebola outbreak hits West African countries	Virus carried by bush meat, and transmitted through human breath, sweat or faeces	20 000 deaths forecast. Losses in Guinea, Liberia and Sierra Leone could total $1.17 billion

To overcome that, we take a peek into all the major departments of an organization. We analyse the typical risks in each; and checklists at the end of Chapters 5–24 help you identify the most important risks for your own business.

What the Words Mean

As we've said, risk management used to be about managing hazards. But in recent years it has been widened to include positive risks, such as acquiring a business or launching a new product. Those activities could have a positive or negative outcome.

When Chevron drills for oil in turbulent Kazakhstan, the oil company is undertaking mighty risks. But it would be wrong for the oil business to abandon exploration simply on the grounds that it poses a risk. Instead it should understand the risks of extracting that oil (whether political turmoil or environmental spills). Then it can try to control, share or reduce the risk. If all else fails it can avoid the risk by giving up and going home.

Some aspects of risk don't really have a positive connotation. Fire officers are unlikely to view fire as anything other than causing harm. Similarly, health and safety consultants won't see any advantage in broken bones. So while the newer definitions will please marketing and finance people, they aren't attractive to those who deal with harmful risk.

There are many definitions of the words used in risk management, many of which are hard to grasp. Here are some definitions that seek to combine clarity with brevity (Table 1.2).

Table 1.2   Old and new definitions

Phrase	Old definition	New definition
Risk	The possibility that a hazard will cause loss or damage	1. The impact of a currently unknown event on the business 2. A potential problem
Risk assessment	Defining what can go wrong	Defining the likely outcome of uncertain events
Risk management	A discipline for controlling losses	A discipline for dealing with the organization's risks

The risk management standard ISO 31000 defines risk as 'the effect of uncertainty on objectives'. In simple English that mean risk the impact (the effect) of a currently unknown event (the uncertainty) on the business (its objectives).

Meanwhile some definitions are unchanged from the earlier era of risk management (see Table 1.3).

Thus, a staircase is a *hazard*. People could suffer *harm* if they fell on it. You can reduce the risk of accidents on the stairs by providing strong handrails or anti-slip treads. A risk assessment could identify the risk of someone falling down the stairs. Auditing the safety of all stairs might form part of your RMS.

In the newer world of risk, setting up an export office in a foreign country is a risk, both as a hazard and an opportunity. If the venture loses money, the business will suffer harm. If it finds willing customers, it will bring rewards.

There is a more detailed glossary at the back of this book, and the ISO has a glossary: 'ISO/IEC Guide 73: Risk management. Vocabulary'.

Table 1.3   Unchanged definitions

Hazard	A source of potential harm
Harm	Loss or damage to people, assets, the environment or the organization

Note that risk management is not the same as continuity or disaster planning. Although continuity planning is dramatic and interesting, the greatest risks to a business are more insidious. The mundane risks of poor quality goods, loss of computer data, fraud and weak corporate governance are more likely to affect the business.

Which Risks are Important?

Numerous surveys have asked managers which risks are the most important. The results vary according to the questions asked, who completes the survey and what has been in the news. The viewpoint of senior financial people differs from mid-ranking risk managers, while those in the financial services have different concerns from people in manufacturing. And if new regulatory legislation is impending, governance risk rises to the top.

A study by the Association for Financial Professionals (AFPonline.org) is shown in Table 1.4.

Table 1.4   Risk factors expected to have the greatest impact on your organization's earnings in the next three years

	%
Customer satisfaction and retention	44
Regulatory risk	37
GDP growth	35
Political risk	28
Energy price volatility	18
Labour and HR issues	16
Natural catastrophe	7

Source: AFP

Despite their financial orientation, respondents put customer retention at the top of their list, perhaps reflecting the difficulty of winning sales in a tough climate.

In an international survey of 161 companies, the Conference Board of Canada (conferenceboard. ca) found that regulation was the greatest risk to business performance, followed by operations and strategy (Table 1.5). It also found that HR was the worst managed internal risk, despite the impact that employees have on the business.

The importance of different risks changes over time. New legislation, the economy, trends in the market and world events jostle for management's attention. In the early 1980s, the environment was not debated in the boardroom. In the late 1980s it became one of the most important items on the management agenda. Today, it has taken its rightful place as one of many risks that companies have to manage.

Table 1.5   Threats to business

Which risks have most impact on business?		How effectively are these business risks managed?	
1	Regulation	1	Finance
2	Operations	2	Operations
3	Strategy	3	IT
4	Human capital	4	Supply chain
5	Reputation	5	Regulation
6	Finance	6	Reputation
7	Supply chain	7	Strategy
8	Political/country	8	Natural hazards
9	IT	9	Crime, terrorism and physical security
10	Natural hazard	10	Human capital
11	Crime: terrorism and physical security	11	Political and/or country risk

Source: Conference Board of Canada

International terrorism became a major issue for many businesses after the 9/11 attacks in New York. Previously it might have been of concern only to companies operating in unsettled countries.

And after the eastern states of America suffered electrical blackouts, risk managers added 'power supply' to their list of worries. All of this goes to show that risk is dynamic – it changes over time; and some big risks are hard to forecast. Companies must regularly re-think their risks.

MISSION-CRITICAL RISKS

Which processes are essential to the survival of your business? Decide which activities are mission critical. Then decide how those activities could be interrupted. You have now defined your major risks. That's important because it's easy to get bogged down in the plethora of risks and issues.

The Dangers of Uncontrolled Risk

A PriceWaterhouseCoopers (PwC)/The Department for Buisiness Innovation & Skills (BIS) survey showed that 35 per cent of firms which suffer a computer disaster lose over £250,000. As each chapter of this book shows, the costs of other risks are equally high.

Table 1.6 opposite shows that all types of risk ultimately result in financial loss. That makes risk an important management topic.

Is Risk Management Relevant to Your Organization?

Risk applies to all businesses, but some are more likely than others to benefit from formal risk assessment. Risk management is particularly necessary to a business with several of the following criteria:

- a number of different sites;
- a size that precludes any individual from knowing the details of every threat;

Table 1.6   The results of uncontrolled risk

Type of risk	Initial effect	Ultimate effect
Quality problem	Product recall; customers defect	Financial losses
Environmental pollution	Damage to the environment. Bad publicity; customer disfavour and defection; court action; fines	Financial losses
Health and safety injury	Bad publicity; worker compensation claims; work-force dissatisfaction; statutory fines	Human suffering; financial losses
Fire	Harm to humans; loss of production and assets	Human suffering; financial losses Financial losses
Computer failure	Inability to take orders, process work or issue invoices; customers defect	Financial losses
Marketing risk	Market share falls. Revenue drops	Financial losses
Fraud	Theft of money	Financial losses
Security	Theft of money, assets or plans	Financial losses
International trading	Foreign exchange losses	Financial losses
Political risks	Foreign government appropriates assets; prevents repatriation of profits.	Financial losses

- a business with overseas operations;
- listed on a stock exchange;
- a range of processes;
- many sub-contractors, suppliers or other business associates who are not under the direct control of the business;
- a site that it has used for more than 30 years. Old sites sometimes have buildings, equipment or work practices dating from times when standards were lower;
- operates in a highly regulated industry such as financial services.

In short, the larger or more complex the business, the more it will benefit from risk assessment.

The Benefits of Managing Risk

Risk management helps a company avoid cost, disruption and unhappiness. Risk analysis also helps you decide which risks are worth pursuing, and which should be shunned.

Risk management can be adapted to meet the needs of each business. It can be used to educate staff, and to give them a deeper understanding of the corporate risks. This turns managers into business people, and makes the business more effective.

But the benefits of risk management are not easy to quantify. It is difficult to claim that risk management in your business has prevented, say, two major fires, a burglary and three serious accidents.

The company should collate records of losses. These are often kept by different departments and classified in ways that makes comparison difficult. But a unified set of information will help the business see how much is currently being lost, and whether investment in risk management has reduced losses. In Table 1.7 below, we consider some advantages of managing risks.

Laggard companies tend to introduce risk management in response to outside factors such as scandals, legislation or regulation (for example, stock exchange reporting requirements). *Forward thinking businesses* introduce risk management because it will help the organization produce better results.

Table 1.7   Advantages of managing risk proactively

Type of risk	Benefits of proactive management
Marketing risks	Maintain market share.
Health and safety risks	Avoid worker litigation; reduce insurance premiums.
Environmental risks	Avoid litigation from regulatory authorities; reduced premiums.
Fire risks	Avoid loss of production, avoid going out of business; reduced premiums.
Bomb threats	Avoid loss of life or destruction of a building.
Computer risks	Prevent inability to invoice, prevent lack of access to information.
Theft and fraud, industrial espionage	Prevent loss of money, assets or market share.
Technical risks	Avoid being left behind with obsolete manufacturing methods or technologies. Avoid production stoppage.
Kidnap and ransom, extortion	Safeguard managers abroad or at home. Prevent payment to criminals.
Product contamination (accidental and criminal)	Avoid harming customers and prevent litigation.

To make risk management work, you need to explain the benefits to other managers and staff. As we have seen from the table above, the corporate advantages are clear. There are also advantages to the workforce. Staff who are aware of risks can prevent them from happening, and can manage them better if they materialize. The most obvious example is fire drills. Staff who have practised a fire drill are less likely to be hurt in a fire. The same applies to other areas of risk, especially those which affect the individual – whether health and safety or kidnap and ransom.

The Risks and Issues Not Covered in this Book

In this book we cover the risks that are most common and most severe. That means ignoring the many smaller risks and issues that won't bring the company down or take management time.

Nor do we focus on the risks specific to the financial services industry, such as stock market risk, hedging or insurance risk. There are other, more specialist books for that.

22 Reasons Why Risk Management is Growing in Importance

At one time, risk management meant buying insurance and having enough fire extinguishers. But several factors have conspired to make insurance and passive deterrents inadequate. Together, they make risk an important boardroom issue. Below we examine some of the issues that are leading management to focus on risk.

Legislation is getting tougher:

1. Legislation is now more extensive. The EU has published 828 edicts on the environment alone.
2. Legislation is more stringent. Company directors can be jailed for corporate offences, and fines can be high.
3. Risk assessment is growing more common in many areas of legislation. The EU now requires companies to carry out risk assessment in health and safety, product liability and finance.

Insurance is more expensive and more difficult to get:

4. Insurance is no longer the cheap option it once was. Insurers are putting up premiums for many categories of risk. This follows years of major claims for environmental and product liability losses. Piper Aircraft was put out of business by product liability problems.
5. Open-ended cover is no longer widely available. For example, it is difficult to get insurance for environmental pollution which has developed over a period of years (for example, long-term pollution of a watercourse). Insurance companies are adding more exclusions, to the point where the insurance won't cover the business for its important areas of risk, which makes such insurance worthless.
6. Insurance companies require their clients to actively manage their risks. More companies have to take action following an audit carried out by their insurance company.
7. Insurance may not recoup the full amount lost. A survey by the Computer Security Institute and the FBI in America found that fewer than 20 per cent of the victims of computer crime even reported security incidents to law enforcers, let alone used external insurance to recoup their losses. This may be due to exclusion clauses and the negative impact on the organization's public image.
8. Insurance payouts can be slow. Later we see how companies can wait over a year for a payout.
9. Many assets cannot be insured. Insurance cannot pay for loss of goodwill and reputation.
10. Insurance is reactive. It does not prevent losses from occurring. That does not detract from the value of insurance, which is a valid fall-back strategy in risk management.
11. Insurance companies are reluctant to pay out. Unless the claim is small or 100 per cent robust, the insurance company may find ways to reject the claim. You may receive a 'Reservation of Rights' letter from your insurer – a notice that even though the company is processing your claim, certain losses might not be covered by the terms of the policy. By this letter, the insurance company reserves its right not to pay out. The letter may also state that the insurance company is not obliged to pay your legal costs, which can be substantial.

Customer attitudes:

12. Corporate customers want to pass legal responsibilities to their suppliers. Many companies look for evidence of risk management in their suppliers (for example, in the form of quality systems).
13. Consumers are more litigious, and less likely to accept product failure. Many examples have occurred in faulty healthcare products that have harmed consumers, who in turn have sought redress in the court. This applies to drugs which had unforeseen side effects, tampons which caused toxic shock or breast implants which went wrong.
14. The internet and social networking sites make organizations' failings more visible, more rapidly.

Shareholders more vocal:

15. Shareholders are more aware of risk. They are seeking more information in the annual report about the company's exposure to risk, because it will directly affect the company's future profits.

A critical public:

16. The public expects higher standards of corporate behaviour than before. It is especially critical of pollution, dangerous products and corporate fraud. This attitude encourages companies to avoid risking the public's hostility.

Management attitudes:

17. Management has learnt from other firms' disasters. Highly publicized disasters, from Bhopal (chemical leak) and Exxon Valdez (oil leak) to Martha Stewart (insider trading) and Hollinger (bad governance), have shown management that risks are a damaging, and sometimes fatal, cost to the business.
18. Companies are becoming more professional: as businesses have started to manage their environmental impacts, they have increasingly discovered that preventing problems is better than trying to cure them.
19. Companies are more global: firms have had to learn how to manage their increasingly international operations. Civil unrest can break out, and suppliers can cause damage to your reputation. Sometimes the solution lies in setting policies and performance standards, while leaving local management to run the business.
20. Growing private sector involvement in national enterprises: governments are withdrawing from the management of national enterprises, such as transportation, healthcare and energy. This means that private enterprises are now running high-risk businesses, and the government will not pay the costs of catastrophe. These companies have to take risk more seriously.

Other factors:

21. The weather has become less predictable, having an impact on the supply chain and companies' ability to deliver to customers.
22. Technology is changing rapidly, putting the organization at risk of falling behind or being stuck with a bad system or weak products.

Risk Management Lacks Maturity

But risk management has a long way to go. Many companies apply risk management only to selected areas of the business, or have varying levels of thoroughness.

Ferma (the Federation of European Risk Management Associations – ferma.eu) assesses risk maturity by asking companies how the manage risk. Figure 1.2 shows the level of maturity according to whether the different areas of risk (health and safety, quality, internal audit and so on) are co-ordinated. It suggests that risk functions are no longer working in silos but co-ordination is limited.

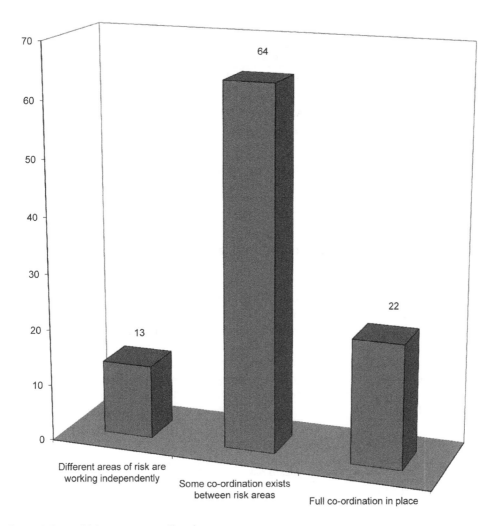

Figure 1.2   Risk areas co-ordination

Many companies still rely on insurance for their protection. For example, 80 per cent of companies have computer insurance, but only 42 per cent have adopted an IT contingency plan. Many computer recovery plans stay within the IT department, and are not discussed with other managers. Nine out of ten companies entrust the preparation of the recovery plan solely to an IT specialist, who understands IT but not the business. This indicates the flawed nature of much risk management.

In Table 1.8, we show three stages of risk maturity: Minimal, Managed and Mature. We indicate the level of activity for each under the headings of Oversight, Awareness, Processes and Measurement. You can extend the headings to provide more detail in the levels of maturity and the aspects reviewed, but this table serves to make the point, namely that you can identify an organization's risk maturity by assessing the way it approaches risk.

Table 1.8   Risk maturity

	Oversight	Awareness	Processes	Measurement
Minimal	Risk is not on the corporate agenda	No awareness of risk	Few controls	Risk is not measured
Managed	Risk is recognized and managed	Awareness varies across the organization.	Controls in place, but not audited	Selected risk measurements are taken
Mature	Board drives risk management	Risk is embedded	Systems are audited	Risks continuously measured and reported on

Risk Appetite: How Much Risk is Acceptable?

Rashness (as shown in Table 1.9) can lead to ultimate disaster. At the other end of the scale, excessive caution leads to missed opportunities. The middle course, involving a proper assessment of the risks, maximizes the company's profit.

This reminds us that the purpose of risk management is not to preclude entrepreneurial flair, but to ensure that it is properly guided.

This is known as *risk appetite*. Some companies are happy to accept new ventures and risky acquisitions. Others want to run a steady course. Young companies with little to lose often take big risks, while mature businesses want to protect their gains. Risk appetite is often a reflection of the CEO's outlook.

Table 1.9   Attitudes to risk, and their implications

Attitude to risk	Opportunity risk	Hazard risk	Result	Effect
Rash (also known as 'risk prone', 'risk seeking' or 'risk loving')	Lack of planning	Inadequate precautions	Harm, loss	Human, financial and other losses
Risk neutral	Proper risk assessment and management	Proper risk assessment and management	Control of risks	Maximized success
Timid (also known as 'risk averse')	Failure to innovate	Excessive caution	Inactivity	Lost opportunities

As shown in Figure 1.3, rash organizations are happy to invest highly for relatively small gains, while Vigorous ones will only invest large sums in projects that promise high rewards. Cautious organizations avoid high risks, preferring to spend small amounts in the hopes of high returns. This can mean they miss opportunities, or have expectations that do not come to fruition. Timid organizations are beset with inertia. They are risk averse: they seek incremental advances and invest little.

Startup and young businesses have a greater risk appetite, because they have less to lose and more to gain. More established businesses tend to be more risk averse. This is shown in Figure 1.4. As organizations grow, they tend to seek more assurance over their risks. As we shall see in

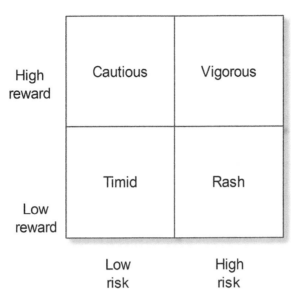

Figure 1.3   Risk appetite

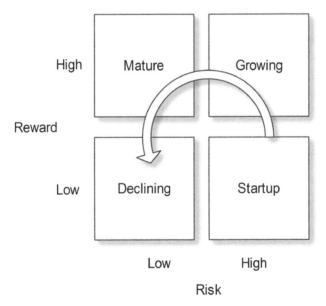

Figure 1.4   Risk appetite by corporate age

Chapter 20, this is particularly true of some market leaders, who want to hold on to their gains and not risk losing them by introducing disruptive products or technologies.

Introducing risk management can help the Board to define what kinds of risks it wants to take. This may never have been explicitly discussed before. Understanding their risk appetite can help Board members achieve more consistent and considered decisions.

There is a saying: 'Don't bet the farm.' This refers to gambling everything you have on a specific outcome. If you lose the bet, the other person takes your farm. In management terms this means avoiding any risk that would put the organization out of business. It particularly relates to investments or acquisitions which, if unsuccessful, would ruin the business. You should be able to put a dollar value on that amount. Bigger organizations can afford to take larger bets. The organization has to match its risk appetite to its *capacity*; that is, your ability to absorb a loss.

Total risk avoidance is an equally dangerous strategy. A freight company says:

> Our strategy is to accept more risks — calculated risk — in order to improve returns. This is being implemented by developing a portfolio of businesses, with a variety of risks and rewards, and continuing to exit from any high risk, low reward businesses.

Primary and basic businesses, like a mine or a pulp mill, often look safe because there is little innovation. Yet these companies are affected by large swings in sales and price, in response to capacity, global demand and raw material prices.

RISK TOLERANCE

You might decide to limit the maximum amount of revenue you earn from any one customer to 20 per cent. Or you might set at 30 per cent the maximum proportion of the company's capital that could be put at risk in any one project. These are sometimes known as *risk tolerance*.

At the other end of the scale are the small but commonly occurring risks. Staff will take home pens and paperclips, a few paperclips and the loss of an employee's life are the upper and lower levels of what management will tolerate.

WHICH TYPE OF TOLERANCE?

Note: there are two types of tolerance, and this can cause confusion:
1.   A range of numbers. A machine might be set to bore a hole, with a minimum and maximum diameter. If the hole size is outside that range the machine needs to be reset.
2.   A synonym for acceptance. 'Our maximum tolerance of new product loss is £1 million.'

COMPARING RISK AND REWARD

The level of acceptable risk depends on the reward. As Figure 1.5 shows, the greater the risk, the greater the reward must be to make it worthwhile.

The best business opportunity (shown with two ticks) is one where the reward is high and the risk low.

A risk greater than the reward (shown by the crosses) is not worth pursuing; while projects that carry low risk and low reward rarely make any impact on the business, and are therefore not worth pursuing.

Companies can affect the level of reward and risk. An oil company would want a greater return on investment from exploring in an unstable country. To achieve that it might negotiate tax concessions and investment subsidies to achieve that aim before starting operations there.

Likewise, a bank would want more collateral for a risky loan than for one that was risk-free. The problem comes in correctly assessing risk. The banks that lent to DisneyLand Paris did not imagine that it could lose money (it made a first year loss of 300m French francs), while those who lent to Eurotunnel did not think its costs would escalate and revenues would be delayed (it lost £925 million in its first year).

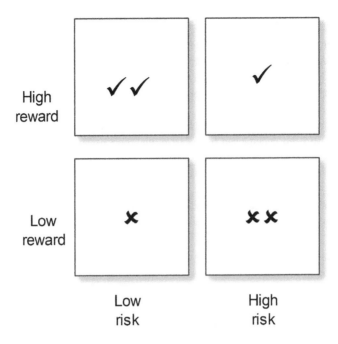

Figure 1.5   Determining acceptable risk

As Figure 1.6 shows, you have to endure a minimum level of risk in order to achieve any reward. Too little risk won't bring any benefit. This applies to marketing risks (spending too little on advertising won't be noticed by the consumer) and to corporate acquisition.

In the middle zone of Figure 1.6, rewards are commensurate with the risk. But at a certain point, the risks outweigh the rewards. That is the rash zone where the CEO refuses to listen to reason or an excessive reduction in customer support serves only to make customers defect.

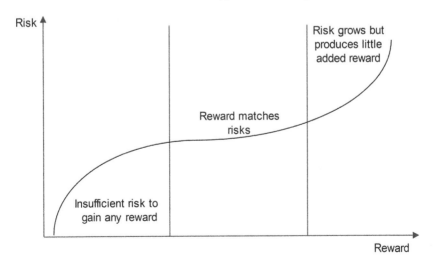

Figure 1.6   Risk/reward zones

CASE STUDY: VIRGIN AND RISK APPETITE

Conventional wisdom says it's wisest to diversify into related markets where you can harness your knowledge to give you a competitive advantage. But Richard Branson has consistently done the opposite, by moving markets he knew nothing about. His Virgin Group, whose core skills lay in publishing and record retailing, has gone into airlines, railways, beauty products, banking and mobile phones. His ventures have even included space flights and umbilical stem cell storage,

If there is a strategy, it has been an attack on mature markets dominated by behemoths.

The launches have been underpinned by the belief that the existing companies are complacent and are failing to meet customers' needs. Branson thereby positions the brand as the consumer's champion. In addition, having no knowledge of an industry has liberated Branson from seeing the problems that limited existing players.

Numbers are hard to find, with Branson locating his companies in the British Virgin Islands tax haven. In many cases he has achieved breathtaking success, as with railways and planes. In others, inevitably, the business failed:

- Virgin Cola failed to make any headway against Coca Cola;
- Virgin Brides folded after making losses;
- Virgin Vie was sold off at a loss of £21 million after disappointing retail sales;
- Virgin Cars was an internet retailer that stopped operating, having sold 'only' 12,000 cars in four years;
- Virginware was set up with 30 stores to compete with Victoria's Secret, but went into receivership.

Unlike traditional business people, Branson is usually photographed in a jumper, which is indicative of his unconventional outlook and his lack of interest in others' opinions.

In recent years he has reduced his risk by doing joint ventures with large cash-rich companies. This has involved him simply lending the Virgin brand name to others' products, which has avoided the risk of huge investment. Thus the Virgin name has become the group's biggest asset, in addition to the founder's restless energy, risk appetite and ability to pick successes.

Branson is a big-picture person, uninterested in numbers or details. He doesn't sit on any of his companies' Boards, but simply sets the scene and then leaves the running of the business to managers.

The author once went to a meeting in Branson's elegant office in Holland Park. It concerned a pitch for a national Sunday newspaper. Branson's lieutenant said our proposal should be no more than one page in length; and he would deliver it to Richard who was living not in the elegant mansion but on his small houseboat moored on the Thames.

There is a clear risk that the failure of one of the Virgin brands could tarnish the others. In some markets the brand name is now in the hands of companies whose values may be different from Branson's. In addition, the Virgin group is a one-man band, rather like Rupert Murdoch's News Corp, and is therefore subject to the founder's strengths and weaknesses. Meanwhile, Branson's risk appetite drives him onwards.

How to Manage Risk

The remainder of this book is about how to manage risk. The process of risk management is relatively straightforward, and is shown in Figure 1.7. The company has to identify risks and set policies. Then it has to take action, and monitor the risks. As the table shows, it is a continuous process.

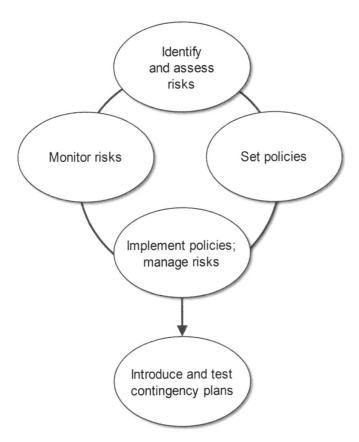

Figure 1.7   A structure for managing risk

Risk management should be done thoroughly. This applies particularly to the task of assessing risk. A company which merely identifies some minor health and safety risks is lulling itself into a false sense of security. Properly used, risk management helps a company to evaluate its strengths and weaknesses. It can help the business to re-engineer itself, and make it more competitive. Risk management is a tool that makes a company grow strong.

2
How to Identify and Assess Risk

There's a virus going about, and 20 per cent of your staff are off sick. The unions are asking whether your workplace is really safe for its members. Incoming calls are going unanswered, and work is not being carried out. Customers are complaining.

At a risk committee meeting, someone sarcastically says, 'If you didn't include epidemics in the register of risks, what else is missing?' You make a mental note to get a more thorough risk assessment carried out in future.

External and Internal Risks

In this chapter we examine how to identify the risks that will affect your business. We start by considering the 'external and internal context'. This curious phrase came into use with the ground-breaking Australian risk standard, and was subsequently adopted by ISO 31000, the international standard.

- The *external* context is the world your organization inhabits. Reflecting on that gives you a wide perspective on risk. If you don't think about the context your business operates in, you might overlook some threats to the business. As we see below, it includes the political, economic and social environment.
- The *internal* context is a broad-brush look at the business itself: its history, ownership, governance, finances, people and operations. Doing that ensures you take all relevant factors into account, rather than just the ones in front of you.
- We start by examining the company's external risks. A common way to do this is to use the PESTLE mnemonic.

PESTLE

Pestle stands for Political, Economic, Social, Technological and Ethical (or Environmental) factors. It is a useful framework for a brainstorming session on risks, especially those that are outside the organization's control. Figure 2.1 contains a list of some PESTLE risks. You may think of others.

We look at a wide range of external risks throughout the book, including later in this chapter and in the chapters on Marketing and Continuity. But for the time being it's useful to assess risk using the categories of the external and internal context.

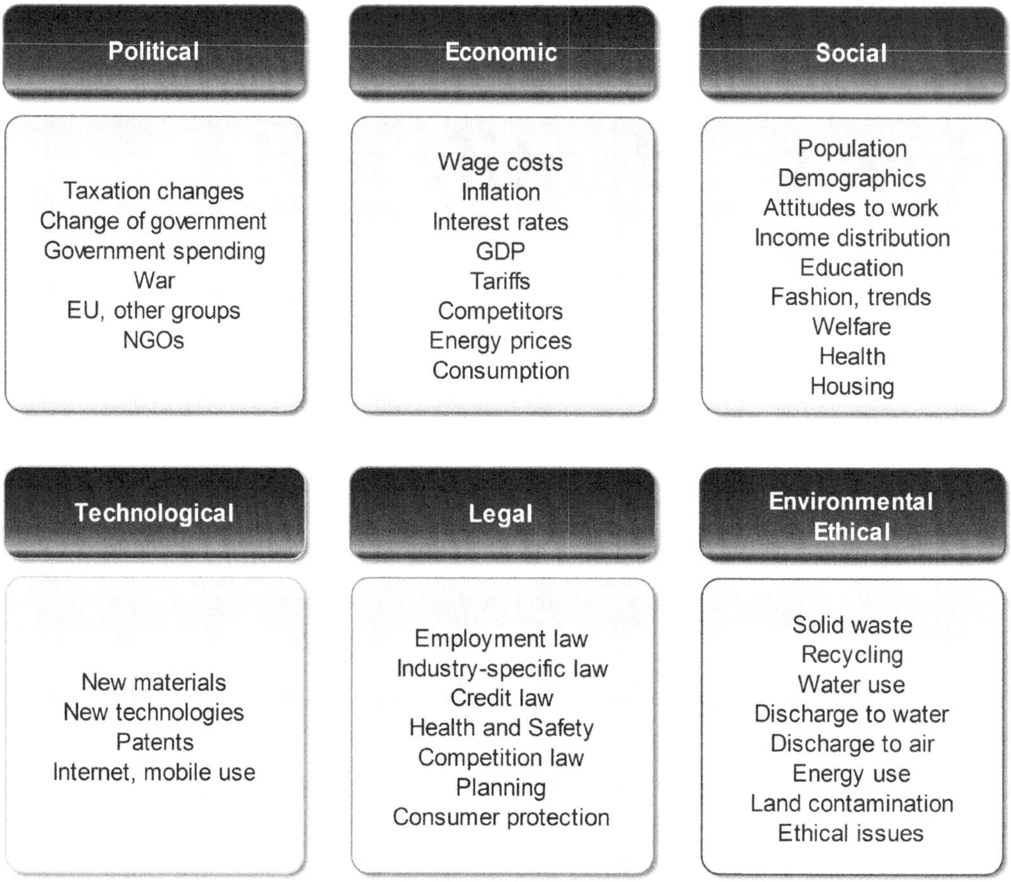

Political	Economic	Social
Taxation changes Change of government Government spending War EU, other groups NGOs	Wage costs Inflation Interest rates GDP Tariffs Competitors Energy prices Consumption	Population Demographics Attitudes to work Income distribution Education Fashion, trends Welfare Health Housing

Technological	Legal	Environmental Ethical
New materials New technologies Patents Internet, mobile use	Employment law Industry-specific law Credit law Health and Safety Competition law Planning Consumer protection	Solid waste Recycling Water use Discharge to water Discharge to air Energy use Land contamination Ethical issues

Figure 2.1   PESTLE analysis

INTERNAL RISKS

Internal risks are those specific to the organization. They include those caused by human, technological, operations, finance or marketing factors. A sample list of risks is shown in Table 2.1. We cover these in more detail below and throughout the rest of the book.

SWOT

Another way of assessing risks, both external and internal is the SWOT analysis. SWOT is a well worn acronym, standing for Strengths, Weaknesses, Opportunities and Threats.

As with PESTLE, the SWOT analysis can help the organization focus on its risks. Strengths and Weaknesses are issues within the organization, while Opportunities and Threats come from outside.

One advantage of SWOT over PESTLE is that it encourages the business to consider its opportunities (which bring risks), rather than just threats. If you're taking part in a new joint venture (JV), this is an opportunity but is risky as well.

In a SWOT analysis, you write down the organization's internal strengths and weaknesses. What are its best features; and what are its failings?

Then you write the threats and opportunities that might affect the company's future. These are the outside influences that could impact on it. A sample SWOT analysis is shown as Figure 2.2.

Table 2.1   Categories of internal risk

Type of internal risk	Example of risk
Human	Industrial action, strikes
	Loss of talent
	Low motivation
	Negligence, dishonesty
	Managerial incompetence, governance problems
	Fraud, theft
Technology	New technology introduced by competitors
	Poor use of technology
Operations	Procurement problems: lack of components or raw materials, poor quality materials
	Environmental damage
	Fire
	Health and safety: trips, slips, employee deaths
	Machinery failure
Financial	Liquidity and cash flow
	Failure to get or sustain credit or overdraft
	Excessive debtors, default by debtors
	Low margins
Marketing	Declining sales
	Loss of customers, churn
	Lack of new product development
Strategic	Acquisition or diversification failure

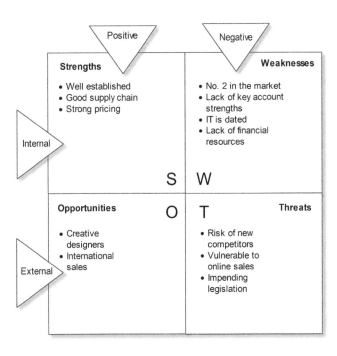

Figure 2.2   Sample SWOT analysis

EVALUATING YOUR STRENGTHS AND WEAKNESSES

After conducting the SWOT analysis, you should assess what the chart tells you. Which quadrant is the fullest? What can the business do to:

- overcome its weaknesses;
- build on its strengths;
- take advantage of the opportunities ;
- pre-empt the risks?

The Six Types of Risk

So far we have looked at risks according to whether they are external or internal. However, some risks span these categories.

In Figure 2.3, we present risks under six broad headings: Strategic, Operational, People, Compliance, Financial and Technology.

> *Strategic risks* are the big issues which require companies to think on a grand scale. These risks should be tackled at Board level and require strategic planning.
> *Operational risks* are those relating to the organization's production or operations. They include faulty raw materials, a supplier going into receivership, or a major customer going into receivership. Unlike Strategic risks, Operational ones are usually implemented at lower levels of management, though they still require Board oversight.
> *People risks* are problems caused by staff. Just as a company's main assets go up and down in the lift, most risks are based on human frailty, whether it's a rash CEO or an alcoholic account manager.
> *Compliance risks* are increasingly important, as regulators and the law often require organizations to report on their financial and operational risks.
> *Financial risks* include external ones such as an adverse exchange rate that reduces the company's exports, and internal ones such as excessive overheads.
> *Technology risks:* With most businesses being driven by technology, failure to use IT properly puts the business at risk. Similarly, technological innovation by competitors may threaten the company.

Most risks fit into one category, but some span more than one. If a major customer goes bust while owing the business money, you could classify it as a financial risk (which could have been managed by better credit control), or as a strategic risk (over-reliance on a few major customers).

In what follows we examine each of these six types of risk in more detail.

STRATEGIC RISK

Some strategic risks and their impact are shown in Figure 2.4. They include:

> *Government* action, such as a recession or new legislation. Companies operate in an increasingly regulated world, and you need to take into account government plans, both at home and abroad.
> *Customers* also bring strategic risk. Their changing attitudes and growing expectations make them less predictable than before.

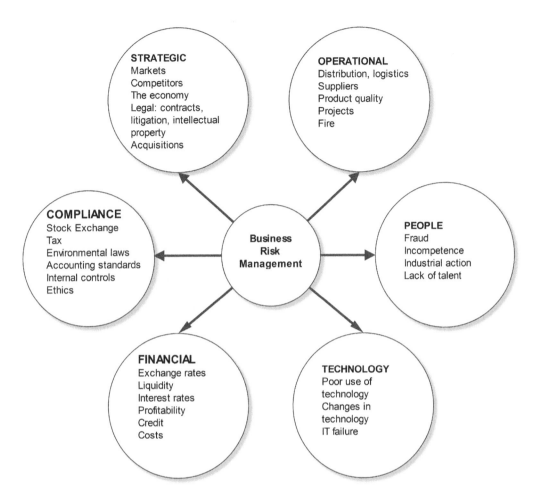

STRATEGIC
Markets
Competitors
The economy
Legal: contracts,
litigation, intellectual
property
Acquisitions

OPERATIONAL
Distribution, logistics
Suppliers
Product quality
Projects
Fire

COMPLIANCE
Stock Exchange
Tax
Environmental laws
Accounting standards
Internal controls
Ethics

**Business
Risk
Management**

PEOPLE
Fraud
Incompetence
Industrial action
Lack of talent

FINANCIAL
Exchange rates
Liquidity
Interest rates
Profitability
Credit
Costs

TECHNOLOGY
Poor use of
technology
Changes in
technology
IT failure

Figure 2.3   The six types of risk

Competition is fluid and has become global for many organizations. No market is immune from new competition. Any market, especially those which are profitable, will attract competition sooner rather than later.

Technology brings new threats and opportunities. If managed properly, it can make the business more competitive; if ignored, it can hobble the organization. Either way, technology brings new unforeseen entrants into the market. We look at technology risks later in this chapter.

These strategic risks have a major impact on the company's costs, prices, products and sales. Figure 2.4 shows some of the solutions that organizations adopt.

We examine these strategic risks in subsequent chapters. In Chapter 20 we review the effects of government action, competitors and customers, while we look at new technology in Chapter 15 and many others. Here we're taking a broad-brush approach to set the scene.

These four strategic issues are ones you have to grasp before you can start to consider the operational risks that affect the company. Management must ensure that it isn't exposed to fundamental risks, like being in the wrong market. Unfortunately, that isn't easy to forecast or assess.

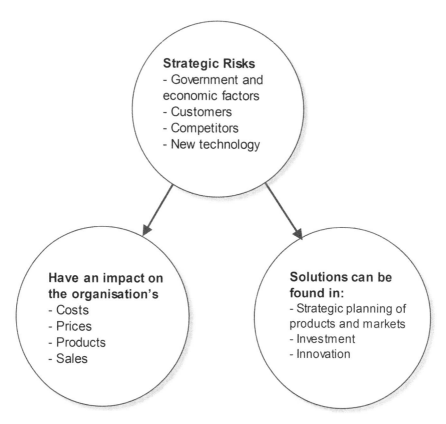

Figure 2.4   Strategic risks

OPERATIONAL RISK

As the names imply, *strategic* and *operational* risk have different levels of management input.

A strategic risk would be overseen by the CEO, rather than a junior manager. However, the reverse is not completely true. That is to say, operational risk does not have minor impacts. In some companies health and safety carries the most severe risk of all – death of employees.

Similarly, if operations come to a halt because of lack of plant maintenance, this will have a substantial impact on the business. Maintenance therefore requires active control by production staff, and can't be ignored by senior management.

As Table 2.2 shows, you can categorize operational risks according to *when* they occur. Some take place at suppliers, and some at the point of production or operations, while others happen in the distribution chain or when the product is consumed.

This table, which also shows whether the risks affect mainly people or assets, serves to remind us that not all risks take place on the shop floor.

There is often some overlap between types of risk. For example, fire often hurts people as well as consuming buildings.

PEOPLE RISK

Virtually everyone in the organization can land the business in trouble. Boards have often been accused of being asleep at the wheel, while at the other extreme, operatives are frequently

Table 2.2   **Operational hazards classified by time**

	< Earlier in time			Later in time >
	Suppliers	**Process and internal risks**	**Distribution**	**Customers**
Risks	Interruption of supplies	Fire	Counterfeiting	Payment problems
		Pollution	Tampering	Competitor activity
	Poor quality supplies	Fraud		
		IT		
		Accidents		
		Labour disputes		
		Terrorism, kidnap and ransom		

forgetful, negligent and even subversive. Stopping people from doing stupid things is part of the risk manager's job.

We devote Chapter 18 to people risks, though every chapter contains risks that are caused by capricious, ill trained or careless individuals.

Some of these risks can be managed through better recruitment, training or motivation. In other cases, such as fraud, solutions lie in putting controls and supervision in place.

COMPLIANCE RISK

Even small and medium sized companies have to deal with the tax authorities, health and safety legislation, and certification bodies. The issue is substantially greater for publicly quoted businesses. In the wake of every scandal comes more legislation to control business; so the days when companies could do as they wanted are long gone.

We look at Compliance risks and how to manage them in the chapters on the Environment (Chapter 8), Governance (Chapter 17), Ethics (Chapter 19) and The Risk Management System (Chapter 24).

FINANCIAL RISK

Most incidents carry a financial penalty, whether it's a loss of sales or an increase in cost. Often financial problems are the result of strategic failings. Excess debt can be caused by buying a company that performs less well than expected. Or it can result from having too many staff for the level of profit. We cover these risks in Chapter 14.

In a few cases, financial problems are caused by financial mismanagement, such as failure to manage debtors. But financial problems are mostly created through business risks created by managers or staff in other departments. Thus the finance department has a strategic role to play in controlling others' desire to spend and their unwillingness to cut costs. And line departments need to be aware of the impact their decisions can have on the future of the organization.

In the financial services industry, namely banks and insurance companies, there are specific risks:

Credit risk: When customers fail to pay their loans.

Market risk: Where the stock market or currency moves in the opposite way that the bank had gambled.

Operational risk: In the financial service industry, operational risk is everything that isn't market or credit risk. This includes insider fraud, lack of controls, the bank's computers failing, incorrect pricing, or any of the myriad risks that businesses in general can suffer from.

TECHNOLOGY RISK

Technology risks are much wider than simply suffering a computer outage at work, serious though that might be.

Technology is everywhere. Smartphones have changed people's behaviour, giving them new ways to pay for services, while new business models are constantly evolving on the internet. Embedded technology is turning even simple items like clothing or fridges into smart devices. Yet as we see in the Marketing chapter (Chapter 20), new technology is often overlooked by existing players in the market; and sticking with a traditional business model can lead to business decline.

Meanwhile, failure to choose the right technology to manage workflow or corporate systems can hamper the organization; getting it right can make the company prosper.

Other Classifications

There are many ways to categorize risk. In the sections below we look at the models used to help assess risk. We start by examining who might be affected if an event should take place.

IDENTIFYING WHO IS AT RISK

As Figure 2.5 shows, the level of risk sometimes ripples outwards from the single worker towards the general population.

On the front line of risk are the individual workers who will be hurt if their clothing is caught in machinery. They are most exposed to many risks: if sales fall, they are the first to be made redundant.

Sometimes, several workers at a plant might be hurt. This includes dispersed workers, as when a train ploughs into lineside rail workers.

Beyond the factory gates, the local population is at risk. A chemical plant might send a fireball into adjoining streets, causing damage to buildings, passers-by and local office workers. Less dramatically, dust, smoke or odour from a plant can extend over the local town.

These risks are of the 'health and safety' sort, but many others are not. A defective product may hurt a customer. A mass-market yoghurt could cause widespread gastro-enteritis.

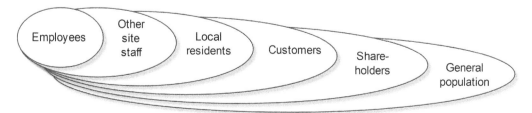

Figure 2.5   The widening effects of risk

At the extreme is a risk to the general population. This is not a problem facing many companies, but it is found in energy (where power stations and their emissions cover the country). The Chernobyl accident affected populations 1,700 miles in radius, all the way from Ukraine to Welsh hill farmers who were restricted from selling their sheep.

Eventually, these risks will affect the shareholder if they cause the share price to fall.

In the case of undetected fraud, the workforce is at risk of redundancy if the fraud is substantial and layoffs take place. If the business closes, the local residents lose a source of employment, and ultimately the customers lose a service or product.

One group which is not especially at risk but nevertheless closely involved are the organization's regulators. They will want to know that the organization is managing its risks properly.

These then are the organization's stakeholders. Good practice requires the business to seek their views, understand their expectations, and communicate with them. The full range is shown in Figure 2.6.

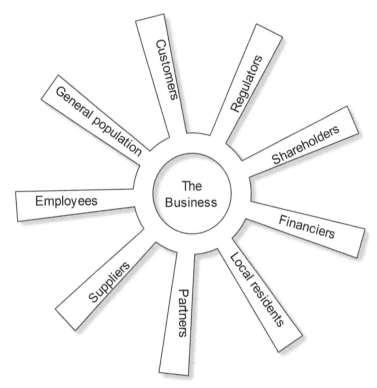

Figure 2.6   Stakeholders

MATCHING RISKS TO CORPORATE MISSION OR OBJECTIVES

Risks should be linked to corporate objectives. This ensures that you identify your *significant* risks, those that are critical to the future success of the business. It is nice to discuss the risks to the wellbeing of the ducks in the head office pond, but they are unlikely to be relevant to your corporate mission, unless the organization is a wildlife charity.

By contrast, if the business claims to operate with high ethical standards, then risks associated with the corporate culture must be a priority.

In Figure 2.7, we show how you might define risks according to the mission, corporate objectives, business plan or other factors including core processes (which we examine in Chapter 5). What matters is to identify risks that are significant to the business.

Figure 2.7   Mapping risks to corporate mission and objectives

DEFINE THE 'REVENUE DRIVERS'

In a survey of large companies, senior managers voted their production facilities, logistics and IT equipment as their top 'revenue drivers' (Table 2.3).

The survey asked CFOs, treasurers and risk managers what contributed most to corporate earnings. The survey wanted to know what managers were most concerned to protect – items whose disruption would have the greatest financial impact on the business.

It comes as no surprise that they listed the elements needed to deliver their product or service. Manufacturing will concentrate on production facilities while service businesses will be more concerned with operations.

Put simply, there is no point in creating a risk strategy until you know what things are important. You need to guard against the risks that count.

KEY SUCCESS FACTORS (KSF)

What factors must the organization be good at? To succeed, what must you do well? These are the Key Success Factors (KSF).

Table 2.3   Top revenue drivers

Revenue driver	% most important
Manufacturing plant, equipment	21
Delivery, logistics	19
IT, telecomms systems	19
Personnel and customer support	19
Raw materials/ inventory	15
Intellectual property	7
Total	100

As with the 'top revenue drivers', risk management must protect these KSFs. In Table 2.4 below, we look at the KSFs for an online business.

The business must ask, 'What are our KSFs? What risks could affect them? And what controls should we put in place?'

Since people visit this hypothetical online company's website 24 hours a day, there must be no downtime. If the internet service provider (ISP) (or the company's server) fails, this would damage the business. To protect itself, the company would carefully select an ISP with proven reliability, and monitor the server's availability. Already a risk management assessment is emerging, together with the elements of a strategy.

Table 2.4   KSFs for an online business

Key Success Factor	Risk	Control
No downtime	ISP has excessive outages, or even ceases trading.	Use a business-oriented ISP with a track record, and a service level agreement. Monitor uptime.
Effective website design	Dated design or poor usability will reduce sales.	Hire professional web designer. Use a content management system to maintain quality when site content is updated. Test usability.
Fast fulfilment	Slow fulfilment will increase returns and causes loss of reputation.	Introduce written procedures. Train staff. Monitor courier company. Automate the stock ordering process.
Effective or distinctive product range	Uncompetitive range or weak offer will reduce sales.	Test new products or format. Monitor competitors, both online and 'bricks and mortar'. Conduct benchmarking.
Profitable customer acquisition	Excessive costs will render business unprofitable.	Monitor promotional costs. Identify referrers.
Effective customer support	Weak support will reduce sales.	Train staff. Use automated systems.
Adequate profit margin	Wrong cost and price mix will ultimately put the company out of business.	Ensure that the business has high-quality management information.

CLASSIFYING RISKS BY THEIR CAUSE

Table 2.5 shows risks by category, and highlights the assets they affect. This could help you decide what assets are vulnerable, and how they might be protected. It also reveals that all business risks ultimately affect the company's profits.

Table 2.5   Classifying risks by cause

Assets	Risk						
	Natural disaster	Government. action, economic forces	Supplier Problems	Customer Problems	Production problems	Theft and fraud	Vandalism and revenge
Examples	*Fire, explosion*	*Tax changes*	*Late delivery*	*Bad debts*	*Labour dispute*	*Theft of stock*	*Computer virus*
Land	✓	✓					
Buildings	✓						✓
Plant and equipment	✓		✓				
Raw materials	✓		✓			✓	
Stock	✓				✓	✓	
Vehicles	✓					✓	
Computers	✓					✓	✓
Staff and visitors	✓						
Local residents	✓						
Cash				✓			

SHORT-TERM OR LONG-TERM RISKS

An explosion or fire takes place in a matter of seconds. But exposure to a carcinogen only reveals its effects after many years. Long-term risks are sometimes overlooked because they aren't uppermost on people's minds.

Short-term risks cause instant disruption. A building gets flooded or the server breaks down. And there are often immediate solutions that can be applied, such as bringing in pumps or IT guys.

By contrast, it takes longer to judge whether a project is a success or a failure. Longer-term risks can also relate to insurance claims, product liability issues, polluted land, and employee or partner contracts. The harm often occurs long after the risk assessment is carried out, and after the current management have moved on. They are harder to quantify and cost. Sometimes insurance is the answer for these risks.

HAZARD AND OPERABILITY STUDY (HAZOP), HAZARD ANALYSIS AND CRITICAL CONTROL POINT (HACCP) AND OTHER SYSTEMS

Many risk managers use flow charts or fishbone diagrams showing processes and the movement of material through the business. This may reveal points of vulnerability.

In the Operations chapter (insert Chapter number here)we examine the leading systems for assessing operational risk. These include Hazard and Operability Study (HAZOP), Hazard Analysis and Critical Control Point (HACCP) and Failure Mode and Effects Analysis (FMEA). We review them in the Operations chapter (insert Chapter number here) because they are specific to manufacturing or operations, rather than the general overview which we are looking at here.

HAZARD RISKS VERSUS OPPORTUNITY RISKS

Some risks have a big upside. They include new product development, medical research programmes and export markets. If they succeed, the company prospers. These are known as opportunity risks.

Hazard risks carry no such advantages. The threat of an oil leak is only ever a negative one. The danger of being taken to an employment tribunal has no accompanying benefit.

INTRINSIC, RESIDUAL, CURRENT AND TARGET RISK

There are various states of risk, depending on whether the risk has been controlled:

*   *Intrinsic risk* is risk in its original state, one without controls. This could be a piece of machinery with no protection in place.
*   *Current risk* is the risk as it currently is, given the current performance and design of controls. The machinery could be surrounded with guards.
*   *Residual risk* is the risk that remains if all controls were fully applied. The organization may plan to carry out training, and install fume extraction. This would reduce the risk to its minimum.
*   *Target risk* is the level of risk that is acceptable to the organization. It might decide that the piece of equipment is still too dangerous, and it needs to be replaced with safer piece of kit.

This is important because the scale of the risk could range from very minor (if fully controlled) to enormous in its raw, intrinsic state. So an assessor may need to define exactly what state of risk is being described.

SITE BY SITE

You may need to assess risk on a site-by-site basis. Where a company owns several sites, they will vary in age, level of investment and production capability. A manufacturing site with old buildings, a large unskilled workforce and combustible raw materials is more risky than a modern office building (though no building is completely risk-free). Figure 2.8 shows that risk priorities will vary from one site to another.

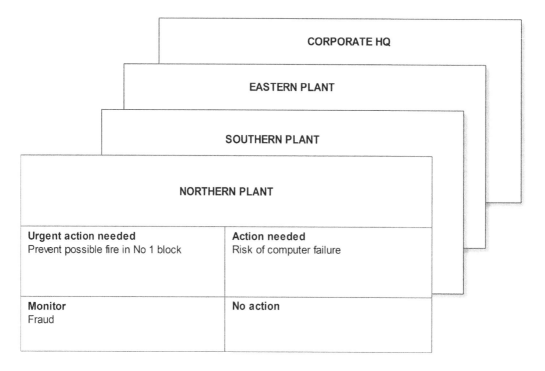

Figure 2.8　Site-by-site action plan

RISK CATEGORIZED BY THREAT

Figure 2.9 shows that you can examine risk by the types of threat the business is exposed to. In the example, the company is examining the threat of fire.

The first column shows the resources that could be affected by the threat.

The assessment then considers the Severity and Probability of the risk. In the table, fire could cause loss of life as well as loss of assets. This produces an overall scale of risk.

Next we consider the key problems. These are areas of special risk. They amplify the information contained in 'Resources'.

Finally, we consider the strategy which should be adopted (or has been adopted) to prevent the risk from materializing.

This chart can be adapted to different requirements. It can look broadly at the issues involved, or it can examine a topic in detail. For example, you can draw up separate charts for different areas of the business.

THE SWISS CHEESE MODEL

The discussion thus far has implied a single source of risk, such as a fire or a labour dispute. But events can have many causes, prompting British psychologist James T. Reason to introduce the Swiss Cheese model (Figure 2.10).

This says that events can have several causes, represented by the holes in the cheese, such as a weakness in the system or a human error. Each factor is insignificant on its own, but taken together they build up to a problem waiting to happen. Eventually they all line up, and an accident takes place.

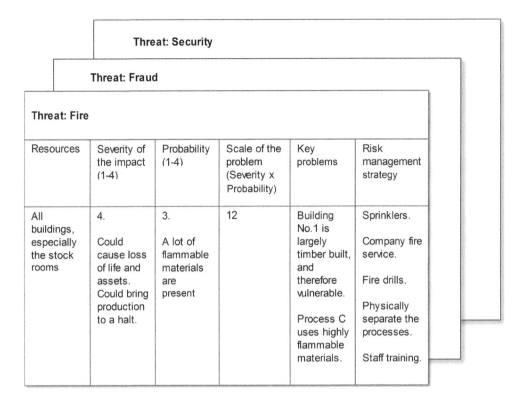

Threat: Security

Threat: Fraud

Threat: Fire

Resources	Severity of the impact (1-4)	Probability (1-4)	Scale of the problem (Severity x Probability)	Key problems	Risk management strategy
All buildings, especially the stock rooms	4. Could cause loss of life and assets. Could bring production to a halt.	3. A lot of flammable materials are present	12	Building No.1 is largely timber built, and therefore vulnerable. Process C uses highly flammable materials.	Sprinklers. Company fire service. Fire drills. Physically separate the processes. Staff training.

Figure 2.9   Risk assessment and management

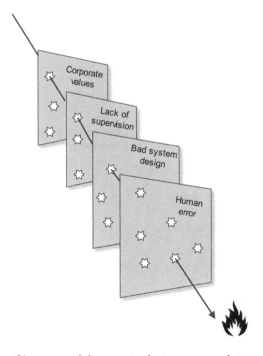

Figure 2.10   The Swiss Cheese model suggests that numerous latent factors lead towards a crisis

The Swiss Cheese model distinguishes between *latent* risks, which might be a wet floor; and *active* risks, for example a badly placed ladder. It is the latter that sets off the incident.

The model is often applied to aircraft accident investigations, with problems having lain dormant for some time (a leaking suspension unit) followed by an active problem taking place (a heavy landing).

Latent risks are also useful as a preventative concept in medicine (if different intravenous fluid containers all look identical, a nurse may use the wrong one).

ACCOUNTING FOR THE UNEXPECTED

As Figure 2.11 shows, some risks are largely manageable. These include fire, product quality and IT viruses. But many other types of risk are hard to predict – a new competitor, a new technology or even terrorism.

A good RMS must allow – and encourage – 'blue sky' thinking. It must avoid being mechanistic and being overly focused on the company's internal processes.

An extremely serious and improbable event is one that is hard to plan for. It is what Nassim Taleb calls a 'Black Swan' event. Predicting major change, he says, is almost impossible. Like generals preparing for the last war, we never expect the surprise event when it occurs.

Taleb refers to the turkey that is fed every day for 1,000 days by the farmer. Every passing day confirms to the turkey that the farmer loves turkeys. But on day 1,001 the turkey is in for a surprise. For the turkey this is a Black Swan event. But not for the farmer.

Meanwhile a 'Grey Swan' event is rather more predictable than a Black Swan one, but is hard to know whether and when it might happen. This might include a recession or a political upheaval. Though possible, it is unlikely and therefore hard to plan for.

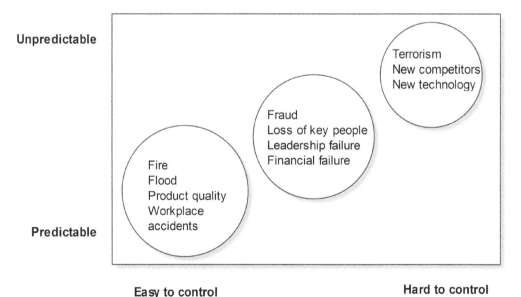

Figure 2.11   Predictability of risks

Gathering the Data

We can categorize data as having *external* and *internal* sources. In the following sections we look at each in turn.

GATHERING EXTERNAL INFORMATION

Data gathering should start with published economic or industry-specific information. What are the main risks associated with the industry? What reports are available? This may provide useful data against which the company can benchmark its own records, and assess the quality of the company's risk management.

You need to gather information from as many sources as possible, and on a regular basis. External sources of information include those shown in Table 2.6.

The data can be collected through conferences, workshops, sales force visits, market research questionnaires, and an online news service such as Google's.

Each of these methods has strengths and weaknesses. Informal methods such as chats have no statistical value, while quantitative questionnaires are weak when there are many unknowns. You should try to include a range of methods.

The risk manager should be the central store for all the information relating to risk.

Table 2.6   External sources of information

Source	Type of information
Suppliers and vendors	Raw materials, process improvements, technology risks and opportunities
National media	Information about risk in society
Trade media	Competitor activity
Bloggers	Insights into the trade, profession or risk in general
Workforce and trade unions	Operational risks
Partners	Risks in the industry, process improvements
Distributors, customers, and end users	Marketplace risks, customer needs, product quality, market opportunities
Trade body	Industry risks, legislation and regulation
Government, EU	Legislation and regulation, cross-border collaboration
Local authority, police	Crime, health and safety issues, environmental health
Shareholders	Industry, investment, corporate and governance risks
Pressure groups	Single issues that might affect the business, for example construction companies and the anti-road lobby

INTERNAL INFORMATION

Once you have gathered all relevant external data, you can begin to collect internal data.

As with external information, a range of methods are available, including informal chats, workshops, brainstorming, staff surveys, audits and business process modelling with flow charts.

Figure 2.12 is simple questionnaire which can be given to departmental managers. Staff often have detailed knowledge about the risks that affect them, and the questionnaire involves them in the process of risk assessment.

RISK QUESTIONNAIRE

Name   Department   Date

What risks exist in your department?	How should these risks be managed?

Any other comment about risks in the business?

Please return to: By date:

What are risks? Risks are anything that can affect you or other members of staff, customers, visitors, shareholders or the environment. Examples of risk include fire, slips or trips, chemicals, dangerous machinery, pollution, IT failure, fraud and theft, or unsafe products.

Figure 2.12   Risk questionnaire

CONTROL SELF-ASSESSMENT

Some organizations also use a questionnaire for auditing purposes. This is known as the 'Control Self Assessment' (CSA) or 'Control and Risk Self Assessment' (CRSA). Here the word 'Control' refers to procedures or methods that you put in place to reduce risk. A sample extract is shown below as Table 2.7. These are often issued by an audit committee, which may choose to audit different issues each year.

The questionnaire is used by the Board to bulk out the work done by internal auditors, and serves to build risk awareness among operational departments. It is also useful to compare results across different sites. By completing a questionnaire and testifying to its accuracy, local staff and managers have to think about their risks and take responsibility for them. The questionnaire is either paper based or electronic. The latter is easier to analyse, and various CSA software programs exist.

The results can be put on to a heat map (see below) and sent to the Board.

CSA questionnaires are effective in reaching more parts of the organization than auditors can. But they are sometimes criticized for taking up staff time in compiling, responding to and analysing yet more paperwork. Operational staff don't always take them seriously or don't have sufficient knowledge to answer the questions. And if the survey is repeated each year against a background of unchanging risks, both the response rates and the survey's usefulness will decline.

Other organizations undertake CSA as a workshop, but this is usually as an initial fact-finding mission, to identify risks and discover what controls are, or should be, in place.

Table 2.7   Control Self-Assessment form (extract)

	Yes	No	Don't know
If there are problems with cash discrepancies, accounts receivable or payable, do management and employees report the matter in accordance with company policy?			
Are the recording and payments of invoices separated so as to prevent fraud?			
Is there a written job description for everyone in your department?			
Are there plans in place to cope with sudden changes in personnel?			
Are performance management measures in place for members of staff who are under-performing?			

PARETO ANALYSIS

You could also carry out a Pareto analysis. This is also known as the 80:20 rule. It often shows that a few issues are responsible for most of the problems. Figure 2.13 shows that 10 per cent of customers account for 70 per cent of the company's unpaid invoices.

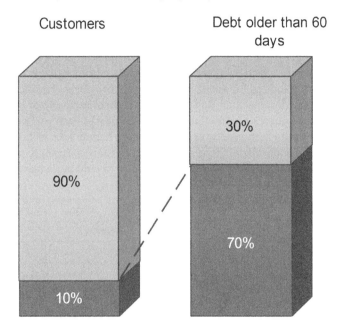

Figure 2.13   Pareto chart

Other tools for gathering data

Other qualitative methods that aid analysis include:

* focus groups;
* brainstorming;
* advice from experts.

Other quantitative methods include the following:

* analysis of historical data;
* influence diagrams;
* fault tree and event tree analysis;
* lifecycle analysis;
* test marketing and market research;
* Monte Carlo analysis.

Circulating the information

Having received the information, the business needs to circulate it. The risk manager may issue an internal newsletter that keeps relevant managers up to date on matters of risk.

Managing the information also takes effort in this age of information overload. The risk information may be fed into the company's enterprise information system, and displayed on the management 'dashboard', if the business has one. The use of red, amber and green colours (for 'major problem', 'minor problem' and 'no problem') quickly highlights key areas.

How Likely is the Risk? How Severe Would its Impact Be?

Probability and severity are two important factors in measuring risk. Severity is also known as Impact or Significance, while Probability is also known as Likelihood.

Organizations suffer *small* problems *frequently*, and encounter *major* problems *rarely*. The more severe the event, the less likely it is.

Some risks are low in severity and happen a lot, such as a worker slipping in the rain. Minor accidents are common in construction and, while unfortunate, they don't bring the building site to a halt. These risks are very probable but have a small impact, and are in the bottom left-hand corner of Figure 2.14. Because they are common, the company should guard against them, but they usually cause little loss.

Catastrophic events tend to be very improbable, such as deaths resulting from a terrorist bomb. This is a catastrophic but very improbable risk, located in the top right-hand corner of the table. Since for most companies the probability is very small, there is no point in trying to manage it. If, however, the business had an office in a known terrorist area, the probability would be higher. The company might then decide to manage the risk by moving the office.

But some catastrophic events happen frequently, such as fire. The risk of fire in a textile plant is both quite probable and catastrophic, so management should spare no effort to manage the threat.

It is worth compiling a grid similar to Figure 2.14. This will help you prioritize your risk management programme. Risks should be prioritized in order from top left of the grid to the bottom right. A risk which has a high impact and is probable should be tackled urgently.

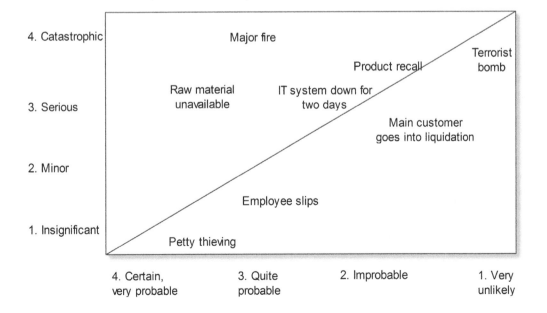

Figure 2.14   Risk severity and probability

There is no point in protecting the business against risks in the bottom right-hand corner of the table. Those are risks which are unlikely to happen and will not be serious. However, if the company has many of these risks, they may add up to a serious problem. For example, the risk of fire in individual rooms A, B and C might be slight. But the risk of a fire in any room could be quite high.

The British Government uses this kind of risk assessment to target fraud prevention. It has tried to work out where social security fraud is most likely and most serious. For example, UK citizens claiming to live in Canada but actually resident in the USA may be a more major source of fraud than citizens living in the Indian sub-continent.

PUTTING NUMBERS TO THE RISK

It is useful to quantify the impact of a risk. By multiplying the Probability by the Severity, you can quantify the importance of the risk. You can either use a simple 3x3 grid (giving 9 as the most major risk), or make the analysis more sensitive by using a 5x5 or a 10x10 grid.

In a 10x10 grid, the terrorist bomb would rate as 10 (probability 1 x severity 10). The risk of a major fire could be 54 (probability 6 x severity 9).

The 10x10 ranking has the added advantage of ending up as a percentage, which people readily understand.

Compiling a Risk Register

At this point you can start a register of risks. This can be quite simple – just a list of the greatest risks, a note on how each is to be controlled and who is responsible for managing it. A simple risk register is shown below as Table 2.8. Note that each risk has been given a unique number.

Regulators often want to see a risk register, though the author has yet to see a regulator query its content.

The risk register should be put on the intranet. This will help to foster risk awareness. Equally, once a risk has been identified, it potentially becomes a smoking gun. In other words, if you identify a risk and don't do anything about it, you could be legally liable. That is why each serious risk needs the following:

• a risk owner: someone who has been delegated to manage it;
• written procedures for controlling the risk;
• a regular audit to check whether the procedures are being followed.

At the same time, having a risk register can also be a protection in law. If you can show that you assessed the risk, and took reasonable steps to prevent it from happening, you are more likely to be absolved of responsibility for loss or harm suffered.

It is sensible to keep the risk register simple. This means restricting the amount of detail in it. Otherwise it becomes cumbersome and burdensome.

Table 2.8   Risk register

Department: IT Services		Location: Bridgend						
Assessed by		Date assessed		Reviewed by			Date reviewed	
Risk No.	Risk Description	Probability (1-5)	Severity (1-5)	Impact (probability x Severity)	Owner	Controls in place	Action needed	
47	Data loss due to corruption or hard drive failure or virus	3	5	15	J Sims	Nightly backup Quarterly restore exercise	-	
48	Loss of access to data due to denial of service attack	1	5	5	J Sims	DDoS plan	-	

A simpler register might look like this:

Risk no.	Risk description	Probability	Severity	Overall rating	Owner	Controls
49	Personal data revealed	3	3	9	J Sims	Data protection plan

COMPLEX REGISTER

Below are the points that you might include in a more detailed register:

- risk identification number (each risk should have a unique number);
- risk type (where a classification would help in planning the response);
- risk owner (person responsible for the risk);
- date identified;
- date last updated;
- description of the risk;
- causes;
- dependencies – are there upstream events that might trigger it?
- consequences;
- cost if the risk materializes;
- probability;
- unknowns. By definition, a medical research project has an unknown outcome. Similarly, people's future behaviour is unknown;

- severity;
- proximity;
- existing controls, chosen action;
- possible additional controls
- target date;
- action by the owner or custodian;
- closure date;
- cross-references to plans and associated risks;
- risk status and risk action status (what are you doing to control the risk?).

USING A 3X3 GRID AND HEAT MAP

You can make the register more visual putting each risk on a square on a 3x3 grid, as shown in Figure 2.15

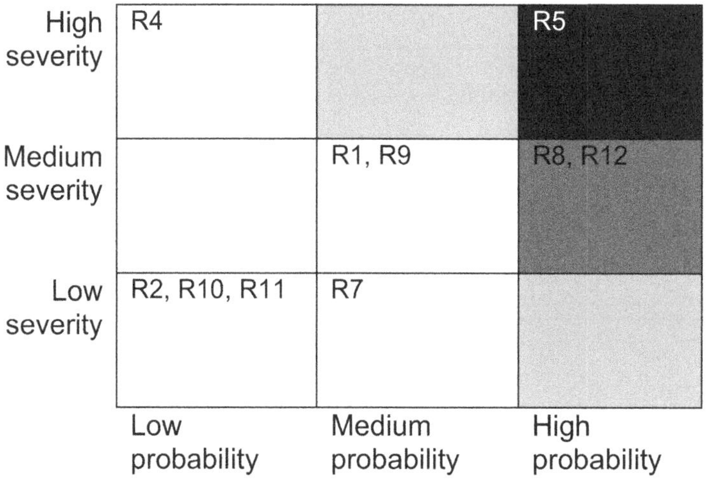

	Low probability	Medium probability	High probability
High severity	R4		R5
Medium severity		R1, R9	R8, R12
Low severity	R2, R10, R11	R7	

Figure 2.15   3x3 matrix for risk assessment

You can colour code the grid, by darkening the squares the nearer they get to the High Severity and High Probability quadrant, indicating the greater the impact they will have.

Or you colour low risks green, moderate risks orange and high risks red. Known as a heat map, this draws the eye to the issues that need attention.

Figure 2.16 looks at risks in a company fleet, with each risk being assessed for its likelihood and impact. This results in an overall risk factor. Again, the darker squares indicate a greater problem.

QUANTIFYING SEVERITY

You can make the judgement more scientific still by quantifying severity. This can be done using the headings in Table 2.9, to which we have added some examples. This is known as 'risk criteria'. It helps managers know how much risk is acceptable, and what the scale of a problem is.

We can measure the scale of an impact through either financial costs or from non-financial impact (such as the duration of a server failure).

Issue	Description	Likelihood	Impact	Risk
Fleet	Driver death	Unlikely	Severe	Moderate
	Vehicle theft	Possible	Moderate	Moderate
	Theft from lorries	Almost certain	Minor	Low

Figure 2.16   Departmental likelihood and Impact grid

Once you have quantified a risk, you will know where it slots into the table in Figure 2.15.

The analysis in Table 2.9 will help to clarify the company's thinking. For example, the business might judge that damage to its local reputation will have only a minor impact. This in turn will direct attention and resources to areas that would have a serious effect.

It also tells us that minor and insignificant impacts can be handled by less senior people, allowing top management to concentrate on the bigger risks.

The headings for this grid are not comprehensive; and you can add categories that are more relevant to your organization. Similarly, the impacts need to be scaled up for a large business, and reduced for a small one. Under the heading of 'health and safety', a large multinational business might re-classify one death as 'serious' and multiple deaths 'catastrophic'.

Table 2.9   Defining the scale of impact on the business (with some examples)

	Catastrophic	Serious	Minor	Insignificant
Financial impact				
Loss of revenue	Over £10 million	£1 million–£10 million	£100,000–£1 million	Less than £100,000
Costs and penalties				
Non financial impact				
Distribution	De-listed by two or more major customers	De-listed by one major or three to four minor multiples	De-listed by one to two minor multiples or 50%+ of independents	Loss of a handful of independent accounts
Reputation – consumer				
Reputation – local				
Environmental damage				
Output				
Product quality failure				
Health and safety	Death of employee	Broken limb or hospitalization	Bruising, sprain, minor cut	Trip or slip
Legal				
IT				

IS IT MATERIAL?

'Materiality' is an ungainly word that is frequently found in auditors' reports, particularly in financial auditing but it has increasingly crept into risk management. At its most basic it simply means: 'Something worth reporting on'. This helps us decide whether a risk is significant or not. An example is shown in Table 2.10.

Table 2.10   Material risks for Amcor (extract)

• The lifecycle impact of packaging continues to be a highly material issue, with increasing interest from our customers.
• Several governments have introduced new extended producer responsibility legislation for packaging in the markets Amcor serves.
• The Australian Government prepared for the introduction of its Clean Energy Legislative Package.
• Supply chain issues, including the sustainability of raw materials, and the environmental and human rights performance of suppliers, increased in importance to Amcor and other stakeholders.
• Food/product safety was identified as a material issue due to an increased awareness amongst our customers, particularly in Asia.
• The continuity of supply to key customer sites was identified as a material issue for the first time.
(Amcor.com)

In more advanced assessments, materiality is used as a specific number. As in Table 2.9, breaching that number makes a risk or problem material. Some examples of material risks are shown below. But note that numbers vary for each industry and according to the organization's risk appetite.

- 10 per cent of average pre-tax income, on a three-year average;
- 5 per cent of gross profit;
- 2 per cent of total assets;
- 1 per cent of total equity.

DEALING WITH UNCERTAINTY

All this makes it sound as though we can accurately quantify risk, and that is dangerous because it could mislead people. The chances of any risk occurring is at best an estimate. In fact the only risks we can fairly accurately forecast are common events in large organizations, such as slips and trips. Everything else is largely guesswork.

One way of emphasizing uncertainty is to draw risks as circles or ellipses on a matrix or heat map, rather than an exact point.

Known unknowns

Former Defense Secretary Donald Rumsfeld made a much derided speech about 'known unknowns'. Most people thought he was talking gobbledegook. But for anyone in risk management, it makes sense:

- **Known knowns:** These are things we know we know. We know our processes, our staffing and our revenue. We also know exactly what will happen if the oil tank leaks into the nearby river.
- **Known unknowns:** Some well-known risks have uncertain outcomes. We can only guess at what our competitors will do, or how the price of raw materials will change. These are risks we have identified, whose outcome is unknown.
- **Unknown unknowns:** These are the risks we are unaware of. They are 'Black Swan' events. They are completely unexpected, and could have major consequences. These risks do not appear on the risk register because we never thought to include them. They are the most dangerous of all.

What if...?

The risk assessment should include alternative outcomes. What if sales don't achieve their forecast levels? What if the costs of the project over-run? What if a specific chemical is banned in an important export market?

These 'what if' scenarios can alert the company to potential problems. Some risk scenarios include pessimistic, optimistic and probable forecasts. This is particularly useful for marketing or project risks. We look at scenario planning in Chapter 23.

Doing the Assessment

Anyone carrying out a risk assessment needs to understand the context in which it takes place. The assessor needs to be briefed on the following issues:

- What is the organization's external 'context'? As we have seen, that is the environment it operates in, for example stakeholder expectations. An engineering business is different from a publicly owned hospital.
- What is our internal context? What is the history, governance and structure of the business?
- What are the organization's aims or values?
- How much risk is acceptable? What should we do about unacceptable risks?
- Are there standardized techniques the auditor should use to assess risk?
- How does this risk assessment fit into the overall RMS?
- Who is responsible for performing the risk assessment? Who do they report to?
- Should the assessor provide feedback to the department being audited? If so, in what form and timescale?
- What freedom of movement does the assessor have? What happens if the assessor unearths unpalatable truths?
- What resources are available for carrying out the risk assessment?
- What will happen after the risk assessment? How will it be reported and reviewed?

As we have seen, each kind of risk has its own assessment methods. But there are common headings, which are shown below. In the subsequent sections, we pose the kind of questions which need to be asked.

1. purpose of the assessment;
2. nature of the risk, or description of the project;
3. who has responsibility for the risk?

4. resources involved or affected;
5. scale of the impact;
6. benefits of the hazard;
7. controls in place;
8. recommendations for improvement;
9. uncertainties;
10. continuity plans;
11. limitations of the assessment.

Let's look at each of these points in more detail.

1.   Purpose of the assessment.
Why is the assessment being undertaken? What will happen as a result of the assessment? How will it help the company?

2.   The nature of the risk.
You need to describe and then classify the hazard. It might be operational or technical, management or customer-related, or a political or financial risk.

- Is the hazard a continuous or catastrophic risk? For example, inhaling carcinogenic fibres is the former, while a terrorist attack is the latter.
- Immediacy (how soon is the event likely to occur?).
- Is the risk present at all times, only occasionally, or at specific times? An overseas factory is probably only at risk of attack at times of extreme industrial unrest, civil disorder or war.
- Has the threat materialized before? A company which has suffered a flood or industrial unrest is likely to be a victim again. A small fire is likely to be followed by a bigger one if no action is taken. Big accidents are usually preceded by lots of smaller ones.
- Is the hazard man-made (such as an internet denial-of-service attack) or natural (such as an earthquake)?
- How predictable is the risk? In some places, storms and typhoons are seasonal and you can predict them.
- Is it internal (under our control) or external (beyond our control)?
- Have similar hazards resulted in loss? Are there comparable examples in other industries or countries?
- What is the attitude of interested parties (such as the local or national government, opinion formers, local residents, employees and trade unions)?
- What is the view of experts in the field? The auditor may refer to consultants' reports. What trust does the company place on the opinion of the experts?
- What level of competence do staff have? Could they recover lost computer data or put out a small fire?

3.   Who has responsibility for the risk?
It is good practice to delegate ownership of the risk to the department where the hazard exists. And there should be a named manager who will be responsible for it.

4.   The resources affected.
Some exercises require a separate examination of the resource. In other cases, it may be already defined in 'Purpose of the assessment' above.

The auditor should state what resources will be affected. Some risks affect finance, while others affect people. In turn, people can be categorized as staff, customers, visitors, local residents, the trade (distributors and retailers), legislators or pressure groups.

- How would people be affected? This could be through disease, or through being burnt.
- Is the whole population at risk, or just certain groups (such as maintenance staff)?
- For plant and equipment, is the resource static or mobile? For example, an oil tanker is mobile while a pipeline is static. Monitoring and controlling a risk are more difficult for mobile assets, and they therefore need to be managed differently.
- What is the scale of the asset at risk? A pipe line may extend over hundreds of miles.

5.   The scale of the impact.
What effect would an accident have on the business? What is the scale of the disaster – would it wipe out the business? As we have seen, you can apply an impact rating to any risk, from insignificant to catastrophic.

- Would fatalities be instant or delayed? Some cancers take many years to develop.
- What effect would an accident have on opinion formers or pressure groups? And what effect might this have on the business? For example, altering the financial structure of the business might be risky, because its shares could become less attractive to investors.
- Can the risk harm future generations? This would apply in the case of radioactive materials being released.
- What effect would the hazard have without an accident? Could its very existence harm the company's image? Toxic waste incinerators are often unpopular among local residents.
- What is the company's experience in this area? It may be less risky for an oil company to drill for oil in a new geographic area than to explore for gas (if it has no experience of this). New developments are inherently risky.
- How reversible would the problem be? The effects of even a major oil spill on local eco-systems (like the Prestige tanker off the coast of Spain which cost $42 million) are now less permanent than once thought.

6.   Benefits.

- How important is the risk? What benefits does it bring?
- Would its absence reduce the company's ability to compete? For example a factory might provide essential raw materials to another part of the business.
- What benefits will it bring to:
  - the company (especially in terms of profit);
  - its workers (from wages);
  - the surrounding population (indirect employment).
- Are there any non-financial benefits (such as the creation of amenities, or the stemming of population loss)?

7.   Controls – Risk treatment.
Following the assessment, the business should decide how to manage the risk by avoiding, minimizing, spreading or transferring it. We look at this in more detail in Chapter 4.

- Can the risk be minimized (perhaps by engineering design)?

- What can the company do to prevent the risk from becoming a disaster? Mitigating factors might include building fire walls or installing alarms.
- Priority: As we see below, risks need to be prioritized since they cannot all be tackled at once.

8.   Recommendations for improvement.

These recommendations are likely to involve future plans, and longer term solutions. They are not things that can be implemented straightaway.

- What might be done in the future to minimize the risk? For example, could the raw materials or energy source be changed?
- What alternatives exist? How great are their risks?
- Could the technology be altered, for example from a combustible plastic to a lightweight metal?
- Ease and cost of mitigating factors (The risk of polluting a river may cost less (in fines) than the cost of installing expensive equipment). How might mitigating factors affect the scale of the risk and the viability of the project?

9.   Uncertainties.

As we have seen, almost all risk management is subject to unknown factors. These uncertainties might include:

- what competitors will do;
- how staff might act;
- what legislation the government might introduce;
- the strength of feeling in the local community;
- the scale of future weather events, such as storms;
- what future technologies might introduce.

10.   Business continuity plans.

- What is the organization's ability to deal with an incident or accident? Are emergency procedures and equipment in place?
- Have contingency plans been put to the test? When did the last test take place? What were the results?
- How long would it take to re-start the business?

11.   Limitations

- What are the constraints of the assessment? Were some areas inaccessible?
- How much trust can be placed in the assessment of risk? How reliable are the numbers? What uncertainties exist?

12.   Conclusions and recommendations

The final section of the assessment will include recommendations for action.

13.   Action taken.

Added to the report should be a note of the decisions taken, and the actions ordered. This section may also include a date for a review.

How Complex Should the Analysis Be?

Academics have mathematical formulae to assess risk. This type of assessment is rarely understood in boardrooms, and still less so by shop floor supervisors. Such calculations are of little help to management (unless they can be clearly explained and have practical implications for the business).

Risk management has to be understood by those who will detect or prevent risk. So there is no point in turning it into an academic exercise. What it loses in intellectual rigour, it gains by being understood.

Much risk analysis is common sense, and many hazards can be forecast. A company whose sales are declining will soon run out of money. An old product will eventually face new competitors. New legislation is preceded by years of discussion in the industry. An unguarded machine with blades is likely to hurt an operative.

The company should be aware of its threats; and the major ones are reviewed in the risk assessments at the end of each chapter.

Measurements of risk can and should be taken. Most industries and most companies gather simple statistical information, such as the number of accident-free days. They identify how probable an accident is likely to be in the future. Using historical data shows management which areas of risk to prioritize.

3
Who Does What? People and Their Roles

One of your office workers has just left in an ambulance. She broke her nose after tripping over a piece of wood that had been left on the floor by a builder who was refurbishing the office.

Apparently, no one saw that as a risk; and the office staff seemingly thought that risks were restricted to fire practice and how to lift boxes. 'So what am I paying auditors for?' you ask. You decide that the business needs to embed risk awareness more deeply into the culture. 'Let's start by seeing who's responsible for what,' you say.

Who Does What?

Risk covers everything from slips and trips to new product development. So who should manage these risks? The simple answer is: the manager in charge of the process. But there needs to be co-ordination across departments and oversight. So in this chapter we look at the main players in the organization, and see who does what.

THE CHIEF EXECUTIVE OFFICER (CEO)

Without the CEO's enthusiasm for tackling risk, it is doomed to failure. Therefore the CEO must be committed to risk management, and be seen to be committed.

So the CEO must head up the risk programme. He or she must appoint people to the roles described below, and must be active in developing a risk strategy.

Risk management is in part about ethics. An ethical stance is the only way to behave, and anyway unethical behaviour will come back to haunt the business. In all the topics discussed in this book – whether finance, the environment or treatment of suppliers – there is an ethical and an unethical way to behave. And if the risk management programme is to work, the CEO must demonstrate commitment to behaving as a good corporate citizen.

In starting a risk management programme, some CEOs send a letter or email to all staff, telling them about the company's concern to manage its risk, the steps that are being taken and how the recipients can get involved.

Another necessary action is the CEO's agreement to new appointments or to revised job descriptions, whether for a risk manager or more internal auditors (both of which are discussed below).

CASE STUDY: PRACTISING WHAT YOU PREACH

The chief executive of one major house builder regularly visits his sites, flanked by his top managers. He strides around the site, talking affably to the workers. But neither he nor his senior managers wear hard hats. This contradicts the safety messages pinned up on the sites. It tells employees that safety is not a major issue, and implies that short cuts could be taken.

AUDIT COMMITTEE

The audit committee serves the need of increasing regulation among listed and 'public interest' businesses.

The essence of the audit committee is that although it is a Board committee, it is usually chaired and largely staffed by non-executive Board members. This gives it a valuable independence.

The committee typically receives reports from internal auditors on the effectiveness of internal controls. And whereas once they were firmly fixed on financial controls, their remit has widened in recent years to include all kinds of risk. In many businesses, however, the audit committee remains focused solely on financial risk.

Meanwhile internal auditors are increasingly taking over the role of risk auditing.

Most organizations believe the audit committee should not be given executive responsibility for risk management. It should merely receive reports and provide the Board with information.

However, some risk professionals think that the audit committee should be given responsibility for risk management. This is because the audit committee is senior, and is already responsible for the external auditor and for corporate governance.

But others see this as expanding the role of the audit committee beyond its competency or jurisdiction. It would require the committee members to take on added responsibility, training and activity. And once people get line management responsibility, they can fall prey to all the limitations that managers suffer, including the need to cut costs and deal with office politics.

BOARD MEMBERS

As an alternative to using the audit committee, the overview for managing risk (or for non-financial risk) can go to the Board, or to a sub-committee of the Board named the Risk Committee. And the Board should certainly be actively involved in the strategic management of risk. In the FM Global survey of chief financial officers (CFOs) and risk managers in very large companies, 90 per cent of respondents thought that risk management should be, and is becoming, a Board-level issue. In 93 per cent of the European respondents, risk was a Board-level issue, compared with just 65 per cent of North American companies. This may be because of legislation in European countries requiring companies to disclose their risk management activities.

The Board should discuss the big risk management issues, since that is where strategic decisions are made. Board members should:

- know the organization's attitude to risk;
- be familiar with the corporate strategy;
- be aware of major risk issues in the business;
- receive regular reports on risk, including internal audits;
- know the extent to which risk is managed in the business;

- challenge management's preconceptions, cultural norms or received wisdom on issues relating to risk;
- review corporate governance statements.

Management must be ready to act quickly to changing circumstances. For example, periods of boom need to be managed differently from recessionary times.

The company must also be able to plan its future, rather than relying on its past successes and its current business formula.

The management team must be balanced and experienced, especially in medium-sized companies. A business with a weak or non-existent finance function is unlikely to survive long. One small and medium-sized enterprise (SME) company in the construction industry had survived for many years without a finance director. Recognizing its vulnerability, it brought in an experienced manager. But when it quickly ran into financial difficulty, he was given the job of chasing payment from customers rather than being allowed to examine why the company was failing. After three months of worsening performance, the company got rid of the accountant and went into receivership a year later.

FUNCTIONAL DIVISION OF RESPONSIBILITY

Each major risk must be someone's responsibility. And that responsibility must lie with line managers and the workforce. If staff feel that risk is something managed by an 'expert', they won't take ownership of it.

In recent years the chief risk officer (CRO) has come to the fore, taking responsibility for all risks. In medium-sized organisations, that role may be performed by the CEO.

But within that, individual risks get parcelled out to the relevant department. And there has been movement and amalgamation as to who is responsible for what. As Figure 3.1 shows, nothing is tidy about risk management.

Traditionally financial risk and fraud has been the responsibility of the finance director. Today the internal audit department has taken a remit that is often wider than just internal financial controls.

Some areas of risk are more likely to have a specifically appointed manager. This is especially true of the quality manager, the health and safety manager, or the environmental manager.

In Figure 3.1, some risks have been amalgamated, as with the safety, health and environment (SHE) manager.

In other cases, the quality manager has taken over responsibility for environment, reflecting the increased importance of environmental management.

At the same time, the operations manager or production manager is increasingly given responsibility for preventing various threats to the business.

Fire is often a separate responsibility, sometimes combined with security. IT risks are typically managed by the IT department.

And business continuity is handled by either the CEO, the IT department, Facilities or Human Resources (HR), depending on the nature of the organization.

On the one hand, this separation is a good thing, in so far as it hands responsibility to functional line management. But it also prevents the business from seeing all its risks together. Only by looking at risks as a whole will a business be able to weigh them and judge which are the most dangerous. The alternative is to end up with a messy chart like Figure 3.1.

CEOs and regulators would prefer to have just one person to turn to when it comes to risk. Therefore some roles are likely to be amalgamated as time goes by.

Every industry and each organization is different, so the contents of Figure 3.1 will vary considerably. The chart reflects a manufacturing operation. Retail organizations will place more emphasis on estates and purchasing. Local authorities will be more concerned with governance. Construction companies will be focused on project management. Figure 3.1 merely provides a structure for examining risk management responsibilities.

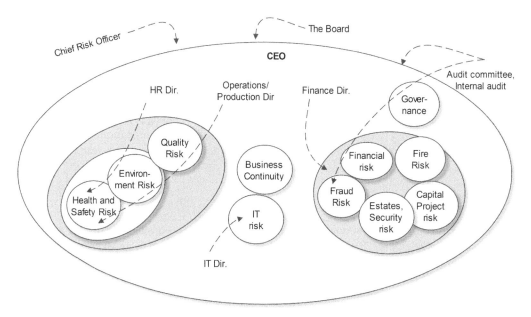

Figure 3.1   Management responsibilities for risk

THE CHIEF RISK OFFICER (CRO)

Companies are increasingly allocating the management of more of these risks to one person, the risk manager, also known as the CRO. First introduced at GE Capital in 1993, CROs (or someone fulfilling that role) are now found in 84 per cent of major corporations, according to Accenture.

If the organization has a risk management system (RMS), which we examine in Chapter 24, it is very likely to have a CRO.

According to a survey by FERMA (the Federation of Risk Management Associations), the majority of risk managers report to the president/chief executive or the CFO (see Figure 3.4).

Medium-sized companies are less likely to have such a person exclusively devoted to risk management. In such cases, the task is likely to be given to the CEO, company secretary or (in regulated industries like financial services) the chief compliance officer (CCO). However if these people already have a full- time job, it is unlikely that they will be able to devote sufficient time to risk management, and so risk management could founder. Hence many organizations have several risk managers, sometimes embedded in different divisions or plants,

It is impractical to expect one individual to have a detailed knowledge of too many subjects. At the same time, there is a risk of duplication if similar areas are the responsibility of different people.

This emphasizes the need for a risk management plan. It will ensure that everyone knows what the major risks are and who is responsible for them.

Every company is different, and size plays an important role. The large firm can have specialists in each area, while the small business cannot afford that luxury.

The risk manager will have a broad understanding of corporate risks. By contrast, an environmental manager or health and safety officer may feel that their subject is the only major risk.

The problem with grouping risks is that managers who are expert in one of the fields often know less about the other areas, which are often thrust upon them as an added responsibility. Someone trained in health and safety rarely knows about environmental risk, let alone financial risk.

Major companies are, nevertheless, increasingly allocating the management of all their risks to one person. The risk manager does not have functional responsibility for the risks – that remains with the line manager. Otherwise, staff will not feel personally liable for the risks they might engender. Thus the risk manager could be more correctly titled risk advisor, though that might seem to demote the individual and therefore demoralize them. The responsibilities of this person include the following:

* embed an awareness of risk in the business;
* evaluate the risks of the business, and introduce plans to minimize them;
* advise employees how to manage risk;
* create or administer a RMS;
* enable the directors to fulfil their statutory or regulatory responsibilities;
* keep the business up to date on changing risk issues.

The risk manager will understand risk, but not every process, while process owners will know the process but may not understand all its risks. This highlights the importance for the two to work together.

The risk manager will also liaise with internal audit. The relationship can be fractious, because internal audit reports to the Board, whereas the risk manager is more likely to report to a senior manager.

Where the organization has many sites, each should be visited once a year where possible. Individual sites should have their own risk advice staff.

RISK COMMITTEE

If the organization is sufficiently large or disparate, there may be a risk committee or risk council, as shown in Figure 3.2. This can take several forms. In some cases it is a sub-committee of the Board. In other case, it is formed from representatives of the main departments who get together to learn from each other and build a company-wide appreciation of risk.

So whereas the audit committee is non-executive (it receives reports and comments on them), the risk committee is actively engaged in ensuring that risk is properly managed. This also reduces the risk that the audit committee will be pressed into taking action, for as soon as it does, it is no longer independent. Depending on its constitution and membership, the risk committee will:

* ensure that staff adhere to the company's risk policies;
* make sure that audits are carried out;
* review the findings of audits, and implement controls;
* develop risk awareness among their staff;
* control risks within their own departments;
* make recommendations to the Board for risk-related changes;
* support the risk manager in facing the challenges presented by business risk.

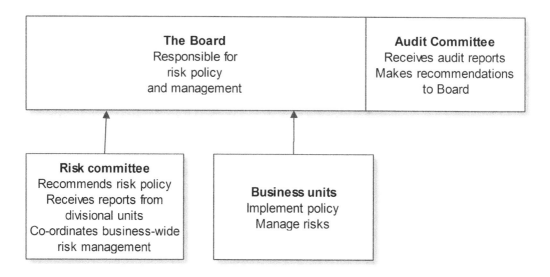

Figure 3.2   Simplified accountability structure

At Aventis, the pharmaceutical business, members of the risk council are drawn from each business area (for example, science-driven research, regulatory affairs and industrial operations). The council is comprised of those people who are responsible for risk reporting in their area. Each council member provides a quarterly risk report, updating the others on the major risks in their business area. The reports are edited and presented to the audit and advisory Board once a year. This format provides a unified corporate view of risk, rather than the opinion of individuals, and therefore carries more weight.

LINE MANAGERS

Line managers must have responsibility for managing the risks within their areas. Their staff must understand the kind of risks in the business, and know how to manage them.

It is easy but dangerous to over-manage line managers, by presenting them with ready-made risk management solutions, by condescending to them, or by sending in inexperienced auditors. In many cases the line managers are running multi-million pound operations. They must take control over their risks, and be given support and advice. If line managers contact the risk manager for advice, you will know the system is working.

Line managers are likely to be the 'risk owner', that is to say, the person responsible for controlling a risk. This emphasizes that every risk should have an owner, and that person should have the authority to manage the risk. This means it should be someone with line responsibility, rather than a risk officer.

SHOP FLOOR EMPLOYEES

Production employees usually face the greatest level of personal risk. They are also most likely to cause environmental or operational failure, for example by emptying chemicals into a river or by allowing a fire to start. Therefore they need to thoroughly understand the nature of risk as it applies to them, and be involved in deciding how to manage it.

SUPPORT STAFF

Most of the company's staff have a role to play in risk management. They include the following:

- Company secretary or compliance officer: Keeping everyone abreast of regulatory changes.
- IT: Helping users to understand IT risks, including the risk of downloading viruses and opening email attachments.
- HR: Providing training in risk awareness and risk assessment; and organizing new staffing appointments.
- Research and Development (R&D): Advising the business on potential threats from change in technology, raw materials and processes.
- Marketing: Helping staff recognize threats from competitors and new channels.
- Sales: Alerting people to new competitors and their strategies.
- Facilities: Warning employees about risks inherent in buildings and plant.

INTERNAL AUDITORS

Internal audit, and the people who carry it out, is the means by which the directors can satisfy themselves that the business is sound.

As explained by Dejan Kosutic (http://blog.iso27001standard.com), internal auditors are of three types:

1. *Employee, full-time auditor.* These are found only in larger organizations, ones that have enough regular work for this activity.
2. *Employee, part-time auditor.* This is more usual in medium-sized organizations. You can use your own employees to perform internal audits in addition to doing their normal jobs. Since auditors cannot audit their own work, because there would be a conflict of interest, you need at least two auditors, so they can audit each other's normal job.
3. *External consultant.* Although not employed by your organization, they will be regarded as an internal auditor because you are commissioning and managing them to do an internal audit. They will operate to your rules and standards, unlike someone performing an external audit.

Internal auditing tends to be uncomfortable for auditees. The auditor is seen to be checking their work and looking for problems. Therefore you should present the audit in a positive light. Internal audit should be seen as a strategy to improve processes, to forestall bigger problems, and to avoid problems from being uncovered by external auditors or the media.

You should encourage internal auditors to be sensitive to the fears that their presence brings, and help them acquire emotional intelligence and people skills. This is particularly true for people brought up within the finance function, where accuracy is valued more highly than interpersonal relationships. One auditor in a government department responsible for giving business grants was known to be arrogant and used his role to bully members of staff.

At the author's distance learning business, the auditor role is circulated among all tutors once they have gained experience. Because it is a regular event the audit becomes routine rather than anxiety-laden. It also builds expertise and the skills of self-reflection and critical analysis among employees.

On the other hand, internal auditors are at risk of being 'captured' by the departments they audit. It is hard for a relatively junior employee to criticize senior management's departments. There is a risk, therefore, that problems will be swept under the carpet. Your internal auditors will be strengthened if they have written questionnaires, or specific questions to address. They also need to be trained to recognize intimidating or hostile behaviour and how to manage it.

Internal audit need not be solely top down. You should encourage department managers to request audit on specialist topics, if only to confirm that the department is working properly. One manager did just this, to verify that her department was adhering to 'right to work' legislation. The auditor discovered that lower-level employees were not carrying out proper checks, which would endanger the organization's right to sponsor the immigration of overseas workers. However the discovery meant that the business was able to quickly rectify its procedures and pre-empt any findings by a government audit.

The growing scope of internal audit

With risk becoming a strategic issue for management, internal auditors are taking on responsibility for auditing areas they might not have done in previous years. They are being asked to audit areas outside their traditional competency of financial controls. Management therefore risks being lulled into believing that audits are being successfully carried out when in fact the auditors are either over-stretched and under-resourced, or are insufficiently trained to understand operational risk.

Internal auditors' backgrounds

The word auditor means different things to different people. There are at least three types of auditor:

1. *Accountancy.* Financial auditors have traditionally carried out financial audits on corporate subsidiaries and divisions. In recent years, as corporate governance has risen up the agenda, these auditors have acquired wider responsibility for risk audits. Such individuals are expensive, are likely to have high acumen, and are able to think and communicate clearly.

    Such auditors tend to be ACA (Associate Chartered Accountant), or in the USA CPA (Certified Public Accountant). Lesser accounting qualifications include the Certificate in International Auditing from ACCA, the global body for professional accountants..

2. *Quality.* These auditors have come from inspection and manufacturing.
    From their origins in quality, supplier, and health and safety audits, they have moved into environmental issues and thence into risk management. They are especially suited to production-oriented risks, rather than risks relating to fraud or financial issues.

    Such internal auditors may need to conform to the international standard ISO/TS 16949 on Internal Quality Audits – Auditor Qualification. You can also find advice on choosing auditors in *ISO 19011: Guidelines for Quality and Environmental Management Systems Auditing.*

3. *Internal auditors.* It is worth distinguishing people who have specialized in internal auditing, rather than being accountants or production people. They may have a CIA (Certified Internal Auditor) qualification from the Institute of Internal Auditors (IIA), This requires a first degree, 24 months internal auditing experience, an exam and continuing professional education. They are likely to be cheaper than a qualified accountant, but less financially adept; and they may be more expensive than someone with a shop floor quality background, but possess a wider view of risk.

    Many people use auditing as a step to greater responsibility. It gives people a much wider view of the organization, as well as the opportunity to travel, and to have contact with management. Thus audit departments can include people who are in training for their accountancy exams and who are required to have a sponsoring organization.

Internal auditors must be trained to carry out their audits, and given clear terms of reference for their audit. This must cover the objectives of the audit, audit review and the relevant dates. And to break down suspicion this information must be shared with auditees. The aim of the audit should be to add value, as well as ensuring that internal controls are working. The auditors' role is to:

• conduct audits to check that the company's risk policy and controls are being followed;
• evaluate whether the business is exposed to uncontrolled risks;
• provide assurance to management that its risk management policies are being met;
• give advice to line management and employees on how to control risk (if they have the experience or responsibility to do this).

Internal audits must be independent of the department they audit and, if the organization is large enough, independent of the division.

Auditors should not do any of the following:
• make policy;
• manage risks;
• design controls;
• implement controls.

Figure 3.3 shows a comparison between the roles of senior management, line management, the risk manager and internal audit.

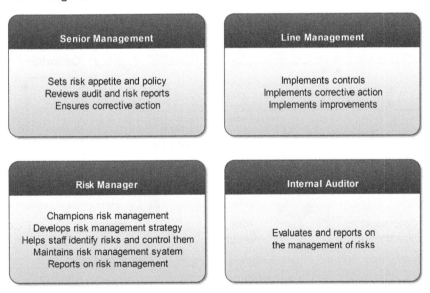

Senior Management

Sets risk appetite and policy
Reviews audit and risk reports
Ensures corrective action

Line Management

Implements controls
Implements corrective action
Implements improvements

Risk Manager

Champions risk management
Develops risk management strategy
Helps staff identify risks and control them
Maintains risk management syatem
Reports on risk management

Internal Auditor

Evaluates and reports on
the management of risks

Figure 3.3   Responsibilities of management, the risk manager and auditors

OTHER STAFF MEMBERS

All staff members should be involved – no matter how briefly – in considering the risks that might occur in their department, and how they can be controlled. This will help to embed risk awareness within the business. You might add responsibilities for risk management into job descriptions, where appropriate.

THE THREE LINES OF DEFENCE

To recap: responsibility for risk must lie with operational managers and their staff. The risk manager should help employees become aware of risk, and implement procedures for managing it. And internal audit will check that all controls are in place and are working properly.

This provides what is known in risk management as the 'three lines of defence', and is shown in Figure 3.4. The external audit provided by the organization's accountants then provides a fourth line of defence.

The dotted lines show that operational departments are being assisted by risk management and checked by internal audit. The latter has a direct line to the Board, while risk and operational management often have a somewhat more lowly status, reporting to senior management.

We can pity the operations people who have not one but two sets of eyes watching them; but at least management knows that risk is being managed.

There are, however, some weaknesses in the 'three lines of defence' approach. Firstly, the military metaphor summons up an image of the Maginot Line – a bulwark that lulled the French into thinking their land was secure – until the Nazis simply went round it. Moreover, it emphasizes a passive assessment of compliance rather than an active assessment of risk.

Secondly, having two separate monitoring units (risk management and audit) adds cost and complexity. More reports, anyone?

Thirdly, three lines sounds like 'more is better'. And just as razor companies add more blades as a way of 'improving' their products, it may only be an illusion.

Finally, since all of these managers and auditors take their information from the same source, if that data is misleading, fraudulent or limited, all will come to the same wrong conclusions.

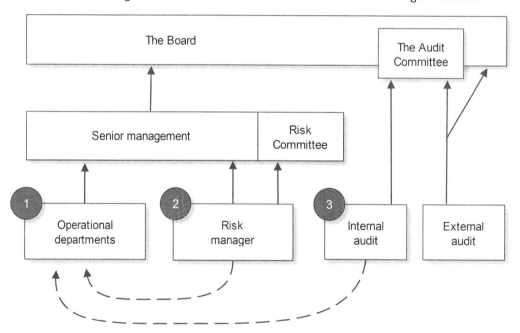

Figure 3.4   The three (or four) lines of defence

Who Will Assess Risk?

In some organizations, the risk manager carries out risk assessments. However it is better that line managers and their staff should do it. This ensures that they get involved in the process of risk management.

These assessors should be trained for the job. This means they should know what risks might exist, and how to measure them. They should also know how to minimize the risk.

The assessment process should be looked upon as a learning experience for everyone involved. Through discussing the risks, the company will learn how to manage them.

Note that this is a different job from that of 'auditor'. The auditor should be someone who does not work in the department that is being audited. We discuss this work below.

EXTERNAL AUDITORS

The company will be used to dealing with accountancy auditors. But using external auditors for risk auditing may be a new development, especially if regulations require the business not to use the firm that is hired for the financial audit.

Although the annual audit may refer to the company's risks, it is not a major part of the audit. Therefore the business may decide to conduct a separate exercise to validate its view of its risks.

There is also a second kind of external auditor, namely the type that audits quality or environmental management systems (ISO 9000 or 14000). These are more concerned with operational risks rather than (for example) fraud; but they may have more experience of shop floor risks than the accountant-auditor.

CASE HISTORY: HOW IBM MANAGES RISK

IBM UK has a Project Assurance and Risk Assessment group which aims to increase awareness of risk among the company's project managers. The Industry Business Division provides IT solutions for medium and large firms. The group uses risk management to ensure that its jobs are profitable and effective. Used in this way, risk management has become a powerful tool for gaining competitive advantage and winning more business. In IT projects, there are many risks. They range from internal failures, such as not making a 'business case' (assessing the project's profitability), to client-based risk. For example, a project where the client thinks IT is a nuisance is more risky than one where IT is seen as useful.

The risk group works with IBM project managers, and helps them to understand the risks inherent in each contract. This lets the project manager alter the design of the project to reduce the risks. It also lets him set a price that reflects the risk.

The group does not seek to gain a detailed understanding of the risks. In fact it deliberately tries to avoid this, in order to provide independence and to encourage the project team to gain the information.

Among IBM's risk management tools is an 87-point checklist, which is broken into six categories (such as Technical and Implementation). The checklist allocates points for 'more risk' and 'less risk'. This helps the project manager see the scale of risk and where it is concentrated.

RISK MANAGEMENT QUALIFICATIONS

Qualifications in risk management are only now emerging. Alarm, the public sector risk management forum, has been promoting a Registered Risk Practitioner (RRP) qualification, in conjunction with the IRM and The Association of Insurance and Risk Managers (AIRMIC).

There are other qualifications that demonstrate that the individual has worked in an area relating to risk, such as NEBOSH, the health and safety qualification, or Certified/Chartered Insurance Professional (CIP).

In the UK, the Universities of Leicester, Nottingham, Portsmouth and Glasgow Caledonian provide degrees in risk management. In the USA are NYU Stern, Columbia, and St John's.

Some university courses are founded in older or more specific types of risk management such as insurance or health and safety; while others are more focused on business risk.

Short courses are available from universities such as Harvard as well as many commercial trainers.

Risk Culture

The risk-aware organization is a safer one. All staff should be aware of hazards, and the practical implications of an incident. It is particularly important in high-risk environments.

This comes from the top. Senior management needs to make people aware that it takes risks seriously. It must also foster an environment that is open to ideas, committed to learning from failure, one that does not stand on ceremony, values knowledge rather than pecking order, is low on bureaucracy and has a bias to action.

In Chapter 5 on operational risks we look at the research into Highly Reliable Organizations (HROs), which are those businesses that have low failure rates for their industry.

The risk manager also carries the responsibility of communicating the importance of risk. People need to know that risk is everyone's responsibility. It doesn't belong to the risk manager. This means raising the profile of risk management by the following means:

- getting departments to do their own risk assessment workshops;
- providing training. This can include formal and informal sessions;
- maintaining a flow of information, including through the intranet and company email.

This also means that the risk manager has to possess good interpersonal skills. It is not enough to possess in-depth knowledge of company processes. The manager must also be able to motivate, inspire and charm people. He or she will encounter managers who have other priorities, and who are unsympathetic to yet another initiative, or who will want the manager to do risk management for them.

The risk manager must therefore possess a degree or charm, charisma and some animal cunning. He or she must be able to hobnob with the lowest of staff, and command respect among top management. The manager must be able to articulate his or her thoughts clearly, and present findings with confidence and presence.

TRAINING PEOPLE TO BE AWARE OF RISK AND TO REDUCE ERROR

Managers must receive training in managing their risks, and they must train their own staff.

It is not enough to hand out risk management manuals to all managers. This will have no effect on the business, even though it may superficially appear that everyone has been briefed.

A more painstaking approach is needed. The risk manager should provide a training programme in risk awareness, and records should be kept. The training will not be as thorough as the auditors' training (above); it may be segmented into the risks that affect the specific business unit, or else specific types of risk (such as environment, or health and safety).

CASE STUDY: NATIONAL HEALTH SERVICE (NHS)

Total legal liabilities for the NHS (the theoretical cost of paying all outstanding claims immediately) are estimated at an astonishing £7.9 billion. Risk management is therefore an important topic for the health service.

The Clinical Negligence Scheme for Trusts (CNST) provides a means for NHS hospitals to fund the cost of clinical negligence litigation. Membership of the scheme is voluntary for all hospital and primary care trusts in England, and there are advantages to membership – Trusts get a discount off their Scheme contributions where they can demonstrate they are complying with the Standards.

Table 3.1, an extract from the NHS risk manual, discusses a specific requirement – that staff who operate equipment should be trained to do so. This is set at Level 3, which is a measure of difficulty. Trusts are assessed first at Level 1, and in subsequent years will seek to achieve the more demanding Levels 2 and 3.

The audits are carried out by an independent firm of auditors, and there is a maximum of one audit per year. This case study highlights the merits of a scheme which:

1. is voluntary and therefore has only willing participants;
2. has financial advantages for participants;
3. is graded in difficulty – this means it is accessible to all operating units; yet higher standards can be attained over time;
4. is limited in duration – with a maximum of one audit per year, and with the level one audit taking only one day.

Table 3.1   Extract from NHS Clinical Negligence Scheme for Trusts (CNST) Manual

Criterion 5.3.3	Staff who operate diagnostic or therapeutic equipment are systematically trained to do so safely and effectively.
Level	3
Source	*Code of Professional Conduct* NMC April 2002 *Devices in Practice A Guide for Health and Social Care Professionals* Medical Devices Agency June 2001 *Equipped to Care The Safe Use of Medical devices in the 21st Century* Medical Devices Agency April 2000 *Provision and Use of Work Equipment Regulations* 1998
Guidance	This is a further development of criteria 5.1.5 and 5.2.6. All equipment that requires operator training should have been identified, and appropriate training programmes should be in place. The Trust should be able to produce evidence of all staff having received appropriate training, including Maternity Services where applicable. Frequency of training updates should be considered. The effectiveness of the system should be monitored by the Trust Board.
Verification	Training programmes. Records of attendance must cover all departments and professions, including the Maternity Services where applicable.
Links	5.1.5 5.2.7
Scoring	10

Most problems occur through human error, especially when new methods are being introduced. The company should identify where catastrophic error could take place, and train staff to avoid it.

Training helps a company meet its legal obligations. One major organization invites staff from its divisions to attend seminars on subjects like waste management. By providing the training the group shows staff how to stay within the law. If staff fail to attend the training, or if they break the law, the company believes that it can reasonably plead due diligence.

In addition to training, you can put up posters on notice boards, and issue leaflets and brochures. Staff meetings where managers are giving briefing documents are also useful, especially if the meeting encourages involvement by doing risk assessments.

DEVOLVING RESPONSIBILITY AND A CULTURE OF OPENNESS

According to Kenwood Appliances, 'The devolution of responsibility is critical to making things happen. This does not happen often enough because of the desire to control risk.'

Coca Cola in the UK echoes this view. 'If corporate value systems place too much emphasis on penalizing failure than rewarding success, people will not take risks,' they say.

Empowerment is vital. The company must let staff question existing views and criticize lax safety standards. There must be a willingness to debate issues in the business. This may require a change in corporate culture and the development of some form of Total Quality Management (TQM).

AVOIDING GROUPTHINK

Does everyone in your organization agree what needs to be done? Is there a strong degree of unanimity? If so, you may be a victim of 'groupthink'

The term was coined by the psychologist Irving Janis in 1972. He noted that some individuals made decisions they felt would be acceptable to the rest of the group. Janis defined it as 'decision-making characterized by uncritical acceptance of a prevailing point of view'. Members of the group suffer an illusion of invulnerability and morality, and construct negative stereotypes of outsiders.

A business that suffers from inadequate internal debate may stultify, leaving it exposed to danger. Innovation is an example where groupthink often holds true. For years, PC manufacturers put computers in beige boxes because 'that's how computers look'. All new models were 'safe' products of groupthink. Eventually Apple broke the mould with its colourful, rounded iMac computers, and took advantage of that sales opportunity.

One method of preventing groupthink is to use an impartial chairperson, who can lead decision-making sessions and encourage participants to 'think outside the box'.

Another idea is to include a lay person in technical decision-making sessions. Alternatively, you could use an 'embedded alien', according to Cindy Barton Rabe, a strategist for Intel Corp. She suggests putting a non-expert into the team for a time to act as a catalyst for new ways of thinking. As the name suggests, this person should be foreign to the project; but also a strong innovative thinker with the capability and experience to understand the key issues at hand.

Using Risk Management Consultants

There are many kinds of consultants who could help the task of risk management. Most specialize in a specific area of risk. For example, some are skilled in international risks and kidnap and ransom. Others concentrate on building security.

There are also environmental consultants who can help prevent pollution. The same applies to health and safety, and fire consultants. Public Relations (PR) consultancies are expert at media relations, but may be less good at understanding how to introduce management systems.

The first approach should be to organizations which provide free information or help, and which have no vested interest in selling a particular solution. This includes government departments, trade bodies and possibly the company's insurance company. There are also magazines and books on different aspects of risk.

Only when these sources of information have been exhausted should the company consider hiring a consultant. You should consider the following guidelines:

- Find a consultancy whose skills match your needs. Be cautious of a consultancy which claims to be an expert in all kinds of risk. It probably has a bias to a particular kind of risk.
- Beware of consultants who market software or some proprietary solution. The consultant will discover that his product miraculously meets your needs.
- Always ask more than one consultancy to offer a proposal. This will allow you to compare price and expertise.
- Find out how much experience the consultancy has in the company's area of concern. What clients has it worked for? Talk to those clients, to get their opinion of the consultancy.
- Does the size of the consultancy affect the work? If, for example, the work involves many plants around the world, could the consultancy handle it?
- What will it charge for the work? The cost of the project should be less than the benefits or savings the consultant achieves. It is best to agree a price for the project, rather than have an open-ended commitment. If the scale of the job is unknown, you can commit to a project that just assesses what needs to be done.
- What would be the outcome of its work? In what way would this help the company?
- Meet the consultants who would work on your business. How much experience do they have?
- Beware of a consultancy which wants to do too much of the work. Its aim should be to teach your staff how to cope with business risks. How will they ensure that your staff take ownership of the new ideas?

Risk Assessment

This is the first of many checklists which cover the main areas of risk in the business. The checklists, which are at the end of each chapter, reveal the areas in which the business is most at risk. They indicate the kinds of risk that you should prioritize for action.

You can assess your organization's risk preparedness by answering the questions below. Score one point for each box ticked.

Question	✓
Is risk management the responsibility of a Board member?	
Has the chief executive initiated any discussion about risk in the last three months?	
Has the business appointed a risk manager?	
Does the company have external risk auditors?	
Are risk audits reviewed by senior management regularly?	
Have staff received risk awareness training?	
Have departmental staff conducted risk assessments?	
Do major risks have an 'owner'?	
Are internal auditors independent of the operations they audit?	
Does the organization have a culture of openness?	
Total points scored:	

Score: 0–3 points: high risk. 4–7 points: moderate risk. 8–10 points: low risk. The Appendix has a summary of the checklists. By entering the results of this one, you can compare your organization's preparedness against other more specific categories of risk.

4
Treating Risk: Avoid, Accept, Control or Share?

The national press are ringing. Customer records from your organization have been found in a recycling centre. And instead of coming to you, the person who found them has gone to the media. HR says it has a clear policy that old records should be shredded. But apparently the department got a school leaver to do it, and no one checked the bin bags. You sigh. It was down in black and white: 'Shred all old client records.'

'Summon the risk committee,' you tell your assistant. 'We have to manage these risks properly.' Wearily you pick up the phone to talk to the Daily Mail reporter.

Four Ways to Manage Risks

Risk management is about priority. You start with the most pressing risks, those that will have the biggest impact and probability. Then you deal with lesser risks, and finish with those that are insignificant and unlikely to happen.

There are four ways to manage risks. They are as follows:

- avoid them;
- share them;
- accept them;
- control them.

You adopt one of these solutions for each risk, depending on how likely the threat is, and how severe its impact will be. This is shown in Figure 4.1.

These four actions used to be known as the 4Ts: Terminate, Transfer, Treat and Tolerate. They were a neat set of words which were easy for people to understand. But ISO 31000, the risk management standard, uses the phrase 'risk treatment' to mean any or all of the four actions. This change in terminology could lead to confusion with the 4Ts, which is why many organizations have stopped using them. But where they are part of an organization's risk vocabulary, it makes sense to keep them. Let's look at each of these actions in turn.

AVOID THE RISK

'Avoiding the risk' means choosing not to accept it. You might decide to stop offering a high-risk service, one that could lead to expensive litigation. You might choose not to acquire another firm

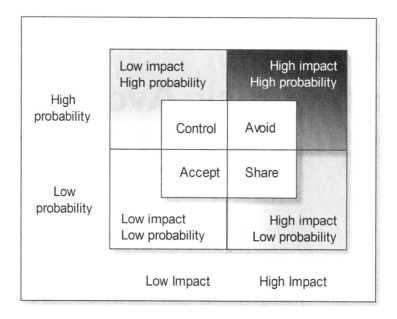

Figure 4.1   The four actions of risk management

because the risks of its failing are too great. Or you might sell a division that has large peaks and troughs in its profits.

A serious threat that is highly likely to occur is one you want to get rid of. It might be an old piece of critical equipment that is on the verge of dying. You would probably decide to replace it before that happens.

Similarly, some processes and activities are more risky than others. You should dispose of any non-core high-risk divisions unless they provide high rewards.

Some organizations, especially those in the public sector, operate with high-severity, high-probability risks, and management may have no choice but to retain them. In such cases, you will have to adopt measures that reduce the probability or severity of those risks.

For example, a homeless charity would be unable to turn away homeless people even though there is a high chance of fights. A hospital cannot refuse to treat very sick people even though there is a chance that they will die during surgery. Instead, it must use the best available staff and techniques to heal such patients.

Figure 4.2 shows a flow diagram for the treatment of risks. It shows how you identify whether a risk is acceptable; and if not, whether you can control it. If you can't, you have to get rid of it.

Since we're discussing highly probable risks that have a big impact, the organization also needs a business continuity plan for these eventualities, something that we discuss more fully in Chapter 23. This lets the company plan for the worst case scenario. If disaster happens, the company must be able to survive and start up again. Having emergency plans will minimize the scale of the problem. However, the cost of the plans should be proportionate to the risk: no business can avoid risk completely.

SHARE THE RISK

A problem shared is a problem halved, as the saying goes. And there are many risks which you can share with others to lighten the load.

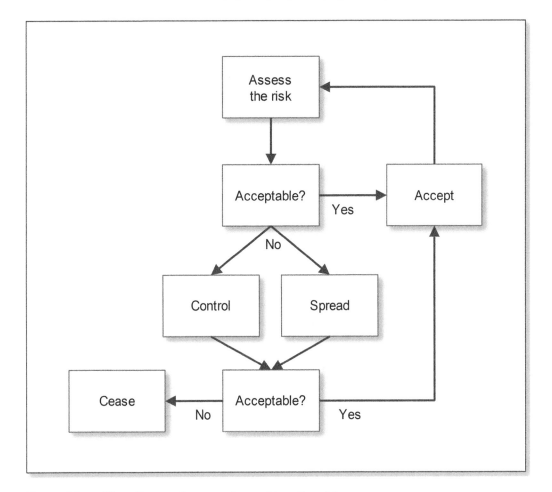

Figure 4.2   Flow diagram for assessing and treating risk

Sharing the risk is also known as transferring or spreading risk. Techniques include JV, redundancy, sub-contracting, outsourcing, dual sourcing, offsetting risk to suppliers, hedging, diversification or buying insurance.

> *Joint venture (JV)*: JVs are relevant where the organization is bidding for a project beyond its resources or where the project requires skills it doesn't possess. We discuss JVs in Chapter 21 on Acquisition and Divestment.
>
> *Redundancy*: A RAID drive that writes to two hard drives at once is an example of redundancy. And some organizations buy redundant space on servers to cope with unexpected demand or a denial of service attack (DDOS).
>
> *Sub-contracting*: With its wide range of processes, the construction industry is well known for using sub-contractors. They are often used to even out the peaks and troughs of the work, but also solve the problem of having to manage specialisms such as planning consultants, or high-risk work such as roofing or asbestos treatment. If part of your product requires manufacturing processes that you lack, you might sub-contract it rather than acquire your own plant and experts.
>
> *Outsourcing*: Governments increasingly outsource their work, from privatizing prisons to outsourcing back-to-work schemes. It means they get a more flexible and more disposable workforce. Equally, much software development is now outsourced, often to low-wage economies

like India. This is usually cheaper (though sometimes of lower quality) than using experienced in-house staff, and allows the business to manage its peaks and troughs. We cover outsourcing in Chapter 7 on Procurement.

Dual sourcing: This involves buying the same product from two suppliers. On the one hand the price might be higher, and you double the workload; but it ensures you never find yourself in a position of being without the product. Dual sourcing is important if there is uncertainty over supply.

Offsetting risk to suppliers: Companies can push responsibility upstream to their suppliers, by requiring them to be responsible for product quality and safety. This is becoming more essential as laws on product liability and pollution become more widespread and stringent.

Note that involving partners, whether through sub-contracting, outsourcing or JVs adds further risk.

Hedging: Hedging is a way of setting a price for a vital raw material, commodity or currency in advance. This avoids the cost fluctuations of the open market, and ensures that the price won't jump in the coming months. With fuel accounting for 35 per cent of an airline's costs, Virgin Atlantic spends £1 billion each year on fuel. Hedging helps the firm contain its costs. In one recent year, the price of oil varied from $38 to $147 a barrel.

Diversification: Founded in 1925, Metal Box diversified out of making metal boxes and became Europe's largest supplier of central heating radiators. It then became a highly diversified business selling sanitary ware, plastic pipes and electronic controls, before ultimately being sold in 2005 to Honeywell who wanted its fire detection and environmental controls products.

Hence, if you need to expand out of a declining market or find new growth, diversification is a way of spreading risk. You can develop products that use new technologies and these can co-exist with their present ones. Or you can sell complementary products.

Similarly, when pension funds buy shares in many companies, they are seeking to lessen their risk through diversification.

But every change creates a new risk. If you neglect your core market, or invest too heavily in a new one, you could suffer more than by staying in your existing market.

CASE STUDY: A SPORTING CHANCE

With many UK football clubs losing money, Airton Risk Management recommended hedging. The company suggested that clubs put their players on a variable wage with incentives for success. Thus a player earning £80,000 a week might be offered a basic wage of £40,000, but would receive £90,000 a week if the team were to qualify for Europe. Clubs would get a lower wage bill and could hedge the risk of paying out in the event of success.

Quoted in *Sports Pro* magazine (sportspromedia.com) Rod O'Callaghan, Director of Airton, reckoned that with a 9 per cent chance of being crowned champions, Liverpool would pay out a £10 million bonus for winning the league, but would only have to invest only £909,000 with Airton to cover that potential payout.

INSURANCE

The biggest and most obvious method of spreading risk is though insurance. Common forms are as follows:

- Employer and public liability, designed to pay out to staff and the public in the event of an accident.
- Professional liability (known in the USA as Errors and Omissions or E&O) insurance.

- Professional indemnity insurance, also known as professional liability insurance. It protects professionals against claims for negligence and malpractice.
- Vehicle insurance, for vehicle fleets and managers' cars.
- Health insurance: compulsory in some countries but not in the UK.
- Various forms of property insurance, including theft, fire and flood.
- Trade credit insurance or factoring, which ensures you get paid if your customer defaults.
- Business interruption insurance, which is relevant to business continuity plans.
- Key person insurance, which covers the loss of an important executive. At one point, the giant US grocery chain Walmart was taking out insurance on the lives of its employees, including low-level ones such as cashiers and janitors. Critics called it 'Dead Peasants Insurance', while the US Internal Revenue Service said the company was seeking to profit from the deaths of its employees, and take advantage of the tax law which allowed it to deduct the insurance premiums. Walmart halted the policy after the federal government closed the tax deduction.
- Kidnap and ransom insurance, used for operations in unstable parts of the world.
- Defamation insurance, which is used by the media to cover the risk of libelling someone.

Many people fear that the insurance company won't pay out when presented with a claim. In some policies, a 'basis clause' makes your declaration the basis of any contract. In other words, it governs what will be covered. An innocent non-disclosure of some fact may allow the insurance company to avoid paying a claim. One method of mitigating this rule is where you answer a question in the proposal form 'to the best of my knowledge and belief'. To avoid paying out, your insurer would have to prove that you were dishonest.

CASE STUDY: IS IT WORTH IT?

In assessing risk, you need to define:
- which risks the organization is exposed to;
- which of those it could buy insurance for;
- whether insurance would be a useful way of mitigating the risk.

You need to decide the following:
- Is the risk sufficiently severe?
- Is the insurance comprehensive? It may have exclusions and small print that would allow the insurance company to reject the claim.
- Is the insurance cost effective?

When seeking insurance against the loss of its computers, one company found that its insurance company demanded so much security as to make the building a virtual prison. The company also knew that the cost of the hardware was insignificant in comparison with the value of the data. The business therefore decided not to buy insurance but to concentrate on additional efforts to protect and back up its data. 'I don't care if burglars take the PCs,' said the CEO. 'They're just cheap metal boxes. The backups are the only things that matter.'

CAPTIVES

Larger companies set up their own captive insurance companies. This keeps the cash in the business and provides insurance at reasonable rates, thereby saving the profit margin that would have gone to an outside insurance company. The risk is retained inside the company, but the business pays sufficient money into the captive to pay out in the event of claims.

This assumes that the business has sufficient scale and expertise to manage its own insurance company subsidiary, though third-party businesses can provide the expertise. Captives allow the business to operate a more comprehensive risk strategy and gather data on its risks. There are also tax advantages. Many are located offshore to avoid paying tax.

Other companies have set up their own healthcare insurance. This has the same advantage of providing insurance at modest cost. The disadvantage lies in the company's lack of core skills in what are entirely new business ventures.

ACCEPT THE RISK

A low-impact, low-probability hazard is quite rare, if only because we don't really notice them. Organizations can reasonably fail to plan for an infestation of mites, moles digging some holes in the lawn or fluorescent light tubes needing to be replaced. The business can safely ignore (or tolerate) these and deal with them when they happen.

With no shortage of serious risks to manage, you may decide that a risk is within agreed risk tolerances. This will relate to the small risks that happen rarely. It is a relevant topic, because chasing every small risk wastes your time and resources. Deciding to accept small risks allows you to concentrate on the major ones, and prevents the risk system from becoming a behemoth.

You might also decide to accept some entrepreneurial risks. Launching a new product or acquiring a business is a risk, but after weighing up the pros and cons you may decide the risk is worth it. Theoretically, you might even decide to increase the risk. This happens when publishers enter a bidding war for a book, or when a business increases its offer to the reluctant shareholders of a company it wants to buy.

Later in this chapter we examine how you can actively incorporate opportunity into your risk management strategy.

CASE HISTORY: HOW A FASHION STORE GROUP MANAGES FASHION RISK

Fashion retailers face two major risks, according to McKinsey, the management consultancy. The first is having unsold stock that consumers don't want to buy. The second is being out of stock in lines that consumers do want. Both these risks cost a lot of money.

One store group tackled these problems by imposing more discipline on the buying process. The company drew up a profile of the typical customer, so that all buyers knew who they were targeting. This encouraged the company's buyers to get stock that had the right appeal in style and price.

The company then divided its offering into four styles, covering the spectrum from everyday wear to high fashion. It also defined how much of each type would be bought. This ensured that buyers bought the full range of goods that the customer wanted.

The change meant buying low-risk (and therefore high-profit) items like jeans, not just the high-fashion (and high-risk) kind that the buyers preferred.

As a result of this process sales are up substantially, while mark-downs have fallen by a modest 5 per cent.

CONTROL THE RISK

Having got rid of unacceptable risks (those in the 'Avoid' category), Shared others for various reasons, and decided to Accept or ignore a third set, we are left with the rest. These typically make up the majority of risks, and are the ones you need to actively manage.

Specifically, you need to reduce the scale of their impact or the probability that they will happen.

The standard way of managing these risks is through 'controls'. Controls can take many forms: they can involve a process, practice or policy, for example:

- Changing a policy: you might decide to stop offering credit to certain kinds of customer.
- Altering a process: you could substitute hazardous chemicals with safer ones.
- Changing a procedure. For example to reduce fraud, you might split the work of approving invoices and making payments.

The overall aim is to minimize, reduce or control the risk. For example, staff might be at risk of slipping on wet floors. You can mitigate that by installing anti-slip flooring, putting up 'Caution: wet floor' stands, or even providing slip-resistant footwear – the later being common on ships.

Controls don't always work as expected and unforeseen factors come into play. So you should avoid being lulled into complacency by the mere existence of controls.

Take cross-infection in a hospital. Just because you require doctors to wash their hands before and after touching a patient doesn't mean that all doctors will do this, all the time. Similarly, the handwash may not actually kill the bacteria. And cleaners may bring bacteria into a ward on their mops, which the hospital might have omitted to include in its controls.

There are several ways to specify how you will control the risk. It can be put in a risk register as discussed in Chapter 2, or put into a risk treatment plan which we discuss later in the chapter.

First, though, let's examine what kind of controls we can put in place.

HOW TO CLASSIFY CONTROLS

There are several ways to classify controls, and this can aid understanding. Three common classifications are:

1. preventative, directive or detective controls;
2. physical, management or technical controls;
3. manual or automatic controls.

Preventative, directive and detective

> Preventative controls stop the risk from occurring. An online form you can't submit until you have completed all fields prevents incomplete records. Preventative controls stop problems before they occur, so they are the best type, but people can circumvent them (by breaking a door lock, for example). Directive controls are usually aimed at getting people to do things. They include policies, procedures and training. You might direct people not to run on escalators, but some children will ignore that instruction. Detective controls give feedback, letting staff know whether a system is working properly, or alerting people if a problem has occurred.

Table 4.1 overleaf has some examples of each of these types of control.

Physical, management, or technical controls

Examples of these controls are:

- non-slip flooring (physical);
- a policy of using protective footwear (management);
- password-protected access control (technical).

Table 4.1   Preventative, directive and directive controls

Type of risk	Preventative	Directive	Detective
Burglary	Locks	Require staff to close windows when leaving	Intruder alarms, CCTV
Fraud	Numbered order forms, having two people sign cheques	Train people not to give out passwords on the phone	Audit
Hard drive failure	Raid drives, redundancy, strong password protection	Educate users to watch for data errors occurring	Software to monitor and diagnose drive wear

Manual or automatic controls

Audits require someone to do an inspection, while automated backups don't require human interaction (though getting a staff member to check that the backups are working is just as important, and reminds us that there is often more than one control for any process). The expression 'belt and braces' refers to using two ways to hold up your trousers; and this can usefully be applied to risk controls.

CHECKLIST: THE RIGHT TYPE OF CONTROL

Any control should have the following characteristics. This checklist ensures you institute controls that are right for the risk.

☐ Appropriate   ☐ Complete
☐ Consistent   ☐ Cost effective
☐ Meaningful   ☐ Practical
☐ Reliable   ☐ Simple
☐ Timely   ☐ Usable

BOW TIE DIAGRAM

The bow tie is a graphical way of showing a hazard, its causes and consequences. It can also show the controls put in place to reduce the risk. A sample is shown in Figure 4.3. It shows the consequences of a river ferry collision, together with the possible causes, such as staff taking alcohol. It also outlines the consequences, such as passenger deaths, plus the measures in place to reduce its impact, such as the use of life jackets.

The bow tie is a blunt instrument, suitable mainly for initial assessments and staff training. Its value lies in its simplicity. Non-specialist staff quickly understand how to use it, and it serves to prompt ideas, making it useful at brain storming sessions and for identifying current practice. You can also use it after an event to learn lessons.

You can use the bow tie diagram to show that risks have causes which can be controlled, and that the severity of an event can be reduced by putting measures in place.

The more complex a bow tie diagram becomes, the less helpful it is. It is easy to smother the idea with extra lines and boxes.

Originally used by Royal Dutch/Shell to assess operational risks, the bow tie diagram has spread to many industries, and is used for many kinds of risk including security, business continuity and project risk.

The bow tie diagram is not an end in itself. You need to move the information gained into practical results ('Let's give all passengers life jackets') and into written procedures. Bow tie diagrams can easily be drawn using flip charts or sticky notes, and you can also get bow tie software from suppliers such as Risktec (risktec.co.uk).

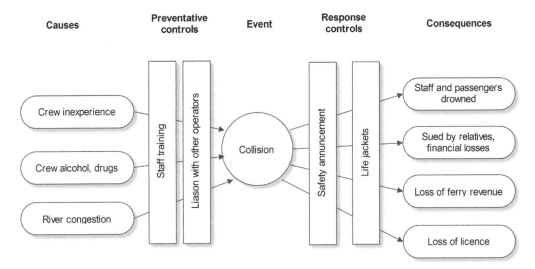

Figure 4.3   Bow tie diagram

THE RISK TREATMENT PLAN

Having decided what controls to put in place, you need to document this. We have already begun to look at creating a plan. In Chapter 2, the findings from risk assessment began to merge into a plan of action. This is often called a risk treatment plan, an example of which is shown in Table 4.2 overleaf.

Measure the Losses

The company should measure its losses and impacts. Taking measurements lets you see where the organization is vulnerable, and where losses are occurring. It also ensures that action is taken. As the saying goes, 'What gets measured gets done.'

Once you are managing your risks, they should reduce in frequency and severity. And in an ideal world you should be able to quantify that. Figure 4.4 is a chart produced by a transportation business that demonstrates the outcome risk management.

Prior to implementing risk management, the company suffered most incidents in the 'high' and 'very high' risk categories. After implementing risk management plans, most of the company's incidents were in the 'low' and 'moderate' risk categories.

Table 4.2   Risk treatment plan

Risk Treatment Plan
Risk name :
Risk no.
Use a unique identifier for easy reference.
Risk description
Describe the risk. Risk classification (for example, environmental). Location. Likely timescale of its impact. Hazard or opportunity?
Assessment
Scale and probability of the impact. Has it happened before? What were the consequences?
Stakeholders
Who would be affected, both inside and outside the organization? What are their expectations?
Risk treatment
What controls are in place? What other controls might be added? What would that cost?
Responsibilities
Who is the 'risk owner'?
Tolerance
What level of risk is the organization willing to accept for this risk? What warning signs might alert us to its likely occurrence?
Audit and monitoring
How will the risk be monitored?
Potential for improvement
What extra could be done to reduce the risk? What would the cost and benefits be? Who would do that?

Assessed by:	Date:
Reviewed by:	Date:

Comments by reviewer:

Date of next review:

You can see the effect of controls in Figure 4.5. Fire becomes less likely if flammable solvents are removed from the workplace. And if sprinklers are then installed, the impact of a fire is much reduced. Successive controls will further reduce the probability or impact of a fire.

Documentation

Managing risk means keeping records and having information to hand. The risk manager will need to keep a range of records. Prime among them will be the risk manual which we consider next.

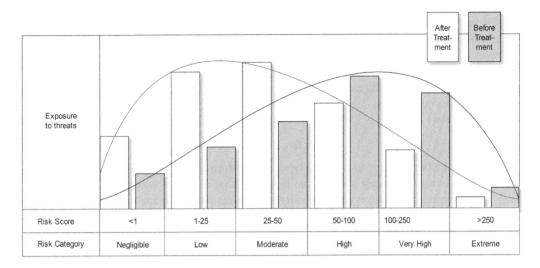

Figure 4.4   Effect of reducing risks

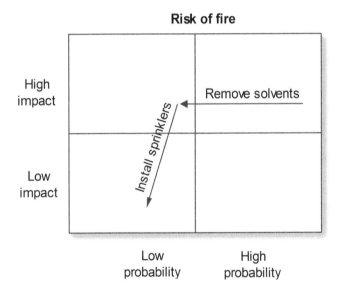

Figure 4.5   The effect of controls on a hazard

THE RISK MANUAL

All facts about risk should be contained in one document, which should be 'controlled'. That is to say, it should contain a version number and old copies should be withdrawn when an update takes place. The risk manual should contain the following items:

- risk policy, CEO's commitment to risk management;
- purpose of the manual;
- scope;

- risk register;
- risk management procedures;
- roles and responsibilities for managing risk;
- procedures for managing major risks;
- audit and review procedures;
- timetable for actions;
- issue number;
- date.

Workforces are sceptical of manuals of all types, and with good reason. They are often excessively large. They use arcane language and contain platitudes that everyone knows are not adhered to. They are produced under duress at the insistence of a regulator, and are then allowed to gather dust. So it is incumbent on those who produce the risk manual to keep it brief, to engage staff in its development, and to ensure it is kept up to date.

As a matter of policy the manual should be reviewed and updated each year, even if nothing obviously major has changed. This ensures that the manual doesn't get out of date and encourages senior management to be aware of risk.

RISK MANAGER'S DOCUMENTATION

In addition to the manual, the risk manager will need more documentation. This includes the following:

- detailed risk procedures;
- insurance policies;
- risk training plans and records;
- improvement plans;
- audit reports;
- management reviews;
- continuity plans and the results of simulations;
- incident and event reports, including complaints and near misses;
- key performance indicators (for example, health and safety incidents);
- control self-assessment results (see Table 2.7).

The risk owner will also need a sub-set of these documents, specifically those that relate to the threat for which that individual is responsible. Depending on the risk, they might include:

- maintenance records;
- plans of specific premises, plus maps, photographs, flow charts and other information recording drains, stopcock, hazards and so on;
- loss records.

STORING AND MANAGING THE DATA

Storing the risk assessment and treatment information on a database gives the company three advantages:

1. It lets the company quantify the data. Using the scoring system in Table 2.8, the risk manager can put a value on each type of impact.

2. The database can perform analyses. For example, by identifying specific buildings or work processes by number, the database can be asked:
   - Which building is most at risk, or which group of workers?
   - What is the descending order of threat, by severity of risk?
3. The data can easily be updated and retrieved.

AVOID THE SPREADSHEET; BUT BEWARE THE BEHEMOTH PROGRAM

Too many organizations rely on that simplest form of electronic storage, the spreadsheet, and this can lead to errors. Power generator TransAlta lost $24 million after it bid too much for a hedging contract, due to an Excel error.

Spreadsheet data looks authoritative, so people rarely question it. Yet the formulae behind a spreadsheet are usually opaque to all but the spreadsheet owner. And even that person can forget how the formulae were designed. In his own organization the author recently saw a spreadsheet that produced erroneous data. A former employee, entrusted with producing monthly figures, but inadequately trained in Excel functions, had started over-riding the formulae, without telling anyone. As a result, data was being omitted, and the results were useless – but they looked quite sensible on first glance.

At the other extreme, some organizations use an ERP system that includes single risk management. This carries the advantage of unifying data and avoiding systems that cannot talk to each other. However, the disadvantage lies in the cost and difficulty of integrating the company's systems.

An alternative is data warehousing where the organization plugs its data from different sources into one repository that presents the company's information in a consistent manner.

An alternative is to use specialist risk management software. We examine several risk management programs in Chapter 25.

Monitoring and Auditing the Risks

Mature audit programmes require the auditor to use a checklist to ensure that risks are identified and managed. The audit may take a specific threat, such as health and safety, and check it throughout the plant.

Alternatively, the audit can involve a specific resource, usually a department or building, and carry out a comprehensive risk audit. The audit should look at all aspects of risk, or else a selected area. For example, is the organization complying with 'right to work' legislation?

The final stage is to monitor risks. This includes regularly measuring the risk (to ensure that it remains within stated tolerances), and auditing (to ensure that the procedure is being followed.

The auditors should report findings to the CEO or risk manager; and the findings discussed. This review is the time to consider how the company's risk exposure could be reduced. A programme of continuous improvement will help to keep the company abreast of best practice, reduce its risks and lower its costs.

Because there is so much information in the business, monitoring should focus on the most important risks. Managers should examine:

- trends that indicate a growing danger;
- data that shows variances from the norm, or is outside pre-set limits (known as 'management by exception');
- key performance indicators;

- one-off reports on new areas of risk;
- information from a range of sources;
- key findings from audits.

CASE HISTORY: RISK AND THE PROFESSIONAL FIRM

For any service company, there is always the risk that an assignment might go wrong and the customer would sue the firm. A human resources consultancy knows that in recommending a new pay structure for a client, it is fundamentally affecting the client's business.

A system that gives too much money to the workforce could bankrupt the client; while a miserly system could cause key workers to defect.

Before submitting a proposal, the consultancy evaluates the potential impact of an assignment on the business, and charges accordingly. Says one executive, 'It's crazy to charge just £9,000 for a job which exposes us to ruinous legal costs if it goes wrong.'

TAKING CORRECTIVE ACTION

As part of the review process, managers need to agree on what corrective action should be taken. Some of this will already be suggested in the internal audit reports, such as investing in new equipment, or introducing new controls in a specific area.

The options for corrective action are the same as for risk treatment as a whole (as discussed in the earlier section, 'Reduce the risk'). Management can avoid, minimize, spread or accept the risk; and there are many ways to do each of those.

The corrective action should be minuted, and an individual made responsible for this action. At the next review meeting you should check to see that the corrective action has been carried out.

EXPLOITING OPPORTUNITIES

So far, this discussion has been slightly negative, even timid. We've seen risks as something bad. They're dangers to be feared and controlled.

But you can also seek out risks and set them to work. In order to prosper you need to identify and positively exploit opportunities, for the alternative is stagnation. Here are some typical opportunities:

- Launch a new product
- Expand into a new market
- Invest in new facilities
- Acquire a business
- Implement an R&D programme
- Take on more sales people
- Do deals, bid for contracts, or provide finance to others
- Take on debt to finance any of the above
- Give more freedom to your talented employees

This is an offensive rather than defensive strategy. In sport it's called Play To Win (PTW), rather than Play Not To Lose (PNTL). In business, organisations are increasingly recognising the limitations of risk management by referring to it as 'risk and opportunity'. But how do you manage it? A structure for exploiting opportunities is as follows:

- Identify your opportunities. This can involve brain storming, qualitative market research, using management consultants or introducing employee suggestion schemes. This can be in response to challenges in the market (such as digital disruption) or just recognising the need to stay ahead.

- Sort, evaluate and prioritize the opportunities. You'll need a system for collating ideas, so they don't get overlooked or forgotten. Then you need to assess the risks and rewards of each, as discussed in Chapter 2.
- Implement your chosen opportunities, using project management techniques (see Chapter 16).

Opportunities are rarely managed as neatly as this structure implies. They're usually seized opportunistically, often on the whim or intuition of the CEO. Senior management often fails to weigh up the pros and cons of each opportunity.

To summarize, if you add opportunity to risk management, and are methodical about it, you can help your organization to flourish.

Implementing a Management System

A management system involves writing down the company's main procedures, and then ensuring all staff adopt them.

There are many well-known management systems. ISO 9000 is used to maintain the consistency of a product or service, while ISO 14001 seeks to minimize environmental damage.

The use of such systems can help to avoid crises, and we examine them further in various chapters including:

- Chapter 5: ISO 9000 and others.
- Chapter 8: The environmental management system ISO 14001.
- Chapter 15: IT and ISO 27001.
- Chapter 23: Continuity and ISO 22310.

There are a plethora of management systems. And although ISO is introducing harmonized wording and structures, senior management may not be sure that all is working as it should.

There are two solutions to this issue. One is to use a RMS such as ISO 31000. We discuss this in detail in Chapter 26. Another is to adopt a compliance management system (CMS) – yet another management system! One such standard is AssS 980.

USING A COMPLIANCE MANAGEMENT SYSTEM (CMS)

The German standard AssS 980 (www.idw.de) helps businesses comply with legislation, and ensures that company policies are being adhered to. To misquote JRR Tolkien, it is one standard to rule them all.

An assurance audit relating should provide 'reasonable assurance' about whether the assertions contained in the CMS about the company's CMS policies and procedures are working. Specifically the audit would check that the company's policies and procedures:

- are fairly presented in all material respects. For example, all important elements should be included, and should not be presented in a misleading way. The areas to be covered should be appropriate and their selection should not be biased;
- comply with the applicable CMS principles;
- can identify in due time and with reasonable assurance risks of material non-compliance, and can prevent non-compliance.

The audit should also assess whether that the policies and procedures have been implemented, and are effective.

The standard involves an audit from an independent organization, such as a management consultancy, following the principles of ISAE 3000, the international standard on assurance engagements (www.ifac.org).

How Much Should You Spend on Risk Management?

Every investment needs to produce benefits, and the same is true of risk management. The returns from a risk management programme should be greater than its costs. By returns we mean:

*   A reduction in costs caused by losses (projects that over-run, financial penalties imposed by regulators, level of refunds, amount of unplanned downtime, staff absenteeism, and so on).
*   An increase in revenue from profitable ventures (referring to opportunity risks, such as sales and marketing activity or acquisitions).

These benefits have to be greater than the costs of running a risk programme. The costs of risk management are as follows:

*   cost of controls, including investments;
*   cost of insurance;
*   cost of management and staff time.

This analysis is useful if seeking to put a figure on how much should be spent on risk management.

But it is hard to identify how much risk management has saved the organization since you can't run the business in two different universes and compare the results.

Moreover, there are many variables which would confound any analysis; and blind chance plays a role, too.

However, there are examples where management can show reductions in injuries or improvements in quality which translate to the bottom line.

Companies without a risk management strategy are more likely to suffer the costs of problems and crises. There are the costs of mopping up after a pollution incident, paying to settle an industrial accident or the cost of recalling faulty products. This is represented at the left edge of the chart in Figure 4.6.

As the company becomes aware of the need to manage corporate risk, the company starts to invest money in prevention. This includes the cost of audits, the cost of preventative maintenance and the salary of a risk manager.

As the prevention costs grow, the number of incidents falls, and so do their costs. As a result, total costs also fall. The company is now at the middle of the chart.

The company can continue to invest more money in prevention, seeking to reduce ever further the likelihood of disaster. If it does so, however, the prevention costs will continue to rise.

Eventually, at the right-hand edge of the chart, the total costs are the same as they were before the company started managing its risks.

The chart suggests that there is a maximum level of investment to be made in risk management. Too great an investment will burden the company with costs and render it uncompetitive, while insufficient attention to risk will make it liable for heavy incident costs. Somewhere in the middle of the chart is the optimal position.

The term ALARP refers to 'As low as reasonably practicable', meaning you need only reduce a risk to as low a level as is practical. A near identical term, SFAIRP, means 'So far as is reasonably practicable'. The terms involve weighing a risk against the trouble, time and money needed to control it.

The UK's Health and Safety Executive (hse.gov.uk) says:

*   to spend £1 million to prevent five staff suffering bruised knees is obviously grossly disproportionate; but
*   to spend £1 million to prevent a major explosion capable of killing 150 people is obviously proportionate.

In the past, auditors and inspectors have pushed to have written procedures for ever more processes. An over-zealous risk environment not only hampers the business but also ends up with the workforce ignoring it.

Risk Awareness

It is worth adding that risk management takes time, usually measured in years. It cannot be installed overnight. The time needed to carry out risk assessments and design procedures and audits is only the starting point. Risk management then needs to be embedded into the organization, and you have to encourage people to develop an awareness of risk.

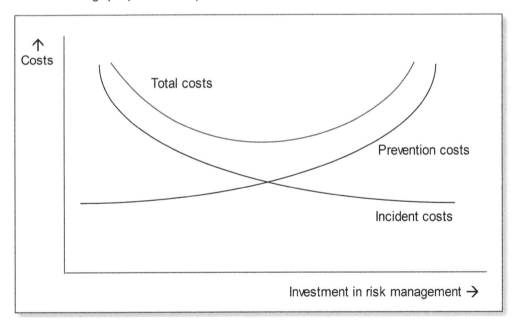

Figure 4.6   The effect on cost of managing risk

SETTING A RISK MANAGEMENT BUDGET

The size of the risk management budget will depend on the size of business, the complexity of its operations, and the responsibilities of the risk manager.

A survey by Deloitte among large financial institutions, particularly those based in North America, showed that 39 per cent have more than 250 full-time employees in their risk management function.

A survey among visitors to the continuityinstights.com website showed that the majority spent less than $500,000 on business continuity management (BCM), and we may assume that many firms spent nowhere near that amount (Table 4.3).

Table 4.3   **What is your organization's annual budget for BCM products and services (response, resumption and recovery)?**

Budget	Number of responses	%
Under $500,000	224	62
$500,000 – $999,999	60	17
$1 million – <$5 million	52	14
$5 million+	23	7
Total	359	100

Source: Continuity Insights

There is another reason for treating this with caution. A question in the same survey showed that in only 8 per cent of cases were BCM funds allocated as part of a risk management budget. The greatest number, 41 per cent, said that funds were allocated on a case-by-case basis. Apart from anything else, this suggests that risk management and BCM are treated as separate issues, or that companies opt for either one or the other.

Some companies make the company's risk management services available free of charge to plant managers – unless the plant suffers a loss, in which case the charge is allocated to their plant. This encourages plant managers to seek advice and support rather than wait for problems to occur. Thus each plant carries a notional budgeted cost for risk. If not used, the money is allocated to the plant's profits.

Risk Assessment

You can assess your organization's ability to manage its risks by answering the questions below. Score one point for each box ticked.

Question	✓
Has the organization categorized its risks into those it will Avoid, Reduce, Share or Accept?	
Can you name any specific risks the organization has explicitly chosen to avoid within the last three years?	
In the last year, has the organization investigated any risk sharing strategies other than buying insurance?	
Can you name at least four detective risk management controls currently in place?	
Does the organization have a risk treatment plan or similar system?	
Do you measure the cost of your risk management programme (the cost of controls, insurance and staff)?	
Is a risk manual or similar source of information in place?	
Does the organization store its data in software that is readily accessible to managers?	
Is there a process of corrective action in place for reducing risks?	
Do you identify and manage opportunities?	
Total points scored:	

Score: 0–3 points: high risk. 4–7 points: moderate risk. 8–10 points: low risk. The Appendix has a summary of the checklists. By entering the results of this one, you can compare your organization's preparedness against other more specific categories of risk.

5
Product and Service Problems: Operations and Production Risk

It's all a bit of a mess. Some employees are sitting around, reading magazines. Others seem overworked. 'It's a scheduling issue,' shrugs Blake, the departmental manager. 'The work is unpredictable, so we can never get it exactly right.'

'But the clients complain that tickets aren't being answered,' you say. 'Yes, but it's not a fully integrated system,' says Blake. 'Things fall between the cracks.' You pause. Should you mention the problem about clients not being invoiced on time? Maybe you'll keep that for another time.

Somehow you have to clean up the mess. It needs better process automation, better reports, maybe a dashboard. Looks like you'll be home late again.

Causes of Production Problems

Production and service problems have many causes, some of which are shown in Figure 5.1. Whatever their origin, they give rise to product and service failings – whether in terms of cost, reliability or human interaction.

When the product or service gets to the market, the problems can become real, which result in loss of customers and other problems.

The solution, as shown in Figure 5.1, is to effectively manage all the elements of production – the people, suppliers, equipment and systems. It is the systems in particular that we emphasize in this chapter. Effective information, auditing and feedback will identify problems and suggest solutions long before they become a threat to the business.

The Consequences of Mundane Failures

Most operational and production failure is banal – a lack of staffing, lack of stock availability and communication failures. The results are equally banal – products that has to be sent for re-work, a patient who has to be booked in again or a customer who defects to another store.

Yet these failures matter cumulatively. For commercial companies, they can make the difference between profit and loss. For the public sector, it can mean a constant wearing down of employees' enthusiasm, or a regular shortfall in the number of clients seen. In all cases, it means financial losses.

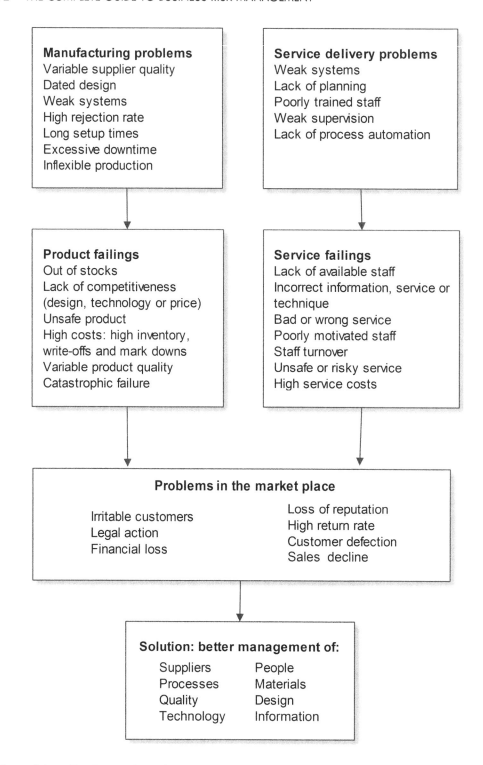

Figure 5.1   Product and service risks and solutions

Nor is the issue confined to product failure. Twenty-four million out of 30 million jobs in the UK are service-based. They include:

- public administration, education and health;
- retail, hotels and restaurants;
- transport and communication;
- finance and business services.

As Table 5.1 shows, there are wide variations between complaints among UK energy suppliers. EDF has five times more complaints per 100,000 customers than Scottish Power.

Table 5.1   Complaints for 100,000 customers

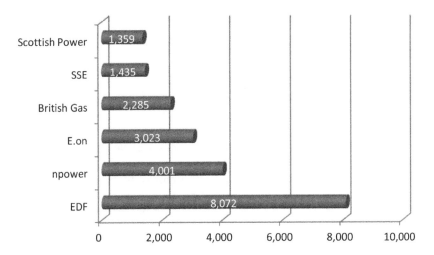

The difference will be attributable to many factors, all of which are under the control of the business, including investment in IT and other systems, corporate culture and internal communications, number of staff allocated to jobs and staff training.

These complaints relate to payment issues, billing, debt, metering, waiting times to call centres and other customer service issues. Inevitably it costs money to bring down complaints, but this should cut customer defection and raise profits.

Service companies are also prone to disaster. Computer software is often delayed, and is notoriously and routinely faulty.

In short, production and operational risk is all around us. Quality failures can be minimized by understanding where they might occur, and by adopting techniques that will forestall them, as we see later in this chapter.

The Impact of Catastrophic Failure

Production and service failures that lead to recall or litigation are very expensive.

- In the US, 1.1 billion newly printed $100 bills had to be put in storage when some were found to have blank areas, caused by the paper creasing. A government audit found that the Bureau of Engraving and Printing should have done more testing before printing the notes.

- Battery maker A123 Systems spent $55 million recalling defective vehicle batteries. The company said a welding machine was calibrated wrongly. This caused a component in the battery to pierce electrical insulating foil. And when the car was driven, a short circuit would occur and the battery would fail.
- Though famed for its manufacturing excellence, Toyota had to recall 9 million vehicles. This was due to an unsecured rubber floor mat trapping the accelerator pedal, and also separately by the accelerator pedal sticking (though this was exaggerated by fraudulent claims). Sales of cars were suspended for several weeks, as the company waited for replacement parts. The pedal problem was alleged to be responsible for 37 deaths, though driver error may have been responsible for some or all. It was estimated that the crisis cost each US Toyota dealer US$1.75 million to US$2 million a month in revenue during the crisis.

Catastrophic Failings in Service and Government Organizations

Service organizations are also prone to catastrophic failure.

- A virus swept the factory of a biotechnology company that produced palm trees for Middle East plantations. It takes months to multiply and grow the crops; and this production failure lost the company much of its annual profits, as well as a loss of reputation among its worldwide export markets.
- Meanwhile government services deal with the big issues of life and death, children and the elderly, roads and travel. So there is more at stake.
- The US army ditched its camouflage uniforms at a cost of $5 billion, after finding that they made soldiers stand out rather than camouflage them. Reports suggested that the army wanted 'cooler' uniforms than the Marines, and that decisions were made by top management rather than subject matter experts (that is, the soldiers who used them).
- The British Home Office scrapped a £423 million fire service reorganization that was to reduce 46 emergency response centres down to nine. The work had progressed so slowly (six years) that the IT systems were no longer fit for purpose. There were fears that the planned 30 per cent reduction in staff would result in lost calls, and there would be a loss of local knowledge.

Newspapers routinely feature scare stories of smear test errors, social work failures and killer nurses. And although these are statistically rare, no manager wants to make the headlines.

One of the biggest risks for local authorities and municipalities is a shortage of money. With much of their annual budget coming from central government, and with many areas of cost, such as pay awards, being outside their control, local authorities risk sudden deficits, as well as the consequences of a lack of personnel to care for the vulnerable. Risks for local authorities and government bodies include:

- civil disaster such as flooding;
- failure to perform – for example, in fostering, environmental health, tendering or in managing housing;
- litigation – by employees, tenants or citizens;
- legionella;
- occupational road risks;
- school trip risks;
- health and safety risks;

- workplace risks;
- risks to reputation;
- environmental issues;
- security;
- risk financing alternatives;
- harm to patients, or death.

OPERATIONAL RISK: TWO MEANINGS

Operational risk means different things to different people. It is most commonly used by banks and finance companies. To them it means problems such as internal fraud, data entry errors and rogue traders who make bad bets. The term excludes *credit risk* which is the chance that a borrower will default; and *market risk*, which is the possibility that interest rates or share prices will go the wrong way.

If you pick up a book or article referring to operational risk you could get confused by the references to Basel III, Solvency II and financial regulations. This is because there the writer is referring to operational risk as it relates to banks and financial institutions.

The book you are currently reading is about business or enterprise risk, rather than specifically the financial sector. When we talk about operational risk, we mean the core processes of any organization. For a local authority, it means the way that planning applications are handled and the roads are swept. For an oil business, it relates to the drilling of holes and extracting of oil. For an advertising agency, it's the way that advertisements are designed and placed.

HEALTHCARE OPERATIONS

According to a survey by the health union Unison, frontline NHS staff are so busy, they don't have time to help explain treatments, keep proper records, or help patients eat and drink. Unison said this risks a care scandal.

One in five respondents said failings were as bad as at a Stafford hospital where 400–1,200 patients were alleged to have died over a three-year period due to bad care. Patients there were left in their own urine by nurses, and resorted to drinking from flower pots. Of those surveyed, 30 per cent said their Trust was at risk of a crisis like that of the Stafford hospital, and 10 per cent said it was already happening.

The fact that the Stafford hospital tragedy was not repeated in hospitals with broadly similar staffing levels indicates that staff numbers is not the only issue. A review of early warning systems in the NHS said, 'Systems and processes [are] almost entirely dependent on the values and behaviours of the staff. Strong leadership at every level is needed to ensure that values and behaviours that put patients first can prevail.'

Criticisms levelled at the Stafford hospital serve as risk factors in healthcare generally:

- clinical governance was poor;
- clinicians did not raise concerns about the poor quality of care for patients;
- the Board became focused on promoting itself as an organization, with considerable attention given to marketing and public relations;
- the Board lost sight of its responsibilities to deliver acceptable standards of care to patients;
- the Board and senior leaders did not develop an open, learning culture, inform themselves sufficiently about the quality of care, or appear willing to challenge themselves in the light of adverse information.

DRIFTING TOWARDS FAILURE

According to Sidney Dekker (*Drift into Failure*, published by Ashgate), systems tend to drift in the direction of failure. As Figure 5.2 shows, each single step is small, so it isn't noticed. A new lower 'normal' is established, and this journey continues downwards. The drift may take the form of small shortcuts, none of which cause any problems.

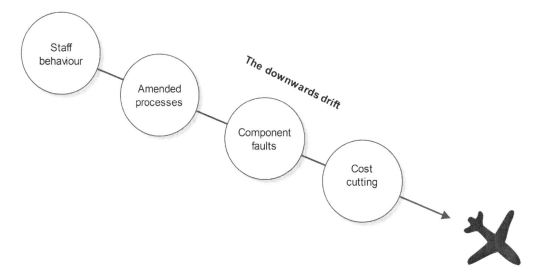

Figure 5.2   The drift towards failure

But careless staff behaviour and amended processes gradually reduce the margin of safety and create more risk.

Eventually, the problems line up, as with the Swiss Cheese model (Figure 2.10), and an accident occurs.

Drift can also occur as a business tries to cut out flab and introduce efficiencies. This seems sensible, until a problem happens. People get tired and make mistakes. Machines wear out and break down.

Failure, says Dekker, is often due to unanticipated and non-linear interactions between components, rather than actual component failure. For example, two parts of a software program might not relate properly. You should also look out for weak systems that are being tolerated or treated as normal. What looks wrong after an event often seemed quite normal at the time, with even abnormal data being written off as a fluke or something that would right itself in due course. There are several solutions to the problem of drift:

1. Failures should be investigated and the lessons learnt. These often help to prevent other future failures.
2. Built-in redundancy (such as duplicate machinery) and slack (having too many people at low periods) can help, because if an element fails the system can deal with it.
3. The organization should adopt the style of the 'Highly Reliable Organization' (HRO), those with low failure rates, which we look at in more detail below.

Today's more regulated world tends to discourage drift. Being audited encourages management to maintain standards. It also focuses employees' minds; and the audit itself helps to spot problems before they become serious.

HIGHLY RELIABLE ORGANIZATIONS (HROS)

HROs are ones that avoid accidents and catastrophes in complex and risky environments where such problems would be expected. Complex and risky areas include nuclear power plants, chemical plants, firefighting and air traffic control. Studies of HROs have also compared high-profile catastrophes such as the Bhopal chemical leak, airplane crashes and the Challenger space craft explosion.

In *Managing the Unexpected*, Weick and Sutcliffe concluded that HROs have five characteristics. All relate to openness and all that flows from it. The five characteristics are shown in Table 5.2:

Table 5.2   The five characteristics of a HRO

Characteristic	Practical implications
1. Preoccupied with failure	The organization is constantly aware of the possibility of trouble, and is alert to the early signs of failure.
2. Reluctant to simplify interpretations	People in HROs look beyond first impressions, labels and established beliefs.
3. Sensitive to operations	Senior people in HROs remain closely in touch with the organization's daily operations.
4. Committed to resilience	HROs value continual learning. People adapt as circumstances change.
5. Defers to expertise	The HRO values relevant knowledge, skills and observations, irrespective of the individual's position in the hierarchy.

In short, safety in high-risk organizations depends on creating a culture of openness. People must be willing and able to exchange information at all levels, and without fear, even when that means challenging authority figures. Developing such a culture allows the organization to be resilient when faced with unexpected problems.

Products in Service Organizations

Service operations and government departments often contain a mix of product and service, as can be seen in the examples in Table 5.3. This has an effect on the kinds of risks that the organization is prone to.

As these examples show, many service organizations engage in activities that have physical end products.

These organizations need to be able to manage both the product and the service. Where the service organization has a product, it will also have all the paraphernalia of suppliers, raw materials and production units. Some managers may be less familiar with how to control such elements, and in such cases greater risk accrues.

Table 5.3   Risks in three types of service operation

Services that involve products	Sample operational risk
Restaurants and catering – food needs to be supplied and served.	Food poisoning
Tyre and exhaust fitting – this is more product-oriented, with physical removal of the old tyre and fitting the new one. Physical distribution requires the movement of goods around the country, vehicle management, as well as route planning and warehouse management.	Wheel nuts come loose Vehicle accidents
Housing management – this requires the construction and maintenance of buildings, as well as the processing of applicants, financial control and record keeping.	Repairs backlog
Active services	
Surgery – involves physical activity of meeting patients, diagnosing them, moving them, the surgical procedure and post-operative care.	Anaesthetic error
Education – activities, events and trips involve increased risks.	Physical accidents
Retail – in addition to the checkout activities, the store has to organize products to a warehouse, thence to the store, and then on to the shelves.	Out-of-stock problems
Service only	
Insurance services require visits to clients, information gathering, credit risk analysis and proposal writing, and paperwork for compliance.	Faulty decisions
Banking processes involve activities such as data processing, paper transactions and credit risk assessment.	Data entry errors
Software writing is predominantly a desk-based activity. (This does not mean to say it is risk-free: software project failure is common.)	Project failure

Other organizations have an active service component which, if not delivered correctly, will lead to customer dissatisfaction. Purely service businesses have their own risks, which are no less hazardous.

ASSESSING PROCESSES

From a risk perspective, there are three types of process: strategic, tactical and operations.

- strategic processes set the organization on a path;
- tactical processes create change;
- operational processes are the day-to-day actions of the business.

Table 5.4 shows some staff levels and processes for a retail organization.

Each type of process carries its own risks. Strategic processes are few in number and high in risk, whereas operational processes are many and low in risk. That said, if an organization has disaffected staff at an operational level, this can in time bring the organization down.

Table 5.4   Strategic, tactical and operational processes

	Strategic	Tactical	Operational
Staff involved	Directors	Area managers, store manager	Sales assistants
Typical tasks	New store locations	Training	Checkout operations
	Diversification	Hiring and firing	Shelf stacking

MAPPING YOUR CORE PROCESSES

Organizations should 'map' their processes to see exactly what goes on. Until this is done, people often have only a hazy idea of the components parts of the system, nor how the system fits together. Mapping the processes can lead to business process re-engineering (BPR), where an organization decides to change its processes. It might simplify them or make them more customer-focused. A business might choose to cut out several steps in its tendering process, so that tenders are produced faster.

A development of BPR is Target Operating Model (TOM), where an organization decides how it would like its services to be delivered. For example, it might seek to simplify customer access to the organization, including one point of contact for all customers. From that it draws up a plan of how that might be implemented.

The point about these models is that organizations gradually ossify over time and become less fit for purpose. Private sector organizations get overtaken by their competitors. In the public sector, dissatisfaction grows until a politician decides to shut the business, transfer it to private ownership or replace the management. These are all risks that models like BPR or TOM can prevent.

The Wider Impact of Operations Risk: Lack of Competitiveness

We have seen that businesses need to avoid operational disaster or crisis. A more insidious problem occurs when the company's service or products don't match the customer's expectations. Customers rarely complain: instead they buy from a different supplier. In the case of not-for-profits, the organization eventually gets a bad grading by regulators or negative headlines in the press.

Customer satisfaction can mean speed of delivery, design quality or value for money.

A survey by YouGov found that most people agree that their schooling left them ill-prepared for the problems of everyday life and work. Most blamed this on traditional text-book learning and inadequate careers advice. Eight out of ten felt their school, college or university could have given them more help in identifying their strengths so they could find a career that would suit them. Inevitably, organizations never ask some kinds of fundamental questions, because it never occurs to them to do so.

Good practice blends many issues: technology, people and systems. On their own these topics do little. But when combined, they can revolutionize the organization.

For example, excessive costs can drag down the organization. They can be caused by many factors: lack of technology, lack of process control, failure to control suppliers or demotivated staff. Some of the solutions in this chapter help to remedy that.

How to Achieve Reliability and Consistency

As Figure 5.3 shows, there are three ways of approaching quality. The most risky is the *Failure* area where complaints and recalls take place. This is rarely very dramatic, because often the customer simply decides to buy from another supplier in future. Failure mode, does, however, include the dramatic product recalls and catastrophes we saw earlier.

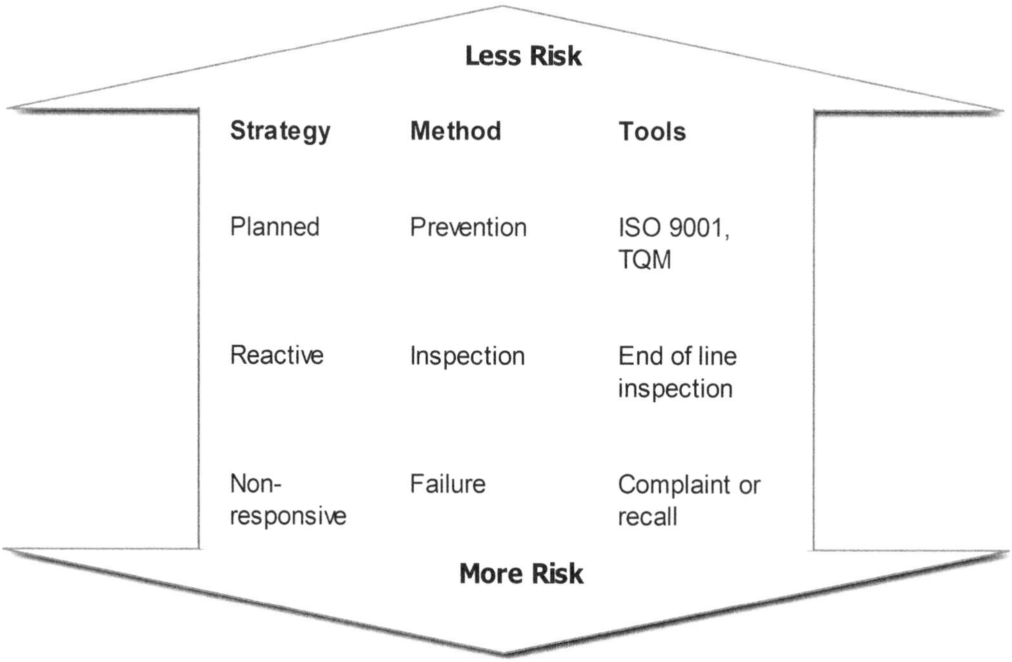

Less Risk

Strategy	Method	Tools
Planned	Prevention	ISO 9001, TQM
Reactive	Inspection	End of line inspection
Non-responsive	Failure	Complaint or recall

More Risk

Figure 5.3   The three production modes

Somewhat more successful is the *Inspection* mode. This was the way many goods were produced in the past. In Inspection mode, quality is the responsibility of inspectors, and their job is to 'inspect out' faults. This makes it easy for ordinary production staff to ignore the quality of the product, because it is not their responsibility. Likewise, production managers come into conflict with the quality department, whose staff are seen as meddlers.

The least risky area involves a *Prevention* mode. Here, the organization's processes are managed in a way that ensures consistency, and stops things going wrong. For example, hospitals use Standard Operating Procedures (SOP) for medical procedures. There is a best way of carrying out every surgical technique, and this will be written down.

But it is important to know that everyone is adhering to the SOP, and that the SOP is fit for practice. This results in the *quality management system.*

The Quality Management System

The basic tool for managing reliability and consistency is the quality management system; and the best known one is ISO 9001 (sometimes known as ISO 9000).

As we see below ('How to use a management system to control your risks'), ISO 9001 is based on sound ideas of inspection, control of important documents and written procedures.

Companies which have implemented ISO 9001 often go on to adopt *total quality management,* or (especially in manufacturing) *statistical process control.* These are all ways of reducing risk.

A management system like ISO 9001 encourages the organization to identify a fault or hazard through audits and checks. It gets people to investigate the root cause, to correct the fault and to prevent it happening again.

The system helps to prevent catastrophe because it aims to catch problems before they turn into a crisis. Getting bakery staff to check their own work before it leaves the factory reduces the chance of a diner finding foreign bodies in the croissant.

CASE STUDY: FORESTALLING PROBLEMS IN A MATERNITY HOSPITAL

Alarmed at a spate of baby snatching from maternity hospitals, St Mary's Hospital in Manchester straps a tagging device to the wrists of the 4,000 babies born there each year. The tag, which cannot be removed with conventional scissors, sounds an alarm if the baby is taken from the ward by an unauthorized person.

The tag, which looks like a teddy bear, was developed to guard prisoners in the first Gulf War. It now helps to minimize one of the hospital's operational risks.

HOW TO USE A MANAGEMENT SYSTEM TO CONTROL YOUR RISKS

The majority of organizations that implement ISO 9001 do so to get on a tender list. Automotive suppliers are one such example.

Others do it because they have complex production processes, in which a small failure can have catastrophic effect. A maintenance department fitted a windscreen to a British Airways jet using bolts that were slightly too small (they were short by a matchstick's thickness). As a result, the pilot's windscreen blew out at 17,300ft over Oxfordshire, dragging the pilot half way out of the plane. Remarkably, no one died.

THE ELEMENTS OF A QUALITY MANAGEMENT SYSTEM

There are several elements to a quality management system:

- *Understand your main processes* – unless you can see how transactions occur, you can't begin to improve them or manage them. Mapping the company's processes often yields valuable and surprising information. BHS, the chain store, analysed its supply chain and found that there were 63 unconnected parts to the process, many of them operating in parallel and duplicating each other. In some cases it was taking three months to agree a sample because it went to and from the supplier several times. Now the company has four parts rather than 63: create, make, move and sell.

- *Have an organizational structure* – so that everyone knows who is responsible for each area.
- *Have written procedures* – so that everyone knows how a job should be carried out. For example, pilots use checklists for every procedure, something that is credited for keeping accidents down. According to Atul Gawande, author of *The Checklist Manifesto*, surgeons' use of checklists is also saving lives in A&E.
- *Keep records* – so that if anything goes wrong, you can trace faulty products or the source of the problem.
- *Do regular checks* – so that faulty goods are not allowed into the market, and so that the whole system keeps running smoothly.
- *Identify faults and correct them* – so that the company learns from mistakes and does not keep making them.
- *Seek continuous improvements* – so your product or service is regularly enhanced.
- *Communicate well* – the design department must talk to manufacturing; and everyone must have the information they need to do their job.
- *Allow an external organization to regularly assess the system* – so that you get an independent view on its effectiveness.

The well-organized business does all these things as a matter of habit; but ISO 9001 ensures that they are formally adopted. Without a proper structure and without inspection by an outside firm, parts of the system may quickly lapse into disuse.

VARIANTS OF ISO 9001

ISO 9001 has many offshoots for different industries. Most industries have similar systems or guidelines.

Aerospace and Aviation AS9100

Automotive industry ISO/TS 16949, VDA 6, QS-9000

Food   ISO 9001 HACCP (ISO 22000)

Information Security Management ISO 17799

Laboratories ISO/IEC 17025

Medical devices ISO 13485

Oil and gas ISO/TS 29001, API Spec Q1

Information security ISO/IEC 27001

Telecoms industry TL 9000

ISO 9001 IN SERVICE ORGANIZATIONS

There is much in ISO 9001 and ISO 9004 that service organizations find useful, including the following:

- level of service to be provided;
- the organization's image and reputation for quality;
- objectives for service quality;
- approach to be adopted in pursuit of quality objectives;
- role of the people responsible for implementing the quality policy.

Management needs to do the following:

- define customers' needs, with appropriate quality measures;
- design preventive actions to avoid customer dissatisfaction;
- minimize quality-related costs to achieve the required performance;
- create a collective commitment to quality within the service organization;
- undertake regular reviews of service requirements and achievements in order to identify opportunities for service quality improvement;
- prevent adverse effects by the service organization.

As an example of the need for consistency in services, a child died after a doctor used a 'naked decimal point' (one without a preceding zero) in a prescription for morphine. The child was given 5mg, ten times the amount, rather than 0.5mg. This shows that staff should be trained to use standardized notations, such as 'mm' rather than 'cm', in mission critical documents.

CRITICISMS OF ISO 9001

People criticize ISO 9001 for being a system that organizations get only under duress. They say it's cumbersome, bureaucratic and old fashioned. And that once implemented, it's ignored.

Like any system, ISO 9001 only works if management recognizes its advantages and is committed to it. And the more recent versions of ISO 9001 have made it less complex. There are only six mandatory procedures.

ISO 9001 reduces the likelihood of those 'uh oh' moments. The combination of written procedures, audits and management reviews makes the organization more consistent. An independent external audit keeps people alert. And so there is less 'flying by the seat of your pants'. However, ISO 9001 isn't the 'be all and end all'. Many organizations put additional systems in place, which we consider next.

Eleven Other Systems for Reducing Risk

Apart from ISO 9000, there are other systems that help you to assess and manage risk. Many have originated in a specific industry and have been found so useful that they have been adopted by other organizations. Below we look at MCP, HAZOP, HACCP, FMEA, Lean, Six Sigma and several others.

THE MANAGEMENT CONTROL PLAN (MCP)

Not every business wants to install ISO 9000. An alternative is the Management Control Plan or Programme (MCP) which is a more lightweight and focused tool.

The MCP comes from government and the military, and seeks to ensure that the organization has a plan. The MCP sets out annual objectives, usually no more than five to seven. These stem from the company's vision, mission and objectives. From this the MCP derives a sales or revenue forecast.

The MCP is also likely to have a marketing and communication plan, and a plan for human capital.

Each department will have an action plan, and the results will be measured with scorecards. This includes holding the most senior officers accountable for what they should deliver. Depending on the size of the business, you can add refinements such as risk assessments, audit and reviews.

All this might be packaged in a Management Control Manual that sets out what the organization aims to do and how it will achieve that.

HAZARD AND OPERABILITY STUDY (HAZOP)

A Hazard and Operability study (HAZOP) is a method of analysing risks in a business process. Originally used in chemical plants, it has been extended to manufacturing and risky environments such as nuclear power plants.

In a HAZOP study (Figure 5.4) you map out your processes, and examine what can go wrong at each stage. What would happen if a thermostat fails to prevent temperature from rising? What if the freezers broke down? What if gas escaped? What if pressure doubled? Ideally you quantify the scale of the problem. What temperature could an oven reach? How much oil might leak?

HAZOP looks for weaknesses in the system, using 'guide words' such as More, Less, Temperature and Time. It looks at the way the system is operated and managed, including the use of controls. It looks at what happens in different modes of operation, such as startup and closedown, standby, and normal and abnormal operations.

An industry has grown up to sell the HAZOP methodology, and while it is better undertaken by someone with HAZOP experience, the process is not overly complicated. You can apply various levels of thoroughness, from using it as a simple method to aid brainstorming all the way through to statistical analysis.

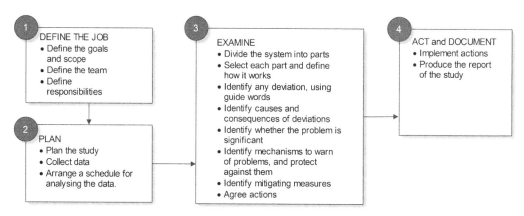

Figure 5.4   The HAZOP methodology

HAZARD ANALYSIS AND CRITICAL CONTROL POINTS (HACCP)

Hazard Analysis and Critical Control Points (HACCP), and pronounced Hasup, is a systematic way of reviewing process risk in the food industry, but as with HAZOP it has been extended to other businesses. It is intended to prevent problems from occurring rather than finding them at the end of the process. This is because bad food can kill people and end-of-line quality checking is inappropriate for food. HACCP is based on seven principles:

1. *Conduct a hazard analysis.* Determine the food safety hazards and identify the preventive measures you can apply to control them.
2. *Identify critical control points.* A critical control point (CCP) is a point, step or procedure in a manufacturing process at which a control can be applied and, as a result eliminates, prevents or reduces a hazard.

3. *Establish critical limits for each CCP.* This is the maximum or minimum value needed to prevent, eliminate or reduce the hazard to an acceptable level.
4. *Establish CCP monitoring requirements.* Monitoring activities ensure that the process is under control at each CCP.
5. *Introduce corrective actions.* These are actions to be taken when monitoring indicates a deviation from an established critical limit.
6. *Establish procedures for ensuring the HACCP system is working as intended.* This involves audits to make sure that the plant is operating as intended.
7. *Keep records.* You should maintain documents, such as (a) the hazard analysis and (b) a written HACCP plan, plus records documenting how you CCPs, critical limits and verification activities; and how you handle deviations.

ISO 22000 has incorporated the principles of HACCP, and may eventually overtake it.

FAILURE MODE AND EFFECTS ANALYSIS (FMEA)

Failure Mode and Effects Analysis (FMEA) is a tool for systematically eliminating errors and cutting risk.

Typically it is used in aerospace, automotive, food and technology. But it can be applied to any large organization, such as hospitals.

An FMEA team will log failures and rank them according to three factors:

- Severity (how big a problem is it?)
- Occurrence (how often is it happening?)
- Detection rate (how likely is the failure to be spotted during manufacture?)

This is shown in Figure 5.5 overleaf. Note that a failure could relate to a process, a design or an item, which makes FMEA universally applicable across all types of organization.

Each of these factors are graded on a 1–10 scale, and then multiplied together to arrive at a Risk Priority Number or index. Thus a small problem that occurs all the time and escapes detection is likely to be prioritized over a serious but visible problem that happens rarely. The team then tries to apply a corrective action, such as a different part, different design, more frequent maintenance or better training.

Like ISO 9001, FMEA has its critics. Multiplying three numbers together is a crude technique and can produce nonsensical weightings. The numbers for severity, frequency and detection are often just guesses, and are only as good as the people who choose them. It doesn't work well for new products which have no track record, though FMEA can be used to identify problems during testing at the design stage.

And finally major failures happen due to causes that FMEA doesn't spot. HMS Sheffield was sunk during the Falklands War because the captain was talking on a satellite phone which blocked the ship's radar. As a result the ship failed to spot an incoming Exocet missile. Meanwhile, David Forster, a radio operator on the nearby HMS Invincible who saw the rockets, said a senior officer ignored his warnings and accusing him of 'chasing rabbits'.

LEAN

Lean is the term applied to removing waste from business processes. Originally a Toyota system, it has since been applied to IT, government and service businesses. Lean sees waste as anything

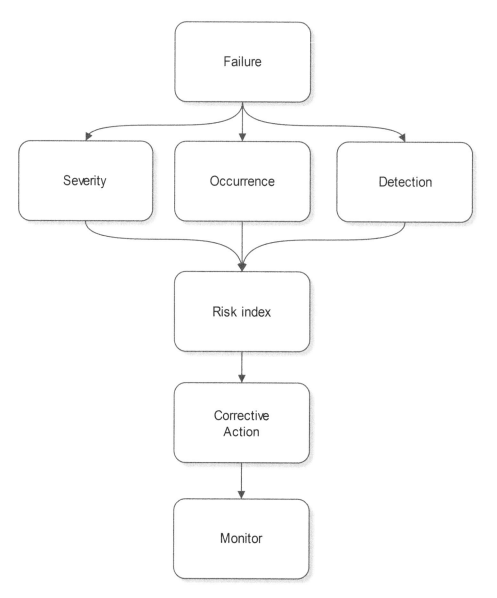

Figure 5.5   FMEA method

that isn't contributing value to the customer. In other words it is much more than just the physical waste that occurs in a process. The main types of waste are often defined as:

Inventory: Unnecessary raw materials or stock in a warehouse represents waste because it represents a cost.

Motion: Motion relates to wasteful activity on goods being processed. This includes wear and tear on equipment. Repetitive stress injuries and inefficient production are also caused by undue effort on the part of the workforce.

Transportation: Nothing should move more or further than it needs, because each time it happens, there is a cost, as well as the risk of damage. This relates to transporting goods in manufacturing and warehousing, and transporting goods to a customer.

Waiting: Goods and services should either be in the process of being worked on, or else despatched to the customer. Often, however, parts are awaiting further processing, or are held up due to lack of parts or staff.

Over-processing: Companies often produce goods and services that are over-engineered or over-specified for the purpose.

Over-production: Over-production happens when the business produces more than is needed. This leads to the risk of excess inventory which may later need to be written off. Smaller batches and Just in Time (JIT) can overcome this.

Defects and scrap: Defects either have to be binned, or require re-working or recall.

Companies sometimes use the acronym TIMWOOD as a mnemonic for the Lean elements.

5S

Workplace risks include lost production, defective output and injuries to the workforce. Many of these can be overcome by a 5S campaign. 5S is a way of organizing the workplace for maximum efficiency. It boosts consistency, which leads to higher productivity and improved safety.

5S: Comes from the five Japanese words: *seiri, seiton, seiso, seiketsu* and *shitsuke*. These translate into English as sort, stabilize, shine, standardize and sustain.

Sort: Identify all tools, materials and equipment. Keep only what is required. Put them in accessible places.

Stabilize or *Straighten Out*: There should be a place for everything and everything should be in its place. The place for each item should be clearly indicated.

Sweep or *Shine*: Everything should be clean. At the end of each shift, staff should tidy and sweep the work area, and put everything back in its place.

Standardize: The business should have standardized procedures for all processes. A member of staff should be able to work at any station, and find the same tools in the same location.

Sustain: Management must ensure that the previous 4Ss are maintained, and people don't slip back into old ways. When improvements are made, they should be applied throughout the organization.

3P

3P or 'Production Preparation Process' aims to eliminate waste through re-designing the product or process.

A 3P team starts with a 'clean slate', examines how a process is carried out and comes up with many alternatives. The team uses fishbone or flow chart diagrams to map the process. It then chooses and implements the best alternative. The outcome is often a simpler product, using simpler raw materials, or involves a process that has fewer steps or less environmental impact. 3P has similarities with BPR.

TOTAL PRODUCTIVE MAINTENANCE (TPM)

Total Productive Maintenance (TPM) is a way of keeping equipment working properly. This reduces operational failure and downtime, which is clearly a business hazard.

TPM works by engaging all members of the workforce in maintaining equipment.

Traditionally, this work was done by a maintenance department, which meant that operators sometimes felt uninvolved. As a result, equipment tends to go out of calibration or fail.

TPM trains equipment operators to carry out autonomous maintenance on a daily basis. In service organizations, staff can be trained to install computer updates, thus reducing the risk of virus infection.

MISTAKE PROOFING

'Mistake proofing' is another Japanese concept ('poka yoke' in Japanese, pronounced poh-kah-yoh-keh). It is a powerful stand-alone tool for preventing operational error, and works across all types of organization.

Mistake proofing applies fail-safe mechanisms which makes mistakes either impossible or easy to detect and correct. There are two types of mistake proofing: Prevention and Detection. The process is shown in Figure 5.6.

Prevention

A *prevention* device stops machines or people making a mistake. Here are some examples.

- An automatic car won't start unless it is in neutral; and some cars won't let the driver select reverse unless their foot is on the brake.
- Lawnmowers shut off the power when the handgrip is released, preventing the operator from being hurt. Food mixers won't operate unless the lid is on. Guillotines don't cut unless the guard is in place.
- Hospitals use different connectors for different IV lines, so that patients can't accidentally receive dangerous fluids.
- Some legal practices cut down mistakes by having a second lawyer inspect all documents.
- Pre-dosed syringes prevent over dosing.
- Typed prescriptions prevent errors resulting from handwriting.
- Computer programs can check that a postcode or zip code is accurate.
- In the service sector, mistake proofing is often applied through the use of checklists or appointment reminder texts.
- Retailers use computer screens that require staff to ask questions such as 'Do you have a loyalty card?', or Would you like cash back?'

Detection

A *detection* device tells the user if a mistake has been made, so they can correct the problem. Here are some examples.

- A car buzzes if you open the door while the key is in the ignition.
- Rumble strips alert a driver to slow down before an approaching roundabout.
- Painted marks on the airport tarmac tell staff where to place airplane chock blocks.
- In surgical operations, a 'surgical count' requires two members of the surgical team to count the items out loud and in unison. This ensures no swabs or other items are left inside the patient. Any shortfall alerts the team that something is amiss.
- In the old days, a coal miner would place his brass tag on a board at the top of the shaft, to show he was in the mine. He would then remove the tag at the end of the day. An empty board showed that no one was left below.

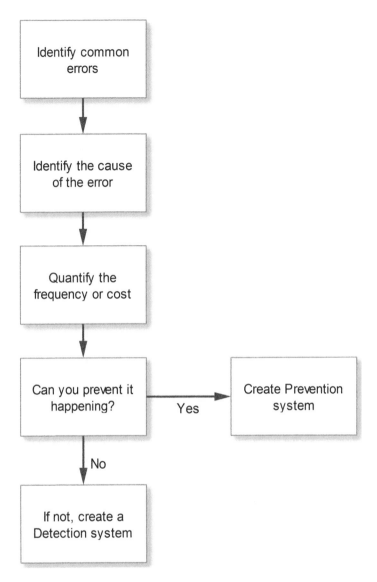

Figure 5.6   Mistake proofing method

In short, mistake proofing is a powerful tool. You simply identify where errors happen, and work out what you need to do to either prevent or detect them. The best types of mistake proofing are simple and cheap.

SIX SIGMA

Six Sigma is another quality management tool primarily suited to mass manufacturing. The name sounds somewhat scientific, but it's a fairly simple idea: if you improve the quality of your output, your production costs will fall, due to the reduced need for repair or re-work.

So the goal is to get as near as possible to zero defects, by identifying mistakes and correcting them as early as possible. Thus Six Sigma aims to reduce the risks that come from poor quality operations.

Six Sigma aims to make your output consistent, and to do that on the basis of factual evidence, by measuring and analysing manufacturing data. Six Sigma uses tools such as identifying the root causes of problems, conducting experiments and applying statistical process control (where a computer identifies variations from the norm).

CASE HISTORY: HOW ONE HOUSE BUILDER STAYS OUT OF TROUBLE

Not every company manufactures a standard product, nor do its problems take place on a shop floor. A case in point is house builders who face many areas of risk. Paying too much for land, and allowing part exchange to get out of control, are two major risks. One house builder, a £200 million business, has systems designed to prevent the company from being exposed to excessive risk.

Paying Too Much for Land

House builders need land; and they sometimes pay a premium for a plot, based on rising house prices. When the houses are built in 18 months' time, houses may be selling for 10 per cent more than they are today.

On the other hand, in a recessionary period UK house builders can buy expensive land, only to find that house prices fall. They are then stuck with land on which they can't make a profit.

The house builder above only allows itself to buy land priced at today's house prices plus the cost of one year's expected inflation for building materials. This means it has to work harder to find land; but it makes it less likely to own unsalable sites.

The Part-Exchange Risk

'Part-exchange' deals are another hazard for house builders. Today, many house buyers trade in their old house when buying a new one, just as they would their car.

Some house builders pay too much for second-hand property, and are then unable to sell the old houses at the price they have paid. Thus they get stuck with unsalable property. The builder minimizes this risk by getting three independent estimates of the house's value.

These procedures reduce the likelihood of the company buying land and second-hand houses on which it cannot make a profit.

Reducing the Scale of the Impact

Another construction and quarrying company found that 40 per cent of its capital was invested in new house building. Since the housing market suffers from peaks and troughs, this made the group's profits volatile. In one year, profits leapt from £2.5 million to £23 million. To solve the problem, the company has taken a policy decision not to invest more than 25 per cent of its capital in housing. The two companies' different approaches demonstrate that there is more than one way of tackling the same risk.

CREW RESOURCE MANAGEMENT (CRM)

With the primary cause of most aviation accidents being human error, Crew Resource Management (CRM) has developed as a programme to help crew make better decisions under stress. It encompasses leadership, team work and interpersonal skills, as well as liaising with other organizations such as a ground team.

A case in point is the moment when the Costa Concordia cruise liner ran aground. According to *Vanity Fair* magazine, video of the chaos on the bridge that night gives insight into the captain's state of mind and his inability to make decisions. John Konrad, an American captain and the editor in chief of the maritime-news site gCaptain says: 'You can tell he was stunned. The captain really froze. It doesn't seem his brain was processing.'

Also known as Cockpit Resource Management, CRM has been adopted by other industries where all crew members need to be able to intervene, and where calm action is required in an

CASE STUDY: RISK IN THE MOVIE INDUSTRY

The movie industry finds it hard to forecast which films will make money and which won't.

A single major failure can bankrupt the studio, so the risks are high. 'John Carter' cost $350 million to make, and poor sales caused Disney to take a write down of $220 million. Insiders said the director, Andrew Stanton, had no experience of live action movies, his previous films having been animations such as Wall-E and Finding Nemo.

Top 15 film losses, adjusted for inflation

1. *John Carter* (2012) $220 million
2. *Cutthroat Island* (1995) $146 million
3. *The Alamo* (2004) $134 million
4. *The Adventures of Pluto Nash* (2002) $134 million
5. *Sahara* (2005) $133 million
6. *The 13th Warrior* (1999) $125 million
7. *Town & Country* (2001) $115 million
8. *Speed Racer* (2008) $106 million
9. *Heaven's Gate* (1980) $104 million
10. *Final Fantasy: The Spirits Within* (2001) $99 million
11. *Inchon* (1982) $89 million
12. *Treasure Planet* (2002) $83 million
13. *The Postman* (1997) $83 million
14. *Red Planet* (2000) $83 million
15. *Soldier* (1998) $78 million
16. *Gigli* (2003) $66 million

Source: CNBC

Many of these films failed because they were boring and failed to attract an audience. Some, like *Sahara*, attracted a big audience, but were sunk by their unexpectedly heavy costs. In several cases, the studio went bust or had to be taken over.

With movies now costing $100 million it's vital that each makes money. However, 80 per cent of the industry's profits over the last decade were generated from just 6 per cent of the films; and eight out of ten movies lost money over the same time period, according to *Sloan Management Review*.

Movies are a mercurial blend of screenwriting, acting, directing and editing, with big egos and strong personalities at work. And the mix is too complex to easily forecast. However, with so much at stake, and with the theory understood (narrative arc, believable characters and so on), studios need to eradicate the obvious weaknesses that the flops listed above suffered from.

Publishing shares similar characteristics with the movies, with 20 per cent of titles producing 80 per cent of the profits.

emergency. Hence it has spread to the armed services, space industry, firefighting and maritime (where it is known as Bridge Resource Management). It is also found in air traffic control, aircraft maintenance (Maintenance Resource Management) and railways.

CRM includes assertiveness training where a risk may require intervention by junior staff. A five-step statement process includes the following:

- *Open the conversation* by addressing the individual. Call out 'Captain, sir,' or 'Matt,' to get their attention.
- *State your concern* in a direct manner. Be open about your emotion. 'I'm worried that the craft isn't rated for this type of weather.'

- *State the problem as you see it.* 'The forecast is for severe weather, and we might find ourselves unable to deal with the storm.'
- *Offer a solution.* 'Let's hole up in the creek until it passes.'
- *Gain agreement.* 'Is that good, Captain?'

CRM primarily involves so-called soft skills. This stands in contrast to much of the rest of risk management which tends to be fact-driven and procedures-based. Although aimed at stressful environments, it is suitable for any workplace where management wants to reduce risk by breaking down hierarchies and instilling team working principles.

Risk Assessment

You can assess your vulnerability to operational and production problems by answering the questions below.

Question	✓
Does the organization operate in high-risk or high-profile areas, such as oil exploration or care of the elderly?	
Does the business lack a quality management system such as ISO 9001?	
Does the organization lack regular feedback on key quality measures?	
Have you suffered from a major failure in the last three years?	
Does the organization suffer a high number of complaints or low satisfaction ratings?	
Has the business failed to invest in up-to-date technology?	
Does the organization lack effective controls over its services?	
Are staff inflexible, lacking in skills or paid for piecework?	
Are the organization's products, services or designs worse than the average in the industry?	
Are costs higher than others in the industry?	
Total points scored:	

Score: 0–3 points: low risk. 4–6 points: moderate risk. 7–10 points: high risk.

The Appendix has a summary of the checklists. By entering the results of this one, you can compare operational and production risk against other categories of risk.

6
Managing Health and Safety

Helen in the Legal Department has emailed you to say the business is being sued by a keyboard operator who works from home.

'She's using a computer supplied by us,' says Helen, 'but she worked at her kitchen worktop on a stool. She's been to her GP complaining of back pain. She says we didn't give her any ergonomic assessment, advice on set-up, seating, posture or breaks.'

You sigh. You've a lot of work on. This is a really unwelcome distraction. And you can't help but think it's a try-on. Should you stop people working from home? Should you conduct assessments in everyone's home, for everyone who ever works at home? As if you don't have enough to do. 'Let's call the health and safety experts,' you say, wearily.

The Real Cost of Accidents

Accidents are hugely expensive, both in terms of personal suffering as well as corporate losses:

- Over 200 people are killed at work in the UK each year. In the USA, there are 13 workplace fatalities each day, while Canada's workforce suffers 1,000 deaths a year.
- Many thousands more people die each year from occupational illnesses, including some cancers and respiratory diseases.
- UK organizations lose around 30 million working days each year due to occupational ill health and injury.
- About two million people suffer from an illness that they believe to be caused or made worse by work.
- This ill-health, injury and death causes further costs in organizations, such as uninsured losses and bad PR.

But it isn't simply rogue businesses that are affected. The Health and Safety Executive (HSE, www. hse.gov.uk) examined five firms with lower than average safety bills. Over a 13-week survey period, all lost large amounts of money due to accidents which caused injury, damage or disruption. A transport firm lost 37 per cent of its profits, an oil platform dropped 14 per cent of its output, a creamery lost £243,000, and a construction company forfeited 8 per cent of a project's value. Below we look at these cases in more detail.

- The transport firm suffered 296 accidents, costing £48,000. One bill, for £2,000, was the result of damage done by lorries manoeuvring in tight spaces.

- The building contractor (part of an international civil engineering company) suffered 3,626 preventable site accidents over the 13-week period. This cost the firm £245,000. Accidents included the collapse of a five-story column, 20 cases of vehicles or cranes hitting property, and six accidents when fork-lift trucks dropped their loads.
- The creamery lost £3,000 when food-handling equipment was bacterially contaminated. A further £1,800 damage was caused when a tanker drove away while still connected to the factory pipe. During the 13 weeks, there were 926 accidents, and the bill was £243,000.
- The North Sea oil firm lost £940,000 because of 299 accidents. They included a £2,000 health bill after a worker hit his head with a 7lb hammer. Firefighting foam was also accidentally set off. This caused workers to assemble at fire stations, and the launch of a standby vessel.

Only a small proportion of these costs were covered by insurance. During the survey period, no accidents involved death or large-scale loss through fire or explosion. In other words, large losses occurred from everyday accidents.

Risks for Service Businesses

Health and safety risks affect service companies just as much as manufacturers. The HSE made a 13-week study of the health and safety costs sustained by a cheque clearing company. The workplace was an ordinary office environment where the majority of work was clerical.

In the 13-week period, nine injuries were sustained and there were four work-related ill-health absences. These incidents cost £23,000, including equipment replacement or repair, lost working hours and reduced output. Insurance paid out £5,000, leaving over £17,000 worth of costs that the company had to pay.

Advantages of Managing Health and Safety

- Improved staff morale and lower staff turnover
- Improved standing in the local community
- Better working conditions
- Less absenteeism
- Increased productivity.

Disadvantages of Not Managing Health and Safety

- Injury or death of employees
- Damaged machinery
- Lost output
- Poor industrial relations
- Damage to reputation
- Bad publicity; PR in local press or radio stations or TV
- Increased insurance premiums.

Legal Requirements

UK health and safety law states that organizations must:

- provide a written health and safety policy (if they employ five or more people);

- assess risks to employees, customers, partners and any other people who could be affected by their activities;
- arrange for the effective planning, organization, control, monitoring and review of preventive and protective measures;
- ensure they have access to competent health and safety advice;
- consult employees about their risks at work and current preventive and protective measures.

Corporate Risks in Health and Safety

The health and safety risks to an organization include:

- being sued by an employee or his family for causing death, injury, loss of hearing, cancer or loss of limb. These are insurable; but preventing accidents is better than dealing with the problems they cause;
- being prosecuted by a regulatory authority, and suffering fines or even a prison sentence;
- loss of output caused by loss of a key worker;
- increased staff costs (including the training and replacement of the key worker, for example, using agency or contract staff).

Next we examine the major hazards in health and safety.

SLIPS, TRIPS AND FALLS

Slips and trips are the most common accidents at work, with over 10,000 workers suffering serious injury each year, according to the HSE. This costs employers £500 million a year, due to production delays, insurance costs and clerical work.

> *Floors* should be smooth and level, kept dry, free of grease, and free of obstructions, especially cables and pipes.
> *Stairs* should have handrails.
> Good *lighting* is important, especially on stairs.
> There should be vision panels in *doors*, and there should be no unprotected *floor openings* such as inspection pits.

People working on upper floors or heights should be protected from *falling* by the use of railings or scaffolding. Great care needs to be taken when working on ladders, scaffolding or roofs. Ladders should be anchored and at the right angle (one length out for four lengths high, about 75 degrees). People should not remain on mobile tower scaffolding when it is being moved.

The business should prevent materials from falling by enclosing scaffolding with sheeting, and by having solid floors without holes. Mobile tower scaffolds should be fitted with guard rails and toe boards.

Appropriate footwear should be worn, for example, steel toe cap boots or trainers, non-slip soles. Sandals, open-toe shoes and high heels should not be worn in hazardous places. You should make this clear in site rules or in a dress code for offices.

DANGEROUS MACHINES

Any machine which could cut a worker or catch clothing is hazardous, and should have safeguards. Workers should be *trained* to be aware of the hazard, and know how to use it properly. They should also have the right to stop a dangerous machine. *Fail-safe devices* should be used: for example, a guillotine should not operate unless the guard is in position. Emergency *cut-off buttons* should be prominent and in easy reach. Regular *auditing* should be carried out, using a written checklist.

Many accidents occur during cleaning and maintenance. Only *trained staff* should be allowed to clean or service machinery. They should wear proper *protective clothing* (including glasses or footwear), and the *machine should be disconnected*. A Permit to Work system should be instituted and a lock-out tag-out procedure introduced. Sufficient time should be allowed for and planned for the cleaning/maintenance.

Workers sometimes ignore safety procedures or over-ride switches to make the cleaning job simpler. Some machines are easier to clean (if a lot more hazardous) when in motion. If workers are found to be taking short cuts, you should find ways to make cleaning simpler.

CASE STUDY: PAUL WHITE GETS 12 MONTHS IN PRISON FOR MANAGING AN UNSAFE MACHINE

Paul White, the director of MW White Ltd, a Norfolk recycling firm, was sent to jail for 12 months and fined £30,000 after an employee was killed when he was crushed by a paper shredding machine he was cleaning. Later the employee's widow won £400,000 compensation from the company.

The employee, Kevin Arnup, aged 36, was killed after climbing into a paper shredder to unblock it when the machine started up, crushing him to death.

Mr Arnup left behind three children: Claire, Kelly and Jason. His teenage son Jason was also in the shredder when the accident happened but he managed to escape uninjured.

A police investigation revealed that the machine was not securely isolated whilst work was being carried out. There was no safe system for such work. And because the electrical controls were contaminated with dust, the emergency stop button failed to work. The judge who sentenced Paul White said the safety breaches were 'chronic'.

NOISE AND VIBRATION

Noise can be measured by metres, and there are laws governing noise levels. If you can't clearly hear someone talking two metres away, you need to take action. If you can't hear someone talking only one metre away, then you have a problem and must take action. That includes noise reduction, isolation of the activity, provision of appropriate hearing protection and appropriate warning signage.

Noise can often be reduced by *improved maintenance* (including lubrication), by running the machine at a slower speed, by *fitting exhausts* by surrounding machines with *sound deadening panels*, and by *separating* noisy machines from the rest of the plant.

New machines should be evaluated for the noise they produce. Giving workers ear protection should be the last resort. It tackles the symptoms not the causes, and is often put aside by workers because it is uncomfortable. Take care that any hearing protection is appropriate to the situation and to the individual worker. It should fit properly and workers should be shown the correct way to wear it.

Vibration from continued use of hand held tools can lead to 'white finger', caused by a reduced flow of blood to the fingers. Vibration can be reduced by better installation and maintenance, and by using better quality equipment. Damage can be permanent, but in the early stages it can be reduced by job rotation and by keeping the hands warm ('anti-vibration' gloves are largely ineffective and expensive: gloves which keep hands warm and dry are more useful).

Noise also affects local residents, and the company should seek to stay on good terms with its neighbours. This may require sound-deadening doors, earthworks and trees to reduce the volume. The company should also avoid revving lorries early in the morning or late at night, or letting vans roar through residential streets when people are asleep.

Failure to be a good neighbour can make a company the loser if local residents gang up on it. This may also bring you to the attention of the Environmental Agency as well as local press and pressure groups, which leads to bad publicity.

ELECTRICAL SAFETY

In the UK, according to the Electrical Safety Council (esc.org.uk), nine to 12 people die from electrocution at work, and around 100 suffer major injuries (many of which are in service industries). Electrical faults also cause many fires, resulting in more injuries and death.

Electrical safety starts with proper planning of installations, with maintenance and alterations being carried out by qualified electricians, and with the use of an electrical audit.

Some installations are more dangerous than others. *Outdoor use*, where cables and workers can come into contact with water, is most dangerous. Overhead electric lines are responsible for many deaths. Even when there is no direct contact, electricity can arc across to a nearby ladder, scaffolding or crane. You should pay attention to the materials used, such as using fibreglass ladders when near anything electrical, not metal ones or wooden ones which may be damp or contain metal parts. The HSE has guidance available on the correct ladders to be used in different situations. A full risk assessment should take this into account.

The use of *circuit breakers*, *correctly rated fuses* and adequately sheathed cables is essential. In construction work, care should be taken against cutting through electrical cables by checking with the local service providers (obtain maps if possible) and/or the use of Cable Avoidance Technology (CAT).

However, legislation designed to prevent major problems has sometimes been extended to trivial risks. Many organizations are duped into believing they must have their kettles tested each year and paying for this unnecessarily. The HSE has recently issued guidance on Portable Appliance Testing and the frequency required to ensure compliance and safety.

HAZARDOUS SUBSTANCES: DANGER TO LUNGS OR SKIN

Hazardous substances are those which can be swallowed, which can harm or burn the skin, or whose vapours can be inhaled. They can be in the form of liquids, gases, powders, solids or dusts.

Hazardous substances can be produced while being poured or mixed, during processing, as waste, or when disposed of.

They occur not just during the process, but also in research, cleaning, repair work and maintenance. Hazardous substances are found everywhere, in factories, farms, swimming pools and offices. For example legionella can lurk in unused taps or shower heads. This is particularly important to schools and colleges with long holiday periods.

A review and risk assessment of the entire water system should be undertaken by an appropriately qualified company and any necessary remedial works undertaken including the removal of dead-leg pipe-work. They should check the condition of the water storage tanks ensuring there is no access to birds or rodents, and no rust. The HSE has a publication on legionella.

The importance of these hazards cannot be underestimated. In the UK at least 50 people die every week from long-term occupational cancer alone. Asbestos is the biggest single cause of work-related deaths in the UK, killing 4,000 people a year. Mesothelioma, a cancer that can be caused by asbestos, has a latency period of 20–40 years; and people who carry out improvements to old office buildings are still contracting the disease when they disturb long-dormant fire insulation material, or drill into Asbestos Containing Materials (ACM) with appropriate equipment.

If necessary, you should commission an asbestos survey, record its findings and act upon its recommendations. You should also share it with any contractors who are undertaking work in the building. Failure to do so may result in prosecution and fines. There is a legal requirement to manage asbestos: see www//hse.gov.uk.

CASE STUDY: MARKS & SPENCER FINED £1 MILLION FOR ASBESTOS FAILINGS

Marks and Spencer was fined £1 million and costs of £600,000 for putting members of the public, staff and construction workers at risk of exposure to asbestos-containing materials during the refurbishment of its stores in Reading and Bournemouth.

The judge found that the company did not allocate sufficient time and space for the removal of asbestos-containing materials at the Reading store. Its contractors had to work overnight in enclosures on the shop floor, with the aim of completing small areas of asbestos removal before the shop opened to the public each day.

The HSE said Marks and Spencer had failed to ensure that work at the Reading store complied with the appropriate minimum standards set out in legislation and approved codes of practice.

M&S had produced its own guidance on how asbestos should be removed inside its stores, but this guidance was not properly followed by the contractors.

Table 6.1 shows how to assess hazardous substances, and incorporates an action plan. It asks the assessor to state the basic information (such as the name of the product and its known effect). It also asks who will be affected, and to what extent. This leads the assessor to provide an assessment of the risk.

This is followed by a plan of action. It will include work procedures, precautions to be taken, training and emergency action.

At the famous Guinness brewery in Dublin, staff who cleaned the empty vats had a rope tied around their waist. When the cleaners started singing, staff knew that the beer fumes had reached them, and it was time to haul them out. That was long ago. Today the company has more contemporary methods of cleaning.

CASE STUDY: WHAT HAPPENS WHEN GOOD INTENTIONS ARE BADLY MANAGED

As part of its corporate social responsibility (CSR) programme, a well-intentioned university gave work experience to a 17-year-old boy. He was the son of one of the organization's manual workers.

He was sent to clean out a paint store. There he inhaled fumes and was temporarily overcome by them. An investigation later found that the organization hadn't carried out a risk assessment, nor gave him any personal protection equipment such as a face mask. The department had simply sent him in to 'do a bit of tidying'.

As a result of this incident, his application to work at the RAF is on hold to see whether the damage to his lungs is permanent.

Table 6.1   **Hazardous substances assessment**

Assessment	Name of substance:
	Toxic or irritant effect:
	Process involved:
	Staff involved:
	Level of staff exposure:
	Extent of risk to health:
Action plan	Method of safe working, and controls to be applied (e.g. measurement, protective clothing):
	Training and information needs:
	Action in the event of an emergency:

CONFINED SPACES

In the UK, 15 workers die each year when working in confined spaces, and many more are injured. Confined spaces are common in all kinds of industry, from vats and silos to sewers. Workers can be overcome by fumes or lack of oxygen, drowned in liquid or asphyxiated in grain. Others die in fire or explosions, and others die trying to help them.

This kind of dangerous work needs to be more stringently controlled, using a *Permit to Work* system. The dangers should be carefully forecast, and safety measures put in place. Checks should be made for gas or fumes, and fresh air must be kept flowing into the confined space.

In construction, similar problems occur with trenches, which are prone to collapse. Care should be taken not to drill through electricity or gas cables, and service plans should be consulted before work starts.

LIFTING, HANDLING AND CARRYING

More than a third of major injuries reported to the HSE each year arise from manual handling – the moving, pushing, pulling, lifting or carrying loads by hand or bodily force.

Back injuries are just one of the problems that arise from moving goods around. Back trouble accounts for many days lost; and even allowing for malingerers, the problem is all too common.

In fact back pain is now the most common cause of work-related ill-health and accounts for one in eight sick days according to Arthritis UK.

CASE STUDY: BANK CASHIER WASN'T SHOWN HOW TO CARRY HEAVY BAGS

A bank cashier won £18,500 from NatWest Bank after injuring her back lifting heavy bags of coins, according to the *Daily Mail*.

Mary Deller said she was asked to carry two bags of cash – each containing 500 £1 coins – without being given health and safety training at a branch of NatWest in Kent,

The 26-year-old said that as she lifted the bags she felt a 'click in her back' and 'intense pain'.

Doctors said that Deller's injury had aggravated a previously undiagnosed genetic degenerative spinal condition. To that extent the case is abnormal. However, she said that she wasn't given any training. 'I wasn't given any health and safety induction or manual handling training or even just any general hints and tips about how to lift things without hurting myself.'

Where possible, *lifting aids* should be used, ranging from trolleys and fork-lift trucks to conveyor systems. Apart from any other consideration, they should reduce the amount of damaged goods.

Another strategy is to make the load lighter: one manufacturer found that sales were suffering because their products were packed in twenties compared with the competitors' boxes which were packed in tens. The trade disliked handling the heavier boxes.

While important, mechanical handling devices add further hazards, because they can fall or break lose. Loads must be balanced and securely packed, and hooks firmly attached. With hindsight, most people can predict an accident: the aim of the risk assessment is to recognize it before it happens.

Accidents involving roll cages have become common, partly due to their prevalence. The accidents occur on slopes and around lorries and their tail lifts; they involve the roll cage toppling over, and tend to affect feet and ankles.

REPETITIVE STRAIN INJURY (RSI) AND COMPUTER SCREENS

Once seen as newsworthy, repetitive strain injury (RSI) no longer captures the headlines. Google shows number of press reports on RSI has fallen by 80 per cent in the last ten years.

Whether this is due to effective training about computer screens, or the failure of court claims, it is difficult to say. The courts have been reluctant to agree that the workplace causes the pain, and RSI itself is not a recognized condition but rather an umbrella term for various musculo-skeletal conditions. There is some doubt of the extent to which claims are based on resistance to change, normal fatigue, malingering or the compensation culture.

However, there have been successful cases. An RAF data input clerk in her twenties successfully sued the Ministry of Defence for a RSI claim that affected her thumb, and she was awarded compensation of £484,000.

In another case, two masseuses successfully sued Virgin Airlines for £300,000. They had been employed to give massage treatments to First Class passengers. They suffered pain in their thumb joints, as well as wrists, shoulders and back. As a result, they were unable to continue with their careers, despite Virgin reducing the number of massages they could give per hour and providing additional breaks.

Legal claims may come from workers who have suffered from carrying out a repetitive job. The most common complaint has come from journalists or secretaries who have typed for extended periods. But anyone who carries out a repetitive movement of the arm, wrist or neck, or one that involves awkward movements, is at risk.

RSI can be avoided by altering the job, by broadening it, by allowing rest periods and by varying the work. Improving the workstation, so that the worker does not have to bend or twist, can also help.

In financial services and many other service industries, computer screens are the heart of many employees' work, and so RSI is a hazard.

Employers should check that there is no glare or flicker from the screen, that the chair supports the small of the back and is at the right height, and that both legs are on the floor. The screen should be moveable, and breaks should be allowed (ten minutes in every 60). Workers should perform a workstation self-assessment; you should address any identified problems.

RADIATION

Each type of radiation has its own method of safe working: the company should get advice from the equipment supplier. All hazards should be clearly marked, equipment should be carefully maintained and workers should be informed about the risks.

Table 6.2 below shows the effects of the main forms of radiation.

Table 6.2   Radiations: causes and effects

Hazard	Emitted from	Possible effects
Microwaves and radio frequency	Plastics welding, some communication catering, drying and heating equipment	Excessive heating of parts of the body; headaches, eye pain
Infra-red	Any glowing source, for example, glass production, and some lasers	Reddening of skin, burns, cataracts
Visible radiation	All high-intensity visible light sources. High-intensity beams, such as from some lasers, can be especially damaging	Heating and destruction of eye or skin tissue
Ultraviolet (UV)	Welding, some lasers, mercury vapour lamps, carbon arcs, the sun	Conjunctivitis, arc eye, skin cancer
Ionizing radiations (X-rays, gamma rays and particulate radiation)	Radiation generators, some high-voltage equipment, radiography containers, gauges, other radioactive substances, including radon gas	Burns, dermatitis, cancer, cell damage or blood changes, cataracts

INJURY CAUSED BY VEHICLES

In the UK 32 people a year die and over 2,500 are injured at work in accidents involving transport. Workers are crushed by moving, reversing or runaway vehicles, including HGVs and fork-lift trucks.

Good practice includes *separating people and vehicles*, by marking separate routes, and by adding *ramps* to reduce the speed of a vehicle. The company can provide *mirrors, audible alarms* and *visible clothing*. Vehicles may also require an *escort* when reversing. The company should make sure other companies' vehicles adopt its safety procedures when on company premises.

DRIVING

Employers have a responsibility for the health and safety of their employees when they are driving in the course of their work. This means lots of van drivers, as well as lorry drivers and sales people.

In many cases, the employee is 'out of sight, and out of mind', but the statistics show that driving is a dangerous activity. More than 1,000 employees are killed each year in work-related motor accidents – far more than the 249 deaths reported annually to the HSE. Employees driving more than 25,000 miles as part of their job have the same risk of being killed as a construction worker. And there are 77,000 injuries to employees every year because of 'at work' road accidents. High accident rates will also increase the insurance premiums.

Tiredness can be a factor in road accidents – you should monitor tachograph records to ensure that drivers are taking sufficient rest breaks.

Employers should carry out an assessment of the driver's behaviour, and implement training. This can include induction training, and on-the-job training, especially for drivers with high accident rates. Advanced driver training will reduce premiums and also accidents.

PRESSURE SYSTEMS

Steam boilers can scald workers, and compressed air can damage eyesight. All kinds of pressurized containers can explode, maiming or killing those nearby.

It is better to avoid the use of a pressured system where possible. If it is essential, it must be properly maintained and used appropriately.

SMOKING

There are government bans on smoking in the workplace in many countries, including the UK, Canada, Australia, France and Mexico. In the USA, the situation is more complex. Some US companies refuse to hire smokers, while 30 states have legislation protecting smokers' rights.

In countries where a smoking ban is in place, the risks of workplace smoking are much smaller. However, the business needs to be its guard against illicit smoking in fire escapes and on roofs, due to the threat of fire and the risk of prosecution.

WASTE MANAGEMENT

We examine waste in Chapter 8, on environmental risk. But the disposal of waste also has health and safety implications. If the organization has clinical waste, sharps or used sanitary protection, staff are at risk from injury or infection. Similarly, incineration can cause toxic emissions. The Environment Agency has guidance on this (environment-agency.gov.uk).

STRESS

Work-related stress currently costs UK employers 10–12 million work days a year, and represents 40 per cent of all work-related illnesses.

Stress can be caused by changes at work, including increased work pressure and reduced job security. This can be due to mergers, takeovers, downsizing, outsourcing, and the growth of agency working and subcontracting.

People can have an unsympathetic or controlling boss, feel they lack control over their job, believe that management is remote and unsupportive. They can also find that clients are more demanding.

Days-off from stress is particularly prevalent in healthcare, social work, education and public administration.

Reported stress is much higher in large organizations, where alienation can set in, with employees feeling remote from management decisions, where unionization is higher, and where the scale of the organization can give an individual a degree of anonymity.

SIGNS OF STRESS

- Physical signs of stress include fatigue, muscular tension, insomnia and gastro intestinal problems.
- Psychological signs are depression, irritability, loss of self-confidence and difficulty in making decisions.
- Behavioural changes include aggression, mood swings and demotivation.

However, what some individuals call stress may be regarded by others as the normal challenges of work.

Some individuals may be unsuited to their job function, level of responsibility or career.

And unlike physical illness, stress is hard to assess, which makes it a convenient label for people who want time off work.

Doctors also find it convenient to give in to an assertive patient who says they're stressed. It gets them out of the surgery faster, by giving the patient what they want, namely a 'fit note'.

Stress is sometimes based on accusations of bullying. These are sometimes well founded. At other times they are the result of management attempting to manage an individual's performance. It is not unknown for an individual who is asked to carry out specific tasks by a certain time to respond by saying he is being bullied.

Increased absenteeism and staff turnover may be signs of a stressed workplace, and this costs the employer lost output and reduced productivity.

Management can reduce workplace stress by ensuring that employees are well supported and managed, with everyone having agreed workloads and priorities. Annual reviews with clear goals will also help (though many employees find such reviews stressful in themselves).

As for the individual who takes days off due to stress, the employer can help them with a managed return to work. This can include the following:

- seek to understand the causes of the individual's stress;
- provide counseling;
- reduce the weekly hours worked;
- coach the individual in stress management;
- restructure the work;
- provide additional training.

In order to reduce stress in the workplace, you can:

- provide adequate training;
- reduce overtime hours;

- help individuals take control of their work;
- rnsure good management practices;
- discuss issues with employees and respond to grievances.

Safe Systems of Work

In Chapter 5 we looked at how a management system such as ISO 9001 can improve quality. The same process applies to health and safety, with ISO 45001:2016 and OHSAS 18001. Using a management system is particularly relevant for tasks which are dangerous.

Written procedures are the basis for staff training, because they represent best practice. Written procedures should never be handed to staff in the form of a thick manual. Staff should receive only the procedures that relate to them, and the procedure should be read aloud and demonstrated.

Risk Assessment in Health and Safety

Carrying out a risk assessment is better than preventing a recurrence after the event. You should write down the findings of the risk assessment, and identify any significant risks, and make it available to health and safety inspectors. It need only be a simple document: simplicity aids understanding. The risk assessment should include the following elements:

IDENTIFY THE DANGERS

To identify the dangers, you can:

> *Ask staff about health and safety risks.* They often know what hazards exist (though they still manage to suffer accidents).
> *Review all purchases.* Check suppliers' data sheets, especially for toxic substances. The sheets are not always very helpful, often being vague about the danger while also suggesting excessive precautions, probably to avoid legal liability.
> *Audit the workplace for dangerous processes.* For example, hardwood dust can cause nasal cancer, making it as deadly as any chemical; while staff in buildings with wet cooling systems may be exposed to legionella.
> *Check a government list of dangerous substances* (in the UK available from HMSO, hmso.gov.uk). Any substance listed as 'irritant', 'harmful' or 'corrosive' is hazardous to health.
> *Read trade publications* or information published by trade associations.

ASSESSING THE HAZARD

As we saw in Chapter 2, the organization should consider how grave the accident could be, how probable it is and who might be affected by it. This analysis will help the business to concentrate on the biggest risks. In most firms there are a small number of obvious dangers.

In a metal-working plant, workers are at risk from flying metal; while in a board-processing plant they may be at risk from glue fumes. Most accidents can be forecast and are avoidable. They happen to staff who open bags with a knife, lift heavy weights or walk over floors covered in granules.

There are also ancillary processes which can harm workers. Workers can be hurt by fork-lift trucks, and by poorly stacked goods.

However, some hazards are not immediately obvious, and many are specific to an industry. Cabin crew are likely to suffer viruses and bacterial infections from germs being re-circulated. Crews are seeking an increase in the amount of fresh air pumped into the plane.

Table 6.3 categorizes the main hazards into groups. Not every hazard can be listed here. You can add other risks at the bottom of the chart.

DECIDE WHO MIGHT BE HARMED

It is easy to omit people from the assessment, so you should include everyone who might be affected. Apart from staff, this comprises visitors, members of the public, cleaners, contractors or other companies which share the workplace.

LIST CONTROLS AND PRECAUTIONS

The company should show how the hazard is managed. What systems exist? Are there written procedures, Permits to Work or training? If procedures are listed in a safety manual, the chart should state where the manual is located. We consider relevant precautions below when looking at the main types of hazard.

Anyone included in the previous section ('Who Might be Harmed?') needs to be made aware of the risk. That is why many firms give visitors brief health and safety booklets. It will help visitors stay safe, and it may mitigate the company's liability if a visitor is hurt.

ASSESS WHAT FURTHER ACTION IS NEEDED

The auditor should now assess whether the controls are adequate. In the course of the assessment you may discover some unforeseen problems, perhaps relating to a new machine.

In Table 6.3, the heading 'Further action needed' is an action list. It shows what needs to be added to the 'Controls and precautions' list. The action list may include ways of making the hazard less risky, such as fitting extra guards, upgrading machinery, or installing monitors or alarms.

ACT ON THE FINDINGS

There is no value in doing an assessment and not acting on it. The real purpose of the assessment is to make sure that the company rectifies any problems that are unearthed. The ideal solution is to get rid of a hazard completely, but this is often not possible.

Many improvements are inexpensive or even without cost, such as putting a non-slip coating on a slippery surface, or marking separate lanes for pedestrians and vehicles.

REVIEW THE ASSESSMENT PERIODICALLY

Assessments should be done at least annually. This will ensure that safeguards are still in place and are being used. Each assessment also serves to update the assessment chart (Table 6.3). Each new piece of equipment or change to a process makes the old assessment out of date. The updated assessment should ensure that safety system remains current and relevant.

Table 6.3   **Risk assessment chart**

Auditor			Date	
	What is the danger?	**Who might be harmed?**	**Controls and precautions**	**Further action needed**
Dangerous machinery				
Pressure systems				
Noise and vibration				
Electrical safety				
Hazardous substances (fumes, dust, toxic chemicals)				
Lifting and handling				
Slips, trips and falls				
Computer screens				
Radiation				
Vehicles				
Fire				
Other risks				

You can adapt this chart to meet the specific needs of the organization.

Minimizing the Risks

As Table 6.4 shows, some solutions are better than others. In a noisy environment, it is easy to give ear protectors (personal protective clothing) to staff, but they may not use them.

A better solution is to substitute or alter the process, or improve the design. A commonly used phrase is, 'Re-design the job, not the person.' We look at each of these strategies next.

REMOVE THE HAZARD

Getting rid of the hazard is the only sure way to prevent accidents. Often this is not possible, but sometimes processes can be changed or even discontinued.

Table 6.4   **Strategies for minimizing risk**

Most effective	Quite effective	Less effective
Remove the hazard	Control the hazard. Isolate and	Train staff
Change the activity to remove or	enclose; ventilate	Personal protective clothing
minimize the risk	Audit regularly	Control staff and exposure
	Adapt; use in a safer form	Use work methods that minimize
	Good housekeeping	the chances of spill or escape

You should substitute safer processes for hazardous ones. This might mean using less dangerous chemicals, or substituting mechanical handling devices for manual lifting.

CONTROL THE HAZARD

Modifying a process or making it safer is one of the more common solutions, which include fitting guards. Isolating and enclosing a process, or ventilating the area, are similar examples. Improving a process might involve reducing machine vibration, a subject which is discussed further below.

Good housekeeping is essential. Companies that have wet floors, flammable materials all around and obstructed exits are more likely to have a poor health and safety record. There should be nothing on the floor which might trip a worker, and adequate space for movement.

Regular audits help to prevent accidents from happening. If devolved, they encourage the workforce to become aware of dangers. Audits should be professionally planned and executed. This means making sure that all areas are audited during the course of a year, and that staff do not audit areas for which they are responsible. The results should be written down and acted upon.

CONTROL STAFF AND EXPOSURE

'Sticking plaster' solutions leave the process as it is, and adapt workers or their conditions. In the case of the noisy machine above, the 'sticking plaster solution' would be to hand out ear plugs.

Training workers and making them aware of the dangers is a continuing process. As processes change, and new staff are hired, it is important to prevent standards from slipping. Staff should sign their training records to show that they have been trained and have understood what they have been told. This will emphasize to staff the importance of safety, and it may reduce the employer's liability in the event of an accident.

Dangerous work should be controlled by using Permit to Work documents which state how the work is to be carried out, and what the dangers are. The permit is signed by the worker, though liability stays with the company.

Isolating and enclosing equipment helps to minimize its effect on other staff.

Ventilation can be either general (to reduce the likelihood of fumes building up), or local exhaust ventilation (to take dust or fumes away from a specific process).

Minimizing exposure can be implemented by using devices which switch off a computer for ten minutes an hour, to ensure that a worker takes a break. In a conventional factory, there may be a written procedure stating the maximum length of time to which a worker may be exposed to a process.

Personal protective equipment can be as simple as providing overalls to ensure that employees do not have clothing which could catch in moving equipment. Personal equipment also covers ear, toe and head protection.

Analysing Accidents

You should investigate all accidents. A report should be written and action taken. Many companies post a sign showing the number of days that have passed without an accident. This encourages people to be aware of the possibility of injury.

In the UK, Reporting of Injuries, Diseases and Dangerous Occurrences Regulations (RIDDOR, hse.gov.uk/riddor) requires companies to report serious workplace accidents, occupational diseases and specified dangerous occurrences (near misses) where an incident has prevented someone from working for seven days.

Human Factors

As we have seen above, people often over-ride or ignore corporate systems. A busy theatre tells staff to wear hard hats when lighting is being rigged, in case equipment should fall. The hard hats are neatly hung on a rack in full view of staff, but are never used.

Staff think that an accident could never happen to them. So unless management ensures that staff obey the rules, someone will inevitably be hurt. Senior management must get involved in risk management.

The Efficiency-Thoroughness Trade Off (ETTO) Principle

ETTO stands for 'Efficiency-Thoroughness Trade Off'. Originated by Dr Erik Hollnagel, ETTO highlights the eternal conflict between being slap-dash on the one hand and overly protective on the other. The location of the Goldilocks or 'just right' position depends on each circumstance. Below are some aspects of ETTO.

Efficient	Thorough
Fast	Slow
Careless	Cautious
Failure to review enough possibilities	Burdensome procedures
Too little control	Too much control
Organization operates in an ad-hoc manner	Organization becomes over-rigid and hide-bound
Staff have too little appreciation of a hazard	Staff take shortcuts to speed matters up
Insufficient thinking	Insufficient output
Decisions made too quickly	Decisions made too slowly
More action	Less action

Management must be aware that staff usually try to achieve efficiency over thoroughness, by taking short cuts, especially when familiarity reduces risk awareness. When on a ladder and something is rather too far way, most of us will reach out rather than descend the ladder and move it closer.

The ETTO principle is used in safety management where practitioners need to be alert to situations where staff are inclined to take the path of least resistance. Equally you may be able to find efficient solutions that do not require excessive thoroughness, such as the 'mistake proofing' examples we looked at in Chapter 5.

Examples of efficiency over thoroughness include failing to check your work, making an assumption without verifying it ('It's only making that noise because it's a bit old'), being careless ('It'll be OK if it's left there') and expecting someone else to finish a job with the result that it is left dangerously incomplete.

Management often introduces ETTO conflicts. A business that values cost reduction over safety is prioritizing efficiency over thoroughness. So is one that claims 'safety is paramount' but staff know output is more important.

We have placed the ETTO principle in this chapter, but it could go anywhere, and everywhere, in this book. There must be enough risk assessment without it requiring an army of assessors. There should be sufficient continuity planning but not for every small risk. And there needs to be adequate security but not so as to make the place a prison.

Risk Assessment

You can check the organization's vulnerability to health and safety risk by answering the questions below. Score one point for each box ticked.

Topic	Question	✓
Machinery	Do staff work with dangerous machinery or pressurized systems?	
Hazardous substances	Do employees work with hazardous substances?	
Electrical safety	Do staff work with electrical tools or equipment?	
Slips, trips and falls	Is the work environment ever wet? Do employees work at heights? Can staff trip over wires or other obstructions?	
Lifting and handling	Does the business store a lot of stock? Does anyone manually lift goods?	
Computer screens	Do any staff spend long periods working at computer screens?	
RSI	Do employees carry out repetitive tasks?	
Noise and vibration	Is the workplace noisy? Is vibrating equipment in use?	
Confined spaces	Does anyone work in confined spaces?	
Vehicles	Does the work involve vans, lorries or fork-lift trucks?	
Total points scored:		

Score: 0–3 points: low risk. 4–6 points: moderate risk. 7–10 points: high risk.

The Appendix has a summary of the checklists. By entering the results of this one, you can compare your health and safety risks against those in other categories.

7
Procurement Problems

Steve is shouting at you. 'The container is held up at Jakarta. The customs documentation has discrepancies. I need you to sort that paperwork and get my stuff over here, pronto. Bunch of amateurs.'

When he pauses for breath you tell him there's nothing wrong with the customs documents. It's actually a bribe they want. It's what they term a 'fine'. He can either disobey the Bribery Act and pay up. Or he can return the container to Brisbane and resend it with 'perfect' documentation (that is, the same documents). Steve starts shouting again. You look out of the window. It hasn't been a great start to the day.

Categorizing Supply Risks

Raw materials and components are an essential input in any organization, accounting in some industries for up to 70 per cent of costs. And the threat of lost production, with customers fuming and orders lost, weighs heavy on the minds of CEOs. All this puts purchasing at the heart of risk management.

Procurement risk can be divided into two broad categories: *supplier* and *non-supplier*.

Non-supplier issues include the following:

*   depletion of natural resources, leading to price rises;
*   disruption of supply due to weather or a natural or manmade catastrophe;
*   price fluctuation of currencies and commodities
*   disruption of supply caused by political upheaval.

As we've seen elsewhere, the business has to weigh up the probability and impact of such risks happening, and take steps. These can include amending the production process by substituting different raw materials, and avoiding long supply chains.

As shown in Figure 7.1, supplier risks include:

*   regulatory issues, especially those affecting the organization's reputation;
*   financial concerns;
*   quality problems;
*   disruption of supply caused by a supplier (an unreliable or slow supplier, or a one who goes bankrupt).

These supplier risks can be managed by monitoring and auditing suppliers. But purchasing risk is also increased by certain types of business practice, specifically:

*   outsourcing;
*   offshoring;

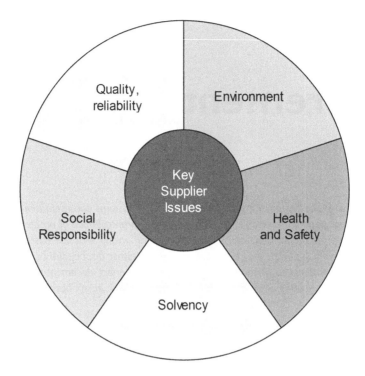

Figure 7.1   Key supplier issues

- single sourcing;
- JIT;
- overstocking.

This is further made more difficult still in situations of supplier oligopoly (few suppliers), or when price hikes keep occurring.

A different type of risk emerges when the organization, typically supermarkets, is accused of abusing its own oligopolistic power, as we shall see.

Businesses are also at risk from people issues in procurement, such as corruption and conflicts of interest.

Meanwhile, other strategies for managing procurement risk include:

- supplier auditing;
- electronic data interchange (EDI);
- partnership sourcing.

This is the complex range of interwoven issues that will be tackled in this chapter.

Non-Supplier Issues

As we have seen, many problems caused in the supply chain don't have anything to do with the supplier. We look at the main ones next.

DEPLETION OF NATURAL RESOURCES

As Brazil, Russia, India and China (known as BRIC countries) have grown, so has their demand for raw materials such as iron, steel and copper.

Similarly the need for rare earth (the lanthanides, scandium and yttrium) has grown. Used in renewable technologies such as wind turbines, 90 per cent of rare earth comes from China which has warned that its supplies are diminishing.

So any business that relies on raw materials is now competing for it with more countries; and the availability of some of these resources is declining, which in turn pushes prices higher.

Substituting with new materials (where possible) is one solution, or even to move into other markets.

The future of oil and gas

Oil and gas might be one exception to rising raw materials prices. Though production of oil has long been expected to decline, there has also been a growth of alternatives such as ethanol and synthetic oils.

More importantly, the extraction of shale oil and gas through fracking means that shale oil reserves now equal to that of conventional crude oil. Many countries are seeking to exploit their reserves, including the USA, Canada, China, South Africa, the UK and Argentina

Apart from keeping oil prices from rising, there are other major implications of fracking, albeit tangential ones as regards procurement:

- Thanks to its shale reserves, the USA could become the world's biggest oil producer, according to the International Energy Agency (IEA).
- The new found oil will make the USA and other countries less interested in carbon reduction. This will lead to greater climate change. Extreme weather events will occur more frequently, something that businesses will have to deal with. We examine climate change in the following sections.
- With US gas prices 20 per cent of those in the EU, US businesses will gain a price advantage for its exports, which poses a risk for non-US businesses.
- With the US no longer thirsting for Middle East oil, it will no longer need to ensure continued supplies of its oil from that area. Therefore its attitude to Middle East politics may change.

IMPACT OF CLIMATE CHANGE

As climate change becomes more obvious, the patterns of agriculture are changing. Different parts of the planet are experiencing increased storms, temperature rises, increased rainfall and drought. This is having a major effect on food and agriculture industries. Supplies long assumed reliable are no longer so.

But while agriculture has always had to battle with the uncertainty of the weather, most observers agree the range of weather conditions and the frequency of freak events is growing.

Mankind is ever resourceful, and will find new places to grow crops. Farmers will also seek to reduce their reliance on the sun and the rain, relying instead on irrigation and artificial heating, even though this will push prices up. Nevertheless, organizations need to take climate change into account when evaluating their risks.

Disruption from weather and natural events

Leaving aside changes in climate, weather and natural conditions cause regular disruption. This comes in many forms such as:

- earthquakes, tsunamis, volcanic eruption;
- weather incidents: storms, flooding, lightning strikes, drought.

Any of these can stop a supplier from producing a company's components or delivering them.

The business therefore needs to consider how reliant it is on supplies that are affected by weather events, and what it should do to minimize that vulnerability.

CASE STUDY: THE FUKUSHIMA TSUNAMI

When a massive 9.0 earthquake struck the east coast of Japan, it was followed by a tsunami with 100ft (30m) waves. Fukushima's nuclear power plant was destroyed, and for some weeks the world worried whether a nuclear blast would happen. It didn't, but the tsunami killed 20,000 people, destroyed 138,000 buildings and caused $360 billion in economic losses.

There were major knock-on effects in the car industry. Honda halved production at its Swindon car plant due to a shortage of components. Toyota halted daily overtime and cancelled shifts at its plant in Burnaston, Derbyshire, where 2,600 staff produced 137,000 cars a year. Ford Motor idled three factories in Asia and South Africa due to lack of parts.

The area hit by the tsunami wasn't highly industrialized, accounting for only 4 per cent of the country's manufacturing output. It does, however, produce much of Japan's intermediate output, including 33 per cent of car heaters and 32 per cent of camera parts.

The car industry recovered quickly, but the loss of output reminds procurement specialists of the risks they run from distant suppliers in areas of risk.

PRICE FLUCTUATION OF CURRENCIES AND COMMODITIES

As soon as you buy from outside your own currency area, you face the challenge of currency fluctuations. And with Western companies getting their goods made in the East, this has become a common problem.

Similarly, if you buy commodities, you face the challenge of changing prices, affected as they are by supply and demand.

These fluctuations are the concern of many departments, including procurement, finance and marketing. We discuss some solutions for this risk in Chapter 14. Suffice it to say at this stage that purchasing needs to stay in close contact with finance or treasury over these matters.

DISRUPTION FROM POLITICAL OR ECONOMIC UPHEAVAL

With wars even in the developed world (Bosnia Herzegovina in 1992–1995), and financial instability throughout the world since the 2008 economic meltdown, we can't assume that any country is immune from a future crisis.

On the one hand we have to assume that most of our normal activities will continue. If we thought otherwise, there would be no point in getting up in the morning. The point is that the more unstable the country from which we get our raw materials, the greater the threat to our supplies.

MAN-MADE (ANTHROPOGENIC) DISASTER

When Japan imposed a 12-mile exclusion zone around the stricken Fukushima nuclear plant (see above), a natural disaster (the tsunami) turned into a man-made one (the losses caused by the radiation risk).

Wars tend to have the biggest impact on supplies, in terms of the magnitude of impact and duration of the disruption.

The biggest man-made disasters (such as oil spills and chemical plant leaks) tend to affect nature more than businesses, and organizations can usually source from other suppliers.

Supplier Issues

So far we have examined risks outside the control of the organization. But what about problems caused by the company's suppliers, over whom the organization has more influence?

An Aravo poll of Fortune 1000 companies about their suppliers in the emerging world indicated that regulatory compliance was their top concern (42 per cent). This possibly reflects media coverage of labour abuses in the developing world. It was followed by the suppliers' financial viability (33 per cent). Only 25 per cent of those polled were concerned about product quality.

SUPPLIER QUALITY

Poor quality supplies is a major risk for any business. We discussed quality in the previous chapter, but buying quality goods and services requires different processes from ensuring in-house quality.

Suppliers can be troublesome when their deliveries become unreliable or the quality of their goods becomes variable.

Supplier quality is best managed by assessment and audit, including site visits and receiving inspection. We review this later in the chapter.

CASE STUDY: SUPPLIER PROBLEMS AT A COMPUTER MAKER

In his autobiography *What You See is What You Get*, Alan Sugar narrates how his Amstrad PC2000 computers suffered disc drive errors, which caused the company a major loss of reputation.

Since the errors occurred on drives from both his two suppliers, Seagate and Western Digital, Amstrad assumed the fault must lie in his own factory. It had to be something Amstrad was doing wrong; and the suppliers asserted that the problem must be caused by Amstrad's non-standard hard disc controller.

Amstrad changed the controller card on both replacement and new machines at high cost, but the problem continued. Eventually the company found that, remarkably, both Seagate and Western Digital were supplying faulty drives,

Sugar said the fiasco ruined his company's credibility in the market. 'It pains me to talk about this situation,' he said. Sugar also admitted that the company didn't have enough in-house engineering expertise, believing it would lead to 'having non-productive people hanging around'.

SUPPLIER INSOLVENCY

Having a key supplier going bankrupt is another key issue. A sudden cut in supply can cripple production.

The best ways to avoid the problem are to not become too dependent on a supplier and to institute a warning system. Gap, the international clothing retailer, uses a scorecard that tracks delivery failures, a leading indicator of supplier viability. A delivery failure automatically triggers the exploration of alternative suppliers.

In good times, most suppliers survive. But a recession can have a major effect on smaller or undercapitalized suppliers.

Companies also need to be cautious about extracting excessive discounts from suppliers, which can lead to the supplier going bust.

REPUTATION ISSUES

Activists have long targeted leading clothing companies such as Nike and Gap, protesting about the companies' suppliers. Employment practices, notably child labour, indentured labour (using unfree labourers), and bad conditions of work have been the main complaints.

Conflict minerals report

US Securities companies and any foreign business that trades in the US have to report on their purchase of 'conflict minerals', including tantalum, tin, tungsten and gold (known as 3TG), if they come from the Democratic Republic of Congo and its neighbours, including Angola, Burundi and the Central African Republic. This is to prevent the funding of armed conflict in Africa.

The businesses have to conduct due diligence, have that independently audited, then publish it on their website. To some degree, the rule affects all companies in the minerals supply chain, including manufacturers of cameras, computers and mobile phones.

Auditing suppliers in the emerging countries is an important solution, and one that we examine later in the chapter.

CASE STUDY: CHILD LABOUR

Around the world, 120 million children aged under 14 work full time, especially in Honduras, Guatemala, Haiti, Bangladesh and India. Employers use child labour because it is cheap – the children will work long hours for low pay and without overtime.

Often their parents hand them over to work as indentured labour to pay off debts. Human Rights Watch reckons that 15 million children in India work as bonded labourers, and are often cheated through fraudulent accounting. From their cramped repetitive work in dusty conditions (especially in weaving carpets), the children suffer spinal deformities and breathing problems.

The use of such labour is controversial – as well as being unethical – and could damage the company's reputation if a pressure group decides to target the business.

Risk-Creating Supply Strategies

CEOs love strategies that lower costs and create flexibility. These include outsourcing, offshoring and single sourcing. And while they have many advantages, they can also create risk. We examine some of them next.

OUTSOURCING

Sourcing materials and services from other companies carries advantages, such as lower costs, more flexibility and greater expertise. For public sector businesses it is often easier to create change when dealing with an agency, largely because the work moves to a non-unionized environment, and employees lose their ability to resist change. Outsourcing is especially effective for non-core processes, such as office cleaning. It is also ideal for work that is not needed on a full-time basis, such as PR.

Risks and disadvantages include the possibility of poor quality work or inefficiency. Just days before the start of the London Olympics, the organizing committee discovered that its contractor, G4S, had recruited only 7,000 recruits out of the required 13,400, leaving a shortfall of 6,400 personnel. The games succeeded only because the government drafted in the police and army to provide the manpower.

The business also stands to lose specialist knowledge, such as engineering, if it outsources manufacturing or design work.

After outsourcing, companies can also find they lose management control and flexibility. Hospitals discover that cleaners won't move to another ward or do additional work, because it isn't in the contract and because the hospital is no longer their boss.

However, these risks can and should be managed; for as we've seen, there are many advantages to outsourcing.

OFFSHORING

Sourcing from low-cost countries (or BRIC-sourcing) such as India or China gives businesses a price advantage, but only until your competitors do the same. At that point, sourcing from low-cost countries becomes the industry norm. Then, however, there is no merit in sourcing from the home country due to the difference in price.

As long as the difference in service or production costs remains wide, it makes sense to get things done in countries with lower costs, assuming you can maintain quality.

The Organisation for Economic Co-operation and Development (OECD) reckons that 20 per cent of all jobs in the OECD countries could be offshored, and that those most likely are in the financial, insurance and ICT sector.

The OECD says that jobs that are most likely to be offshored are those:

- with an intensive use of ICT;
- whose outputs that can be traded and transmitted by ICT;
- with a highly codifiable knowledge content;
- low face-to-face contact requirements.

At one point, accountants and software engineers in India were costing only 13 per cent – 17 per cent of what they would be in the West. However costs have been rising and the gap narrowing.

Moreover the literacy rate is only 74 per cent, and lower for women, according to the United Nations Educational, Scientific and Cultural Organization (UNESCO); and India is running out of skilled people says The Work Foundation.

As wage rises in China and India have grown, the differential has reduced. However, observers such as Hackett (thehackettgroup.com) believe that a reduction in the cost differential this will merely prompt manufacturers to move to other low-wage economies, rather than inshore the work (move it back to the West).

For manufacturers, China also has other advantages such as weak unions, weak planning laws and large production units (Foxconn employs 1.2 million people in its factories, with workers producing iPads under conditions of military discipline. It would be hard to replicate that in the West).

There are risks, however. Sourcing from abroad carries a clearly higher risk. Apart from foreign exchange risks (discussed in Chapter 14), suppliers often require bigger orders, and the lead times can be longer.

Suppliers also sub-contract to their own suppliers, leading to an extended supply chain which adds risk as well as being ever harder to monitor.

For niche markets, rather than volume suppliers, manufacturing in your local market may be an advantage. Organizations may have to make a trade-off decision between speed of logistics and cost.

CALL CENTRES: ONE OF THE FIRST TO GO, AND THE FIRST TO RETURN

Consumers are very aware of the offshoring of call centres, because they can detect cultural and language differences. And this can have a negative effect on customer perceptions of the business.

It is partly in response to this, and partly because of the difficulty of having an offshore staff handle complex questions that some call centres have been inshored or reshored (returned to the West).

A survey by ContactBabel revealed that one in seven of Britons who have knowingly come into contact with an overseas call centre responded by taking their business elsewhere. Three out of four people have a negative view of companies who route their enquiries abroad. However, this only applies where offshoring is visible to the customer.

New Call Telecom moved work back to Burnley, UK, from Mumbai, because of staff turnover and increased costs, an example of what has been called 'northshoring', referring to the growth of call centre in the relatively poorer north of the UK. Aegis, a call centre business, forecasted that more voice contact work would return to the UK, while data-processing work would move abroad.

With the advance of technology, more businesses now employ home agents, working from home. This is said to produce higher levels of customer satisfaction due to higher quality staff, itself the result of a higher applicant pool.

Quoted in *Harvard Business Review*, General Electric CEO, Jeffrey Immelt said outsourcing is 'quickly becoming mostly outdated as a business model for GE Appliances'. Just four years after he tried to sell GE's vast mile-long Appliance Park factory complex in Louisville, Kentucky, believing it to be a relic of an era GE had left behind, he has now spent $800 million to resurrect it, and bring appliance manufacturing back to the US. According to *Atlantic Magazine*, several factors have come together:

- the price of gas has fallen due to rising availability;
- US labour relations have improved;
- robotics has made manufacturing less labour intensive;
- the cost of Chinese labour has risen substantially;
- the price of oil has grown, thereby boosting shipping costs.

Immelt said: 'I think the era of inexpensive labor is basically over. People that are out there just chasing what they view as today's low-cost labor – that's yesterday's playbook.'

Note, however, that inshoring mainly applies to specific kinds of products, especially technology ones. Mass-market clothing such as jeans that require a human eye, and where labour is the main cost, may still need to be outsourced to overseas markets.

SINGLE SOURCING OF SUPPLIERS

Many businesses prefer to source raw materials from a reduced number of suppliers, with some aiming to have only one supplier for each product they buy. Others have found that cutting the number of suppliers saves time and money, avoids duplication, and gives them greater control and an improved relationship with suppliers. But it can also make the business more vulnerable to an interruption of supplies, and is prone to corruption. Transparency International claims that the high level of single sourcing in national defence spending is to facilitate government bribes.

According to the Procurement Strategy Council (PSC), which is comprised of large businesses, 40 per cent of members rely on a single supplier per product, with the rest using several. Those who use several suppliers are likely to have between two and five.

The choice is driven largely by the size of the spend in that category, and the importance of the component or category. Each of these dictates the scale of risk to the business.

With single sourcing, the company no longer sees bids from competitors, so there is a risk of higher prices. Some companies use target pricing, where the supplier is asked to supply a product at a price determined by the customer. At the very least, the supplier may be asked to quote a firm price for 12 months.

So while management theory may preach the benefits of restricting the number of suppliers, or of sourcing goods from cheap and distant lands, the risk-aware manager should seek to avoid over-reliance on suppliers.

It is especially wise to keep more than one supplier for products whose price fluctuates, such as products made from commodities. Companies which buy, for example, copper, will buy on the futures market of the London Metal Exchange or a similar financial institution.

Where single sourcing is used, it's important to have systems that discourage corruption. They include assessing whether there is sufficient reason to use a single supplier, and having an independent reviewer assess bids, especially losing bids, to ensure that the winner was properly chosen.

According to Ralph Szygenda, retired Chief Information Officer of General Motors, it's important to be able to change suppliers quickly. 'You should, when they fail to deliver the right service, be able to move to another company quickly,' he says. This means ensuring they don't set up proprietary systems that lock you in. 'And you never give it all to one company,' he says. That's an argument for dual sourcing, where one supplier gets most of the work and a secondary business receives a small amount. It lets you compare the two, and you don't become reliant on one business.

JUST IN TIME (JIT)

The principles of JIT seek to reduce bulging warehouses and to avoid having parts in stock for upwards of two months.

However, like any good idea it can be taken to extremes, and many companies have scaled back the intensity of their JIT processes, because JIT increases supplier risk. There are three main problems:

1. *Out of stock risk:* Where JIT is implemented, a traffic hold-up on the motorway can bring a factory to a halt. Some businesses now have only two days' parts in stock, and receive daily supplies from the supplier. This makes them vulnerable to having their plant halted if suppliers fail to deliver the goods. The Kobe earthquake virtually halted some motor manufacturers.
2. *Supplier resistance:* Suppliers are increasingly unhappy about providing a high level of flexibility (and the costs that go with it) unless they are paid for it. For many sub-contractors, JIT is simply a means to shift cost to the small company. And dominance by a major customer can lead to problems for the supplier if orders are reduced.
3. *Pollution:* Some companies believe that the extra pollution is undesirable.

OVERSTOCKING

At the opposite end of the risk spectrum from JIT is overstocking. Not a business strategy, it is usually the result of lack of management information and/or a sales downturn.

For some businesses, stock kept in inventories can represent the largest asset on the balance sheet.

This is especially true for retailers, where not having products in stock translates instantly into lost sales. However, inventories can easily grow to an unacceptable size and this will often result in obsolescent stock, followed by heavy markdowns, driven by a desire to generate cash flow.

The solution to these stocking dilemmas is somewhere between the risk of JIT and the waste of overstocking. While it is impossible to forecast demand exactly, managers should be able to estimate future sales. This can either be done with sophisticated software, or with a simple spreadsheet that shows historical sales data, depending on month, events or other important factors. It is important to avoid making inventory purchases too far in advance. Generally the goal is to be flexible enough to respond to most changes in demand, whilst not being wasteful.

Some retailers have adopted scan-based trading (SBT). Even though they are on a retailer's shelves, the goods are owned by the manufacturer, and the retailer is not invoiced until the goods are scanned at the checkout. But the trade-off is that the retailer pays faster than using traditional methods.

OLIGOPOLY

An oligopoly is a market with few suppliers, and this poses risk to downstream businesses.

- In many countries, the banks are few in number, and therefore there will be little competition or innovation.
- The UK's gas and electricity suppliers are dominated by British Gas, EDF and Eon, with the result that their profits are high.
- The same is true of academic publishers,
- In electronics, D-ram chips are made by just three firms, Samsung, Hynix and Elpida (which went bust).
- There are only four big accountancy audit firms: PwC, KPMG, Ernst & Young and Deloitte.

This issue presents the same problem as we saw before in single sourcing, namely a vulnerability over supplies.

THE EFFECT OF A COMPANY'S OLIGOPOLY POSITION

With only a handful of buying points for food, supermarkets are often criticized for exerting downward pressure on prices of their suppliers, especially growers and food producers, or imposing unfair terms on them. This is a reputation risk issue.

Walmart controls 30 per cent of many US markets, while in the UK the big four, Tesco, Sainsbury, Morrison and Asda, take 75 per cent of grocery sales.

This may be the inevitable the price of success, as individual companies become increasingly successful. However, it can also lead to growing resentment.

Caught between the consumer's need for low prices and suppliers' need for profit, the supermarket is always likely to favour their customer.

Companies in this position might want to foster local suppliers, support local businesses and contribute to the local economy.

It is not inconceivable that legislation could be passed, limiting a company's grocery market share, in the same way as media ownership is restricted in Germany and the UK. A UK code of practice was established by the UK Competition Commission, affecting any supermarket with more than 8 per cent of the market.

Managing Procurement Risks

We've looked at the risks that procurement brings. Now let's see how those risks can be managed. Many strategies involve getting closer to the supplier.

PROACTIVELY MANAGING SUPPLIERS

Organizations can get better results by proactively managing suppliers and introducing penalties for failure and bonuses for meeting quality targets.

- Some businesses ask their suppliers to agree targets for quality, in some cases to as low as 20 defective parts in a million.
- Others require suppliers to accept responsibility (and the costs) for warranty work which can be attributed to their parts.
- Partnership sourcing, which we discuss below, is also used to maintain quality. This can include educating suppliers, providing advice and investment loans.

SUPPLIER AUDITING

The solution is to ensure that suppliers (and their sub-contractors) are independently audited, and by ensuring that adults are paid a fair wage.

But effective auditing can be hard to adequately achieve, particularly if suppliers are in distant locations or if they use sub-contractors. Moreover, a company's internal audits are rarely credible to non-government organizations (NGOs) or protest groups.

Self-declarations or statements such as the following are unlikely to carry much weight in the eyes of the companies' critics:

'We ask all suppliers to sign an agreement stating that they understand HP's expectations and that they agree to work with us towards conformity with our Supplier Code of Conduct' (www. hp.com).

'We have a process of 'self-declaration' when new suppliers provide information on their production facilities' (Monsoon Accessorize).

INDEPENDENT AUDITING ORGANIZATIONS

The Fair Labor Association (FLA, www.fairlabor.org) provides independent audits.

The FLA is a non-profit organization that promotes adherence to international labour standards and improves working conditions worldwide. Seventy-five per cent of its audits are in clothing companies.

Participating companies adopt the FLA Code of Conduct, and implement a compliance programme throughout their supply chain. The FLA then uses accredited monitors to monitor each company's high-risk facilities, works with companies to resolve problems identified in their facilities and independently verifies internal compliance programmes.

However, the FLA lost credibility after its CEO praised Foxconn, a business with plants in China supplying Apple products, where workers had committed suicide. Critics also say it suffers a conflict of interest since its revenue comes from the companies it audits.

The Worker Rights Consortium (www.workersrights.org) is another mainly US-based independent labour rights monitoring organization, focusing primarily on clothing.

Achilles (www.Achilles.com) provides the independent logging and auditing of 77,000 suppliers and their sub-contractors. It helps businesses find and qualify potential suppliers, then audit and monitor them, thus helping companies to identify the risks in their suppliers.

Aravo (www.Aravo.com) offers similar services and has 1.8 million suppliers on its database.

PARTNERSHIP SOURCING AND SUPPLIER DEVELOPMENT

At the opposite end of the traditional 'arms-length' purchasing model is partnership sourcing, a model that gets suppliers deeply involved in their customers' businesses.

Partner sourcing entails giving more information to suppliers, involving them at an early stage in development projects, and giving them advance notice of production plans.

Using supplier expertise also helps to boost the quality of the final product.

Nearly three-quarters of major UK purchasers and 61 per cent of suppliers use it. Of those that have adopted partnership sourcing, 78 per cent say that it has improved product quality and 81 per cent say that it has improved the service they give their customers.

Working closely with a supplier can reduce risk, because the supplier better understands your needs. In addition, there is the opportunity to harness the supplier's skills to improve your product or service.

There are attendant risks, however. The supplier can use the knowledge acquired to win contracts with competitors; but this kind of risk can be reduced by the wording of contracts, non-disclosure agreements (NDA) or a non-compete agreement.

A worked-out plan can focus attention on important issues such as how to add value through collaboration, how to share that value and what exit strategy should be agreed.

Partnership sourcing is a good example of how to use risk as an opportunity, rather than simply treating it as a threat.

SUPPLIER RELATIONSHIP MANAGEMENT

Supplier Relationship Management (SRM) is an extended version of partnership sourcing. It usually involves collaborative software (an online database involving SAP or Oracle, connected through EDI or an XML portal) that links the organization and its suppliers, using EDI or XML. SRM allows both to have a clear view of the relationship. It's particularly useful in cases where the business outsources much of its production. Typical content includes:

- Data about the supplier. This can include its financial results, renewal dates for the contract, the relationship framework, contact details supplier risks and the organizational structure.
- The product and process: lifecycle management and ERP.
- Contract performance. You can have a scorecard showing key performance indicators and survey score results for the relationship.
- Status of the project. This can incorporate status reports, outstanding actions and ideas from suppliers.

HORSE BURGERS, ANYONE?

A pan-European investigation started when Tesco, Lidl, Aldi and Iceland were discovered to be stocking beefburgers contaminated by horse DNA. Although the DNA is perfectly harmless and many societies eat horse, the issue raised concerns about traceability of supplies, and whether low price had paid a role.

Of the beefburgers tested, 85 per cent contained other animal meat, and value beefburgers were 29 per cent composed of horse. The meat is thought to have originated from the Netherlands and Spain.

Previous food scandals had included illegal 'Sudan 1' dye being found in many supermarket ready meals, and pet food containing contaminated rice protein from China causing kidney failure in US pets.

Some food markets have supplier chains stretching across the world. Add a concentration of buying points among supermarkets, grocery price competition, rising grain prices and shrinking margins among producers, and it isn't surprising that corners would be cut.

To avoid this kind of problem, businesses have to have a clear oversight of their suppliers, conduct regular audits and carry out independent tests. Ensuring suppliers have an adequate margin is also important.

BS11000

Billed as the world's first collaborative relationship standard, BS11000 is designed to help organizations create effective partnerships. It moves the debate away from computer databases to a more dialogue-centred strategy, and turns partnership sourcing into a controllable system.

BS11000 was originally trialled by the UK's National Air Traffic Service (NATS) which is responsible for managing 2.2 million flights through UK airspace each year, it also involved its partners EMCOR UK, Lockheed Martin, NATS, Raytheon Systems and VT Group.

It is relevant for large organizations with a big reliance of suppliers, such as in engineering, infrastructure, aerospace or automotive markets.

At the heart of the standard is the Relationship Management Plan (RMP) which documents how the partners will work together.

The initial phase, dubbed 'Strategic', involves preparation on the part of the organization (see Figure 7.2). The business needs to know what it plans to do, and must equip managers with the relevant knowledge. It needs to create systems and prepare a risk analysis. It should allocate resources, set partner selection criteria and write an action plan.

In the middle, or Engagement, phase, the business will nominate potential partners, and then select the right ones. And it will establish a collaborative strategy with the partners.

In the final, or Management, phase, the business will seek to maintain delivery and performance, along with innovation. It then needs to implement its exit or disengagement strategy, and evaluate future opportunities.

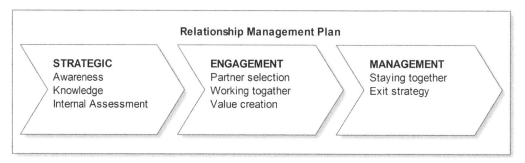

Figure 7.2   ISO 11000

Security and ISO 28000

Security of the supply chain is ever more important as high-value goods are moved around the globe. These products share similar characteristics of being:

*   High-value items, and so worth the effort;
*   Of a manageable size, and therefore easy to steal;
*   Non-traceable and therefore easy to sell.

ISO 28000 is a management standard for the security of your supply chain. It focuses on threats to cargo security such as theft, piracy and terrorism, as well as hacking and identity theft. The standard is especially suitable for organizations which warehouse and ship high-value items across long distances, such as electronics or designer goods.

The Transport Asset Protection Association (TAPA) (www.tapaonline.org) has published two standards aimed at warehousing businesses and logistics companies:

*   Freight Security Requirements (FSR)
*   Trucking Security Requirements (TSR).

These aim to ensure cargo security when transporting goods through the supply chain. They specify the minimum acceptable security procedures for HVTT (heavy vehicle transport technology) goods, as well as how to maintain those standards.

C-TPAT is another supply chain security standard, for manufacturers shipping to the USA.

Hacking and data loss is also a thereat since large organizations exchange data with their suppliers. Those risks are covered in the Security and IT chapters of this book.

SUPPLIER ASSESSMENT

In the Aravo poll of Fortune 100 businesses mentioned earlier in this chapter, over half of respondents said they actively manage risk for fewer than 20 per cent of their suppliers. Part of this may be the extra work it entails, but it also indicates a weakness in carrying out supplier assessment.

Supplier assessment often starts with a form completed by a prospective supplier, such as the extract in Table 7.1. This requires references, history and evidence of quality control procedures such as ISO 9001.

Table 7.1   Extract from a suppler checklist

	Fully implemented ✓	Partly implemented ✓	Not implemented ✓	Evidence
1 The company guarantees the rights of freedom of association and collective bargaining (ILO fundamental Principles and Rights at Work)				
2 The company does not practise forced labour or inhumane treatment (ILO fundamental Principles and Rights at Work)				
3 The company does not practise child labour (ILO fundamental Principles and Rights at Work)				
4 The company does not practise discrimination in hiring and employment (ILO fundamental Principles and Rights at Work)				
5 The company has taken measures to protect occupational health and safety (ILO International Labor Standards)				
6 The company observes the national (host country's) laws regarding working hours. (ILO International Labor Standards)				
7 The company observes the national (host country's) laws regarding minimum wages (ILO International Labor Standards)				

Forms like this can be sent as a questionnaire to an organization's suppliers; staff can use a variant as a checklist when visiting the supplier and auditing it. This helps the business assess the suitability of a new supplier. It also provides regular factual reviews of the supplier's delivery reliability, its management's ability and flexibility, and the quality of its products.

Subsequently, the client visits the supplier from time to time to audit their premises and processes. Finally, receiving inspection takes place at the client's premises when goods arrive.

This is the minimum requirement for any major supplier. It ensures that the customer checks the supplier in a methodical way. Beyond that lies greater involvement with the supplier and partnership sourcing, which we discussed earlier.

Given that each supplier can be inundated with hugely detailed questionnaires from dozens of clients, and even more prospective clients, best practice includes the following. Dale Neef, author of *The Supply Chain Imperative* suggests the following:

- Use outline broad-brush questionnaires to identify vendors and for existing low-risk, low-importance suppliers. Use them to simply identify issues such as what certifications they hold and whether they employ minors. Reserve more detailed questionnaires for key suppliers.
- In the questionnaire include a request for permission to inspect the supplier's site to verify the answers.
- Avoid the need for long narrative answers. This will increase the speed and response of suppliers.
- Make questions straightforward to avoid misinterpretation.

Figure 7.3 is a simpler form used by Cow & Gate to assess supplier suitability. It reminds the auditor to check that the supplier keeps adequate process records, quarantines non-confirming products, and maintains its equipment adequately.

GRADING SUPPLIES BY THE SCALE OF RISK

There are several ways to grade supplier risk. PRC, as shown in Table 7.2 suggests three straightforward categories. Supplies are graded into three groups, with different purchasing strategies and managers allocated to each.

Table 7.2   Grading suppliers

Level	Items	Level of procurement staff
A	High-value items valued above $5 million dollars Includes critical mass items such as raw materials	Commodity teams with strategic skills
B	Middle-range items valued between $1 million to $5 million dollars Includes supplies such as specific packaging items	Commodity teams with mid-level strategic skills
C	Low-value items valued under $1 million dollars. Includes MRO (maintenance, repair and operations) items	Commodity teams with transactional and tactical skills

Source: PRC

Wells

Quality Assurance

Subject PURCHASING PROCEDURES	Date 16.09.93	Nr. PUR/06.22
SUPPLIER AUDIT CHECKLIST	Page 9 of 11	Modification Nr. 1

5. MANUFACTURING PROCESS CONTROL	COMMENTS
5.1 Processing Equipment – condition cleanliness	
5.2 Equipment Checks – documents frequency	
5.3 Process Documentation – document checklist	
5.4 Process Systems – GMP HACCP	
5.5 Process Records – temperatures dwell times	
5.6 Non Conforming Products – area documentation	
5.7 Weighers – area checks	

5.8   Packaging Handling – area
             checks
            frequency	

5.9 Personnel Training – equipment documentation	
5.10 Train Review – scope frequency	
5.11 Maintenance logs used?	

COMMENTS

Drawn Up:	Approved	Q A Verification

PUR/06.22.8

Figure 7.3   Supplier assessment at Cow and Gate

Another method of grading suppliers is the Kraljic Matrix, shown in Figure 7.4. This is more widely used in northern Europe.

This is a simple 2 x 2 grid which categorizes supplies according to their availability and their impact on profit.

For a motor manufacturer, examples of each quadrant might be as follows:

- Routine: stationery.
- Leverage: extruded plastic parts.
- Bottleneck: door locks.
- Strategic: engine, airbags.

It results in four types of purchases: Routine, Leverage, Bottleneck and Strategic. This allows for a strategy for each category, ranging from standardizing the purchasing for unimportant purchases to developing strategic long-term partnership for critical ones.

The matrix also lets the organization determine how much management time (or visibility) each product should receive.

The Kraljic Matrix is mostly common sense, but it is a useful tool for reducing complexity and imposing order.

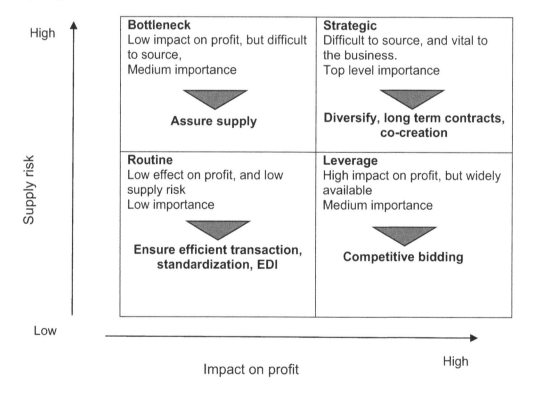

Figure 7.4   Kraljic Matrix

PORTFOLIO ANALYSIS

Having completed the Kraljik Matrix, companies should take the products defined as Strategic, and place them in one of the nine boxes in Figure 7.5.

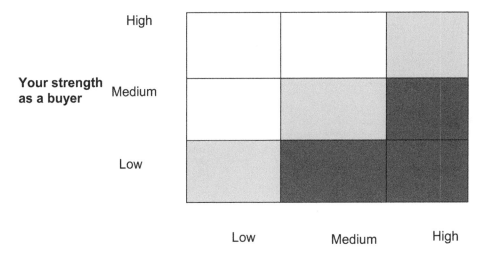

Your supplier's strength

Figure 7.5   Portfolio analysis

This Portfolio Analysis will then provide a purchasing strategy for each of the three colours.

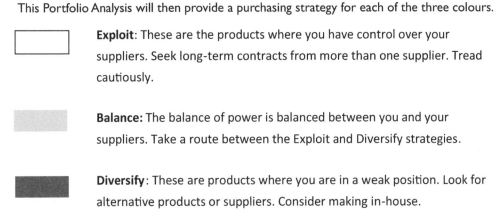

Exploit: These are the products where you have control over your suppliers. Seek long-term contracts from more than one supplier. Tread cautiously.

Balance: The balance of power is balanced between you and your suppliers. Take a route between the Exploit and Diversify strategies.

Diversify: These are products where you are in a weak position. Look for alternative products or suppliers. Consider making in-house.

MSU AND MONCZA

Other organizations, especially in the US, use the Moncza/MSU model, as shown in Figure 7.6. It operates as follows:

1. Make or Buy? The organization decides whether the product is best made in house or should be outsourced.
2. Group similar products together, so you don't have to analyse each one separately.
3. Locate a range of suitable suppliers.
4. Develop and manage suppliers, by using SRM.
5. Work with the supplier at the earliest opportunity, to avoid confusion and misunderstanding.
6. Involve the supplier in your production process. The more they know, the most supportive they will be.
7. Monitor and aid your supplier to improve their quality.
8. Push costs down to remain competitive

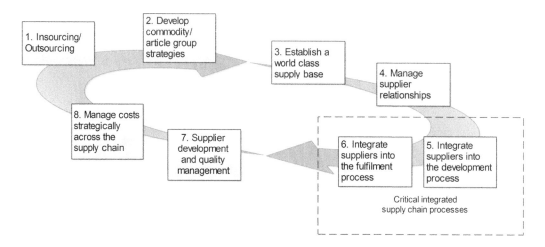

Figure 7.6   Monckza/MSU model

Source: Robert Monckza

Each of these systems has their own uses, and will be appropriate in different situations.

Brigitte Faber at the HAN University for Professional Education has reviewed both Monczka and Kraljic, and concluded the following. Kraljic stimulates thought and creativity. By contrast the MSU model represents a checklist rather than a model. Kraljic is easy to use for a one-off purchase. But Monczka is better for strategic purchasing decisions. Monczka requires many questions to be answered, and is therefore slow to implement. Thus each has its own strengths and weaknesses.

Procurement Fraud

We discuss procurement fraud in more detail in Chapter 13. But along with the finance department, procurement is an area with more opportunity than most to suffer from corruption both from within and from supplier fraud.

Procurement fraud is especially prevalent in construction where large sums are spent on many contractors and sub-contractors. Purchasing officers and specifiers are at risk in government bodies that commission public works or grant licenses. And it is especially liable to happen in developing countries, and territories that involve intermediaries or subsidiaries.

As Figure 7.7 shows, procurement fraud occurs in the following areas:

1. When projects are put out to tender or quotes requested. Tender documents can be skewed to favour one bidder or another. And where suppliers can conspire to bid in such as way as to ensure one of them is chosen.
2. When the winning bidder or supplier is chosen, sometimes as a result of bribery.
3. When the work is carried out, with substandard goods or services being supplied.
4. When invoices are submitted. This can include fake invoices from non-existent suppliers.

You need to have systems in place to prevent these frauds from happening, and to identify them if they should occur.

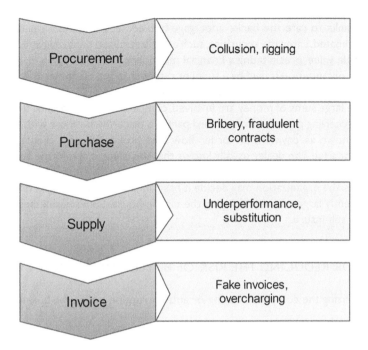

Figure 7.7 **Four moments of procurement risk**

CORRUPTION RISKS

With suppliers desperate to get orders, the buyer is vulnerable to corruption. Aware of this, many companies have stringent codes of conduct about accepting gifts and inducements from suppliers.

Sometimes the inducements are subtle, and take the form of company events and hospitality. The insurance industry, for example, entertains brokers or big company clients with golf days, outings to motor racing circuits and industry awards ceremonies. GlaxoSmithKline was fined $3 billion in the US for failings which included offering golf lessons and fishing trips to doctors in the hopes that these would persuade them to use its drugs.

CONFLICTS OF INTEREST

The International Organization of Securities Commissions (IOSCO) has warned that some banks have been channelling clients' exchange traded fund (ETF) business through in-house trading desks, which keeps the profit in-house, with the margin being wider than might be obtained from another dealer. Is that a customer, not a supplier?

And in the past, record distributor Sony BMG was fined $10 million in New York to settle charges of bribing radio stations to play the company's records (known as payola).

Auditors get a fee for approving a company's accounts, so there is a clear benefit on both sides for the auditor not to inspect too closely. The auditor may receive further fees for consultancy. Hence investors may not receive fully independent information.

Similarly, in the run up to the sub-prime mortgage crisis, when banks gave mortgages worth $3.2 trillion to people who couldn't afford them, rating agencies such as Moody's were paid by the investment banks to rate the banks' mortgage-backed securities. As a result the risks were severely underestimated. Later the securities, such as collateralized debt obligations (CDOs), were found to be of little value, precipitating a financial meltdown.

Wherever an organization is paid by a supplier rather than the customer, the company's ability to service the client dispassionately must be in doubt.

Hence where large sums of money are involved, companies need to assure themselves that its advisers are not receiving payment from a third party to place the business with them. In financial markets, this is known as 'payment for order flow': the broker places a client's business with a dealer in return for a fee. The dealer usually knows that the client is ill-informed and therefore the profit on the transaction will be high.

In other cases, the organization may decide it has sufficient in-house expertise, or the sums of money are sufficiently large to merit bringing the task in-house. Some businesses place their own insurance or else self-insure.

STRATEGIES FOR REDUCING THE RISK OF BUYER CORRUPTION

Apart from publicizing the company's buying or anti-corruption policy, the business can:

- routinely switch buyers to different purchasing categories;
- install a purchasing system that makes collusion difficult, such as the use of competitive tendering;
- undertake audits to check for abuse;
- implement a continuous improvement programme. This will increase the extent of change in materials and processes, and thereby prevent fraud due to inertia. Stable or moribund systems can lead to corruption;
- formally warn suppliers that they will lose the business and be reported to the police if corruption is discovered.

Controlling Price Hikes

Industries such as construction and engineering are heavily exposed to raw material price changes, and when prices rise, their profitability is threatened. StoraEnso is an example of a company that weighs its risks, as shown in Table 7.3.

The solutions that companies can adopt include:

- *Negotiating with suppliers* to keep their prices down. Some buyers require a drop of 2 per cent a year, rather than the more usual increase. This has to be found through productivity rises achieved by the supplier.
- *Shopping around for different suppliers* – not everyone puts their prices up at the same time.
- *Hedging* – buying futures in their selected raw material.
- *Add a clause to your customer's contract*, allowing you to increase the price if costs rise. This isn't possible for powerful customers, such as some government contracts.
- *Choosing substitute raw materials or components.* For example the cement industry uses limestone, clay, shale, sand and iron ore, as well as fuels such as coal. Many of these raw materials can be changed. You can add fly ash to the limestone, reducing the amount of limestone needed. Alternatively, you can substitute blast furnace slag or powdered limestone.
- *Re-design your products.* A fork lift truck manufacturer removed 200kg (450lbs) of cast iron from each vehicle by changing the position of the counterweight used to provide stability during lifting. This cut material costs by $200 per vehicle, as well as reducing weight and therefore the vehicle's fuel consumption by 4 per cent.

- *Build alliances and join buying groups.* The French car maker Renault cut 3 per cent of its parts' costs from its Modus small people carrier by sharing half the components with Nissan. The savings could double its profit margin on this class of vehicle.

Supply and Demand Risk

Product prices, raw material and energy costs are cyclical and therefore a period of low product prices or high raw material costs affects profitability.

Reliance on imported wood may oblige the Group to pay higher prices for key raw materials or change manufacturing operations.

Reliance on outside suppliers for the majority of energy needs leaves the Group susceptible to changes in energy prices as well as shortage of supply.

Changes in consumer preferences may have an effect on demand for certain products and thus on profitability.

Exchange rate fluctuations may have a significant impact on financial results.

Table 7.3   Excerpt from StoraEnso's annual report

CASE STUDY: HOW TO SPOT PROCUREMENT CORRUPTION

Deon Anderson, a 48 year old procurement officer at aircraft giant Boeing, seemed a normal guy – until the courts jailed him for taking bribes.

Anderson gave two contractors details of their competitors' bids for $3.5m worth of airplane parts. He got 20 months in prison, while the contractors were sentenced to 18 and 15 months.

William Boozer, owner of Globe Dynamics International, a machine parts supplier, had got Anderson to provide him with confidential information. As a result his company gained seven out of 16 tenders. The two regularly communicated by phone and email in code, with Boozer frequently requesting "Isle 5", a coded reference to a "Price check on aisle 5". This was a request for details about competitors' quotes.

Anderson later gave secret bid information to a second company, a machine shop called J. L. Manufacturing. The contractors paid Anderson $250,000 over the four years that the fraud took place.

TAKE AWAYS

In case of corruption there are often tell-tale signs. In this case, a single bidder won seven of 16 tenders, while another vendor secured seven of nine bids; and that should have raised questions.

A high win ratio should be a risk indicator for corruption. Routine data mining of the procurement system data can identify this kind of anomaly. Covert phone calls and emails with unusual content are also danger signals.

In cases of collusion the criminal's bid must always be delivered last, because they have to know their competitors' quotes before submitting a tender. Therefore a supplier who regularly delivers the last bid could be engaged in fraud. More generally, there is always a risk when a purchasing officer gets to know suppliers. And the risk grows when the same officer can review the bid documents. You can reduce the risk by rotating buyers' responsibilities. This reduces the likelihood of a buyer building a close relationship with a supplier. You should also seal bids until after the closing date, and if possible separate the roles of posting and awarding of contracts.

Be on the lookout for an employee who spends more than they might be expected to. Exotic holidays, expensive cars and big homes are typical give-aways.

And finally you should undertake rigorous audits. The US Defense Contract Audit Agency has a set at dcaa.mil/audit_program_directory.html. They can serve as a useful model in other industries.

REVERSE AUCTIONS

Reverse auctions are a completely different model to single sourcing. Here, suppliers are bidding the lowest price for a commodity or component.

You can set up your own reverse auction site through suppliers such as procureport.com. In the US, the federal government runs a reverse auction for its agencies at reverseauctions.gsa.gov.

The risk associated with reverse e-auctions is that you can make more supplier changes than in the past. And making supplier changes can result in disruptions to your plant operations. But incumbent suppliers often retain the business, not least because they know more about the client, and are able to offer a cogent case. However, this is unlikely to be the case in bids based purely on price.

Other problems in reverse auctions are suppler quality and flexibility. An unknown supplier may have an unproven track record, although suppliers often build up a track record on the site, or can demonstrate their experience through having a quality system in place.

Vertical Integration

Some companies minimize the risk to their supplies by buying their suppliers. For example, Kentucky Fried Chicken bought chicken farms. Other companies extend downstream. Oil companies own petrol stations, while an egg producer started making egg products, including the liquid egg that caterers use to produce omelettes.

But this also contains risks. In order to provide a complete service to clients, advertising agencies have often added direct response, PR or new media departments. But these offshoots have been hard to manage; and if the client is dissatisfied with the direct mail business, this can sour the relationship with the core – and more valuable – advertising account.

If you own the supplier, company policy often requires you to buy from that supplier, rather than seeking the lowest price in the market. This results in higher-priced supplies. And trying to get co-operation between different business units is often hard to enforce.

Vertical integration also reduces the number of suppliers, which in turn reduces competition, which can put prices up. Innovation can also reduce since the supplier now has a captive customer.

Moreover, businesses sometimes find themselves unable to easily manage the very different kind of business represented by the supplier. And if the supplier begins to perform badly, the purchaser can find itself bearing the unexpected losses.

So while vertical integration always looks attractive, it often fails to produce the desired result.

Electronic Data Interchange (EDI)

EDI has been around for many years. It issues orders to suppliers, mostly triggered by the customer's warehouse computer or production planning system. It also provides electronic invoicing.

Sophisticated EDI systems can then initiate production at the supplier's plant, without the need for human involvement. By speeding the information between customer and supplier, and avoiding the dangers of lost paperwork, the customer reduces its risks. Suppliers can also use EDI proactively as a business tool to embed themselves more firmly into the customer's system.

In theory the internet should have superseded EDI, which usually requires costly hardware and software, and is therefore less suitable for smaller suppliers. But EDI has been deeply embedded and therefore hard to change.

Each industry now has its own XML EDI system, and many industries have several. There is therefore a risk of selecting the wrong system and getting locked into one that becomes out of date. The EU has 12 different invoicing EDI systems, according to the *Financial Times*.

However, the Mond platform provides a 'translation' service, converting electronic purchase orders and invoices into a format that can be understood by all businesses, irrespective of which system the company is using.

Eighty-four per cent of all global invoices are still paper-based, which indicates that there is still room for improvement. Tradeshift provides a free online invoicing system, making its money from other services that its 60,000 customers can use.

Risk Assessment

You can assess your vulnerability to procurement risk by answering the questions below. Score one point for each box ticked.

Question	✓
Do raw material prices have a big impact on the company's profits?	
Are you dependent on supplies from distant countries?	
Do you rely on a single supplier for major supplies?	
Could bad PR from your suppliers affect your reputation, for example for the use of child labour?	
Do you rely on 'just in time' supplies?	
Does the business face a possible conflict of interest between suppliers and customers?	
Has the company failed to formally assess its suppliers?	
Does the business lack an e-procurement system?	
Have you failed to introduce partnership sourcing or supplier development?	
Have you failed to test your ability to recover operationally from a supply chain failure?	
Total points scored:	

Score: 0–3 points: low risk. 4–6 points: moderate risk. 7–10 points: high risk.

The Appendix has a summary of the checklists. By entering the results of this one, you can compare your supplies risk against other categories of risk.

8
Preventing Environmental Damage

'The good news is, plant and animal life seem to be unharmed,' says Paul. Your contractors have managed to clog a sewer pipe during building work. 'It was a freak accident,' he says. 'They were filling a trench and some flowable fill got into a crack in the sewer pipe and clogged it up.'

The result was 15,000 gallons of wastewater spilling into the river for a four-hour period before someone noticed. The smell was pretty awful, apparently. It then took the council 12 hours to locate the blockage and fix it.

'The Council say they've taken water samples from different points of the river to test for faecal bacteria, which is what you get in sewage,' says Paul. 'But the numbers are OK. The river has done a good job of dispersing it.'

'But the bad news is, they're going to sue.' 'How much?' you ask. Paul shrugs. 'Five figures maybe,' he says.

13 Changes to the Environmental Background

Most companies have well-established processes for conforming to the environmental legislation. But governments tend to ratchet up environmental standards. This means that some developments are still on the horizon. They create a problem for some businesses and opportunities for others. So, what will the future be like? We examine the changes that are likely to happen.

1. ENERGY COSTS WILL INCREASE

Several factors are likely to push up the cost of energy, while others are set to reduce it. Governments and the EU will want to encourage businesses to reduce energy usage in order to reduce airborne pollution and global warming. Governments will also see it as a legitimate way of taxing less efficient companies. The net result will be a rising cost of energy. On the other hand, renewable energy and a growth in shale oil and gas will serve to push prices down. It would be risky, however, to gamble on prices remaining where they are.

2. WASTE DISPOSAL WILL BECOME MORE CHALLENGING

The costs of waste will increase. The opportunity to send waste to landfill will be reduced. European governments will require firms to recycle, incinerate or microwave its waste. Incineration is a more expensive process than tipping, especially as greater controls will be placed on incinerators to ensure that they do not produce atmospheric pollution. There will be penalties for disposing of non-recylable waste.

3. PHYSICAL DISTRIBUTION WILL GET HARDER

As car ownership increases, city centres will increasingly become polluted and grid locked. As a result, city centres will become increasingly less efficient and pleasant places to work. Physical communications will become more costly and difficult, and deliveries will take longer.

Municipal authorities will take action against cars to clear the air and keep their cities moving. This will include curbs on vehicle access to city centres, as has happened in cities like London, Athens and Hamburg. Companies will have to make choices about the location of their premises and the way they distribute their products.

4. MORE LITIGATION AGAINST POLLUTERS

Regulatory authorities will have increasingly sophisticated monitoring devices in place (for example, in rivers), which will detect whether a company is causing pollution. Legal action will become an automatic response, especially if governments see this as a good source of revenue. Few taxpayers will complain if the government prosecutes polluters. Fines and jail sentences will become heavier and more frequent, as pollution becomes less socially acceptable. Directors will be held liable.

Companies which exceed their consent or which cause a pollution incident will be increasingly prosecuted. Penalties vary around the world. In New Zealand, it is punishable with up to two years in jail. In the US, executives have been jailed, and large fines are common. Hoegh Fleet Services, a Norwegian shipping company, was ordered to pay $3.5 million for intentionally dumping waste oil into the ocean.

5. SUPPLIERS' ENVIRONMENTAL PROBITY WILL COME UNDER MORE SCRUTINY

Corporate customers will want to avoid liability for environment damage. As a result, they will demand evidence of good environmental practice from their suppliers. Customers will increasingly issue questionnaires which will be used to determine the choice of suppliers. They will also undertake audits of the supplier, and will demand evidence of management systems such as ISO 14001.

6. TOXIC OR HAZARDOUS RAW MATERIALS WILL BE PENALIZED

There will be pressure to ban dangerous products. Assessment of product safety will be more public than in the past. Research methods and findings will be available to the public and to pressure groups. Some products now in common use will be no longer available for sale.

7. ECO-LABELLING WILL BECOME MORE PROMINENT

In future, consumers will have more comparative information about the products they buy. This will apply to a varied range of products.

Companies will no longer be able to hide behind witty but spurious advertising; and businesses with environmentally superior products will have an advantage.

8. PRESSURE GROUPS WILL GET SMARTER

Pressure groups have become adept at blocking outflow pipes, hoisting protest banners from tall chimneys and immobilizing contractors' vehicles. Companies will find that pressure groups like Greenpeace will adopt more sophisticated tactics, such as hacking into corporate files, infiltrating companies by seeking employment, and use the annual meeting to publicize their case. Companies that don't relish a fight will aim to avoid controversial products and processes.

9. MORE CONTROLS ON CARBON EMISSIONS

Governments will seek to reduce carbon emissions, whether in energy use in buildings or vehicle and airline exhausts, through increased taxes. Mandatory carbon reporting will take place in some countries.

10. THE COSTS OF COMPLIANCE WILL RISE

There are many pieces of national and international legislation on the environment that push up the costs of compliance. They include:

- The EU eco-design and energy efficiency directive 2012/27/EU has an impact on any product that uses energy, except transport; and few categories are unaffected.
- The European Liability Directive (ELD) is designed to 'make the polluter pay'. It covers damage to protected species, natural habitats, water pollution and 'land damage'.
- CLP Regulation (the EU directive on chemicals) and Chip Regulations.
- The Restriction of Hazardous Substances Directive (RoHS), which restricts the use of certain hazardous substances in electrical and electronic equipment, and WEEE, which requires companies to collect and recycle electrical goods.

In the US, EU, India and other parts of the world, companies have to buy consents to discharge. In some places pollution is still free, but this is unlikely to continue for very long. In China, for example, city managers in Beijing intend to 'punish' factories that emit toxic emissions to air. For a company with polluting processes, consents can prove costly.

11. RENEWABLES WILL BECOME UBIQUITOUS

Solar, motion power and battery power will extend to more products, including equipment, controls and buildings, as renewables become more efficient. Businesses will find this is a good way to keep energy costs down, and innovative suppliers will be rewarded with greater market share.

12. POLLUTERS WILL INCREASINGLY FIND IT DIFFICULT TO GET FINANCE AND INSURANCE

Banks are less keen to provide loans to companies which could make them liable for cleaning up after a pollution incident. In the US, for example, banks and insurance companies may be liable for pollution clean-up.

13. THE POLLUTING FIRM WILL FIND IT HARDER TO ATTRACT AND RETAIN GOOD STAFF

Staff prefer not to work for a polluting firm. This is particularly prevalent among younger graduate staff. A survey carried out by the National Union of Students found that three-quarters of student job hunters would not work for a company with a poor ethical record and half those surveyed would take less pay to work for a company with a good history.

Without wanting to sound too gloomy, all the evidence points towards organizations needing to be clean and green. Managing down the company's environmental risks makes sense.

In the rest of this chapter, we look at the major risk areas (such as emissions to air) and how to control them. We also look at the common ways businesses nail down risk though identifying impacts, writing procedures to manage them and even setting up an environmental management system (EMS) such as ISO 14000.

Eight Pollution Risks

There are eight ways that a company can cause environmental damage:

1. emissions to air;
2. discharges to water;
3. solid waste;
4. producing or using toxic or hazardous materials;
5. consuming fossil fuels or energy derived from them (that is, non-renewable energy sources);
6. damage to nature (for example, by building projects), through destruction of natural habitats or amenity space.
7. consuming scarce or non-renewable resources;
8. owning (or acquiring) environmental damaging assets, especially contaminated land. Some experts see this as merely the purchase of existing pollution, rather than a separate cause but it is convenient to discuss it here as a separate issue;

In the following section we examine each in more detail.

1. EMISSIONS TO AIR

Air pollution comes from combustion – through industrial chimneys, cars, lorries and buses, and other engines. It also comes from the fumes in petrol stations, dry-cleaners, paints and household products such as varnishes.

Airborne pollution includes sulphur dioxide (SO_2), nitrogen oxides (NO_x), benzene, volatile organic compounds (VOCs) and particulate matter (PM).

This pollution can contribute to respiratory illness and heart disease. It creates acid rain, kills aquatic life, damages trees, reduces visibility and can be carried over long distances before falling on land or water. Major emitters are:

*   coal-fired power stations;
*   acid production plants (producing sulphuric and nitric acid for fertilizer and chemical plants;

- glass manufacturing plants;
- cement manufacturing plants;
- petroleum refineries.

Companies must report emissions. The oil giant ExxonMobil was fined £3.3 million by the Scottish Environment Protection Agency (SEPA) for failing to report over 30,000 tonnes of climate pollution. It followed undeclared emissions at one of its chemical plants in Fife. ExxonMobil later discovered the emissions had not been included in their annual returns, and reported that to SEPA. There is a penalty of £100 per tonne for excess emissions, which resulted in the £3 million fine.

In recent years, major companies have substantially reduced their discharges to air. According to the Chemistry Industry Association of Canada, in the last two decades the Canadian chemical industry has made the following reductions:

- overall emissions of a unit of chemical product are down by 87 per cent;
- VOCs have been reduced by 75 per cent and benzene by 98 per cent;
- the CO_2 output of chemical plants is down by 60 per cent.

It demonstrates that major reductions are possible if businesses put their mind to it.

Way to reduce air pollution and atmospheric emissions include the following:

- switch to more fuel-efficient engines;
- use renewable energies, such as bio fuels or solar power;
- use less polluting fossil fuels, such as low-sulphur or methanol;
- maintain engines and boilers better;
- improve route planning of fleets;
- use thermostatic heating controls;
- apply filters and scrubbers to clean emissions;
- insulate buildings and pipes;
- use vapour recovery systems;
- use low VOC paints and coatings.

2. DISCHARGES TO WATER

Forty years ago, residents of the city of Durham in North East England could tell by the colour of the river Wear what flavour the local toffee factory was producing. At that time, water pollution was accepted; today it is regarded as unacceptable. Most companies have now substantially reduced their water pollution, using 'closed loop' water systems in the painting operation. At Rolls Royce (rolls-roycemotorcars.com), this recycles 53,000 litres per day. The river water downstream of a Volkswagen factory is actually cleaner than the water upstream. The factory returns waste water to the river in a cleaner condition than when it was extracted.

Society values clean waterways, and dead fish or a river foaming with chemicals is an affront. Companies which pollute water courses will find increasing restrictions placed upon them in the future.

Thousands of fish died when caustic soda was spilt into the unspoilt river Ellen in Britain's Lake District. Experts reckoned that the polluted water, which could blister the skin, would damage kingfishers, otters and other important wildlife. Investigators linked the incident to a Dairy Crest creamery, where the chemical was used to clean milk vats.

Here are eight ways to reduce water pollution:

- limit the quantity and toxicity of cleaning materials and pesticides;
- replace solvent cleaners with aqueous ones;
- filter and treat effluent before it reaches the drains or water courses;
- adopt cleaner processes;
- introduce spill control measures, including bunding (watertight surrounds for oil tanks) and above-ground oil storage (that enable visual inspection of leaks and corrosion);
- use of overflow alarms for storage tanks;
- use less chemical fertilizer in landscape maintenance.

Reducing water usage is not strictly speaking a business risk issue, though it is good practice in areas of water shortage. Organizations can reduce their water use by means of the following:

- use percussion taps that turn themselves off (a tap that is left running wastes 10 litres of water per minute, or 14,400 litres a day);
- repair leaks;
- use high pressure, low-volume cleaners;
- reduce the water pressure in taps;
- reduce the amount of water used in processing.

3. SOLID WASTE

The cost of solid waste disposal is likely to grow. Politicians are agreed that solid waste is a bad thing, particularly as its disposal to landfill can lead to methane which is both explosive and contributes to the greenhouse effect. The politicians are therefore likely to restrict the production of solid waste by taxing it, or by passing laws which reduce the company's flexibility in waste disposal.

Companies with substantial amount of waste will therefore find that the cost of their products will rise. The sensible business will find ways of reducing waste, either through substitution of raw materials, better housekeeping or recycling.

CASE STUDY: RECYCLING COMPANY OWNER JAILED FOR 12 MONTHS FOR DUMPING

Mark Watts, a UK waste haulier, received a 12-month prison sentence for illegally dumping thousands of used tyres.

Under the name of Storm Recycling, Mark Watts transported the tyres to rented land at Abercynllaith, Oswestry and also a unit on a Business Park in Whitchurch, neither of which was authorized to accept that sort of waste.

Storm Recycling was paid by tyre-fitting businesses across Shropshire and North Wales to remove and dispose of waste tyres.

When the Environment Agency discovered what was happening at the sites, it served Watts with two notices to remove the tyres.

He failed to properly comply with either notice, which meant thousands of tyres had to be removed at expense to the landowner; and Watts was sent to prison.

Packaging Waste Regulations

Companies with a total turnover, including any subsidiaries, exceed £2 million, which handle more than 50 tonnes of packaging a year must register under the Packaging Waste Regulations. As the case study below shows, companies sometimes fail to meet their obligations.

CASE STUDY: KRISPY CREME FINED

Woking Magistrates' Court fined Krispy Kreme £12,000 for failing to comply with the Packaging Waste Regulations.

Route checks by the Environment Agency had revealed that Krispy Kreme UK Ltd was not registered with a compliance scheme.

The company pleaded guilty to failing to register with the Environment Agency, and failing to recover and recycle packaging waste by purchasing Packaging Recovery Notes as provided by the regulations.

Tackling the major issues in solid waste

Business should distinguish between relevant environmental activities, and those that are 'greenwash'. Recycling plastic cups in a major corporation will not reduce the company's pollution.

In some cases, the energy and monetary cost of collecting, cleaning and recycling used material is higher than making the product from new. However, many valuable commodities can be recycled, including glass, most metals, batteries, tyres and plastics.

4. USE OF TOXIC MATERIALS

Unless handled carefully, toxic materials endanger the workforce and can escape into the drains and rivers where they kill fish and get into the food chain. Others are nerve agents, poisonous to humans, flammable or explosive,

Regulatory authorities are increasingly hostile to companies that create such problems. Table 8.1 shows the world's 20 most dangerous materials, as defined by the US Agency for Toxic Substances and Disease Registry (ATSDR). The EU has its CLP Regulation list, while the USA has a Substance Priority List.

Table 8.1   20 most dangerous substances

1	Arsenic	11	Chloroform
2	Lead	12	Aroclor 1260
3	Mercury	13	Ddt,
4	Vinyl chloride	14	Aroclor 1254
5	Polychlorinated biphenyls	15	Dibenzo(a,h)anthracene
6	Benzene	16	Trichloroethylene
7	Cadmium	17	Chromium, hexavalent
8	Benzo(a)pyrene	18	Dieldrin
9	Polycyclic aromatic hydrocarbons	19	Phosphorus, white
10	Benzo(b)fluoranthene	20	Hexachlorobutadiene

Where possible, you should substitute dangerous chemicals with ones that have less impact, or change the process to one that requires less toxic materials. Table 8.2 contains some examples. More case studies are at www.subsport.eu/.

Table 8.2   Some substitutes

Industry or process	Item	Substitute
Boat maintenance	Marine paint, which is toxic to aquatic life	Use a coating which prevents mussels from attaching themselves to hulls. Or rub down the boat more frequently
Cleaning	Aggressive cleaning products	Use products that have less environmental impact, and/or use micro-fibre cleaning cloths
Timber buildings	Wood coatings containing biocides	Use more resistant types of wood. Or limit the wood's exposure to severe weather conditions
Paint removal	Dichloromethane	Dichloromethane
Paper making	Sursol, used to soften cardboard	Rheocol ACP
Disposable nappies	Acrylic acid, AN absorbing agent	Wheat starch and granulates
Metal working	Chlorinated paraffins	Vegetable and animal oils, or change the metal.
Food wrap	Phthalates as a softener	Castor oil
Equipment cleaning	Hydrochloric acid, nitric acid	Citric acid, acetic acid
PVC cable	Lead to stabilize it, phthalate to soften it	Change the recipes to lower the viscosity, reducing the need for chemical treatment. Change to LSOH cables, or use rubber insulation
Computer manufacture	Lead in tin solder	Abandon use of lead in the solder
Joint filler	Polyurethane	MS polymers
Dry cleaning	Perchloroethylene	CO2, AquaClean or silicone
Plywood fabrication	Epoxy filling	Water-based powder filling
Dessicant	Silica gel	Bentonite clay

CASE STUDY: THE LEGACY OF LONG-ESTABLISHED COMPANIES

A defence contractor unexpectedly found nuclear radiation in the grounds of one of its manufacturing sites. It discovered that the radiation came from World War Two aircraft dials lying just under the grass. The company's staff had painted luminous dials by hand on aircraft instruments, and had thrown faulty dials out of the window (quality control being less scientific than it is now). As a result, the site was still emitting nuclear radiation. It is an example of unforeseen hazards that leave the long-established company with a liability.

But what if chemicals or toxins are the company's core business? Where the business sells biocides or toxic materials (for example, weedkillers or drain cleaners) there is an added risk of product liability. There is a conflict between governments and the public demanding safer chemicals, while at the same time, people are living longer thanks to clean water and sanitary conditions, which in turn require biocides. However, companies can still seek to reduce the toxic load imposed by their products.

Useful actions include the following:

- substitute less powerful substances wherever possible;
- modify or eliminate environmentally damaging raw materials, processes and products;
- use more recycled and recyclable raw materials.

If none of these solutions work, consider exiting from such markets. The loss of short-term profits will be softened by the lack of corporate blame in the longer term.

If the product is not part of the company's core business, you could cease producing the product, or sell the assets to another firm. If the product is central to the company, the problem is less easy to solve, as we see next.

CASE STUDY: BAYER AND THE HONEY TRAP

Companies find it hard to give up a revenue stream; and this can present a management challenges.

Bayer makes the world's most widely used insecticide, a neonicotinoid called imidacloprid. Neonicotinoids have been labelled an unacceptable threat to bees by the European Food Safety Authority, following several studies implicating the nerve agent, which has been blamed for the widespread decline of the honey bee population, along with habitat loss and disease.

Bees and other pollinators are important part of agricultural production, being responsible for one-third of all food. France, Germany, Italy and Slovenia have implemented a partial ban on nicotinoids, but at time of writing it remains in use in the UK and many other countries.

But Bayer (and, it has to be said, the UK Government) has dismissed the research evidence as insufficient. And it claims that a nicotinoid ban would cost UK farmers £620 million in lost food production.

Bayer's Julian Little said, 'We do not believe the … new EFSA reports alter the validity of [existing] risk assessments and the underlying studies. It is very important that any political decision relating to registrations of neonicotinoid-containing products should be based on clear scientific evidence of adverse effects … and not on the basis of an over-interpretation of the precautionary principle.'

In previous generations the same arguments were advanced to defend cigarettes, DDT and other toxic substances. In more recent years, oil companies have vigorously denied the existence of climate change, and the role of carbon emissions in global warming. Conclusive and undeniable proof is hard to come by, and takes time to build; which leaves room for companies to continue to sell those products.

It is true that environmentalists can demand change where evidence is weak. It is equally true that companies don't want to lose a profitable income stream; and most organizations are reluctant to accept unwelcome evidence.

Like many flooring manufacturers, Nairn Floors, once Britain's leading smooth flooring business, used to use white asbestos as the backing layer for its vinyl flooring. The company would write reassuring letters to concerned purchasers, reassuring them that there was no evidence that white asbestos was harmful to health, unlike blue asbestos. No one in the business wanted to hear the growing alarm over asbestos. In due course a competitor, Marley, produced a flooring that had a flexible plastic backing. Consumers preferred it, and turned away from Nairn product.

In summary, you should be prepared to think the unthinkable. You have to be prepared to hear criticisms of your product, and recognize that they might have validity. If you don't, your business may get left behind by more astute competitors.

5. CONSUMPTION OF ENERGY

Smoking power stations and factory chimneys are visible targets for protesters. But for the average company, energy use isn't a high-profile issue.

Apart from major energy users like foundries, the cost savings from energy can sometimes be slight. However, as energy costs rise, companies need to see where they can cut energy consumption. Solutions include the following:

- improve insulation and draught-proofing;
- using energy-efficient lighting and equipment, including timed switches and motion sensors;
- improve boiler efficiency;
- reduce excessive use of air conditioning;
- increase the use of energy management systems;
- conduct an energy audit, with the aim of cutting energy use;
- turn down space heating and hot water thermostats.

As a case in point, construction companies could see a growth in environmental legislation. This might include taxes on extraction of raw materials, controls over greenfield sites, and an emphasis on higher-density housing with less reliance on cars and higher energy efficiency. However, this shouldn't overly bother the construction industry since it will affect all companies equally, and since other factors affect housing sales such as the economic cycle and availability of mortgage credit.

6. DAMAGE TO NATURE

Mining and quarrying companies, developers and construction companies, and even supermarket chains, are likely to build on greenfield sites. The same applies to petrochemical companies as they explore for, and then extract, oil and gas. The destruction of natural beauty is an obvious and easy target for conservationists, and some companies have been paraded on television news for several years as they battle to develop a site.

In order to minimize damage to nature, the environmentally aware company will take the following steps:

- work on brownfield (urban, developed and possibly run-down) land;
- carry out an environmental impact assessment (EIA);
- introduce mitigating environmental measures (such as maintaining habitats);
- engage in a full consultation process with stakeholders.

CASE STUDY: HOW ONE COMPANY CHANGED FROM VILLAIN TO HERO

Gardeners use compost to improve their soil. In the past, they mainly used peat, a natural substance produced by decaying plants. But peat occurs only in flooded wetlands, a habitat that attracts distinctive species. The peat extractors drain the wetlands, which destroys the wildlife habitat and releases the stored carbon which in turn contributes to global warming.

So when a leading Somerset peat producer found its products coming under scrutiny, it sought alternative products to sell, and found one in its own backyard. Somerset is famous for its apple orchards and cider making, and the latter produces waste in the form of apple peel and pulp. Fruit waste is a well-balanced composting material, containing a ratio of 35 parts of carbon to one part nitrogen, close to ideal.

The company took this waste from local cider makers and turned it into compost. This relieved the cider makers of their waste, and earned revenue for the company from an environmentally sound product.

However, customers often prefer not to live on or shop in brownfield sites, which is one of the basic conflicts in planning land use.

CASE STUDY: LAND POLLUTION FROM OIL EXTRACTION

Over an eight-year period, as Texaco drilled for oil in eastern Ecuador, 17 million gallons of oil escaped into local water supplies. In comparison, the Exxon Valdez disaster involved only 11 million gallons.

The native peoples of the region brought a billion dollar suit against Texaco. But lack of support from their own government meant that only $5–$10 million was offered in compensation.

This kind of threat emphasizes the need for companies to ensure that they have proper *systems for protecting the environment*. These systems need to be documented, to ensure that they are implemented and act as insurance against future litigation.

---

7. USE OF SCARCE OR NON-RENEWABLE MATERIALS

Extracting rainforest timber and even quarried stone have become major environmental issues. Conservationists can easily brand a company which uses scarce materials as a corporate vandal.

There are often substitutes which the company can use, and such substitutes pre-empt the company from law suits and protesters at its gates.

Ensuring that the local population gets benefits from the activity will substantially reduce opposition. The UK's Sellafield nuclear plant, while controversial nationally, is locally supported because of the massive employment it creates. Local support does not, however, mitigate the risks.

The business should adopt the following measures:

- audit its use of scarce or non-renewable materials;
- substitute with materials that have less environmental impact;
- reduce overall use of materials (for example through lightweighting or waste reduction);
- ensure proper controls for protecting the environment are in place.

8. CONTAMINATED LAND

The UK's Environment Agency estimates that there are between 5,000 and 20,000 contaminated sites in England and Wales. This amounts to over 300,000 ha of land, equivalent to one and a half times the area inside the M25 motorway.

Many of these sites may require decontamination in future years as the Government looks to eliminate the risks they pose and increase the percentage of new housing built on brownfield sites. Therefore businesses that purchase land must remain vigilant.

Property developer Mountleigh, one of the largest international property groups, put one of its sites, the Merry Hill Centre at Dudley, on the market for £125 million. Several potential purchasers carried out site surveys and discovered that the land was contaminated. All decided against going ahead. This stopped Mountleigh paying back debts of £75 million to its Swiss bondholders. The land was finally reckoned to be worth £35 million less than Mountleigh had estimated, and the company went into liquidation.

Even government bodies are not free from risk. The Ministry of Defence is one of the UK's biggest land owners. Over the years various processes have taken place on its land, ranging from weapons testing to vehicle maintenance, Partly because of the scale of its ownership, the Ministry lacks a complete record of the contamination.

Like any landowner, the Ministry needs to know the extent of contamination so that it can carry out remedial works, especially as government regulations on contaminated land is likely to become progressively more stringent. In other words, remedial works are likely to become more costly.

The business can undertake the following measures:

- conduct a site assessment;
- establish the scale of contamination, if any;
- implement a plan to reduce the contamination and improve the ecological value of the land.

Dismissing Legislation

Some organizations, especially service businesses, are relatively unaffected by environmental legislation. However, some service industries have operations that impact on the environment, such as supermarkets' lorries.

Others find it cheaper to produce emissions and, in the case of solid waste, dump it rather than invest in pollution control. Companies that adopt this strategy need to be alert to changes in legislation, taxation or other economic factors that would change the equation.

Equally, there are many advantages of improved environmental performance:

- lower costs – from less tax, less waste, less packaging, less energy use;
- less risk: from bad PR or environmental catastrophe;
- less likely to be targeted by environmentalists;
- increased sales from improved reputation;
- easier to recruit and retain staff.

DOES BEING GREEN GIVE A COMPANY AN ADVANTAGE?

Green issues are now established headline issues across the Western world. A DEFRA survey (defra.gov.uk) found that when asked whether 'prices and jobs are more important than protecting the environment for the future,' 55 per cent disagreed and only 30 per cent agreed. And 20 per cent of people regularly buy organic foods, despite the higher costs.

In short, the consumer continues to be concerned about the environment. But do these attitudes really affect purchase behaviour? Will the consumer (or the corporate buyer) prefer to buy from a green supplier? Some markets are more environmentally sensitive than others. Some consumers buy toilet paper made from recycled fibres out of concern for the environment. But others do so because the product is cheaper than luxury non-recycled toilet tissue. So the price and performance of products that claim environmental superiority should at least match that of traditional products.

Consumers will become hostile to companies which are revealed to disregard the environment. This could have a long-lasting effect on the corporate brand image and therefore sales.

OR IS ENVIRONMENTAL CONFORMANCE A THREAT?

Some businesses fear that the cost of environmental protection is excessive and therefore threatening. Such companies can take a hard-nosed approach to environmental protection by adopting the following steps:

- gather information about the company's environmental impacts, and quantify them;
- assess the environmental gains against the cost savings;
- rank the most profitable measures;
- implement low-cost and no-cost measures. They will bring an immediate return;
- plan for higher-cost measures in the longer term (for example, replacing a boiler with a more efficient system);
- seek innovative solutions which replace environmental hazards with cheap and safe solutions;
- pass the costs of pollution control on to the customer.

STRATEGIES FOR REDUCING ENVIRONMENTAL RISK

There are ways to avoid these risks. This involves the action shown in Figure 8.1.

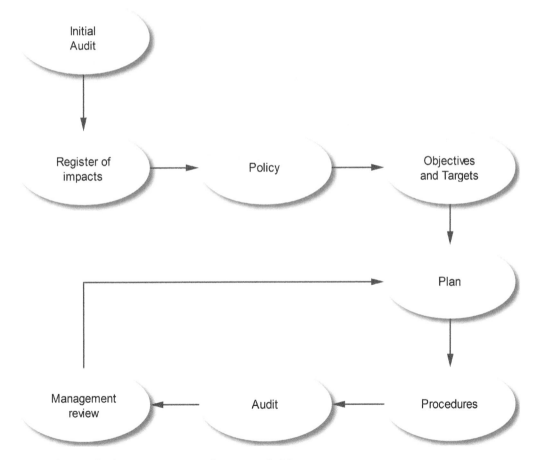

Figure 8.1   Action to prevent environmental risk

1. CARRY OUT AN ENVIRONMENTAL AUDIT

The environmental audit is a tool for finding out the extent of the company's impacts. It should bring together all relevant information about the way the company affects the environment.

Someone who is independent of the activity being audited should conduct the audit. A production manager should not audit their own plant.

By carrying out an environmental audit, companies can assess the environmental risks associated with each of its divisions. This results in a 'risk register', which we discuss later. The audit lets the corporation determine whether a division is likely to cause environmental problems.

2. DEFINE A POLICY ON ENVIRONMENT ISSUES

Once management has information about the scale of its impacts, it can make an informed decision about them. How important are they? How much energy does the company want to spend on them?

An environmental policy tells employees and the public about the company's intention to manage its impacts. But how strong should the company's commitment be?

It can decide to become an environmental Leader (a company against which others measure their progress), or a Conformer (a company which simply obeys all statutory obligations). See Figure 8.2.

It should avoid being a Laggard (a company which falls below the environmental standards set by the rest of its industry), or Punished (that is, a Laggard which continues to take no action, and eventually loses customers to other businesses or is taken to court by regulatory authorities).

A Leader is committed to being foremost in its field, for example 3M. A Conformer aims to obey the law, but doesn't seek environmental excellence.

Leaders that rest on their laurels become Conformers, as the rest of industry catches up. Similarly, Conformers can slip into the Laggard category, and Laggards can easily become Punished.

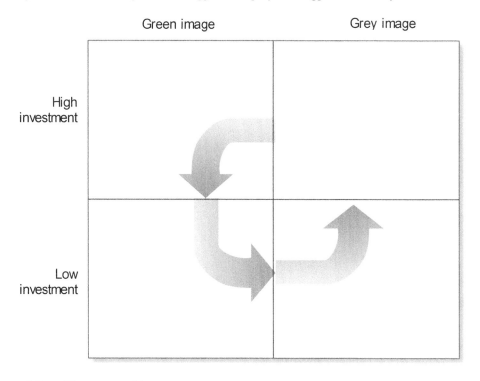

Figure 8.2   The green grid

The reverse is also true. Punished companies can become Leaders in response to an environmental crisis. One such example is Shell, which changed course following criticism over its plan to sink the Brent Spar platform in the Atlantic, and anger over its alleged damage to Ogoni lands in Nigeria. Most of the Nigerian oil leaks were later found to have been caused by local people stealing oil. Subsequently however, Shell planned an oil and gas drilling site beside a coral reef off the coast of Australia, indicating the industry's permanent conflict between earning revenue and risking damage to the environment.

CO-OPERATIVE BANK'S ETHICAL POLICY

To communicate its environmental stance to employees, the business should produce a policy outlining its attitude on green issues.

As environmental and ethical issues meet at various points, the Co-operative Bank's ethical policy includes its policy on ecological impact, animal welfare, genetic modification and other topics. The bank declares that it will not invest in any business whose core activity contributes to:

- global climate change, through the extraction or production of fossil fuels;
- the manufacture of chemicals which are persistent in the environment and linked to long-term health concerns;
- the unsustainable harvest of natural resources, including timber and fish.

And it will seek to support businesses involved in:

- recycling and sustainable waste management;
- renewable energy and energy efficiency;
- sustainable natural products and services, including timber and organic produce;
- the pursuit of ecological sustainability.

An environmental policy should include a commitment to:

- comply with legislation;
- prevent pollution;
- set objectives and targets for managing its impacts;
- conduct audits and review their outcomes.

The business may also commit to continuous improvement in this area.

3. SET OBJECTIVES AND TARGETS

Next the company needs to quantify its policy statement. It can do this by producing an annual target for reducing its solid waste or for reducing the BOD (biochemical oxygen demand) level of its waste water. Setting targets allows the business to determine whether it is achieving change.

4. PRODUCE A PLAN THAT WILL MEET YOUR TARGETS

Targets do not in themselves produce results; so the company needs a plan. This might include investing in a water treatment plant or undertaking a study to reduce the company's solid waste.

It is important to select the right action to take. Some companies cherry pick their environmental actions. One UK electricity company boasted that it had given £200 to a wildlife charity, and had sponsored a children's painting competition. Meanwhile it was bulldozing areas of outstanding natural beauty, evicting wildlife and disfiguring the countryside with its pylons. Such behaviour is risky because it may be seen as window-dressing.

Businesses should take action on their most significant impacts. An agricultural business found that water pollution was its major impact, and installed a water treatment plant. Other, less rigorous, firms might have chosen instead to recycle old packaging, which was a less serious issue.

The organization has to ask itself how it can meet its targets. If the business wants a 5 per cent reduction in energy use, how might that be achieved? For example, it might require:

- staff training;
- changes to processes;
- new materials;
- better monitoring;
- new equipment.

5. INVOLVE PEOPLE

The company must ensure that all its staff are motivated to manage the company's environmental impacts. This means undertaking a period of consultation, and delegating action to the lowest appropriate level.

The company also has to manage its external communications. Corporate customers are increasingly asking their suppliers to show how they are managing their environmental impacts. The media, pressure groups and local residents are also keenly interested in a company's environmental activity.

6. WRITE PROCEDURES

Left to their own devices, staff carry out processes in ways that suit themselves. They may take short cuts, or fail to make checks. That is why the business needs to have written procedures for any activity that could harm the environment. The company also needs procedures for dealing with emergencies.

7. UNDERTAKE AUDITS

An audit will assess whether the procedures are being followed. Does the tanker driver use safety slings when transferring liquid carbon dioxide, to prevent the hose flailing? Does he turn the engine off when stopped? Does he use wheel chocks when emptying and filling the tanker?

ISO 19011 contains guidelines for auditing quality or EMS, as well as guidance on assessing the competence of auditors.

8. MEASURE AND RECORD IMPACTS

Companies keep lots of financial and production information, but many have little environmental data. This is because in the past the environment was not seen to affect the business, and nor was it a cost.

Things have changed; and companies need to start collecting environmental data. But it isn't easy. RTZ, the mining company, found that all over the world its mines used different measures, and had different production methods. Trying to harmonize the data wasn't easy. But recognizing the problem and identifying the goal is a good starting point.

It is also important to standardize the measurements. The amount of pollution will grow if production rises, so the company should adopt a measure such 'kilowatts of energy per tonne of output'.

Software such as Eco-footprint (ifsworld.com) allows businesses to record impacts and aid compliance reporting.

9. HOLD A MANAGEMENT REVIEW

Records of impacts should go to a management review meeting, along with non-conformances observed during an audit. In turn the review should ensure that corrective or improvement action is taken.

The Environmental Management System (EMS) – ISO 14000

By the time you've undertaken points 1–9 above, you'll have implemented a full-scale EMS.

The best-known EMS is ISO 14000. 188,000 certificates for ISO 14001 have been issued globally, of which almost 48,000 have been in Europe.

ISO 14000 is very similar to other management standards, especially ISO 9001. Any company which has ISO 9000 is already half way towards getting ISO 14000.

Among the tasks listed in 1–9 above, ISO 14000 requires a company to produce a register of significant environmental 'aspects'. These are activities, products or services that interact with the environment.

DEFINITIONS

Environmental Aspect: Any element of an organization's activities, products or services that can interact with the environment. It might cause an impact.

Environmental Impact: Any change to the environment, whether adverse or beneficial, resulting from an organization's activities, products or services.

A sample blank register is shown as Table 8.3. The chart summarizes the company's environmental impacts in a simple way.

This aids clarity of thought, and helps the business focus on the major effects. It also serves as an environmental risk analysis. Where the business does not have one of the effects listed here, the auditor can state in the relevant box that it is 'not applicable'.

EMAS

EMAS (The EU's eco-management and audit system) is similar to ISO 14000. But it is less popular: only 8,000 companies have registered with EMAS, the majority in Germany, and EMAS is available only in the EU.

Table 8.3   Register of significant aspects

Impact	Current activities			Future planned activities	Past activities
	Normal operating conditions	Abnormal operating conditions*	Incidents, accidents, and emergencies		
Emissions to air					
Discharges to water					
Solid waste					
Toxic materials					
Energy use					
Scarce or non-renewable material					
Damage to nature (incl. contaminated land)					
Noise, odour, dust					
Visual impact					
Other impacts (list)					

* Notably start-up and shut-down activities

Unlike ISO 14000, EMAS requires an open dialogue with the public, thought a Public Environmental Statement.

BS 8555

For businesses that are reluctant to commit to ISO 14000, there is another option: BS 8555. This affordable standard is aimed mainly at SMEs and all types of business activity. It lets the business develop its EMS over time in six separate stages, and demonstrates to interested parties that the company is making progress on its environmental impacts. BS 8555 can be useful in demonstrating the company's environmental probity to large customers. It has six phased stages:

1. Commit to the project. Establish your impacts.
2. Identify legal requirements and ensure that the business complies with them
3. Introduce objectives and targets. Identify how to manage your impacts.
4. Implement the EMS. Have written procedures.

5. Audit your activities, and get management to review the findings.
6. Get external assessor to confirm everything is in place. Get ISO 14000.

You can't get registered to BS 8555, but some organizations offer assessment:

- http://www.iema.net/ems/acorn_scheme/acornhomeAcorn, from the Institute of Environmental Management and Assessment (ems.iema.net/acorn_scheme).
- Green Dragon, run by Groundwork in Wales (groundworkinwales.org.uk/greendragon).
- Seren Scheme, from Tarian (serenscheme.com).

BS 8555 can be addictive. Stage 1 is easy to get, and you receive a nice certificate. Wouldn't it be great to get Stage 2? It doesn't take much effort. But surely you don't want to be just a Stage 2 business? Why not make a little jump to Stage 3? Like an efficient drug dealer, BS 8555 can get you hooked. But at every stage you are not committed to going any further: there is the illusion of free choice.

Note, however, that if the organization already has several of the BS 8555 steps in place, it may be simpler and cheaper to go straight to ISO 14000, because it involves only one external audit, not six.

GETTING ASSESSED

The usual way to demonstrate that the organization conforms to ISO 14000 is to have an independent auditor audit the business. However, this is a major commitment for any business. The company is committing itself to the costs and time of assessment in perpetuity. Alternatively, a company can:

- Make a self-declaration, saying that it operates in a way that meets the requirements of ISO 14000. This is unlikely to satisfy all of the stakeholders, but may be appropriate in markets where certification isn't essential.
- Ask corporate customers to verify that its processes meet the standard. However, they may be unwilling to do that.

Product Certification and Eco-Labels

There is a plethora of 400 competing eco-labels, especially prominent in environmentally sensitive markets. Some labels originate from NGOs, others from government departments, while a third group come from industry associations. The labels vary in how high they set the bar, and in what topics they regard as important, Hence Rainforest Alliance teas sit on a supermarket shelf next to FairTrade packs, with the consumer unsure of what each stands for.

Consumers find the labels confusing in their range but also help them an easy way to choose. Meanwhile it is inevitable that manufacturers and supermarkets will produce products whose labels suggest eco friendly but whose contents are anything but.

In some markets, notably cleaning fluids, the major businesses remain wary of eco-labels, not least because the labels prefer chemicals that have little impact on the planet, and are therefore sometimes less effective at cleaning, and can prevent major companies from innovating. They also hand control of consumer choice to a third party.

SOME LABELS IN SELECTED MARKETS

General: Green Seal (greenseal.org), EU eco-label (ec.europa.eu/environment/ecolabel), Nordic Swan (nordic-ecolabel.org).

Paper: Forest Stewardship Scheme (FSC), Blue Angel (blauer-engel.de) WWW (http://checkyourpaper. panda.org).

Cleaners: Design for the Environment (epa.gov/dfe).

Fishing: Marine Stewardship Council (msc.org), a plethora of 'dolphin-friendly' labels.

Cosmetics and toiletries: Leaping Bunny (leapingbunny.org).

Energy consumption: Energy Star (energystar.gov/), EnerGuide (Canada).

Food: FairTrade (fairtrade.net), Rainforest Alliance (rainforestalliance.org) criticized by some as 'FairTrade Lite', EU organic logo (ec.europa.eu/agriculture/organic), and the USDA organic logo (ams.usda.gov).

As compulsory energy and nutritional labelling has arrived in some countries, so too might uniform eco-labelling, It is both a risk and an opportunity, depending on each company's stance.

There is an ISO standard (ISO 14024) for eco-labelling, and an index of eco-labels at ecolabelindex.com.

Supplier Assessment

Lifecycle analysis is an important concept in environmental management. It means being aware of all your product's impacts, from cradle to grave.

An environmental consultancy gives a plaque to those clients that gain the quality standard ISO 14000. It then discovered that the plaque caused several environmental problems. The engraving process used polluting acids, and the plaque was mounted on a tropical hardwood base. The company switched to a pine plaque which was not only better for the environment but was also a lot less expensive.

The starting point for any lifecycle analysis is a supplier questionnaire. This asks suppliers to state the ingredients or composition of the raw material, and the nature of environmental impacts involved. Most suppliers will respond positively to this request for information. A few suppliers will not; and they may need to be replaced. B&Q, the DIY chain, has de-listed many suppliers who failed to enter a dialogue about the environment.

Choosing environmentally sound suppliers is important, because environmentally aware suppliers are likely to be low-risk suppliers. A supplier that responds positively to the environmental challenge is likely to be concerned about its other challenges, including quality, health and safety, and customer service.

Companies should categorize their supplies into high, medium and low risk. They should then seek to substitute their high-risk purchases. This may involve working with other departments, especially production.

Many companies report difficulty in finding out where responsibility rests. A product has often been processed by several companies in turn. For example, a food processor that buys a puree has a supplier chain which stretches from a farmer who sprays the fruit with pesticide, to a local processor which pulps the fruit, before it reaches a third company which mixes it with other ingredients.

Risk Assessment

Use this chart to determine the scale of your environmental risk. Tick any boxes which apply to your business.

Question	✓
Does the business consume many scarce or non-renewable resources (such as rainforest timber)?	
Does the company make or use many hazardous or toxic materials?	
Does the business produce substantial solid waste?	
Does the company produce substantial air pollution?	
Does the business discharge a large amount of polluted waste water?	
Does the company consume large amounts of energy?	
Does the business impact substantially on nature (for example, through construction projects)?	
Has the business failed to conduct an environmental audit within the last 24 months?	
Does the business lack an EMS?	
Has the company been criticized by environmentalists in the last 12 months?	
Total points scored:	

Score: 0–3 points: low risk. 4–6 points: moderate risk. 7–10 points: high risk.

The Appendix has a summary of the checklists. By entering the results of this one, you can compare the scale of environmental risk against other categories.

9
From Drought to Flooding: Risk and Opportunities from Adverse Weather

Malcolm was reminiscing about the day of the storm. 'It wasn't exactly unexpected,' he said. 'But having the transformer come crashing on to the roof was pretty scary for the people at work.'

Fortunately no one was hurt. The energy company had restored power and removed the debris, but the organization had been idle for three days due to damage. The backlog took weeks to clear. And some clients defected.

'What happened was straightforward,' he said. 'A cable holding the transformer snapped in the high winds, and brought it crashing down on to the office roof. Oil and other chemicals from the transformer also leaked into the building,' he said.

Everyone had accepted that the office location, next to the sub-station, wasn't exactly glamorous. But since it was only for back office work the location didn't seem to matter. And the lease was cheap.

The CEO moved the whole department into head office for a few weeks. Some people had to work in the corridors. Later they asked you to add 'adverse weather' to your risk register.

Overview

In this chapter we look how weather patterns are changing, leading to unexpected storms and heat waves. We consider how this will affect organizations, and what you can do to prevent loss or damage. First, a quick look at some facts:

- The UK has recently been battered by gales and rain, sometimes leaving 27 000 homes and businesses without electricity in the wake of 109mph winds.
- In the US, temperatures are expected to rise between 3 degrees and 5 degrees centrigrade (5–9 degrees fahrenheit) over the next century, according to the Natural Resources Defence Council (nrdc.org).
- Texas has suffered its worst one-year drought in history, which caused billions of dollars in crop damage and record wildfires. Its agricultural losses were the largest in the state's history.

- Sea levels will rise, putting millions at risk of flooding, though there is uncertainty over exactly how much. Estimates vary between 2ft and 12ft (0.6 to 3.6m), with each foot of sea rise putting 100,000 New Yorkers at risk, according to the BBC.

These changes are due to global warming, a phenomenon now accepted by 90 per cent of scientists. It could be due to natural climate variation but the growing consensus among scientists is that it is caused by man-made CO_2 emissions, from burning fossil fuels and deforestation. The effect of global warming is:

- a temperature rise, causing melting of polar ice caps, causing a rise in sea levels, leading to flooding and a loss of coastal land;
- a rise in the number of extreme weather events: storms, floods and drought.

Effects of Climate Change

In Europe, numerous regions have been affected by severe floods. The UK, Germany, Hungary, Poland, Italy, France and Ireland, are among countries that suffered in the recent years.

The European continent has also been paralysed by severe winter storms. Bosnia's Government declared a state of emergency in its capital when hundreds of people got trapped in their homes. In Romania, 200 cities and towns were left without power. In the UK, half of the flights were cancelled at the Heathrow airport, making it extremely difficult for people to get in or out of the country. The UK is also affected by the movement of the jetstream, causing wintery weather.

In Asia, in addition to the rising temperatures, the continent is experiencing increasing precipitation. Tropical cyclones have also become more intense. Areas that already suffer from water stress experience even more severe water shortages, while snow melt leads to floods in other areas of the continent (India and Bangladesh).

The rise in temperatures brings increased threat of drought. Some areas of the globe are particularly vulnerable. In the United States, 57 per cent of the land is exposed to the risk of drought. Warmer temperatures, in combination with lack of rainfall and snow, create conditions in which grasses grow and then dry out, creating fuel for fires.

THE IMPACT ON THE UK OF THE JETSTREAM

For several of the last summers the UK has suffered unusually wet weather. This has been due to the jet stream moving to the south of the UK, where it pulls rain-bearing weather on to the island. In the past, the jetstream was located to the north of the UK, where it dragged the wet weather away from the country.

This change may be due to natural variations in the weather, or it might be due to the melting of the Arctic sea ice and the warming of the Atlantic ocean.

Whatever the cause, the jetstream appears to be less constant than before,

If the jetstream moves north, the UK loses its cloud cover, creating freezing winters and hot summers. This will cause water shortages in summer months, leading to poorer harvests.

If it stays south, the UK will have more rain, leading to flooding. In summer this will cause crop losses. In recent years the UK honey harvest has fallen by 75 per cent as bees had to stay in their hives due to driving rain. More than £600 million of food such as wheat and potatoes has been lost in some years. The effect will be more pronounced in the winter, when the ground will be already saturated and therefore more prone to flash floods, with roads being cut off and deaths from drowning. Floods can also take out electricity sub-stations, leaving some businesses without energy (as we saw at the beginning of this chapter).

Wildfires are often followed by dust storms. These happen when exceptional draught is combined with strong winds. The states across the Great Plains are mostly affected.

The North American continent also suffers from hurricanes that happen every few years. Hurricane Sandy that hit the US and Caribbean killed 165 people and left eight million residents without power. Strong winds, combined with high tides caused $50 million worth of damages in New York.

Tornados are also common in the US. On average, the country receives more than 1,200 tornados each year. Kansas, Nebraska, Oklahoma, Missouri, Iowa and Texas (so-called 'Tornado Alley') experience the highest number of tornados.

How Weather Patterns Affect Organizations

Changes in the weather pattern can affect organizations' assets (buildings, infrastructure and workforce), their production processes, the markets they serve and the natural resources they depend upon. We examine each of these in turn.

1. EFFECT ON CORPORATE ASSETS

The shift in the climate will affect business premises, through the impact on their structure, their fabric and the working conditions inside. This will require changes in the way new buildings are designed and constructed. Existing premises will require a different regime for maintenance and facilities management.

The workforce will be also affected by the new weather pattern. The number of hot days is increasing and workers may have to stay long hours in overheated spaces, in buildings that have not been designed for these conditions.

Heat has a physiological effect on humans: it can lower employee's concentration, productivity and motivation. Staying in overheated spaces can also affect employees' physical health. Employees working outdoors will require better protection against heat as well. These changes may result in new health and safety regulations.

2. IMPACT ON BUSINESS OPERATIONS

Some business activities and industrial processes, for example construction and agriculture, are sensitive to temperature and climate. Consequently, productivity in some sectors will be affected. The new situation may require new equipment or new production processes to be designed in order to maintain business continuity. Positive changes are also possible, for example when a new process become financially viable in a new climate.

Business operations can be influenced by the disruptions in the supply chain and logistics. Floods, hurricanes and other disasters can interrupt the supply of water, energy, raw materials or IT services.

Company finances can be also negatively impacted by the cost of damage, disruption and sales can drop. Increased insurance costs will also follow. And the cost of regulatory compliance is also expected to increase.

3. CHANGES IN MARKETS

Some companies may experience changes on the markets they operate. The demand for some goods and services may decrease, leading to reduced sales (for example, cold weather clothes). The shift will also bring new opportunities, as businesses and consumers will require different products and services. Some renewable energy businesses will flourish, for example.

4. REDUCTION IN NATURAL RESOURCES

The natural environment in which businesses operate is also changing. The most important issues are twofold:

A. The ability to grow crops and raise animals

Atmospheric carbon dioxide has now reached 400ppm. The last time this happened, the Arctic was ice-free and sea levels were 40 metres higher. According to Lord Stern of the Grantham Research Institute, 'Hundreds of millions of people will be forced to leave their homelands because their crops and animals will have died.'

He went on, 'When temperatures rise [by up to 5C] we will have disrupted weather patterns and spreading deserts.' He said agriculture could fail on a continent-wide basis and hundreds of millions of people will be rendered homeless, triggering widespread conflict. He said the trouble will come when people try to migrate into new lands. 'This will bring them into armed conflict with people already living there. Nor will it be an occasional occurrence. It could become a permanent feature of life on Earth.'

B. The availability and the quality of natural resources.

Some areas of the world are already experiencing shortages of water, as a result of higher temperatures. Increasing CO_2 levels in the oceans have negative impact on fish stock. Shellfish production may fall by 10 to 30 per cent due to increased acidification of the oceans, with attendant losses in the food service industry. Businesses relying on natural resources may experience disruption in the supply chain. Others may experience some restriction in the use of natural resources.

Consequences of Climate Change on Businesses – Risks and Opportunities

As a result of climate change some organizations will suffer from the following effects:

1. worsened (and sometimes improved) financial performance;
2. additional costs;
3. disruption of operations;
4. loss of work hours and reduced staff productivity;
5. increased stakeholders' interest and enhanced or damaged corporate reputation;
6. additional regulatory requirements;

7. contractual issues
8. litigation;
9. new market opportunities and product diversification.

In areas most affected by climate change, businesses will see civil unrest and population changes.

However, climate change can also bring positive consequences, such as improved financial results, an enhanced brand image, or a demand for new products and services. We consider these effects next.

1. WORSENED (AND SOMETIMES IMPROVED) FINANCIAL PERFORMANCE

Extreme weather conditions can have a direct impact on a company financial performance. Weather-dependent sectors, such as agriculture or tourism, are the most vulnerable to any weather abnormalities, but other industries can also be affected.

The retail sector can serve here as a good example – when heat waves strike Britain, customers rush to the shops for cold drinks, alcohol and barbecue food. The sales of sandals, summer clothes and ice-creams will also jump.

The opposite situation can happen with cold weather. When snow and freezing winds hit the country, some John Lewis' stores have recorded a 32 per cent decline in sales, comparing with the same period of the previous year. The impact of weather on a business' financial performance can be complex, as difficult to predict as the weather itself.

2. ADDITIONAL COSTS

Some organizations may need to invest in new equipment and infrastructure in order to stay in the market or become more competitive. Farmers may need to buy water tanks, more efficient irrigation systems or build water reservoirs, so that they can provide their crops with an adequate amount of water during extended periods of draught. Haulage companies may decide to supply their vehicles with tyre chains to facilitate driving on snowy and icy roads.

Higher temperatures will create a higher demand for cooling of buildings and, consequently, higher energy costs.

Air conditioning, along with cooling and refrigeration of information and communication infrastructure, accounts for 4 per cent of the total electricity used in the UK. In London, the cost of increased cooling demand is projected to rise to over £1 billion by the 2080s.

3. DISRUPTION OF OPERATIONS

Extreme weather conditions and higher temperatures can pose a serious risk to business continuity, as transport links and supply chains will experience disruptions. This situation is particularly risky for businesses involved in perishable goods.

During cold weather periods, when England is hit by heavy snows and freezing winds, some rural roads become impassable and raw milk cannot be collected from farmers or delivered to processing plants. The plants stand idle or produce limited output, and retailers do not receive dairy products for sale.

Extreme weather can cause interruptions in the supplies of water, energy and communication technology. Losses can be significant. In the UK, 84 per cent of businesses depend heavily on communication technology (ICT). If these systems get disrupted, the scale of losses can be major, but difficult to quantify. The flooding of a power plant at Carlisle in the UK interrupted the electricity supply, and 63 000 customers were affected.

Cyclones, tornados, hurricanes or floods can damage business assets to the extent that further production is no longer possible, or must be stopped for safety reasons.

When a tornado hit Oklahoma, the natural gas processing plant in Canadian County had to be shut down. The damage halted the production of 26 million cubic feet of natural gas and 2,000 barrels of natural gas liquid a day.

4. LOSS OF STAFF WORK HOURS AND REDUCED STAFF PRODUCTIVITY

Staff unable to come to work is the most common problem associated with extreme weather events. Floods, snow, ice, broken trees on roads make it impossible or risky for employees to make their way to work. In some cases, teleworking or working from home may address this issue.

As a result of raising temperatures, many buildings become too hot and uncomfortable for the employees – in extreme cases, overheating may lead to illness or even death. Employees will often take days off, as a result of not being able to cope in such difficult working conditions. The risk of staff days lost due to overheating remains unrecognized by many organizations. In the meantime, the number of days when overheating occurs is growing.

In London, overheating happens 18 days a year on average and this number is projected to grow to even 92 by 2050 and to 121 by 2080. The cost of staff day lost due to workplace overheating was estimated at £0.77 billion and by 2050 it may reach £5.3 billion.

5. INCREASED STAKEHOLDERS' INTEREST AND ENHANCED OR DAMAGED CORPORATE REPUTATION

Brand and reputation are critical company assets. In large FTSE 100 firms, the corporation's brand identity is worth up to one-third of the total company value.

In developed countries, consumers have started paying attention how the companies behave in response to the climate change. Recent surveys show that customers opt for 'responsible' brands and find them more attractive than the alternatives. A loss of reputation may lower consumers' confidence in the company and its products and damage the brand. The company also risk losing the reputation among stakeholders: communities, government, media, employees and so on.

Sectors with significant emissions levels and those dealing directly with customers are particularly exposed to the reputational risks. There is a growing consumer expectation that energy, automotive, aviation and retail businesses should act in an environmentally responsible manner and go beyond the legal requirements.

Some organizations have managed to use this trend to enhance their reputation. Three hundred computer and component manufacturers have joined the Climate Savers Computing Initiative supported by WWF (worldwildlife.org) and committed themselves to make their products more energy efficient.

6. ADDITIONAL REGULATORY REQUIREMENTS

According to the International Energy Agency (iea.org), there are already over 2,000 national policies and regulatory measures aimed at mitigating the effects of climate change. This database is still growing, with thousands of new regulations being considered by governments worldwide. The regulations related to climate change can be divided into two categories:

- traditional legislation (emissions permits and energy efficiency requirements for products);
- market-based regulations (carbon schemes, emission trading schemes and fuel tariffs).

For example if you are building, selling or leasing a commercial property in the UK, you are required to have an Energy Performance Certificate (EPC), a document showing how energy efficient the building is. An EPC certificate is valid for up to ten years and should be provided to the perspective buyer/tenant. The failure to provide an EPC for a building can result in a penalty fine between £500 and £5,000.

The impact of these regulations on businesses can be direct and indirect. In the example of an EPC, a direct impact to the building company is the extra cost incurred for the production of the certificate. But poorly performing buildings are getting more difficult to sell or rent. The EPC score may affect the property's value.

7. CONTRACTUAL ISSUES

Extreme weather can make it difficult for some businesses to meet their contractual obligations. Examples include:

- goods not delivered as a result of transport disruption;
- construction projects not completed due to heavy rains or floods;
- a holiday resort not able to accommodate guests as a result of damage caused by weather to the buildings and infrastructure.

More businesses are now incorporating 'extreme weather events' clauses in their contractual agreements.

8. LITIGATION

As the legislation related to climate change is increasing, there is a growing risk for businesses of litigation. In the US, there are presently three categories of possible climate-change lawsuits:

- *Procedural lawsuits.* In Massachusetts vs. Environmental Protection Agency (EPA), the Supreme Court recognized CO_2 as a pollutant. As such, CO_2 could become a subject to regulations under Clean Air Act. The American States were given the right to force the EPA to act accordingly.
- *Claims against businesses.* These are lawsuits brought by states for climate change-related damages – for example – the State of California sought damages from six car manufacturers for environmental impacts of the greenhouse gas emissions emitted as a result of using their products. To date, most of these claims have been unsuccessful.
- *Claims made by shareholders* against businesses for failing to manage risks related to climate change. This could include claims against senior executives for neglecting fiduciary duties in climate change-related matters.

9. NEW MARKET OPPORTUNITIES AND PRODUCT DIVERSIFICATION

But despite the gloomy narrative over the last few pages of this chapter, climate change could have a positive impact on businesses, by creating a demand for new types of products and services.

Organizations that manage to spot the prospects early will gain a competitive advantage. The main areas of opportunities are:

1. New products, such as:
   - products with increased resilience: products that are heat resistant, moisture- retaining, waterproof or made from permeable materials;
   - products and services not reliant on power, water, communication or transport,
2. New services, such as:
   - monitoring and measuring weather impact;
   - providing advice;
   - managing weather risk (insurance services, clean up or construction of flood barriers);
   - infrastructure solutions (sustainable urban drainage or rainwater harvesting systems).
3. New processes to meet climate change (growing new types of crops or creating new types of outdoor activities).

CAN ANYTHING BE DONE TO REDUCE CLIMATE CHANGE?

The best solution is to reduce the output of CO_2 in the atmosphere, and reduce its concentration. This can be done by decarbonizing the world power generation – that is, reducing our use of oil, coal and gas; and turning instead to energy from solar, wind, wave, biofuel or nuclear sources.

In addition, we have to reduce deforestation and lessen the impact of energy-intensive industries such as transport and cement production.

But there is little political will or commercial imperative to do any of this. Meanwhile fracking is an attractive source of carbon energy for those nations that possess shale oil and gas; and this will serve to increase CO_2 emissions.

Hence scientists are now pushing for 'Plan B' solutions which involve geo engineering. It means being fatalistic about our dependence on oil and gas, and introducing interventions to reduce its effects. This is known as 'adaptation', rather than 'mitigation'. Technological fixes could include the following:

- Placing millions of tiny aluminium mirrors in space. They will reflect light away from the planet.
- Fertilizing the sea with iron or nitrogen to create algae which will soak up atmospheric CO_2.
- Spraying sea water into the sky. This creates clouds that will bounce the sun's light back into space.
- Plant paler crops that reflect sunlight
- Spread olivine on the land. Olivine transforms CO_2 into bicarbonate, which washes down to the oceans where it precipitates as carbonate, a salt.

All of these may have unintended consequences, such as altered weather patterns, so they are not free of risk. Moreover, relying on a 'silver bullet' solution allows governments to prevaricate on the central issue of decarbonizing their energy production.

Governments will also take more practical and less controversial action to protect shores against flooding, increase the capacity of drains, and make new buildings storm proof.

Impact on Specific Industries – Ways to Meet the Threat

Changes in the climate pattern will have an impact on almost every industry sector. However, some industries, such as agriculture, tourism, logistics or insurance are particularly exposed to risk. Detailed characteristics of the threats and opportunities for these sectors are included below. Other industries, even those that do not rely directly on weather, can be also affected by changing patterns of consumption or new regulations.

AGRICULTURE

Agriculture is particularly vulnerable to the weather conditions. Rising temperatures, changing rainfall patterns and sunshine levels, higher concentration of CO_2 in the atmosphere, problems with water availability and extreme weather events – all these factors can severely affect farming operations, productivity and the range of goods offered by the sector. As Table 9.1 shows, there are risks and opportunities.

Table 9.1   Risks and opportunities for agribusiness

Risks	Opportunities
Flood risk to high-quality agricultural land	Higher production levels (wheat, sugar beet
Increased risk of soil erosion	and potatoes)
Drier soils (due to warmer summers)	Increased productivity of grassland and
Increased water demand for irrigation of crops	forage legumes
Extended duration of heat stress in livestock	Greater production in greenhouses
Risk of crop pests and diseases	Increased livestock productivity
	Decreased number of livestock parasites
	Opportunities to grow new crops

The opportunities for agribusiness

Climate change can bring agriculture some tangible benefits such as higher production levels for some crops, increased livestock productivity and opportunities to grown new species.

1. Higher crops production
   Farmers may enjoy increased crops production as a result of warmer temperatures or higher concentration of CO_2 in the atmosphere, or the combination of these two factors. According to the UK Department of Environment, Food and Rural Affairs (DEFRA), the weather changes will bring increased wheat and sugar beet yields. It is estimated that the production of wheat will jump 40 to 140 per cent, sugar beet 20 to 70 per cent, and grass 20 to 50 per cent by 2050. However, these benefits may be reduced by the lower soil moisture and heat stress.
   In some part of Britain, increased levels of CO_2 in the atmosphere may lead to more tor grass and bracken, which are not very nutritious for grazing animals. Poor nutrition may lead to decreased milk production in cows. This calls for physical and chemical control of the weeds.
2. Higher production in greenhouses
   Increased temperatures will have a positive impact on the crops grown in greenhouses, minimizing the requirement for additional heating and making production cheaper.
3. Increased livestock productivity
   Warmer temperatures, increased grass production and a longer vegetation season are beneficial for farmers rearing cattle. This is because it will extend the grazing season and the cost of housing cattle during winter time will be reduced (with less fuel required for heating).

In areas where extending the grazing season is not possible (because of heavy rainfall or other local factors) there are still opportunities to make savings. As grass production increases, the excess grass can be stored as silage and used as winter feed.

4. Decreased number of some livestock parasites
   Hot, dry weather is hostile for some of the parasites, such as roundworms, blowflies and ticks. Some areas may experience a decrease in their numbers.

5. Opportunities to grow new crops
   The milder weather conditions create an opportunity to start growing new crops or expand those that are currently cultivated in small quantities. Signs of this new trend have been already recorded in the south of England.

In Britain, formerly exotic crops such as melons are already thriving in Kent (about 10,000 fruit are harvested each year), along with peaches, nectarines and sunflowers. The first apricot farm has been also established and sells its fruits to one of the national supermarket chains. A farm in Devon has invested in growing walnuts, almonds and Sharon fruits. Grapes have been grown in the UK for some time, but the number of vineyards will grow, as the climate becomes warmer.

By switching from traditional to new crops, farmers can reach new markets and improve their competitiveness on the market.

The risks for agribusiness

Floods, soil erosion, draughts and heat are the most serious risks the industry should address.

1. Agricultural lands can be flooded.
   Agricultural land is at a high risk of flooding from the rivers and the sea. In the UK, high-quality arable and horticultural land is likely to be flooded at least once every three years (according to DEFRA). As a result, some of the affected land may become unsuitable for agricultural activity.

2. Increased risk of soil erosion.
   More frequent and intense rainfall can increase the risk of soil erosion. The implications on the crops have not been fully examined yet.

3. Drier soils.
   As the temperature rises, more water evaporates from the soil, making it drier and less suitable for some species. Farmers will need to change the plant species they grow. Deep-rooting or drought-tolerant species have a better chance of surviving and thriving in these conditions.

4. Increased water demand for irrigation of crops.
   Crops will require more irrigation, especially the high-quality ones. As other industries will also require more water, we can expect that less water will be available for agriculture. This can be mitigated by improved water management and enhanced irrigation techniques.
   - *Rainwater harvesting (water tanks)* – this involves collecting water during periods of rain so that it can be used during dry spells. This is a simple but effective method. There are usually many containers on a farm that can serve this purpose.
   - *Improved irrigation techniques* – these may vary and depend on the region, soil, crops and other local factors, so it is not possible to provide universal guidelines. Farmers may consider reconfiguring irrigation layouts, installing extra infrastructure, such as recycling systems and installing piping, drip or spray systems.

- *Irrigation reservoirs* – these reservoirs are built by a river to store the high winter river flows so the water can be used in the drier periods. This method of water storage is becoming increasingly popular in England. Lindsay Hargreaves, a farm manager from the Elveden Estate Thetford, has built two clay-lined reservoirs of 450,000 m3 (each 100 million gallons). The water stored in the reservoirs is used to irrigate 1,500 ha of crops, including onions, potatoes, carrot and parsnips.

5. Extended duration of heat stress in livestock.

   Animals can suffer from hot temperatures. Prolonged exposure to heat may reduce their physical activity; diminish their fitness, fertility, lifespan and productivity. Pigs and poultry have been found particularly vulnerable to the heat stress. To protect livestock from, heat stress, farmers can take the following steps:

   - make sure adequate quantities of water are available so that animals can replenish their body fluids;
   - when reared outdoors, the animals should have access to shaded areas. These could be trees or man-made structures, constructed for this purpose. Insulated aluminium and galvanized steel are the best materials, as they have reflective surfaces. You may also consider plating more trees;.
   - water sprinkler systems and drip coolers can be used on pigs to cool them;
   - install additional fans in barns or increase the air exchange rate in mechanically ventilated buildings;
   - adapt your feeding regime: animals usually reduce the amount of food consumed to reduce the amount of heat produced due to digestion. The new diet should include higher levels of fats.

CASE STUDY: ICE LOLLIES FOR PIGS

Farmers have come up with new ideas to keep their livestock hydrated. One farm in Devon started giving frozen 'carrot lollies' to its pigs during heat waves.

At Pennywell Farm, near Buckfastleigh, Assistant manager Catherine Tozer said: 'The recent heatwave has been exhausting for everyone, and animals are no exception.'

'All the animals have been struggling with the heat so we have just tried to do everything we can to make them more comfortable. After the initial shock and a bit of investigation, the ice lollies went down a real treat.'

---

6. Increase in crop pests and diseases.

   The possibility that climate change may cause an increase in crop diseases and pests is not supported by strong scientific evidence, but it cannot be entirely disregarded.

LOGISTICS

Extreme weather conditions can severely disrupt logistic operations. In the UK, transport problems caused by snow have the greatest impact, costing the country an estimated £400 million a day. Other weather phenomena affecting the logistics sector are: heavy rain, severe gales, heat and sun, thunderstorms, lightning and dense fog.

With the exception of fog, all these factors can cause damage to the infrastructure. And the longer they last, the more serious the damage can be. The more severe the damage, the longer it takes to repair the roads and train lines and re-establish normal operations. For example, heavy rain can reduce visibility and cause slippery roads (a short-term impact), but when it lasts for a longer period of time, it can cause flooding and landslides, which can take a long time to repair (long-term impact).

In the UK, the total length of roads exposed to flooding is estimated at 12,000km. In 30 years' time, this figure will jump to 14,000 km and by the 2080s by 19,000km. For the railways these figures are: 2,000km, 2,300km and 3,100 km respectively. Over 1,000km of roads are at risk of landslides.

In the UK, the logistics industry is particularly affected by the weather mechanism called arctic oscillation, which used to happen around Christmas. Strong winds from the north-east hit the country and the temperature drops below zero. The cold temperature combined with snow can cause massive disruptions to logistics operations in the busiest trading time of the year. Stobart, a leading UK logistics group, had to spend an extra £1.5 million as a result of adverse weather around one Christmas period.

The number of accidents on the roads during heavy rains and snow is always high. Staff injuries and damage to the fleet are the most serious risks to the logistics industry that should be addressed, as shown in Table 9.2.

Table 9.2 Risks and opportunities for the logistics industry

Risks	Opportunities
Flooded roads and rails	Brand enhancement
Landslides	Sharing and integrating transportation along the
Rail buckling	supply chain
Road traffic disruptions – delivery delays	
Impassable roads – failed deliveries	
Road accidents – staff injuries, delays in deliveries, stock lost	
Damage to the fleet	

The risks for logistics businesses

The immediate implication of the extreme weather is a disruption to the domestic transportation of goods between manufacturers, ports, distribution centres and retailers. These disturbances are particularly serious for the transport of perishable goods such as dairy or bakery products.

Action to reduce risk in logistics

In order to minimize risk and losses, logistics businesses can take the following steps:

- Monitor weather forecasts on a continuous basis, especially during the winter. This step will allow you to identify risks and prepare a plan in advance.
- Keep in touch with retailers to monitor the stock level to avoid under and over-stocking.
- Fit vehicles with tyre chains or AutoSocks to improve traction where relevant. These are particularly helpful where roads are not properly gritted. Tyre chains are common and at times compulsory for commercial trucks in some countries, including Canada. The investment varies between £150 and £300 per drive wheel for a commercial truck, but the chains can be used for many years if properly maintained and stored.
- Make sure your drivers get training in winter road driving.

- Get in contact with the councils responsible for gritting the roads, so you can plan delivery routes in advance. Consider alternative roads for your deliveries.
- If the extreme weather events occur frequently, it may be worth to consider moving distribution centres closer to the end customers.

The opportunities for logistics businesses

Extreme weather events are a real test for logistic companies. If a firm can prove it can perform under extreme conditions, it will build its reputation as a reliable contractor. Good performance may result in new contracts, for example taking over from the companies that failed or performed badly under the same circumstances.

Another opportunity lies in sharing and integrating transport along the supply chain. This might include combining truckloads, or renting farmers' tractors to transport milk closer to the processing plants.

FINANCIAL SERVICES

Insurance companies worldwide have been recording an increasing number of claims resulting from extreme weather events over the last years; and this trend is projected to continue.

Since 1980 there have been 76 weather disasters in the US, including tornados, hurricanes, fires, droughts, heat waves and floods. Each event cost at least $1 billion in damages. In total, it is estimated these events cost the insurers about $500 billion.

Hurricane Katrina cost insurers $45 billion. The economic impact of Hurricane Sandy in New Jersey is estimated at $39 billion, while it cost New York $50 billion. Since 1980 drought is estimated to cost insurers approximately $200 billion, of which $5 billion will be paid by private companies.

The forecasts for the future are far from optimistic. Claims for winter storm damage in Europe are projected to increase by 68 per cent in the next 60 years, which translates to about €11 billion a year.

The risks and opportunities for the industry are shown in Table 9.3.

Table 9.3 Risks to the financial services sector

Risks	Opportunities
Higher number of weather-related insurance claims	Increasing market penetrations and uptake for
Mortgage provision limited due to flood risk	weather-related insurances
Destruction of assets	New financial and insurance services

Action that can be taken by the financial services industry

There are several steps insurance companies can take to minimize the risk:

- Consider raising insurance premiums to balance your books. But if the increases are too steep, the business may lose clients.
- Start a dialogue with local councils: they may consider covering some of the insurance costs in cases where homeowners cannot afford them.
- Consider withdrawing from the worst-affected areas.

When heavy rain caused severe flooding in Worcestershire, Gloucestershire, Devon and Cornwall, major UK insurance companies announced they had to raise sharply the insurance premiums; otherwise they would be not able to cope with the scale of the problem in the future. As UK home insurance includes flood cover as standard, some insurers are prepared to withdraw from the worst-affected areas.

Mortgage dealers are not free from risk either. When Hurricane Sandy caused serious damage to both homes and businesses, many homeowners found themselves unable to keep up with their payments. Despite relief programmes offered, some of the money lent will never be recovered.

In order to minimize the risk, insurance, mortgage brokers and other financial services companies should:

- incorporate weather risks into their business planning;
- use catastrophic modelling to predict the effects of extreme weather events and climate change on business operations.

Barclays and HSBC are involved in research projects aimed at incorporating weather risks into investments.

The opportunities for financial service businesses

In some parts of the world, homeowners can obtain private insurance for extreme weather events. In Australia, you can insure your property against storm, flood, storm surge and actions of the sea, erosion and subsidence and fire. As weather extremes are no longer surprising, this type of insurance product may become even more popular in the years to come.

There is also a potential for a new line of financial products. In the UK, Aon Benfield has introduced a new weather re-insurance to mitigate financial losses associated with adverse weather events.

TOURISM

The tourism industry is one of the most affected by weather, and it is therefore exposed to severe risks. Changes in the climate may cause some of the tourism attractions to disappear. In Australia, the Great Barrier Reef industry is estimated to be worth A$1.5 billion. With a 2–3 degree centigrade increase in the temperature, 97 per cent of the reef may disappear, making the tourism sector fighting to survive.

Table 9.4   Risks and opportunities in the tourism industry

Risks	Opportunities
Physical destruction of natural attractions	Expansion of existing destinations
Tourism assets exposed to flooding	New tourist destinations
Reduction of existing beach areas	All year round visitor traffic (rather than seasonal)
Reduction in snowfall	

As the sea level rises, it may destroy tourism assets and reduce the area of existing beaches.

In the UK, 33,000 tourist and leisure facilities are currently exposed to tidal and river flooding. The loss of beach area due to sea level rise may increase from 12km² to 61km² by the 2080s.

The reduction of snow cover can affect winter tourism. In Switzerland, 85 per cent of ski resorts are now snow-reliant. The change in the climate can move snow reliability from resorts at 1,200m to those at 1,800m, leaving only 44 per cent of the resorts snow-reliable.

While some threats are difficult to mitigate, as they happen on a global scale, there are some steps that can minimize the risks and compensate income loss caused by weather:

- improve business drainage and install flood defend measures to protect your premises against physical damage by flood
- develop all-year facilities and activities that are not dependant on weather (for example, in-house swimming pool, spa, indoor sports facilities or art activities);
- if running a ski resort, consider artificial snow making.

The opportunities for tourism

As the climate becomes warmer, some destinations may become more attractive for tourism. Britain, for example, can capture some of the today's southern Europe tourism market. Moreover, with milder weather, some of the attractions have a chance to get visited all year round, rather than just in summer. New destinations may also emerge as a result of this shift.

GOVERNMENT AND MUNICIPAL AUTHORITIES

All cities, everywhere in the world, will face a range of unexpected challenges: gales, extreme heat, freezing conditions, and storms, with biblical events happening each year. The scale of risk for town halls and municipal authorities is quite extraordinary. Once in 100-year events are now happening every five years or even on a yearly basis.

It was originally hoped that governments would mitigate or reduce the scale of climate change; but with that resolution weakening, cities are being forced to manage the impact on their citizens.

Rising sea levels will mean either having to relocate millions of people from coastal cities around the world, or build sea walls to keep out the projected sea rise of perhaps 6–10ft (1.8m to 3m). Inland municipalities will be struck by extreme weather events, including storms and soaring temperature. Water will either be in short supply, or else arrive as a flood.

Storms and storm surges: In addition to rising sea levels, there will be devastation caused by storms and the kind of surge seen in Hurricane Sandy on the Eastern USA seaboard, or Hurricane Katrina which killed 2,000 people in New Orleans.

National and local governments are major organizations, and they are now having to plan for the risks that are already being visited upon the world.

Strategies for cities

Coastal cities are most at risk. Table 9.5 shows coastal cities at risk of a 1.6ft (0.5m) rise in sea level. Other cities, such as New Orleans, Rotterdam and Amsterdam, which are at or below sea level, have unknown numbers at risk.

CASE STUDY: THE EFFECT OF HURRICANE SANDY

When Hurricane Sandy struck, it was 1,000 miles wide. A 14-foot surge of water hit New York harbour. Eight million people lost power for up to several days. The internet was down, mobile phones didn't work, and the city was plunged into darkness when a power station failed. Vulnerable people were stuck in high-rise buildings with no elevators.

Only 120 people died, compared with Katrina's 2,000 death toll, but the city was paralysed and the impact on businesses and citizens was immense.

As predicted by the scientist Klaus Jacob, the subway system which forms the city's main arteries, was quickly flooded, as were the road tunnels; and this is likely to be replicated in the future. Jacob pointed out to the BBC that sea levels are already one foot higher than when the New York subway was built in 1904.

Hardest hit were the poorest communities living in high-rise public housing such as in the district of Red Hook, Brooklyn. This feature is replicated in disasters elsewhere, such as New Orleans where the poorest were least able to cope with disaster. This has implications for town planners, social services and the emergency services.

Eight million people on America's Eastern seaboard are now at risk from flooding. Yet great cities continue to attract people. One million more inhabitants are expected to come to New York over the next 30 years, and the majority will live in the flood zone.

Table 9.5   Populations at risk from sea level rise

City	No. of people (millions) at risk from a 1.6ft (0.5m) sea level rise by 2070
Kolkata, India	14.0
Mumbai, India	11.4
Shanghai, China	5.4
Bangkok, Thailand	5.1
Miami, USA	4.7
Alexandria, Egypt	4.3
New York-Newark, USA	2.9
Guangzhou, China	2.7
Tokyo, Japan	2.5

Source: BusinessInsider.com

There are two options for planners in coastal cities. You can either defend the city or retreat to higher ground.

If you retreat, you have to persuade, or order, the civilian population to leave; or buy them out. You also have to make building land available in new higher areas.

If you defend the city, you have to build sea walls to defend against storm surges and rising sea levels. In addition, you have to try and protect the infrastructure. The metro system will fill with water, as will the road tunnels. As we saw with Hurricane Sandy, power stations will flood and fail. This will leave the city without heat and power, and lifts won't work. Vulnerable individuals will be trapped for days in high rise buildings. Citizens will die from heat exposure, from freezing, drowning, and being struck by objects.

Non-coastal cities

Inland cities will be equally vulnerable from extreme weather. Extra rainfall will call for permeable pavements, which will allow the rain to seep into drains. New buildings will need building codes that require them to absorb extra heat, and reduce the use of air conditioning.

Rural areas

Rural areas are no less vulnerable, and local governments are facing devastation in many ways.

When super typhoon Bopha struck the Philippine island of Mindanao, it killed 1,067 people and left 800 missing. More than 6.2 million people were affected, 216,000 homes were destroyed or damaged, and the losses amounted to $1 billion. The banana crops were ruined, and it would take the coconut plantations ten years to recover.

Dealing with the aftermath of these storms costs poor governments money they can ill afford, and the events will only gain in intensity and frequency. Local authorities have to take the following measures:

* Protect the city. This will entail building defences and protecting the infrastructure.
* Adapt to the changing climate, and make the city more resilient. Cities can reduce the impact of storms and heat by various measures, as shown in Table 9.6 below.
* Have plans in place to provide disaster relief.

All of this will require major investment. Unfortunately, most politicians are focused on the present; and are unwilling to spend money that will only be of benefit in the long term.

Moreover, the amounts of money are of huge scale, at a time when most economies are short of cash and citizens are unwilling to pay extra taxes. Political will and investment will not be readily available.

Table 9.6   Examples of adaptation

Typical Adaptation Strategies

General
* create early warning systems;
* improve and extend emergency planning.

Sea areas
* abandon some coastal areas to the sea, in order to invest better elsewhere;
* build sea walls;
* create or improve sea marshes or water zones to act as a storm buffer zone.

Buildings
* incentivize building residents to move to safer places;
* ban new building in flood plains;
* create new building codes, for safer properties;
* build houses on stilts or with ground-level parking. Abandon ground and first floor dwellings;
* improve building design to conserve water, and reduce the need for air conditioning;
* install air conditioning in schools and public buildings.

Infrastructure
* elevate critical systems above the ground, such as the High Line park and track in New York;
* improve river flood defences;
* make pavements permeable, allowing rain to soak into drains;
* encourage sustainable transport. Reduce car use, so as to minimize air pollution. Foster renewable transport, including trams, bicycles and walking;
* plant heat-tolerant trees;
* increase the capacity of storm water storage, so that storm water doesn't contaminate rivers.

Risk Assessment

By answering the questions below, you can see how vulnerable your organization is to breaches of security.

Question	✓
Does the business lack an adverse weather risk assessment?	
Are your products or services weather dependent?	
Do you have buildings in an area at risk of flooding?	
Do your workspaces get hot during the day?	
Do you have assets in areas at risk from tornados?	
Do you operate in agribusiness or the food industry?	
Do you operate in the logistics business?	
Does the business operate in tourism?	
Does the business operate in financial services?	
Do you operate in national or municipal government?	
Total points scored:	

Score: 0–3 points: low risk. 4–6 points: moderate risk. 7–10 points: high risk.

The Appendix has a summary of the checklists. By entering the results of this one, you can compare your security risk against other categories of risk.

10
Protecting Against Fire

Alec produces great food. But he isn't a great manager. He likes to concentrate on cooking, and leaves everything else to his staff. Since most staff last only a few months, it isn't surprising that there are occasional disasters.

So now the cafeteria is closed, amid much grumbling from staff. Smoke damage means every surface needs cleaning and repainting. And the business needs a new kitchen. 'It was the grease trap,' said Alec. 'All members of staff know they have to empty it, every Friday. But sometimes these things get overlooked.'

Alec and his staff put out the fire with two extinguishers, no one was hurt, and the fire didn't spread. But there needs to be an enquiry, and maybe a disciplinary hearing. 'Who does Alec report to?' you ask your assistant. 'Er, technically you,' he says. You sigh.

Fire as a Commercial Risk

Fire is one of the most common commercial risks. Most companies believe that fire could never happen to them, and so they focus on more exotic types of risk such as terrorism or environmental issues.

Yet 20 per cent of employers suffer a fire; and 70 per cent of businesses involved in a major fire either do not re-open or fail within three years of the fire.

There are 20,000 commercial fires each year in the UK, 65 per cent of which are accidental. In other words there are a lot of fires (480 non-domestic fires each week), and one in three of these are started by arsonists. Moreover, fires occur not just in risky places like chemical plants, but ordinary ones like shops as well.

So, fire is a major workplace problem. St John Ambulance reported that over a four-year period in the UK 1,425 people were injured in workplace fires and 32 died.

The Association of British Insurers reports that socially deprived areas and schools are especially vulnerable: arson rates are 30 times higher in poorer areas, while 20 schools a week suffer an arson attack, disrupting the education of 90,000 schoolchildren. Factors that give rise to larger fires include:

- open plan buildings, which allow more rapid spread of fire;
- lack of compartmentation (dividing a building into discrete zones);
- out of town developments, where fires can go for longer unnoticed;
- buildings that are under construction or are being renovated.

CASE STUDY: NEW LOOK FINED £400,000

The fashion retailer New Look was fined £400,000 and £136,000 costs following a fire at its Oxford Street, London branch.

150 firefighters using 35 fire engines fought the blaze for three days, and the fire disrupted trade at 50 other shops in the vicinity.

Malik Khan, who owns a souvenir shop opposite New Look, said: 'I heard screams and shouts, and ran out into the street. People were pouring out of New Look. Many of them were screaming and having to dodge large pieces of glass as they came down from above them. It's a miracle no one was hurt.'

After the fire, the fire brigade inspected the premises and found 'substantial' breaches of fire legislation. The company had failed to carry out an adequate fire assessment and staff were insufficiently trained.

Staff had reset the fire alarm after it had gone off, and shoppers continued to walk around the store. Even when the alarm went off again, staff failed to react. By the time evacuation started, windows were being blown out by the heat of the fire.

On one escape route, the corridor was partially obstructed by plastic crates containing stock. This created a fire hazard and restricted access.

New Look had developed its risk assessment in line with government guidance, but it was largely generic in form and effectively a 'tick-box' exercise to be completed by the store management at each of the retailer's premises.

How Fires Start

A fire needs three things to get started: air, fuel and enough heat to ignite the fuel (see Figure 10.1).

Heat comes from many sources in the workplace, including friction produced by a wheel that's turning or a machine that's moving. All mechanical machines are prone to breaking down. Static electricity can spark an explosion, which can lead to fire.

Fuel doesn't have to be the obvious things conjured up in our mental images of oil cans or wood. A spinning part may produce combustible dust, which offers itself as the fuel.

Human error often plays a part, too. So it's easily for industrial and commercial fires to start.

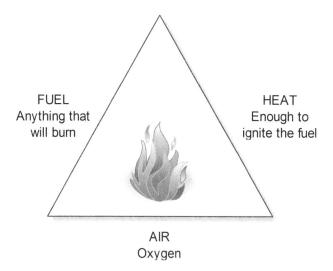

FUEL
Anything that
will burn

HEAT
Enough to
ignite the fuel

AIR
Oxygen

Figure 10.1   How fires start

Designing a Fire Risk Strategy

A fire risk strategy starts with a fire assessment survey, shown in Table 10.1, which should be carried out at regular intervals. This will identify the most likely causes of fire, and indicate how to prevent them and minimize their impact.

The audit should cover all areas of the building, including production facilities, offices, storage areas, basements and roof spaces. Fires often start in unlikely places because those are the areas where fire safety has been overlooked.

In the UK and many other countries, it is a legal requirement for companies to carry out a fire assessment, and maintain a fire management plan. In the next section, we examine how to do that.

The Five Elements of Fire Safety

Fire safety has five elements:

1. Assess, Audit and Review
2. Appoint and Train
3. Detect
4. Mitigate
5. Escape

We look at each of these in turn, starting with the assessment.

1. DOING A FIRE ASSESSMENT

A fire assessment examines the risks that could lead to a fire. It reviews the actions that have been taken to minimize the risk or fire and to reduce its impact. Table 10.1 overleaf is a sample fire assessment. It also identifies the measures currently in place and steps to be taken. It also includes a Review form and an Action Plan.

2. TRAINING

You need to appoint a fire officer for each area. Ensure staff know what to do in the event of a fire.

The most important part of fire precautions is to *protect life*, and the emphasis should be on evacuating the building. You should ensure that staff are familiar with the emergency procedure.

Fire drills should be held regularly (at least annually). The organization should make sure that the building can be quickly evacuated and personnel accounted for. Staff must be trained to be aware of fire risks, such as overloaded electrical circuits.

3. DETECTION

Fire alarms need to be wired into the mains; and they should be tested regularly.

Automatic fire detection (AFD), which uses sprinklers, is the preferred choice in buildings where many people work or live. Sprinklers must be maintained, while smoke detectors are essential in buildings which do not have AFD.

Table 10.1   Fire risk strategy

Fire Risk Assessment	
Organization:	Assessment undertaken by:
Building:	Signature:
Location:	Date:
Floor, area:	

1. What are the major fire hazards?
Fire requires fuel, oxygen and a source of ignition.

Sources of fuel
For example: flammable liquids such as paints, packaging materials, waste products or textiles.

Sources of oxygen
Oxygen is in the air around us. Another source is oxygen cylinders.

Sources of ignition
For example: naked flames, hot surfaces, cookers or electrical equipment.

Are there any hazardous substances?
For example: explosives, poisonous chemicals, irritants or carcinogens.
What control measures are in place? What further steps need to be taken?

2. Who is at risk?
For example: customers, staff, disabled or elderly people, or children. How familiar are they with the building? Lone workers? Cleaners?

3. How do we detect and announce a fire?
Current Measures: *For example: mains wired fire alarms. What further steps need to be taken?*
Action required:

4. How would we mitigate a fire?
How could a fire be fought?
Current Measures: *For example, fire extinguishers (identified on a plan). Staff training. Nominated fire officer/fire marshals.*
Action required:

How do we restrict the spread of fire?
Current Measures: *For example, fire doors. What further steps need to be taken?*
Action required:

How do we segregate areas of higher risk?
Current Measures: *For example, correct storage for dangerous substances.*
Action required:

5. How do people escape?
What escape routes exist?
Current Measures: *For example: signposted fire escapes available on all floors. Are they lit and free of obstructions? Are doors always unlocked? What further steps need to be taken?*
Action required:

What evacuation procedure is in place?
Current Measures: *For example: the plan is shown on notice boards on all floors. It is tested regularly.*
Action required:

Review of the Assessment		
Date:	Reviewed by: *G Fowler*	Signature: G Fowler

Review outcome
A. Stair safety lighting needs to be installed.
B. No rubbish to be stored on the fire exit corridor: AL to advise site manager
C. Construction work in L&G Dept is causing additional risks. Carry out new assessment.

Action Plan				
Priority	Issue	Action to be taken	Date completed	Verified by
1	*Stairs not lit*	*Install emergency lighting*		*M Rowles*

4. MITIGATE

Containing and suppressing a fire can involve fire extinguishers, hose reels, fire blankets for kitchens, and sprinklers. You need a maintenance programme for these devices. The use of hose reels by untrained people isn't recommended, because the force of the water makes the hose difficult to handle and direct effectively.

To delay the spread of fire, the business can introduce *fire-rated construction materials*. For example, some wall panels give an hour's protection against fire. Separating office and production workers from warehousing is one example.

In new buildings, ensure that fire safety is designed into the structure.

5. ESCAPE

In the event of a fire, people need to be able to escape from the building safely. That means having safe escape routes, emergency lighting, escape ladders, clearly signed exits and regularly tested evacuations. Again, you need to carry out regular audits.

You should give thought as to how to disabled people will evacuate the premises:

- *Visually impaired people*: appoint a team member to act as their emergency buddy to ensure they are led to safety.
- *Hearing impairment*: a bright flashing light linked to the fire alarm will visually alert them.
- *Mobility impairment*: ensure people with impaired mobility have an emergency buddy. Ideally they should not be sited above the second floor in a multi-storey building. But if they are, you should provide a refuge near the exit. Mobility-impaired people should not be evacuated at the same time as other occupants; the use of emergency evacuation chairs would slow everybody down. They should use the refuge area until everyone else has evacuated.

Emergency signs (showing the way to the fire exit) should be clear and consistent. In an emergency, misleading signs could lead to death. There are Building Regulations concerning signs.

Escape routes should be constructed from non-flammable materials. Fire exits should be free from obstructions, such as potted plants, furniture or storage. If fire starts, people need to be able to escape easily. Fire doors should not be propped open – this will speed the path of a fire. People wedge doors open to stop the door slamming; yet the noise can be reduced by fitting damped door closers. In the summer, people prop doors open to improve air circulation; but a proper ventilation system will improve the atmosphere.

Major Risks

The organization needs to concentrate on the most significant risks. In assessing the risks of fire, the auditor should look for examples of the main causes of fire:

- electrical hazards;
- hot work;
- machinery;
- smoking;
- flammable liquids;
- bad housekeeping;
- the threat of arson.

We examine each of these next.

ELECTRICAL HAZARDS

Electrical fires can start when circuits overheat, having been wired wrongly or overloaded. Wiring is usually installed properly, but may then be modified by a self-taught handyman. Over a period, wiring may become full of spurs, transformers become overburdened and inadequately sized wiring is introduced. A lack of socket outlets often results in the use of adapters, which can overheat and cause a fire.

As many as one in three electrical fires take place in the office. Overloaded circuits are an increasing problem as companies buy additional printers and computers for use in older buildings designed before computers were invented, bringing in four-way plugboards to overcome the lack of sockets, and then daisy-chaining them, using all the sockets. Some NHS hospitals outlaw their use, allowing only approved ones (where each socket has an on/off switch and a fuse).

Good practice involves using *qualified electricians*, and *planning electrical installations* properly. It is also useful to conduct an electrical audit, to check for unsafe and overburdened wiring.

Portable Appliance Testing: One in four of all electrical accidents involve portable appliances. For this reason the UK's Electricity at Work Regulations require employers to reasonably ensure that they are safe. This means such devices need to be regularly tested, a task that is often sub-contracted.

The frequency of testing required depends on the scale of the risk. As mentioned elsewhere in this book, electric kettles don't usually constitute a sufficiently major risk as to need annual testing. The HSE has issued a guidance note on this, which is available on their website.

HOT WORK

Fires caused by hot work often come from unexpected sources. The painter who burns old paint from a door, and the plumber who tries to thaw a frozen pipe, are just as likely to cause a fire as a welding or cutting operation. Staff may take short cuts when doing a job, and procedures may not be followed.

Good housekeeping is essential for hot work. The work should be supervised, and the area should be checked for four hours afterwards.

The company should avoid using radiant and portable heaters if possible. Hot pipes or lamps should not come into contact with combustible material such as paper, fabric or sawdust.

Where possible, you should avoid equipment that could cause a fire. That can include:

- deep fat fryers;
- welding equipment;
- cutting equipment;
- grinders.

If their use is unavoidable, place an appropriate fire extinguisher within easy reach of the work.

MACHINERY

Machinery should be regularly *serviced*, and kept properly lubricated. Vents should be kept clear to prevent overheating. Oil leaks and drips should be absorbed with mineral absorbents, not sawdust; drip trays should be used where necessary and should regularly be emptied.

Hot equipment, like pressing or soldering irons, should be switched off immediately after work, and placed on rests – not on the work surface.

Extraction systems will remove combustible particles such as sawdust. 'If your facility generates dust, any kind of dust, consider it combustible until proven otherwise,' says John Astad, founder of the Combustible Dust Policy Institute, quoted in Industry Market Trends (thomasnet.com). More than 500 fires and explosions are caused by combustible dust each year in American industrial businesses. Most of the incidents did not result in injuries or fatalities, but some do.

SMOKING

As we saw in Chapter 6 on Health and Safety, many countries have now banned smoking in the workplace. Health risks were the prime motive, but the reduced fire risk is a benefit – 12 per cent of accidental workplace fires in the UK used to be caused by smoking.

However, illicit smoking is still an ever-present risk, and since it takes place secretly it poses a hazard. Knowing the workforce's behaviour and implementing security patrols can avoid this.

FLAMMABLE LIQUIDS

Solvents are a major hazard, causing nearly half of all fires started by flammable liquid. The flammability of solvents is not always recognized, and cleaning fluids are often kept in opened drums which are moved around the plant.

Flammable liquids should be stored kept in enclosed metal containers, and drip trays should be used. A stringent 'no naked lights' policy should in place in their vicinity.

LPG (liquefied petroleum gas) cylinders are a fire hazard, and should be stored safely, as should other compressed gases like propane and butane. They should be kept outdoors in a fenced container, with prominent notices prohibiting smoking and naked lights.

The business should have *ventilation systems* that reduce or disperse flammable gases or vapours.

CASE STUDY: IT COULD HAVE ENDED DIFFERENTLY

As the Asia Pacific manager of a specialist arm of a large publishing business, Miranda was working on the seventh floor of a thirteen-storey Hong Kong building (low rise by Hong Kong standards) when the fire alarm went off.

Initially staff ignored it (as staff do) until someone arrived to say there was a real fire, and they shouldn't use the lifts. Miranda was the first to reach the fire exit door, which she discovered was locked. At that moment Miranda thought she might be facing death.

Fortunately, the fire was a small one and the fire brigade arrived quickly and released the trapped staff. For the rest of her time there, Miranda used to regularly go down the fire exit to check the doors were unlocked, and she had difficulty putting the thought of her near escape out of her mind.

BAD HOUSEKEEPING

Rubbish should not be allowed to accumulate, especially in boiler rooms, under stairs, basements or store rooms. *Fire exits* should not be obstructed.

In offices, piles of computer print outs and sheaves of paper litter executives' desks after they have gone home for the night: a *clear desk policy* can reduce the risks.

Combustible waste should be kept to a minimum, and it should be caged: old pallets are often targeted by vandals. Ensure good work practices.

There should be a procedure for *closing down* the premises when work finishes. All non-essential electrical equipment should be unplugged, including computers and heaters. Fire doors should be closed.

The £36 million fire at Windsor Castle, home of Queen Elizabeth II, began during refurbishment work. A chapel was being used to store paintings while other rooms were being refurbished. Curtains used to screen the altar were closed, hiding spotlights that were used to illuminate the altar. An 'A' frame may possibly have pushed the curtains on to the spotlight. The investigators concluded that the spotlight set fire to the curtains, thus starting the fire, which destroyed some of the most historic parts of the building.

CASE STUDY: FIRE DOESN'T NECESSARILY START INSIDE YOUR PREMISES

Fire hazards don't have to be on your site. A massive explosion and fire at the Buncefield oil refinery made several companies' buildings unusable, including that of Fujifilm, which later had to be demolished.

49 miles away, Addenbrooke's Hospital in Cambridge lost its computerized admissions system for several days, and had to revert to a manual one (the admissions system effectively controls all the hospital's work, so a loss of this sort has profound and rippling effects).

25,000 staff in 600 businesses were unable to get to work. Marks and Spencer had to close a food depot, which disrupted supplies to stores. And ASOS lost stock worth £5 million from its only warehouse, and the company's shares were suspended. The company had to refund 19,000 orders.

This case study demonstrates that you should review nearby businesses and activities, to assess your vulnerability to fire and other threats.

ARSON

In the UK, arson is responsible for 45 per cent of fires in commercial buildings costing an estimated £300 million. In the USA it represents 25 per cent of the fire losses by value, and in Europe the level of arson is growing.

Much arson is either an attempt to conceal a crime, or stems from a personnel dispute. A company with a labour dispute, layoffs or poor industrial relations is more likely to suffer arson. Other likely targets are empty buildings, and political targets (such as oil companies).

Companies can beat arson by ensuring that *fire protection systems* (including sprinklers and alarms) are tested regularly. Improved security is also important. Staff should not be allowed to walk round sensitive parts of the site unless authorized, and *perimeter security* should be enforced (see Chapter 11 on Security).

CASE STUDY: WHO ATTACKED RIVERSIDE OFFICE SUPPLIES?

According to police forensic experts, the fire that destroyed the premises of Riverside Office Supplies had been started deliberately.

Someone had cut a neat hole in a window, having first applied sticky plastic to prevent noise. Then they had poured petrol into a waste bin and dropped a match into it. They had even tinkered with the engine of a van parked inside the loading bay, to make it look as though an electrical fault had started the fire.

Riverside Office Supplies was not the obvious target of an arson attack, yet someone wanted to stop the five-month-old company before it became too big.

An office supplies company stocks a lot of paper, so the blaze was all-engulfing and completely destroyed the premises. Apart from the stock, the fire also destroyed the computers and all the company's records.

Just one thing saved Riverside Office Supplies. June Rathmell, the Managing Director, had made a back-up copy of her computer records and taken it home with her that night. She could find out who had ordered goods, and who owed the company money. Without this single fragile disc, the company would never have continued trading.

'At 7pm on the Friday night, I was pretty tired,' said June. 'I wasn't going to bother with the back-up. But I told myself not to break the habit of a lifetime, and it was lucky I didn't. Many firms take back-ups, but they leave them in the same premises. You have to take the back-up with you.'

Many companies which suffer fire never trade again. Despite June's precautions, Riverside Office Supplies found it difficult to get started again. On the following Monday, June began to trade again from her living room, but sales fell dramatically. Customers believed that the company had stopped trading, because its premises had gone.

The insurance took ten months to pay out, largely due to hold-ups with the neighbouring firm's insurance company.

The case history demonstrates the importance of backing-up computer information, and keeping it off the premises, as a precaution against fire.

Minimizing the Effects of Fire

You need to decide what is necessary to keep the business operational. This is often called the Business Recovery Plan and should regularly be reviewed. We cover this topic in Chapter 23. It may be computer data, certain equipment or specific raw materials. You should then take steps to safeguard them. It is sensible to have remote backup copies of computer data.

Certain products may need to be stored in more than one warehouse. Some equipment might have to be protected by sprinklers.

The company should also check whether it could sub-contract work to a competitor, or another part of the group, while the business gets operational again.

Liaising with the fire service is always beneficial. Plans showing the location of emergency exits, sprinkler system control valves and gas lines should be provided for the fire service, and be kept away from the main building.

After a Fire

Around 60 per cent of every insurance premium goes to cover the cost of rectifying water damage. Thousands of gallons of firefighting water are used to bring a commercial fire under control, leaving the insurers and the occupiers with a big problem.

The sooner that dryers can be brought into a building, the more likely it is that capital equipment on site can be saved and the fabric of the building preserved. Storing goods away from the floor also helps to minimize water damage. In Chapter 24 we look at the steps to take after a crisis like a fire.

Risk Assessment

You can assess your vulnerability to fire by answering the questions below.

Topic	Question	✓
Type of work	Does the company use combustible materials? Does the company carry out flammable work?	
Track record	Has the company suffered from arson in the last two years?	
Fire alarms	Does the company lack an AFD and alarm system?	
Sprinklers	Does the company lack sprinklers?	
Escape routes	Are escape routes combustible?	
Age of occupants	Are occupants aged under five or over 70?	
Experience	Have some occupants not participated in a recent fire drill?	
Familiarity	Are occupants unfamiliar with the building (for example, hotel guests?	
Alertness of occupants	Are occupants ever asleep or not alert, for example, watching television?*	
Total points scored:		

* We aren't talking about dozy employees here!

Score: 0–3 points: low risk. 4–6 points: moderate risk. 7–10 points: high risk.

The Appendix has a summary of the checklists. By entering the results of this one, you can compare the risk of fire against other categories of risk.

11
On Your Guard: How to Maintain Security

'They aren't worth fixing,' says the repair man. 'They're all smashed up. You can see where the thieves ripped out the tubing, the condenser coils, all that copper.'

The air conditioners are looking a mess, and the cost of replacement will run into tens of thousands.

Geoff, the head of security, says the thieves must have got up on to the flat roof via the fire escape. 'The police probably won't even ask the scrap merchants; the components are impossible to trace,' he says. 'But in future, we could try painting their replacements, illuminate them at night, and put them in a locked cage,' he says.

'Which?' says his assistant, 'The units or the thieves?'

What are Security Risks, and What Damage Can They Cause?

The company's buildings, stock, staff and assets are at risk from thieves, vandals, arsonists or even extortionists. Some organizations can easily achieve tight security. But others, such as hospitals, have to be accessible to the public, and are often regarded as 'free supermarkets' by thieves who routinely walk away with handbags and computers.

For all kinds of companies, security is a complex issue. The assets to be protected now include computer data and intellectual property (IP).

For companies with a high profile or overseas operations, security can be a life and death matter, with kidnap and ransom being a real possibility.

Conducting a Security Review

The security process starts with a security review (as in Table 11.1). This entails checking the current security arrangements and assessing the areas of vulnerability.

Table 11.1   Security review

In a security review, include all relevant information under the following headings:

- past incidents; history of break-ins and losses;
- building location, type of perimeter, type of premises, attractiveness to criminals, and likely method of theft;
- external and internal security;
- assets which could be stolen or damaged;
- control of access for employees, visitors and unauthorized people; parking;
- guards, radio contact and internal communications;
- alarms, closed-circuit TV;
- fire and emergency planning;
- management systems and procedures;
- recruitment processes;
- business partners: suppliers, agents and customers;
- company structure and personnel.

Devising a Security Strategy

Once you have completed the security review, you can devise a security strategy. Every set of premises is different, so each property needs an individual strategy. The strategy should operate at several levels, as shown in Figure 11.1.

1. The outermost level is the perimeter, together with the entrances in it.

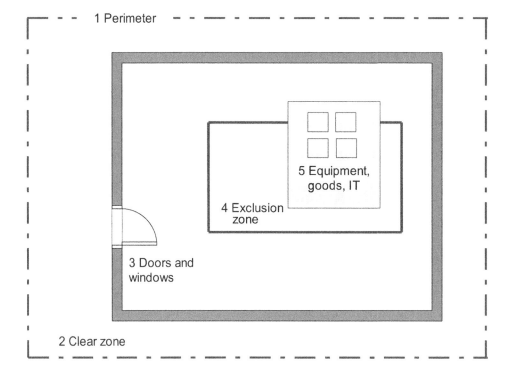

Figure 11.1   The five security zones

2. Within that is the clear zone, the area between fence and buildings.
3. The third area is the windows and doors of the building.
4. Once inside the building, you may want high-security exclusion zones.
5. Within these zones, you may need to secure individual items of equipment, material or information.

MATCHING SECURITY TO THE THREAT

The extent of security must reflect the degree of risk. A company which stocks large amounts of desirable merchandise or has a commercially sensitive R&D laboratory needs tighter security than an office which simply processes information. But even schools and clinics have equipment and computers which are attractive to thieves; and the loss of data could cause either embarrassment or a financial disaster.

The techniques for protecting your premises and goods involve an ascending order of action, each representing an additional barrier to be overcome. As Figure 11.2 shows, they are as follows:

1. Dissuade people from illegally entering the premises with notices and the presence of cameras. Just seeing a camera is sufficient to deter many criminals, so the cameras should be prominent.
2. Observe people with cameras and security staff. Regularly check that the video recorder is working. After a break-in or theft it is not uncommon for management to find that the device was not actually recording.
3. Make it physically difficult to enter, by means of fences and locks. This element of the strategy can be applied all the way from the bolt on the perimeter fence to the keypad on a server room.
4. Announce illegal entry by alarms or lights.
5. Make it difficult to remove items by physically securing them.

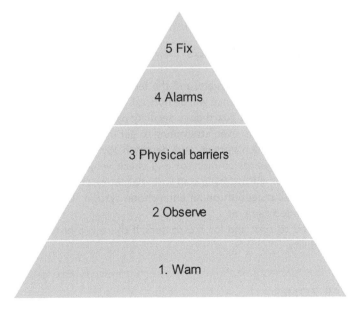

Figure 11.2   The five elements of a security strategy

Not all five elements are always needed. Each simply represents an additional obstacle for the criminal.

SECURING THE PERIMETER

The first security zone is the business perimeter. The premises may be surrounded by lawn, vegetation or trees, or they may be restricted to the building itself.

You have to balance security risks against the needs of employees and suppliers to enter the building. People need to get in and out of your premises without undue difficulty, and without employees feeling that they are under undue observation.

Where necessary, you should protect the perimeter by a fence. This should be high enough to prevent casual burglars from getting over it. A chain link fence is better than a masonry wall, which gives cover to intruders.

You gain extra security by adding razor wire or barbed wire, angled outwards at the top. A masonry wall can be embedded with broken glass. However, these additions can make the organization look like a prison. Thorny bushes dissuade opportunists.

One organization in special need of security is MI6, the British secret service. It moved to new buildings in the centre of London, where the arrival and departure of staff and visitors can now be observed by passers-by. Its building is also overlooked by other office blocks.

STOPPING ARSON

Arson is often committed by bored or malicious young people under the age of 18. Emotional firesetting can be caused by people suffering from mental illness, or can result from arguments with neighbours or organizations, while criminal firesetting can be caused by people engaged in criminal activity such as fraudulent insurance claims.

You should assess the area for arson opportunities. Lock sheds and rubbish containers, and avoid leaving combustible items such as pallets in the open, Do not allow rubbish to build up.

Consider whether parts of your buildings are likely to suffer from arson, such as wooden barns, and whether fire could spread to your building. Your building should be 10m from the nearest source of fire.

MANAGING THE PERIMETER OPENINGS

There should be as few openings as possible, and the fence should be lit at night. It can also be alarmed so that an attempt to cut the fence will be noticed. 'Keep out' signs should be posted, with the possible additional warning about guard dogs, security patrols or video surveillance. Your aim should be to discourage people from even attempting to get inside the property.

Factories or warehouses which receive and despatch lorries loaded with valuable goods require barriers with security guards to monitor and check vehicles. Service companies have a lower level of security need: some firms issue staff with magnetic cards which will raise the barrier. Visitors and suppliers can call reception on an entryphone system. Reception should be able to see the vehicle, if only by video camera.

Staff should be trained to spot and log trespassers. If there is a history of trespass on your site, it indicates vulnerability.

CREATING CLEAR ZONES

A clear zone should be kept all around the fence. Low vegetation and a limited number of sparse trees can be planted to prevent a desert-like appearance. Tall hedges provide places for criminals to hide.

You should mount closed-circuit cameras in a way that allows the whole area to be seen. Attention should be paid to fuel storage tanks, skips or waste paper bins which could be of use to a criminal.

SECURITY OF THE BUILDING

There should be as few openings as possible. Even department store retailers like to make customers enter and exit through a restricted number of doors.

The success of some firms, such as retailers, depends on having many visitors during the day. In such cases, security will be moved to the building itself. The same applies to a company that is based in a business park or in a terrace of buildings.

The vulnerable areas are doors and windows. Doors should be illuminated at night, and windows should be secured. Burglars can easily open old sash and casement windows unless they have been reinforced. You can block sliding doors and windows with a wooden pole in the bottom track, while window frames should be lockable to the window frame. Without such precautions, a burglar can easily jemmy open most windows and doors.

Doors should be solidly made, and fitted with security locks and hinges. Only one door should be openable from outside. The others should be replaced with fire doors openable only from inside. This will give the intruder only one method of entry. Each fire door should be alarmed, so as to sound a warning when it is opened.

INSIDE THE BUILDING

Many organization have coded door locks. These are effective against intruders, but they need to be regularly changed. One Newcastle city centre hotel has codes for each floor, and these remain the same whenever the author stays there.

At another company, it is obvious from the dirt marks around the door lock that only two numbers are being used: 9 and 4. It would not take a criminal long to discover that the code is 9494. A final problem with door codes is that a criminal can watch staff hand movements to ascertain the number.

Similarly, photocopiers at one management consultancy can now only be operated by entering a departmental password. Consultants now do their illegal and personal copying using other departments' passwords. Since the amount of illegal copying is only a small percentage of the legitimate work, the problems created can outweigh the benefits. Targeting heavy users of illegal copying would be more effective.

Sometimes security can rebound to the company's disadvantage. Dunkin Donuts was embarrassed to discover that a member of staff had spied on diners, using its video and audio security system, and had then gossiped about what he had learnt.

SECURING THE INTERIOR

The interior of the premises can be secured by circuit-breaker alarms on doors and windows.

Other kinds of protection include pressure pads, ultrasonic movement detectors and photoelectric beams which sound an alarm when broken. These devices can be wired to sound an alarm or make a telephone call.

They are, however, known to fail or sound a false alarm. Care must be taken to choose a system that will not be set off by passing traffic. Low-cost motion-detection cameras can be linked to the computer network, and can transmit an image of the scene to a distant location, whether reception or (during out of office hours) to someone's home computer.

EXCLUSION ZONES

Departments categorized as security zones will include:

- research and development offices;
- cashier's office;
- finance offices;
- mainframe computer, servers.

The boardroom and directors' offices should be swept for bugs if commercially sensitive information is likely to be discussed.

Ideally, the room should not be overlooked by other buildings where long-distance microphones could be used, nor where information shown on flipcharts could be seen through a telephoto lens. Shredders will destroy sensitive information.

SECURITY OFFICERS

People should not be freely able to enter the building without being questioned and without providing identification.

Large premises will need to be patrolled by a security officer at night, at predetermined frequencies but at irregular intervals, possibly accompanied by a dog. Dogs act as a strong deterrent to criminals.

Security staff are not always well trained. At the BBC in London, members of a pressure group arrived at the studios and asked the way to the newsroom. The security guard courteously directed them to the right place, where they briefly got on air and made their protests on TV.

Security patrols have been found to be effective in reducing crime on large estates; in business premises they may similarly deter criminals, and spot security problems. However, mobile guards also add more cost.

END OF DAY, AND NIGHT TIME

The company needs an end-of-day routine, with one named individual responsible for securing the building. This person should check that:

- all doors and windows are secure;
- no combustible material is left lying around, and flammable liquids are locked away;
- no unauthorized people are; on the premises;
- alarms are on
- outside lights are on.

Night is a vulnerable time for many companies. The periods when cleaners are working are a time when a thief could enter. At one company, a deranged person wielding a knife got into the building late at night, when few staff were around. Security failed to answer the telephone, so the staff had to disarm the man themselves.

Natural Threats to Buildings

Fire is the greatest natural threat to a building, and is examined in detail in Chapter 10. *Water* is an often under-estimated risk. A dripping tap or a blocked drain can flood a basement, and if it happens over a long weekend, it could be four days before the problem is discovered. Mainframe computers, lift equipment and power supplies are often located in basements, and are therefore at risk from water, whether from flooding or from a burst pipe.

Public Events

Conferences and exhibitions are sometimes targets for attack by pressure groups or terrorists. This is either because prominent VIPs or the media will be present, because disrupting the event can cause havoc in the city, or because disruption will harm the government's standing.

It is important to create an exclusion zone before the event starts, with the area being swept for bombs or incendiary devices. Thereafter, everyone entering the area should be registered and badged.

Security should be particularly stringent before the event when stands are being erected or dressed by technicians or sales people. During the event, security guards should constantly check for suspicious packages. Arrangements must also be made to protect VIPs. Protesters should be treated courteously and should be escorted off the premises as quickly as possible but without undue force.

Retail Security

Average shrinkage or 'shrink' (stock loss from crime or wastage) suffered by stores throughout the world is around 1.3 per cent of turnover. While this may sound small, it equates to $171 per household. Figure 11.3 shows the causes of shrinkage, as perceived by retailers, while the most popular items are shown in Table 11.2. Thieves favour well-known brands, especially those which have been launched in the last two years.

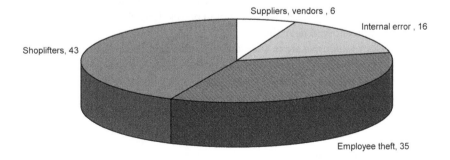

Figure 11.3   Percentage losses caused by shrinkage

Source: Global Retail Theft Barometer/European Retail Theft Barometer

Table 11.2   Items most likely to be stolen

• CDs	• Game consoles, video games
• Cellular phone cards	• Leather belts
• Cheese	• Leather jackets and other leather garments
• Children's clothes	• Leather wallets
• Chocolate	• Lipsticks
• Contraceptives	• Locks and security devices
• Cosmetics	• Mobile/cellular phones
• Costume jewellery and earrings	• Perfume, fragrances
• Designer accessories	• Ready-made curtains
• Designer handbags	• Shaving products
• Designer brands clothing	• Skincare
• Duracell batteries	• Spare parts for electrical items and power drill bits
• DVDs	• Spirits, mainly whisky, vodka, Bacardi and so on
• Electrical power tools (well-known makes such as Bosch)	• Sportswear
	• Sunglasses
• Electrical skin care and toothbrushes, for example Braun and Gillette	• Tea and coffee
	• Trainers, sports footwear
	• Videos
• Female lingerie	• Vitamins
• Fresh meat (supermarkets)	• Wrist watches

Source: Global Retail Theft Barometer/European Retail Theft Barometer

STAFF PILFERING

Staff can take goods and cash either themselves or by collusion with suppliers or customers. A Jack L. Hayes consultancy study showed as many as one out of every 28 US employees were apprehended for theft. Supermarkets are the worst affected, losing around 2–3 per cent of turnover. The two main problem areas are the cash tills and the stock room.

There are many methods of stealing at the tills: staff can fail to ring up purchases, or they can short change customers and pocket the difference (this affects not the company's profit but customer attitudes). In the stock room, goods are piled high, and the room is often empty, making a tempting target for the thief.

Electronic point of sale (EPOS) equipment reduces the prevalence of theft, as does the use of mystery shoppers to check staff honesty. Tills should be regularly changed and checked. Staff coats and bags should not be allowed near the tills.

In the stock room, expensive stock should be locked, while cheaper stock should be delivered shrink-wrapped. Procedures for deliveries should be properly managed, and any goods entrance should be supervised or locked if possible, to prevent goods being removed.

Staff collusion also allows individuals to walk out with stock, or to hide goods in rubbish bags, while collusion with friends and family allows them to make fake refunds or exchanges.

SHOPLIFTING BY THE PUBLIC

Making goods easily accessible encourages the shopper to buy them, but it also makes it easier for them to be stolen. Asymmetrical store layouts lead the consumer around the store but reduce the sightlines and make shoplifting easier.

Electronic Article Surveillance (EAS) tags prevent high-value items from being stolen, while closed-circuit television can deter thieves, and store detectives can identify culprits.

Loop alarms can protect small electrical goods, while some clothes stores operate a tag system to prevent shoppers from taking several garments into a changing room.

Easily pocketable items such as electric shavers may be displayed under glass counters, while small items like batteries can be blister-packed.

PROTECTING THE STORE AT NIGHT

The shop should remain lit at night to discourage theft, and cash should not be kept overnight on the premises. Shop windows should be made of resistant glass, capable of withstanding attack for a specific period. Security grills are useful for higher-value items, and grill alarms can be set. Even toughened glass is vulnerable to 'ram raiders' who attack a window using a stolen car. Bollards or a low metal bar extending the length of the window may be the only way to deter this kind of problem.

THE HOLD-UP

Most robberies are over very quickly: robbers rightly fear being apprehended if they stay too long. For his reason, they rarely get to the safe, preferring instead to grab what cash they can from the nearest till. The store (or bank) should therefore keep only as much cash in each as is needed for the business. Staff should be dissuaded from heroic action: safety is more important than the petty cash.

Where the business takes cash to or from a bank, the timing and route should vary, and two people should make the journey together. The money bag should be unobtrusive.

Transit Security

Stealing goods in transit has always been around, from highwaymen to pirates. Nowadays it is big business, and valued at $130 billion in the US alone. This is because lorries represent an isolated, low-risk, high-reward target, especially compared to retail theft where the gains are limited and security is high.

According to Chubb, attractive goods include computer equipment, high-tech components such as memory chips, consumer electronics, designer clothing and accessories, wine and spirits, cigarettes, cosmetics and perfume and pharmaceuticals.

Desirable goods share the same characteristics:

* high value-to-size ratio (they are small and high value);
* difficult to identify as stolen;
* easy to transport;
* easy to sell because they have a high 'street' value.

To reduce theft, retailers and shippers can use GPS devices and monitoring, locks and seals, training and information sharing.

REDUCING TRANSIT THEFT

Chubb (www.chubb.com) recommends several strategies to reduce transit theft. They include your choice of carrier, packaging and staffing.

Your carrier

- Select a carrier, including all intermediaries, that is reputable; experienced in handling and transporting high-value, theft-attractive goods; uses suitable equipment; follows good security practices and procedures; and maintains an excellent loss record.
- The carrier should verify all data on pre-employment applications and should perform detailed pre-employment screening.
- Avoid carriers that sub-contract the transport of your goods without receiving your prior approval.
- Establish Standards of Care for the transport of your goods and incorporate these into your contract. This might include:
  - direct routing from origin to destination; eliminating unauthorized stoppages;
  - securing vehicles, trailers or containers with high-quality, tamper-evident security seals that are ISO 17712 compliant;
  - real-time tracking of your shipments.
- Require documented accountability (sign-off of count and condition) every time your shipment changes hands during transit.
- If practical, prohibit consolidating your goods with that of other shippers (opt for exclusive-use cargo conveyances).
- Monitor the carrier's performance throughout the contract period.

Packaging

- Ban the prominent display of company names, corporate trademarks or logos or other marks that would identify the contents.
- Consolidate multiple carton shipments to take advantage of economy and added security.
- Introduce tamper-evident features including unique carton tape, banding straps and/or security seals.
- Use opaque plastic stretch-wrap to secure and obscure cartons on pallets or skids.

Transit

- Make your shipment documentation as generic as permissible by law. Use general terms or coded information rather than specific identification of shipper and consignee or description of your goods.
- Plan your shipment departure so that it will arrive at the destination during normal working hours, unless you or your customer has established after-hours receiving procedures.
- Require that your goods are placed in a secure area, one that is under constant personal or electronic surveillance, whenever they are at rest and in the care, custody and control of others.
- Ship your goods door-to-door to minimize, if not eliminate, en route transfers or trans-shipments.

Staffing

- Use security escorts or covert/embedded tracking technology for your high-value shipments transiting to or through known cargo theft 'hot spots'.

- Limit access to sensitive shipment data (values, customer, destination and carrier) and documentation to those within your company who 'need-to-know'.

International and airfreight

- Book your international air shipments on a specific flight. To reduce the time your goods are at rest, they should arrive at the airport as close as possible to the deadline set by the carrier to receive goods.
- Arrange for quick clearance of your shipments from Customs.

Vandalism

Vandalism is usually associated with run-down inner-city locations, and carried out by bored youths. While this is true of vandalism to bus shelters and telephone boxes, many city centre buildings need to be secured against this problem.

Good community relations can also play a part. When race riots erupted in London's inner city area of Brixton, the Marks and Spencer store was unscathed. It is thought that the company's investment in inner city renewal was responsible for its escape.

Vandalism is also prevalent among companies whose property cannot be observed all the time, such as bus and train companies. Vandalism usually comes from a small number of areas, for example where poor housing estates back on to railway tracks. Targeting these areas with observation, video cameras, and good lighting can reduce the problem.

Sometimes vandalism is carried out by staff. A sailor poured sugar into the fuel tanks of his destroyer in order to prevent the boat from leaving port. He had formed an emotional attachment to a woman, and was reluctant to leave her. But more usually, the member of staff has a grudge against the company. Ensuring that door entry codes are regularly changed, that individuals surrender their security pass on leaving the organization's employment and that dismissals are well managed, will minimize the problem.

Public Spaces and Locations That Can't Be Secured

Organizations such as water, forestry and rail companies have acres or linear miles of property that can't be secured. People can break in or trespass at will.

Each industry has its own major problems. With forestry the big risk is fire, while on the railways it's theft of copper cable. In the UK, cable theft costs £16 million annually, leads to the delay or cancellation of 35,000 national rail services and the deaths of ten people.

Similarly, local authorities have seen mounting thefts of iron manhole covers, while churches have lost lead from their roofs,

Theft of metals results from the rising price of raw materials, and the ability of petty criminals to sell their stolen goods to unscrupulous scrap metal merchants. New cable has been stolen from railway depots, indicating that gangs get information from employees.

Remote or unmanned spaces can be protected by warning notices, CCTV, alarms and mobile patrols. Rewards for information can also be productive.

Espionage

Espionage can involve the theft or sale of information, whether records or formulae. Typically, this involves confidential research data being sold to a competitor. It is cheaper for an unscrupulous company to buy competitors' secrets for a few thousand pounds than spend millions in R&D. Industries such as computing, telecommunications and aerospace are particularly vulnerable.

Tender information or *product plans* are also worth money. Detergent companies routinely try to spoil a competitor's launch by rushing out a similar product a few weeks earlier. The success or failure of a launch can have a huge impact on market share and profits.

Active espionage, where one company bugs another's boardroom, is rare. Usually, the perpetrator is a disgruntled employee, and the information is usually freely available to him. Sometimes, the employee has a friend, relative or spouse who works for a competitor. Many precautions can be taken, and these are included in the following section.

According to a survey by International Data Corp (IDC, uk.idc.com/):

• 58 per cent of industrial espionage is perpetrated by current or former employees;
• 48 per cent of large companies blame their worst security breaches on employees;
• 60 per cent of security breaches occur within the company, behind the firewall

SECURITY OF INFORMATION

It is easy for an employee to take confidential information, whether by photocopying customer records, taking drafts of tender documents, or by copying the contents of a computer on to a flash drive. Using a smartphone they can also photograph a screen, leaving no trace. Often, the company never knows that the theft has occurred. The company should make an assessment of:

• What information is sensitive?
• How could it be used or taken?
• How can the theft be prevented?

Paperwork security: Important documents should be shredded, not binned. All departments should have shredders, since even apparently innocuous documents can be risky or have economic value. Sensitive data should be locked away at the end of the day, not left lying on desks. Staff should have lockable filing cabinets if they need them, especially in the HR department. Plans should be restricted to those who need them.

Computer security: In Chapter 15, we look at the problems of computer theft, viruses, unauthorized access and other problems leading to a loss of data. See also 'Natural threats to buildings', earlier in this chapter.

Staff relations: Companies should seek to ensure that staff do not have cause for resentment, thereby pre-empting the thought of espionage or sabotage. This includes treating staff fairly and having transparent and honest procedures for discipline and promotion. Companies should try to resolve complaints and personal problems, so that staff do not grow disaffected.

CASE STUDY: HOW EASY IT IS TO STEAL SOFTWARE

Sinovel, one of China's biggest wind turbine businesses, was charged with stealing software worth $800 million from AMSC, its US software supplier.

The US Department of Justice alleged that once AMSC had developed the software that controlled the turbines, Sinovel stole the source code with the help of an AMSC employee based in Austria.

Sinovel then stopped accepting product shipments from AMSC, which had to lay off 500 workers as revenues fell, and its stocks plunged 93 per cent.

Dejan Karabasevic, the Austrian employee, admitted the fraud and was imprisoned for one year in his native country. In exchange for the software, he had received a $1.7 million contract, an apartment in Beijing and equipment.

If convicted, Sinovel could face fines of up to $1.6 billion (£1 billion) for each count, while Karabasevic and Sinovel's R&D deputy director could each be jailed for 35 years.

This case study illustrates the security risks implicit in employees who liaise with customers, the vulnerability and huge value of software, and the cost savings available to a customer if they don't have to pay for it.

INFORMATION LOSS

If information has been taken, you should assess:

- What information has been taken?
- How was it taken?
- What can be done to stop a recurrence?
- What legal or disciplinary action should be taken?

Before taking action, it is worth checking that the information has really been stolen, rather than misfiled or simply buried in an in-tray.

TELEPHONES

Telephone conversations can be transmitted to a listening post. Some computerized telephones contain diagnostic tools which can check for this.

Cellular telephones are particularly vulnerable, as proved by cases involving transcripts of the UK royal family's private telephone conversations. Some transmit signals which can easily be picked up, though security is improving over time.

Sometimes employees are simply nosy. In some cases, anyone who wanted to eavesdrop on Board meetings could set a boardroom phone to auto answer, and mute the ring. Then they can dial it, and listen in to the conversation.

HUMANS AT WORK

Staff are prone to make comments to the press or to social networking sites, or on their blogs. They are also known to sell information to third parties.

To combat the casual release of information, the organization needs to systems for staying abreast of comments made on the internet (for example, through Google Alerts).

It is harder to discover that competitors whether have gained trade secrets. If suspicions are aroused, the business may need to hire a private investigator.

Staff can ignore or break the most stringent security controls if there is a conflict with workplace needs. A typical example is the emergency door which gets propped open, despite the warning notices. This may be because the building gets hot and airless, and staff enjoy the breeze; or it may be a short-cut from the building to the car park. This can be solved by having the door alarmed; and by improving the air conditioning.

RECRUITMENT

For companies who have secrets to protect, the recruitment process needs to be thorough. References should always be followed up, preferably by telephone (which, being a more informal and instant method of communication, will be more revealing).

Checks may be made on the applicant's current address, former employment, academic background, credit rating (to identify the larcenous or dishonest recruit), and personal references. All information should comply with current legislation.

CASE STUDY: THE SUPPORT WORKER WHO PRETENDED TO BE A BARRISTER

A UK, West Midlands firm of lawyers hired Will as a support worker to liaise between defendants and their lawyers in the mainly criminal practice. The HR department conducted what it thought were reasonable checks, and his CV seemed in order.

Will worked diligently for over a year. But then it was discovered that he had been making false expense claims to the Legal Aid Board. The firm suspended him while they investigated the matter, and Will promptly disappeared. As the legal practice was to discover, Will wasn't who he said he was.

The HR department started ringing everyone Will had cited in his CV, only to discover that no one had ever heard of him. A few months later, the law firm got a call from the Department of Social Security, saying that the National Insurance number Will had provided belonged to someone else.

Will had several female admirers; and he had told them he was a barrister, and kept a box containing a barrister's wig in the back of his car to 'prove' it.

This episode caused some loss of confidence among the practice's clients, not least the Legal Aid Board. As a result of the mishap, the company tightened its vetting procedures for new employees.

SAFEGUARDING INTELLECTUAL PROPERTY (IP)

According to a survey by Symantec (symantic.com), half of employees admit to taking corporate data when they leave a job, and 40 per cent say they plan to use the data in their new job. This means organizations' knowledge and intelligence is getting into the hands of competitors.

A high number of employees don't think that taking corporate data is wrong. 62 per cent of employees say it's acceptable to transfer corporate data to their personal computers, tablets, smartphones and cloud file-sharing apps. And once the data is there, it stays there – most employees never delete it.

Employees don't think twice about taking corporate data because they don't see the harm. Fifty-six per cent don't think it's a crime to use trade secrets taken from a previous employer.

This shows that people don't understand who owns the IP rights. Employees attribute ownership of IP to the person who created it.

Symantec recommends that companies protect themselves by the following means:

- *Educate employees*: Let your employees know that taking confidential information is wrong.
- *Enforce NDAs*: Include stronger, more specific language in employment agreements. Ensure exit interviews include conversations focused around employees' continued responsibility to protect confidential information and return all company information and property (wherever it is stored). Make sure employees know that policy violations will be enforced and that theft of company information will have negative consequences to them and their future employer.
- *Use monitoring technology*: Install data loss prevention software that monitors inappropriate access and use of IP, and automatically notifies managers when sensitive information is accessed, sent or copied.

You should also:

- train people to understand what belongs to the employee and what belongs to the company;
- create an environment that promotes employees' responsibility and accountability in safeguarding business information;
- educate employees that using a former employer's confidential data puts you and them at risk.

Inventions, designs, trademarks and brand names are worth money, especially if a competitor copies them, or if they can be licensed to other companies. Therefore it is essential to establish the company's rights to all IP.

The company secretary or a firm of patent agents should be used to secure these rights. This work needs to be continuous, because the infringement of corporate designs is a continuous threat. We discuss this further, along with identify theft, in Chapter 22.

Risk Assessment

By answering the questions below, you can see how vulnerable your organization is to breaches of security.

Topic	Question	✓
Buildings	Has the organization got manufacturing or warehousing premises?	
	Are your premises easily accessible to the public or visited by many people?	
Information	Could your paperwork or computer data have commercial value to a competitor?	
Espionage	Does the business operate in markets subject to fashion or technological advance?	
IP	Does the organization have inventions, trademarks or well-known brand names?	
Attacks on premises	Does it employ large numbers of people?	
Transport	Do you transport high-value, easy-to-sell goods?	
Review	Has the company failed to carry out a security review?	
Retail	Does the organization have retail outlets?	
Open spaces	Is the business responsible for open spaces or areas that can't be protected?	
Total points scored:		

Score: 0–3 points: low risk. 4–6 points: moderate risk. 7–10 points: high risk.

The Appendix has a summary of the checklists. By entering the results of this one, you can compare your security risk against other categories of risk.

12
Extremists, Terrorism and International Risks

You've been talking to Kehinde Manko, your guy in Nigeria, following his release. Maybe he shouldn't have strayed north, but it happened. He was taken from his car on the Kano by-pass. The kidnappers wanted 15 million Nigerian Naira (£60,000) for his release, a sum your insurers said they would meet.

'My return was miraculous,' he said 'The kidnappers seemed to be college students. They said they didn't want to do it, but they had no jobs. One said he was an engineering graduate, and another said he was a marketing management graduate.'

Helped by a specialist firm of kidnap and ransom experts, plus their local 'black book' contacts, Kehinde was freed after only two days. Kehinde was lucky: it went well. From what the kidnap and ransom people say, other victims have died.

The Risk of Local Violence

In this section, we examine the threats posed by local, as opposed to international violence.

For most businesses, the threat of local violence is thankfully rare. But due to the nature of their work, some organizations are more at risk. These include organizations that:

- deal with the public at large, such as bus drivers, railway staff and healthcare workers;
- open at night, such as employees at bars, off licences, taxi drivers and late night convenience stores;
- deal with the public in conflict situations, such as social workers and staff at social security offices. Teachers, too, get assaulted. In the UK, one teacher a week is taken to hospital;
- are at risk from robbery, such as security van drivers, and staff at banks and post offices;
- are involved with controversial markets, such as armaments or animal experiments;
- work in unstable countries. This applies typically to oil and gas businesses, but any supplier to that industry, along with construction companies, engineering firms and exporters of all sorts are at risk.

Businesses not on this list should be wary of relaxing. People are no longer deferential, and the mob can be stirred up rapidly and unexpectedly. In the US, it is not unknown for employees to go on the rampage, though it is hard to predict 'Black Swan' events.

Angry, criminal or inebriated members of the public represent a risk to staff. To prevent assaults on staff, or reduce their impact, the business needs to carry out a risk assessment, followed by procedures for dealing with an incident.

UNDERTAKE A VIOLENCE RISK ASSESSMENT

The organization should conduct a violence risk assessment, and implement its findings. How likely is the risk of assault? And what steps need to be taken to control it? This might include:

* introduce staff training;
* amend working practices;
* issue personal alarms;
* change the structure of the building, for example put up glass partitions or install panic buttons, something that is suitable in only high-risk situations;
* develop a policy on lone working.

DEALING WITH THE INCIDENT

Staff should not attempt to argue with customers, which would only make the hostile person more angry. You should train them to try and diffuse the problem, by talking it through with the client.

If the problem is getting out of hand, the member of staff should call security and leave them to deal with the problem.

If there is injury, or the exchange of body fluids, extra caution should be exercised, and the staff member should take medical advice.

AFTER THE VIOLENT EVENT

After an event, there should be a review, and action should be taken on the findings. You should notify the police after any violent incident.

The member of staff should be offered counselling. However, early counselling is thought to harden the experience, so it might be beneficial to have some delay between the event and the counselling.

Risks to Employees from Protesters

Employees are sometimes subjected to attacks by pressure groups. One by one, all the UK's ferry operators, including Stena Line and P&O Ferries, stopped carrying research animals to and from Europe. This followed a campaign by the National Anti Vivisection Alliance (antivivisection. info). http://news.sciencemag.org/scienceinsider/2012/03/editor-content.html?cs=UTF-8Directors of the ferry companies received letters at their home addresses, and the protesters sent emails from around the world to company employees.

In previous campaigns, activists had attacked the homes and vehicles of senior managers at firms involved in importing farm animals. Campaigners mailed a letter bomb to a senior executive of Stena Sealink (the predecessor of Stena Lines), which exploded and injured a secretary.

According to Science Insider, Chris Laming of P&O Ferries said, 'Because we have experience of this kind of protest we understand the implications, and have no wish to antagonize people like that,' he says, adding 'The business wasn't worth that much.'

With the growth of social networking sites such as Facebook and LinkedIn and the ability to organize flash mobs through email and texting, employees and directors will increasingly come under attack in future.

As Figure 12.1 shows, activists have spread their attacks to include suppliers, individuals, business partners and investors.

This model of disruption and intimidation can and will be applied to other organizations that offend the activists, including biotech, the extractive industries, energy companies and chemical businesses.

Even theatre companies have come under pressure. The Birmingham Repertory Theatre cancelled 'Behzti', a black comedy depicting murder and sex abuse at a fictional Sikh temple, after a mini riot by Sikh protestors. The building was stormed by Sikhs who damaged doors, set off fire alarms and attacked backstage equipment. They said it demeaned their religion. The theatre said it had to close the play because it couldn't guarantee audience safety.

This means companies with a potentially controversial product or service need to have plans in place to protect the business and its employees. The organization should be ready to:

- increase physical security. Ensure that your buildings are secure, as suggested in Chapter 11;
- support staff arriving and leaving;
- protect the anonymity of employees;
- educate employees about the risks associated with having a presence on social networking sites;
- compile evidence of illegal behaviour by activists;
- liaise closely with the police and emergency services;
- enforce your legal rights, for example against trespassers, including High Court injunctions.

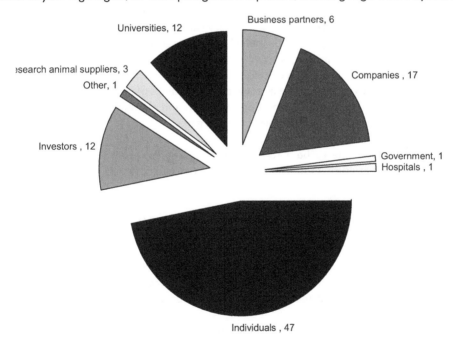

Figure 12.1   Targets of illegal actions by animal rights extremists, USA

Source: Foundation for Biomedical Research

INTELLIGENCE GATHERING

Faced with the threat of violence or direct action, companies sometimes decide to seek intelligence on protest groups.

Scottish Power was revealed as a client of Vericola, a private security firm exposed by *The Guardian* for infiltrating green protest groups.

The Inkerman Group (inkerman.com) monitors protestors as well as surveillance and hostage negotiation services. It is thought to include multinational corporations, companies and governments among its clients. Other similar businesses include Diligence (diligence.com).

And on a lesser scale the University of London was criticized for using security staff employed by Balfour Beatty to forcibly remove students who occupied the university's Senate House, and hand them over to police.

If a company is found to have sought intelligence on a group, perhaps by infiltrating it or 'dumpster diving', the news is usually deemed scandalous. Companies usually seek to distance themselves from being directly associated with these acts, but this activity can rebound to the company's detriment.

Animal Rights

There are three types of people concerned with animal welfare and animal testing.

1. The first is the average consumer, for whom animal welfare is among the nice-to-have issues. Such consumers will not go to the barricades on behalf of animal rights, and mostly don't check to see whether the products they buy measure up to some relevant standard.
2. The second type is the lawful campaigners. They can make life difficult for businesses, but are essentially law abiding. They include animal welfare organizations such as the generic ones such as the RSPCA (rspca.org.uk) and the Vegetarian Society (vegsoc.org) or Viva which campaigns for farm animals (viva.org.uk).
3. The extremists, however, may take direct action, sometimes of a violent kind. They include the Animal Liberation Front (animalliberationfront.com), and vary in their focus, from laboratory animals (buav.org) to generic campaigns (peta.org.uk).

Animal cruelty will continue to be a topic for protesters. The issue occurs in many different markets, especially agriculture, food processing, cosmetics and toiletries, and drug development. The two main issues are:

1. The use of animals in research, for example in medicine or cosmetics testing.
2. The treatment of animals raised for food or for their skin

Many toiletries companies use cruelty-free products as a selling point.

Companies can pre-empt the threat of animal cruelty by *avoiding animal testing* where possible. In many markets where animal testing was thought to be essential, alternative methods have been found.

Where the use of animals is central to the product (such as chicken farming), the company needs to ensure *good husbandry*. Cage sizes and freedom of movement are often attacked as insufficient, and even a modest increase would result in improved conditions. The company should have an animal welfare policy, backed up by effective documentation and procedures.

Good communications are also essential. Some companies become the focus of protest. To prevent this, the company should consider maintaining a dialogue with local environmental groups. However, such protesters may have a fixed mindset, and may not be amenable to debate.

CASE STUDY: THE CAMPAIGNS AGAINST DAVID HALL AND HARLAN

A firm that bred guinea pigs for research was hounded for six years before police put the perpetrators behind bars. David Hall and Partners, a family business which ran the breeding programme at a farm in Newchurch, Staffordshire, closed down after intimidation by animal rights activists.

The protesters dug up the grave of Gladys Hammond, mother in law of one of the directors. Employees received death threats, bricks were thrown through windows and hundreds of malicious letters were sent. Two protesters were jailed for 12 years, and another for four years.

A similar campaign was waged against Halan (*harlan.com*), which breeds rats and mice for medical experiments. Protesters used loudhailers to shout 'Shame on you', 'Puppy killers' and 'Blood on your hands' to employees as they arrived and left, and banged on their cars. In the morning it took employees five minutes to get into the premises. Protesters also smeared employees as paedophiles, and one Harlan worker found that his neighbours had been sent notes claiming he was a rapist.

As a result of this intimidation, the number of UK businesses supplying animals has fallen from 34 in 1981 to three today, and the remaining ones have difficulty in recruiting staff.

Andy Cunningham, a Harlan manager told the BBC: 'We have to have animals if we want to develop new drugs for Alzheimer's and heart disease, and to test products used by the public.'

Extremists as the Leading Terror Threat

In the USA, the FBI regards ecological and animal rights extremists as the country's leading domestic terror threat. And it believes the pharmaceutical and biotech industry is at risk from increasingly violent activists.

Yet those industries have not handled this as a strategic issue, says Phil Celestini, FBI special agent supervising operations against animal and ecological extremists. 'There is a certain amount of denial going on,' he said. 'But until the issue is addressed, it won't go away.'

In addition to harassing and intimidating employees and directors of these businesses, activists now target businesses that have business or financial relationships with the primary target. This means even the most mundane office cleaning business is at risk.

Bomb Threats

Public transport, large offices and companies in city centres often receive bomb threats. Fewer suffer a bombing, though bombs at airports, railway stations and city centres happen.

Ferrylink postponed plans to export British livestock after receiving a bomb threat presumed to have come from animal rights activists. On a trial run from Sheerness in Kent to the Dutch port of Vlissingen, an anonymous caller rang to say that a bomb was on board. This caused the ferry to be held up by six hours outside Vlissingen harbour while Dutch port officials searched the boat.

CASE STUDY: BOMB THREATS AT THE UNIVERSITY OF PITTSBURGH

The University of Pittsburgh received 145 bomb threats over a period of two months, leading to 136 evacuations.

Several threats were written on lavatory cubicle walls in the Chevron Science Center and the Cathedral of Learning. Others came by email, and were routed through an anonymizing server.

The threats were aimed at many buildings at the University, and the nearby educational institutions including Western Pennsylvania School for the Blind.

Police initially arrested Mark Lee Krangle, a former teaching fellow at the university who had written a harassing email to Pitt professors who advised on international terrorism.

The police also interviewed a transgender couple. One member of the couple had been a student at the University until he was expelled following a dispute regarding his use of a male locker room.

The University of Pittsburgh offered a reward of up to $10,000 on for information leading to an arrest and conviction of those responsible for the threats. This was later increased to $50,000.

The University tightened security measures, requiring all people entering university buildings to present university IDs. Only students were allowed into halls of residence, and no backpacks or other packages were allowed into buildings.

Subsequently, the university received an email from 'The Threateners' demanding that it take down its reward, and offering in return not to make any more bomb threats.

On advice from the FBI, the University chose not to negotiate with terrorists, and ignored the demand. A campus newspaper received a similar email from 'The Threateners' The email writer also claimed to have no ties to those the police and University had identified.

After this email, the University of Pittsburgh withdrew the $50,000 reward. At time of writing there have been no more bomb threats.

A US federal grand jury later indicted 64-year-old Adam Busby, a Scottish separatist living in Ireland, with having emailed bomb threats targeting the University of Pittsburgh, three federal courthouses and a federal officer. Busby, who had prior convictions for bomb hoaxes, is not known to have connections to the University, though he may have read about the earlier and possibly separate hoaxes.

WHO MAKES BOMB THREATS?

People have different motivations for making bomb threats. They include the following:

- Former employees and disgruntled current employees. These people want to disrupt the business, and stop it earning money. It's a way of getting back at the company. It gives them power over the organization.
- Employees who want time off work. One Texas company received Friday afternoon bomb threats that coincided with the hunting season, possibly to allow an employee time off to go shooting.
- Activists or terrorists who are phoning to warn about a real or pretend bomb, as part of a campaign of disruption, typically in a city centre

The more detailed the threat, the more likely it is to be real. Threats which contain a recognized code are regarded as more serious than those that don't; but it can be difficult to distinguish between a hoax call and a terrorist call, particularly as terrorists use hoax calls as a weapon.

It is important to have written policy and procedures for:

- managing a telephone or written bomb threat;
- evacuation procedure;
- search procedure.

MANAGING A TELEPHONE THREAT

Staff who take calls should be trained to handle bomb threats. This will help them respond efficiently and without panic. All bomb threats must be taken seriously. The process is as follows:

1. Respond to the caller with appropriate questions.

2. Document the conversation and the exact words of the threat made by the caller.
3. Report the threat to the relevant authorities.

Table 12.1 and the advice that follows comes from advice provided by the State of Texas.

Table 12.1   Telephone bomb threat report

Telephone Bomb Threat Report Form		
Be calm and courteous. Listen and do not interrupt.		
Date		Time
Exact words of the threat		
Ask the following questions	1. When will the bomb go off?	
	2. Where is the bomb?	
	3. What kind of bomb is it?	
	4. What does it look like?	
	5. Where is the caller now?	
	6. What is the caller's name?	
	7. Is there a code?	
Characteristics of the caller and the call	Sex, race, age, accent.	
	Caller's manner (angry, calm, drunk, irrational?)	
	Any background noises?	
Notify your manager		
Phone 999	State the following: • the name of person who received the call; • location of the building • telephone number where call was received; • whether the caller's voice was recognized.	

EVACUATION PROCEDURE

The organization needs to decide whether to evacuate the building. The emergency services are unlikely to make this decision for you. The decision to evacuate must be made on the grounds of probability. It the threat is credible, you must evacuate. And you must be able to evacuate the building within the timeframe given by the caller.

Some organizations have a separate alarm sound for a bomb threats, as compared with the sound of a fire alarm. This creates a different procedure. In the case of a bomb threat, staff might need to evacuate to a further distance, and not use two-way radios in case they trigger an explosive device.

SEARCH PROCEDURE

If there is enough time to conduct a search, you may decide to ask appropriate staff to conduct a search. This process is also useful for routine checks.

The police or bomb squad won't search for a bomb since they don't know what is unusual, suspicious or out of place. Only staff who work in the area will know this.

One person familiar with each area should be asked to look for any object that is unusual, out of place or suspicious.

For common areas such as toilets or break rooms, you should allocate a search team of two or three people familiar with the area. The team should search in the following way:

* look, listen and smell before entering the room;
* divide the room into high, low and waist-level areas;
* use a grid or spiral pattern to conduct the search.

A bomber has only a limited time to plant a bomb, so search areas in priority order. This will typically be:

1. outside the building first;
2. common areas open to the public;
3. private areas last.

Once the area has been searched, the area must be secured. If a suspicious object is found, either evacuate all personnel within 100m (300ft), or evacuate the building, depending on the size of the bomb found and the risk.

WHAT DOES A BOMB LOOK LIKE?

Almost anything could be a bomb. The three questions to ask are:

1. Is it unusual?
2. Is it out of place?
3. Is it suspicious?

If the answer is 'yes' to all three questions, do not move or touch the object. Move people away from the suspicious object or package.

COLOUR CODING

Organizations liable to bomb attack, such as government departments, use a colour to denote the scale of the security alert.

This allows security to be scaled up and down to meet the perceived threat, and helps staff and visitors to take respond accordingly. A graded scale is better than requiring staff to be on a constant 'red alert'.

Hardening the Business

Any terrorist or criminal will select the easier of two similar targets. Organizations should therefore harden their premises. This can mean making it more difficult to get past reception, or adding bollards outside the building to stop a vehicle driving into it.

Since the post room is an easy target for letter bombs or anthrax attacks, experts suggest moving it off-site, so that the main building is protected.

Tampering

The company should assess who might want to tamper with its products. What benefit might they get from doing this? Companies at high risk are those:

- with a well-known brand name;
- whose products are eaten or drunk (and where contamination or poisoning of the contents would damage the company's reputation);
- in a controversial market (for example GM foods or cosmetic companies that involve animal testing).

This makes branded grocery products and supermarkets especially vulnerable.

Tampering can take place at any point along the route from production plant to retail shelves. There are three main types of people who tamper with a product:

- a current or former employee with a grudge;
- a pressure group (especially animal welfare groups, political groups or environmentalists);
- an extortionist who demands money.

The threat of tampering can be minimized by using tamper-evident packaging. With the help of the packaging industry, most industries have developed their own solutions with the help of the packaging industry, such as paper seals, vacuumed lids or transparent collars, all of which demonstrate to the customer that the package has not been opened.

However, none of this prevents tampering in the factory. Security control, vetting of employees, supervision and good management will help to prevent tampering in the workplace.

Extortion

Internet extortionists have threatened to ruin the reputation of Blue Square (www.bluesq.com) an online betting firm by saying they would send out emails containing child pornography in the company's name.

This follows previous blackmail attempts on Blue Square where the gang bombarded the company's website with thousands of bogus emails, in a denial-of-service attack. This made the site unavailable to customers for five hours. The extortionists then emailed a demand for €7,000 (£4,680) to prevent further attacks. Attacks on other betting firms, including William Hill, have sometimes been timed to coincide with an important period for the business, such as Cheltenham Week.

To combat this type of crime, the UK police's High Tech Crime Unit has liaised with their counterparts in Russia, where they have carried out raids on suspected extortionists.

Extortion demands should be immediately reported to the police, and you might decide to call in specialist consultants.

Probability versus Ability to Recover

Yossi Sheffi, head of MIT's Centre for Transportation Studies, says terrorist risk assessment can be viewed on a two-by-two matrix, with the probability of an attack on one axis and the company's ability to recover on the other (Figure 12.2).

US airlines would be on the high end of both scales. 'They are at a high probability of being attacked and if the airplane goes down due to, say, a missile, my guess is that the company goes out of business,' said Professor Sheffi, quoted in the *Financial Times*.

McDonald's has a high probability of attack, he said, because of the visibility of its brand. But because it has thousands of outlets worldwide, attacks on individual restaurants would not put the company out of business.

An own label manufacturer might have a low probability of attack, but if it has a single distribution centre, an attack on that facility might disrupt business permanently or for a long time.

At the low end of both scales would be a cleaning franchise business. No one would want to attack it, it has many outlets and they are all locally operated.

Probability of attack

		Less probable	More probable
Ease of recovery	Easy to Recover	Franchised cleaning business	McDonalds
	Difficult to Recover	Own label manufacturer	US airlines

Figure 12.2   Visibility and ability to recovery

Kidnap and Ransom

There are 15,000 kidnappings a year, mostly taking place in developing countries. The most common are shown in Table 12.2, though the countries vary from year to year.

In recent times a senior banker was kidnapped in Mexico by armed extortionists, and two British MPs were kidnapped in Somalia. Tourists have been held in northern India and a British restaurateur was snatched by Khmer Rouge guerrillas in Cambodia. Unicef paid a six-figure ransom for an employee abducted in Kabul.

Kidnap for ransom is proliferating. Countries destabilized by conflict have seen an increase, as do those where a big increase in foreign investment has resulted in a widening of wealth and poverty, and not matched by enhanced security infrastructure.

Companies can take out an insurance policy that will reimburse the costs of releasing the victim. This includes payment of the ransom, consultancy fees, medical costs, travel and accommodation. Insurance is common in oil, engineering and construction companies where executives frequently travel to high-risk locations.

Seventy per cent of all kidnappings are resolved by a ransom payment, according to Aon, and 95 per cent of victims survive their ordeal. Those who don't die due to a medical condition, shock, or a failed escape or rescue attempt.

In Europe, kidnap and ransom is relatively rare, though taking hostages prior to a robbery is more common. Normally, hostages are only held until the robbery has been carried out. But even in Europe terrorists may seize employees, particularly in areas where political terrorism exists, such as Sicily.

Table 12.2   Most kidnaps by country

1. Nigeria	6. Afghanistan
2. Mexico	7. Iraq
3. India	8. Lebanon
4. Pakistan	9. Colombia
5. Venezuela	10. Guatemala

Source: Control Risks

KIDNAP CORPORATE POLICY

Companies subject to kidnap risk need a *corporate policy*. This should cover:

- attitude towards concessions to kidnappers;
- negotiation strategy, including police and government involvement, and use of specialist consultancy;
- handling the family of the kidnap victim;
- handling the media.

Such companies also need a *contingency plan* for kidnapping. This will set out the responsibilities of a management team and a local team. It should define the procedures for immediate action (including the notifying of responsible bodies). It should identify the policies to be followed. It should enable the company to handle a future kidnap situation.

The plan should ensure that local managers know how to protect themselves when in high-risk countries.

A company which could be subject to kidnap should also carry out an *exercise*, simulating a kidnap event. This should be played over a suitable period, and will serve to familiarize corporate executives with the possible course of events.

In the event of a kidnap, the company should seek to gain as much information as possible. It should evaluate the effects of different courses of action, and the likely actions of the kidnappers. Speed and the quality of communication will be important. It will be important to have a responsible executive in the kidnap country as soon as possible after the event. Experts advise that the company should initially negotiate with the kidnappers, rather than pay out straight away. Offering to pay straightaway indicates that money is no object.

Executives can minimize the risk of kidnap by taking precautions. Gangs employ spotters to identify possible targets. People whom they identify as wealthy or important then become victims, as do their families. Individuals should avoid standing out. They should dress inexpensively and drive a modest car. Around 90 per cent of victims are kidnapped in transit, so travel patterns should be varied, and dangerous areas avoided.

Murder and Execution

In some cases, people are executed for political reasons. The Dutch film-maker, Theo van Gogh, was shot and stabbed to death in seemingly peaceful Amsterdam, before having his throat cut. His murder resulted from a short film he had made, called 'Submission', about the forced marriage of four Muslim women. Lasting only ten minutes, its broadcast caused upheaval in the country when it was broadcast, and van Gogh received death threats from extremist Muslims.

Such incidents can give cause an increase in racial or religious tension, leading to further violence. After van Gogh was murdered, several fires broke out at a new mosque in Utrecht, thought to have been started in revenge, and far-right protesters marched in Amsterdam and Rotterdam in anger at Van Gogh's killing,

Few companies are likely to anger extremists in the way that van Gogh did; but publishers and TV executives are sometimes targeted for producing material that is deemed to give offence. Sponsoring an exhibition or controversial opera could cause the same effect.

After the Danish newspaper *Jyllands-Posten* published cartoons depicting the Islamic prophet Muhammad, there were riots around the world. The violence caused over 200 deaths, there were attacks on Danish and other European countries' embassies, and violence on churches and Christians. A Middle East boycott of Danish goods halved its exports to the Middle East by 50 per cent, causing Denmark's total exports to fall by 15 per cent.

In another case, a businessman, Amarjit Chohan and his wife Nancy, from Hounslow, West London, were murdered so that two villains could take over his freight business, CIBA, importing and exporting fruit, and use it as a front for importing drugs.

But it has to be said that such attacks are exceedingly rare, and executives are much more likely to die of coronary heart disease.

International Security

Doing business abroad carries increased risks. They comprise:

1. *Technical risks*: the problems of running a business far from home, and the lack of technical know-how. Lack of supplies and infrastructure

2. *Economic and political risks*: this involves the problems of working in a climate where riots can happen, inflation undermines the economic stability, violent clashes take place, there are arbitrary changes in tax or the threat of nationalization.
3. *Security risks*: terrorists, war and lawlessness can lead to kidnap, murder, terrorist attack or extortion. We have discussed these above.
4. *Bribery and corruption risks*: in some markets bribery is common. In others, it can be difficult to decide whether bribes are actively sought by the individual awarding the contract, or simply offered by the company seeking to win the contract. Chapter 19 discusses the risks of corruption, and how to prevent it.

These risks stem from political instability, economic underdevelopment, inequality of wealth or personal greed.

Before starting overseas operations, it is important to assess the future political climate. The former Yugoslavia was once a humdrum part of Eastern Europe, so the current state of affairs in any country is not always what it seems.

Nor is it always a gloomy scenario. Some countries which are conventionally seen as hostile offer major marketing opportunities. Even under the mullahs, Iran has provided British and other Western companies with large and profitable contracts. And in many countries with bad reputations, the risks are often limited to specific locations.

The company should therefore assess the likely political developments. Then it needs to assess at a company level the operational risks of doing business in that country. Finally, it needs to be able to manage a crisis, should it arise.

HIGH-RISK REGIONS FOR BUSINESSES

Table 12.3 and Figure 12.3 show the world's major danger spots. In most places, the risks are confined to specific parts of the country, for example on the borders or in rural areas. Large cities are often (though not always) quite safe providing you take proper precautions.

Moreover the high-risk areas quickly alter from month to month as peace breaks out or fighting is renewed. Before sending staff abroad, the business should get an up to date view of foreign risk.

One of the *Wall Street Journal's* correspondents was kidnapped and murdered by a radical group in Pakistan. According to Gabriella Stern of Dow Jones, the newspaper's owner, the 200 journalists in the Middle East and Africa are made aware of the risks they sometimes face. 'But we have to take risks because we want to write about risks,' she said. 'To write about politically risky places, we have to be there. Our customers want to know about risks and opportunities.'

Consultancy firm Control Risks (controlrisks.com) considers a country to be a high risk if:

- it is suffering a civil war;
- law and order are breaking down;
- there is a campaign against foreign businesses.

Note that many of the risks will be confined to certain regions of a country, for example border areas.

Table 12.3   The world's danger spots

Central and South America	
Haiti (political violence, natural disasters, criminal gangs)	Colombia (guerrillas and drug violence)
Venezuela (political instability)	Bolivia
Mexico (murder in the north)	

Europe	
Kyrgyzstan, Uzbekistan, Kajikistan, Georgia	Russia (north Caucasus, Chechnya – political violence, war)

Africa	
Niger, Nigeria (violence, including by police and army personnel)	Sudan (militias, population displacement)
Central African Republic (armed robbery, civil strife)	Chad
Burundi/Rwanda (threat of civil war)	Algeria (hostage taking in the Sahara)
Zimbabwe (political, economic and anarchy)	Eritrea, Somalia (factional fighting)
DRC (fighting, civil unrest)	Guinea (civil unrest), Equatorial Guinea
Cote d'Ivoire (humanitarian crises)	Gambia
Mali, Mauretania (Mali border)	

Middle East	
Yemen (Al-Qa'ida activity risk)	Iraq (political violence, kidnapping)
Syria (civil war)	Iran (tension with Iraq)
Palestine (Gaza)	Libya

Asia	
Afghanistan (civil war)	North Korea (famine, uncertain political future, tension with South Korea)
Pakistan (potential anti-Western violence), Bangladesh	Nepal (Maoist terror activities)
Burma	

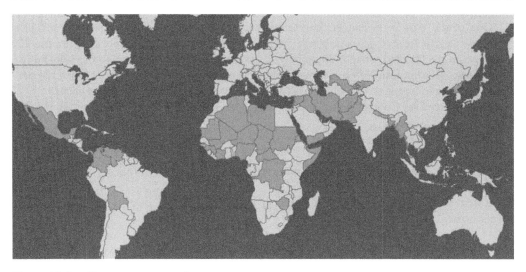

Figure 12.3   Dangerous countries

CASE STUDY: MANAGING RISK IN COCAINE'S HEARTLAND

Colombia is the location of one of the world's big oil fields. The National Liberation Army sabotages the oil pipelines in an effort to get Colombian oil nationalized. At one point the country was losing 250,000 barrels of oil a day, and was able to export on only five days in the month.

Oil pipelines are vulnerable, stretching as they do over such long distances. Therefore the oil companies have to manage the threats to the pipeline.

Before they start prospecting, the companies need to assess not only the geological risk but also the political risk. The damage caused by a country's political instability may outweigh the revenues that accrue from the oil.

Then they need a plan of action to minimize the risk. This will involve regular communication with local inhabitants at all levels. Building up good relations is important, and local people need to see the advantages of the oil, especially through increased employment.

Oil companies also need defensive measures, which will include protecting expensive or especially vulnerable parts of the pipeline, such as pumps. Passive defences, using locked enclosures may need to be augmented by guard patrols.

Like other oil companies, BP explores for oil in some dangerous parts of the world. It has found significant amounts of oil in Colombia, a country with a reputation for drug-related lawlessness. BP hopes to produce up to 600,000 barrels a day from this field.

Because Colombian oil is potentially so important, it has attracted the attention of guerrilla groups who want to see BP's work disrupted. The company has a plan for managing that risk.

Security figures highly, and the company receives help from the Colombian military forces. David Harding, BP Exploration's Chief Executive for the southern hemisphere says that the company invests in staff communications, 'ensuring that they understand the company's policies for handling risks, and training and advising them'.

BP employs local managers. Only one in five employees are expatriates, and that percentage is continually falling. The company takes Colombian graduates and is forging links with local schools.

David Harding says: 'We're managing our risks in Colombia as we do elsewhere. We're listening. We're maintaining the standards of business that we apply throughout the world. We're building contacts with authorities at all levels. And finally, we're communicating with staff – both Colombian and expatriate – so that they know what we're doing, and why we're doing it. That's the right way to manage risks.'

CORRUPT COUNTRIES

Among oil countries, Angola, Azerbaijan, Chad, Russia, Iran, Iraq and Venezuela all score highly on Transparency International's (TI) list of corrupt countries. 'In these countries, public contracting in the oil sector is plagued by revenues vanishing into the pockets of Western oil executives, middlemen and local officials,' said Peter Eigen, Chairman of TI. This is sometimes known as the 'resource curse' where mineral wealth leads to corruption and crime.

As can be seen in Table 12.4 overleaf, Nordic countries ranked highly as being free from corruption. The UK came eleventh, while the USA was seventeenth.

Security Services

Not every task can be undertaken by in-house staff. Some jobs are carried out only occasionally, and so they are better performed by an outside firm.

Table 12.4   Most and least corrupt countries (higher numbers represent greater corruption)

	Least corrupt		Most corrupt
1	Finland	133=	Indonesia
2	New Zealand		Tajikistan
3=	Denmark		Turkmenistan
	Iceland	140=	Azerbaijan
5	Singapore		Paraguay
6	Sweden	142=	Chad
7	Switzerland		Burma
8	Norway	144	Nigeria
9	Australia	145=	Bangladesh
10	Netherlands		Haiti

Source: Transparency International

Note that security consultants have different skills. Some are mere locksmiths, while others want to sell you a uniformed guard service. You should thoroughly check any consultancy's credentials and experience before hiring the firm. Their services can be categorized as follows:

* advice on travel risk;
* protective and security services;
* emergency support;
* kidnap for ransom;
* corporate internal investigations;
* competitive intelligence;
* hostile takeovers;
* computer forensics;
* financial forensic investigations.

Risk Assessment

By answering the questions below, you can see how vulnerable your company is to extremists, violence and international risks.

Topic	Question	✓
The public	Does the business risk violence from the public?	
Employees	Is the organization likely to have disgruntled employees?	
Customers	Does the organization face violence from clients or customers?	
Activists	Is the organization likely to face threats from protestors or extremists?	
Bomb threats	Has the organization suffered a bomb threat in the last two years?	
Products	Does the organization have widely distributed packed products which could suffer a tampering incident?	
Assets	Hs the business failed to harden its buildings against attack?	
	Are the organization's assets centralized?	
International	Do your executives travel to unstable developing countries?	
	Does the company have assets in unstable developing countries?	
Total points scored:		

Score: 0–3 points: low risk. 4–6 points: moderate risk. 7–10 points: high risk.

The Appendix has a summary of the checklists. By entering the results of this one, you can compare your security risk against other categories of risk.

13
Pre-Empting Fraud and Theft

It was unusual to see a police car at the office, and even more of a surprise to receive two officers into your office. Drunken employees? A traffic offence? That they wanted copies of your courier bills seemed odder still.

Later it turned out that the delivery company's finance person was issuing inflated bills to clients, and then siphoning off the cash. 'They targeted companies who weren't scrutinizing their bills,' says Sergeant Pearce. 'You're fortunate,' he says. 'Some companies have lost a lot of money.'

'Well, we pride ourselves on being on the ball,' says Jake, in that slightly smug tone of his.

'The sergeant pauses. 'I've noticed an absence of emergency exit signs, fire extinguishers, and several other health and safety violations here. When I come back next week, it would be in your interests to make sure you're fully compliant, sir,' he says.

The Scale of the Problem

The big fraud cases tend to affect individuals, such as where Bernie Madoff defrauded people of £18 billion and was sentenced to 150 years in prison. Or else they are accounting scandals such as Enron, where a business misleads investors.

But there are countless cases of fraud that don't reach the press, and which can put a company into receivership. In many cases, the fraud goes undetected for several years, and is only discovered when the company runs out of money.

In this chapter we examine occupational fraud, that is, fraud carried out by employees, and also consider the threat of fraudulent behaviour from customers and criminals.

FRAUD IS SURPRISINGLY COMMON

Fraud costs UK businesses £66 billion a year, according to the National Fraud Authority. The Association of Certified Fraud Examiners (www.acfe.com) reckons that companies lose 5 per cent of turnover, which could equate to 10 per cent of profits. Table 13.1 is an estimate of UK losses. Ernst and Young (ey.com) believe that companies suffer from other unreported ways:

- loss of impetus in managing the business;
- loss of business;
- loss of customer and banker confidence;

Table 13.1   Losses from fraud

Total losses from fraud, by sector	£billion
Private	45.5
Public	20.3
Individuals	6.1
Not-for-profit sector	1.1
Total	73.0

Source: National Fraud Authority (gov.uk/government/organisations/national-fraud-authority)

- adverse movement in the share price;
- impaired health and performance of the management team.

Occupational fraud can continue for several years before being discovered, with the median being 18 months. Often the accounts tally, and cannot therefore be easily spotted. The thief often understands the principles of double-entry bookkeeping, and therefore the fraud cannot be easily detected. For example, the thief might insert a false invoice into the system and arrange for its payment.

Detection of fraud not only allows the company to recoup its money through legal proceedings. It also stops the losses from continuing, which could be even more costly.

Even fake solicitors and accountants are busily at work committing fraud. In UK law, anyone is entitled to call themselves an accountant. However, by taking the time to investigate a prospective accountant's credentials, a company can eliminate any risk that might pose.

TYPES OF FRAUD

Fraud is so varied that it is difficult to define. In law it is often treated as a sophisticated form of theft, usually involving deception. It can be undertaken by virtually anyone, whether inside or outside the organization:

- *Blue collar employees.* This could be the theft of goods. A security guard may over-record a vehicle's weight, or an operative may stop a meter running.
- *Clerical employees.* Clerical workers handle paperwork, and are often in positions of trust. It is easy for them to falsify or destroy records.
- *Managers.* A lot of fraud is white-collar. Managers can approve invoices from a fake company which they own.
- *Suppliers.* A supplier can sometimes be involved, typically in collusion with a member of the defrauded company's staff. For example, an office worker may issue fraudulent credit notes to a supplier. Suppliers can also collude to fix prices, or nominate which of them will win specific contracts (cover pricing).
- *Lone customers.* Individuals can commit fraud. Fraudulent insurance claims are one such example.
- *Criminals.* Criminals can buy goods, especially for online or telephone 'customer not present' transactions, using stolen cards, or fake ones using skimmed card information.

How Internal Fraud Happens

For fraud to take place, various preconditions must be in place, as Figure 13.1 shows.

1. *Motive*. The fraudster (often a trusted employee) has a reason for his action. One employee wanted to give his daughter the perfect wedding. Another needed to fund his gambling addiction. Companies with a demotivated workforce may be more at risk. The motive may be cloaked in justification. An employee might want to 'take revenge' on the company for not giving him a pay rise. This is often simply a justification for theft; many employees feel aggrieved without stealing from their employer.

    MI6, the British secret service, appointed a cleaning contractor, Strand Cleaning, to clean its London offices. This was because it cost MI6 50 pence an hour less than the loyal staff cleaners, who were made redundant. Foreign powers now might try to bribe the new cleaners, whose annual earnings average £3,000, to rummage through the MI6 waste paper baskets.

2. *Asset worth stealing*. Products, raw materials, tools and equipment are worth having, particularly if, as we see below, there is an opportunity to sell them. The author's next door neighbour managed a frozen food company's vans, and one of them often stopped outside his house. It later transpired that he was offloading supplies of frozen food which he could use or sell.

3. *Opportunity or access*. Brain injury charity Headway lost £35,000 over a six-month period when its accounts controller David Field used the charity's debit card to pay for impotency drugs, a subscription to a dating website, his council tax bill and a stay at a London hotel. He also used the card to withdraw cash. He was sentenced to four years imprisonment. Blank cheques cost the British Council £520,000, forcing it to cut back its educational programmes throughout the world. An employee simply passed genuine British Council blank cheques through the cheque-signing machine. Field later filled in the amount and the payee's name.

    The person who loads and drives a van full of equipment has an opportunity. So does a manager who is responsible for approving overtime payments. In one case, an IT manager at a magistrate's court wrote computer code that salami-sliced a tiny percentage of every fine paid into the court and paid the sliver of money into his own bank account.

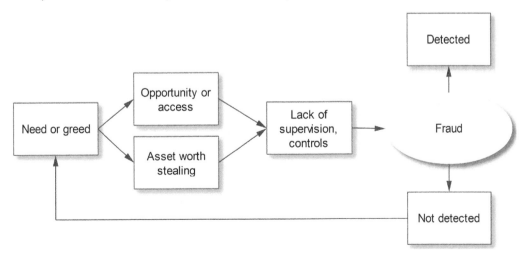

Figure 13.1   Preconditions for fraud

The opportunity must exist for the individual to both steal and sell the goods. In most cases, the fraudster needs an outlet. This might relate to:

- physically taking stock and knowing someone who will buy it;
- falsifying a document and having an acquaintance in a supplier's firm who will benefit, and with whom he will share the proceeds.

Meng Chih-chung, a director of Taiwan's Directorate General of Telecommunications, was jailed for seven years for forging documents and granting illegal favours to Ericsson of Sweden. In the $220 million (£142 million) scandal, five other officials received sentences of five to six years on similar charges.

But the opportunity to steal is pointless unless the thief is not going to be caught. A bank cashier can easily take money from the till, but the bank's systems will soon discover the loss. So the next element (lack of control) must be present.

4. *Lack of control*: The individual must feel able to commit a crime without being discovered. Absence of checks is the way this occurs. An opportunity to steal arises when an accountant is unsupervised. It also takes place when someone handles both the raising of purchase orders and their payment. This indicates the opportunity for fraud for anyone who carries out work on his own, for example in a service department, and who is in a position of trust, no matter how junior or senior.

It is good to separate payroll from HR. In other words, the person who adds staff to the HR database should not be the person who pays their salaries.

In another case, a payroll executive changed a supplier's sort code and bank account details on his company's records, which meant that the payments were made into his own bank account. He then fled before the supplier queried the outstanding bills. It follows that any change to bank details or passwords should be notified to the company or person affected.

The above factors may result in an act of fraud. And if it is undetected, the fraudster will be emboldened and the cycle will repeat itself.

A major bank never queries any event below £500. As a result, junior bank employees can remove £499 off their friends' overdrafts. Spot checks on events under £500 would prevent that from happening.

Restructuring to reduce costs often reduces the level of supervision. This creates greater opportunities for fraud. Ironically, efforts to save money can actually create greater losses.

To prevent fraud, one of the four elements (Asset, Motive, Opportunity and Lack of Control) must be broken. The easiest solutions are making assets less readily available and increasing your controls.

FACTORS THAT MAKE ORGANIZATIONS LESS SUSCEPTIBLE TO FRAUD

Factors that reduce an organization's or industry's risk of fraud are as follows:

Few cash transactions: The more that money is transferred electronically, the less cash that is transferred around the business. This is why stores offer customers cash back on debit cards: it saves them counting and then transferring notes to a bank. That said, there are many examples of where cashless transactions have been abused by fraudsters.

Goods have little street value: If the organization makes screws or bolts, they have little value to the fraudster. Whereas if the company sells jewellery or TV sets, the opportunity is greater.

Few transactions: Where transactions are fewer but higher in value (for example £100 rather than £5) the transactions are easier to record and monitor.

Good systems and staff: If the organization carries out end of day till reconciliations, stocktakes or audits, their business is less likely to suffer fraud. Similarly, anti-fraud awareness and training helps staff to be on the lookout for fraud.

The more closely the business matches these factors, the more-fraud proof it becomes.

But fraud prevention costs money. It's good to introduce anti-fraud procedures, but you have to weigh the costs against the likely benefit. It is best to concentrate on the areas of highest and most probable risk.

CASE STUDY: THE BOOKKEEPER WHO LAVISHED STOLEN CASH ON HIS MISTRESS

Jailed for four years, Andrew Harpur stole £400,000 from his employer to lavish on his home, his wife and his mistress.

Though he earned only £12,000 as the company's bookkeeper, he spent money 'at the rate of someone who had just won the pools' according to Judge Barrington Black.

Over 18 months, he bought a Volvo car with £26,000 in cash, he had helicopter lessons which cost £10,000, and was hiring them to take his family and his mistress on flying trips.

'It's amazing that he wasn't caught earlier,' said Detective Constable David Hanley who led the investigation.

Harpur had been a warehouseman who re-trained as a bookkeeper. He set up bogus companies and siphoned large sums into their bank accounts. The fraud was only discovered when 12 cheques totalling £60,000 were stopped by the company's bankers because the company didn't have enough money to pay them.

Who Commits Fraud?

The majority of fraud is an inside job, according to Kroll (kroll.com). The company estimates that two-thirds (67 per cent) of fraud globally is insider fraud. This makes sense because insiders know where an organization's weak spots are.

And when it comes to employee fraud, about two-thirds of it is committed by directors or senior managers, according to a KPMG survey of British business fraud (kpmg.com).

The survey also found that 32 per cent of fraud was committed by longstanding employees who had been with their companies for between ten and 25 years. About 23 per cent of perpetrators had been with the company for between five and ten years, while 20 per cent had been employed for between two and five years. So longevity is no guarantee of loyalty or honesty, and statistically makes the fraud more likely. In other words, it is your loyal and trusted staff who commit fraud.

Online Fraud

Online fraud mostly affects consumers rather than businesses. Unwise individuals are regularly relieved of their money by online dates or by fraudulent scammers who seek their assistance in getting a dictator's millions out of the country.

But there are exceptions. 70 US lawyers were relieved of $29 million by a wily Nigerian who asked for their help in recovering a debt owed to him in the US, in return for which the lawyers would take a hefty fee from the proceeds. The lawyers collected the debt, and paid the Nigerian, having first taken a share of the proceeds. Later they discovered that the 'debtor' was involved in the scam, and the cheques bounced.

In what is known as phishing, internet users receive emails purporting to come from their bank. The innocents give out their account details and password, and the fraudsters take their money. While the consumer suffers, the bank or organization gets bad publicity.

In other cases, fraudsters break into an internet site, for example an ecommerce site, and steal users' names and passwords. This sometimes gives them access to the users' bank accounts. As with phishing the company suffers reputationally. We discuss how to protect online databases in Chapter 15.

Telephone Fraud

Most businesses are aware of the risk of hackers, but they often overlook one area of vulnerability – the telephone. Ireland's Department of Social and Family Affairs noticed that its phone bill was rising, and found that expensive calls were being made out of office hours, and to Africa and south-east Asia. An investigation found that hackers had gained the password to the organization's PBX system, and were able to make international calls through a part of the system used only for maintenance. These calls cost the department 300,000 euro. This is known as fixed line fraud.

Companies tolerate small levels of abuse from employees making overseas foreign calls, but sometimes this can cost a lot of money. For example, an employee can divert their phone to an international number. At home after work, they dial their office number, which puts them straight through to their friend in Australia or the USA, at the company's expense.

Companies should bar most lines to international calls, and restrict the use of forwarding. Organizations should also monitor any changes in phone use, especially of out of office hours and international calls.

Invoice Fraud and Scams

Companies often receive invoices for services not received. Each year, like thousands of businesses, the author receives invoices from the Domain Registry of America (DROA), stating that his domains need renewing, and request payment. If the invoice is paid, the DROA becomes the domain registrar. The author also receives invoices for renewal of trademarks from what looks like the Intellectual Property Office. These are known as government agency scams. Legal experts say many of these frauds stay on the right side of the law, but only just.

A university finance department received a letter from a supplier, notifying it of a change of address. The department duly changed the address, and sent a cheque for £175,000 – to a criminal's address. Fortunately, the next cheque, for nearly £1 million was stopped just in time. The investigation that followed suggested, rightly or wrongly, that people in finance tend to do what they are told, and avoid conflict or actions that might 'rock the boat'.

Being Used for Fraud

The company might simply be a vehicle for an employee's fraud. For example, certain employees might be colluding with those of another firm to defraud it. Or in local government an employee could be taking money to smooth the path of planning permission.

Other companies are used to launder money. The International Monetary Fund estimates that between $600 billion and $1.5 trillion is laundered annually, equivalent to 2–5 per cent of the world's gross domestic product (GDP). The majority of this is laundered through banks but the increased threat of international terrorism, funded through money laundering, has seen stricter controls being introduced.

CASE STUDY: WHEN USED CAR SALES TURN OUT TO BE FINANCING TERRORISM

A small used car exporter in Florida was delighted to get a sale for one of their vehicles. The money was wired from the Federal Bank of Lebanon, and as requested they duly exported the car to West Africa.

Over the next four years, Tampa-based Mansour Brothers and 30 other used car dealers were wired $329 million, according to the US New York Attorney's office.

But the money came from Hezbollah institutions. And the cars were sold to release clean money back into Hezbollah accounts, a radical Islamic group dedicated to the obliteration (its words) of Israel and the promotion of suicide bombings.

There was no evidence that Mansour Brothers knew anything about the money laundering, but at the time of printing, the Attorney's office was seeking to seize the $3 million sent to the company.

Identity Theft

In Chapter 20 we discuss counterfeiting which is a major problem for luxury goods companies, spirits businesses and many others. But all companies are at risk.

BBC journalists believed that a DowEthics.com website was genuine, and emailed the listed PR contact. The activists who had set up the site then posed as Dow Chemical spokespeople in a BBC TV interview. They apologized on the company's behalf for the Bhopal disaster, and said it would be paying $12 billion compensation to those who suffered. When the news was broadcast around the world, Dow was forced into a 'retraction' and explained that it was not going to give money to the victims.

The same 'Yes Men' pranksters had previously angered toy maker Mattel by swapping the voice boxes of Barbie and GI Joe action figures and putting them back on shop shelves. To the confusion of the children, Barbie wanted to go on the attack, while GI Joe suggesting shopping trips.

In his book, *Defending the Brand: Aggressive Strategies for Protecting Your Brand in the Online Arena*, Brian Murray suggests the following steps to manage the risk of online corporate identity theft:

1. Get your stakeholders to provide early warnings. Make it easy for employees and customers to report any suspicious emails or websites they encounter.
2. Tell your customers that you never ask for their personal or account details by email or on the phone.
3. Make sure you're easy to find online. Promote your web address and keep it simple – to avoid misspellings.
4. Manage your domain registrations, and monitor new registrations that include your company name or trademarks. Register common misspellings of your website address. You can automatically redirect visitors to the correct address.
5. Plan your response to an attack before it happens. Many trade associations have advice on best practice. Also seek advice and establish relationships with the police and other organizations that can help take down fraudulent sites if an attack occurs.

Organizations can suffer identify theft offline as well as online:

- employees can order goods in the company's name, and have them delivered to another address;
- fake references can be used to set up accounts with suppliers;
- if fraudsters know a company's suppliers, they can order goods which will they ask to be delivered to a rented address bearing the supplier's name.

Customer Fraud

Customer fraud is categorized as 'short' and 'long'. Short fraud involves getting goods on credit, and never paying for them. It's a one-off event that is repeated with other companies, or even trying it again with the same firm but under a different name.

Long fraud (or the long game) involves buying goods and paying for them, so as to build up a track record. Then the criminals put in a huge order which they disappear with. The police advise businesses to prevent this kind of fraud by taking the following measures:

New customers

- check the director's credit histories;
- where potentially large sums are involved, visit potential new customers at their claimed business premises;
- for limited companies, check if it has filed accounts. Do an online credit check;
- check to see if the individuals are bankrupt or disqualified from acting as directors. In the UK, this is the Companies House Insolvency Service and Companies House database.

All customers

- ensure all customers provide a physical address and land line phone numbers, not just email addresses and mobile telephone numbers;
- ensure that goods are delivered to identifiable individuals and addresses;
- ask the customer for trade references, and check their authenticity. Fraudsters create fictitious companies that will provide a reference.

Long fraud

- be wary of accepting a large order from a business that has been a customer for a short time.

CUSTOMER FRAUD IN LOCAL GOVERNMENT AND SOCIAL ORGANIZATIONS

According to the National Fraud Authority (gov.uk), housing tenancy fraud is the second biggest drain on local government, only just behind local government procurement fraud which stands at £876 million a year. Social housing organizations in England are reckoned to lose a further £919 million a year.

Housing tenancy fraud occurs when a tenant unlawfully sub-lets a council or housing association-owned property to someone else, when a family passes it on, or continues to claim for the house when not living there.

CUSTOMER FRAUD IN NATIONAL GOVERNMENT

Outside the tax and benefits system, the UK Central Government loses most money in procurement fraud, as shown in Table 13.2.

Table 13.2   Central Government losses from fraud

Fraud Type	Fraud Loss
Tax fraud and vehicle excise	£14.1 billion
Tax credit and benefit fraud	£1.9 billion
Procurement fraud	£1.4 billion
Grant fraud	£504 million
Television licence fee evasion	£204 million
Payroll fraud	£181 million
NHS patient charges fraud	£156 million
NHS dental charges fraud	£73 million
Student finance fraud	£31 million
Pension fraud	£13.7 million
National Savings and Investments fraud	£0.40 million

Source: National Fraud Authority

The Indicators of Fraud

There are often clues that fraud is taking place. Here are a few of them.

Employee behaviour

- an employee gains sudden wealth, which he claims is from a rich relative's will, or a pools win;
- the employee who never takes a holiday (for fear that their fraud will be revealed) –everyone takes their holiday, and that others do their work while they are away;
- an employee who doesn't allow others to see his work;
- the disorganized or chaotic employee – others abandon any prospect of getting information from them, or any hope of understanding their systems. Messiness can disguise fraud.

Computer records

- the physical stock in the warehouse is usually less than computer records show;
- gaps in records, caused by 'computer breakdown'.

Non-standard accounts

- busy accounts used for samples or guarantee claims. Accounts that don't generate invoices make it easier to conceal fraud.

Supplier fraud

- a substantial amount of work going to one supplier;
- a supplier whose additional costs are regularly accepted;
- a supplier who receives multiple orders just below the threshold for tendering.

CASE STUDY: HOW TO GET £750 MILLION WORTH OF BANK LOANS

Allied Irish Banks (AIB), Bank of Scotland (BoS) and Barclays lent a fraudster £750 million on the strength of fake British Virgin Islands (BVI) company accounts and forged references from Credit Suisse.

The fake accounts suggested that Achilleas Kallakis owned billions in shipping assets. He also had forged guarantees from SHKP, a major property developer in Hong Kong, and told the AIB that SHKP would only support the deal if they weren't contacted directly.

When pressed, the banks were granted a meeting at Kallakis' London office with 'Jonathan Lee', who purported to be a director of SKHP's treasury department, but who was in fact an actor.

Kallakis also showed the bank forged references from Credit Suisse and the economist Lord Harris.

With the proceeds Kallakis bought luxury properties in Knightsbridge and Mayfair, which he intended to convert into luxury flats. He also acquired the trappings of wealth: a $44 million private jet, a $8 million helicopter and a £26 million yacht.

He was aided by his long-time business partner, Alex Williams, who was adept at forging documents. The two had been convicted 18 years previously for selling fake manorial titles to unsuspecting Americans. This wasn't picked up because the two had changed their names.

The fraud only unravelled when AIB sought to spread its risk by selling some of its loans to other banks. The German bank Helaba conducted background checks and hired a private investigator, and took their findings back to the horrified bankers at AIB.

At his trial Kallakis said that he was only a humble adviser to the Hermitage Syndicated Trust, the company that had received the loans, and that the real controller was the trust's director, a Swiss lawyer named Michael Becker. However the jury didn't accept that.

In sentencing Kallakis to seven years in jail, the judge said that AIB and BoS had 'acted carelessly and imprudently by failing to make full inquiries before advancing the money'.

Customer collusion

- unexplained credit notes.

Vulnerability Analysis

A company which believes itself to be at risk of being defrauded should undertake a vulnerability analysis. This will assess where and how the company could be defrauded.

It will determine what money or assets are at risk, and which functions could undertake fraud.

Doing this analysis will help to direct fraud prevention resources to where they are most needed, and avoid wasteful checks in unlikely areas.

The company can undertake a vulnerability analysis either as a routine task once a year, or in response to growing losses whose cause is unknown.

As with audits, the test should be done by an independent person. If it is conducted by an internal accountant, they may be the fraudster. A vulnerability analysis will assess:

- what assets might be at risk;
- where they might be stolen
- who might take them and who could benefit;
- how they might take them (the method of theft) and sell them;
- how effective the controls are.

We look at each of these next.

ASSETS

The most easily stolen assets are cash and small high-value stocks. Other assets which could be taken include information, raw materials, and plant and machinery.

Items which are at risk include cash, credit card transactions, intra-firm accounts and credit sales. So are special items which are not billed to anyone (such as samples or test products).

As we have seen, in long and short-firm fraud, criminals set up legitimate businesses and order goods. In the long-firm version, they buy regularly, then put in a very large order and abscond. The short-firm version involves getting just one order, and involves the supplier failing to credit check the new customer properly, or is deceived by them.

FRAUD – THE LOCATIONS

Places where fraud can take place are:

*   locations where money physically changes hands;
*   accounting transactions (where, for example, fractions of a penny can be regularly posted to a wrong-doers account);
*   locations where documents which have monetary value (such as invoices or expense claims) are issued or received;
*   places where poor records are kept, or where much short-term activity occurs (such as suspense accounts);
*   locations where physical goods exchange hands, such as stock leaving the company's premises;
*   departments whose costs are difficult to trace (such as a maintenance department where large numbers of people use spare parts);
*   computer programs where a manager or programmer can divert money to another account.

In its department stores, House of Fraser deploys security tags and uniformed guards, and checks changing rooms for empty hangers in changing rooms – a sign that thieves are operating. In addition, the store video records activities at the point of sale, whereas many retailers video only around the store. This has significantly reduced the level of cash loss, credit card fraud and refund fraud. With its expensive brands, the store believes it will always be the target for thieves. But by presenting a robust and visible deterrence through guards, recordings and changing room controls, the store aims to discourage fraud and theft.

WHO MIGHT TAKE THE ASSETS?

As we have seen, fraud can be committed by staff, customers or suppliers. It can also be committed by organized crime which has a growing involvement in corporate fraud. Fraudulent employees may be those who:

*   are addicted to gambling, drink or drugs;
*   have heavy financial commitments;
*   have close business or personal links with suppliers or customers;
*   are involved with handling assets;
*   display wealth greater than their income would support.

As Figure 13.2 shows, fraud tends to occur in departments where the opportunity for fraud is greatest. The arrows in the table also show that collusion takes places in predictable ways. The table shows common frauds, but does not include all the possible departments involved, nor the full range of frauds.

Note that at director or manager level, frauds may involve inflating expense claims, or 'cooking the books', so collusion may not involved.

HOW FRAUD IS COMMITTED

Fraud may involve any of the following actions:

* theft of assets;
* tampering with equipment or meters to give a wrong reading;
* falsification of records (invoices, cheques, stock control records and so on);
* corruption (for example, where a buyer awards a contract in return for a bribe);
* a customer selling 'seconds' as perfect quality products;
* using today's colour photocopiers or sophisticated printing equipment, which makes forgery easier;
* use of a stolen or fake credit card to make online purchases.

EFFECTIVENESS OF CONTROLS

In analysing the effectiveness of controls, the auditor should examine controls over cash, accounts, purchasing and invoicing. The auditor should also check controls over stock control and production, and supervision of personnel. The checks should assess whether controls are more lax at certain times (such as during night shifts, when the opportunity for theft is greater). An example is shown as Table 13.3.

Table 13.3   Vulnerability analysis

Organization			Site		
Date of assessment			Auditor		
Dept.	Vulnerable assets or areas of potential fraud	Who might be involved?	Method of fraud (for example, forgery)	Effectiveness of controls	Level of risk (low–high)
1					
2					
3 etc					

Doing the analysis

The analysis should check for losses in money or goods, or for areas where controls are insufficient to prove that loss has not taken place. The check should take place over a set period (such as one week), and should be compared with previous records.

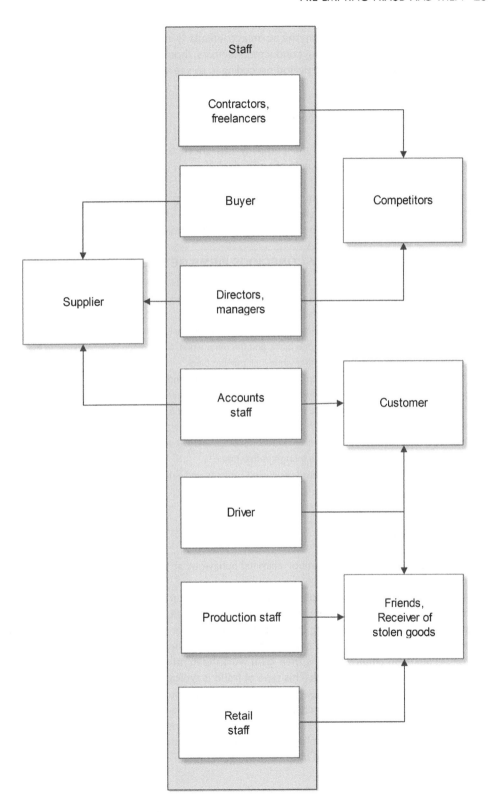

Figure 13.2   Typical areas of fraud

Treat unexplained deviations from the norm as particularly interesting. In some cases, fraud will cease while the investigation is being conducted. Auditors should be able to compare the results of the investigation period with the period preceding it, to see whether the pattern has changed.

After the vulnerability analysis has been done

Once the analysis has been done, the company can check to see whether losses are in fact occurring. This will involve various kinds of audits, including:

- data analysis (for example, examining documents, or comparing ratios over time);
- physical checks (physical stock checks, for example);
- surveillance. This can include electronic tagging to track items which are at risk.

CASE STUDY: EVEN SECURITY-CONSCIOUS FIRMS GET DEFRAUDED

The Bank of England is said to have lost £600,000 through theft by employees paid to destroy old notes. It was alleged in court that over four years, Mrs Christine Gibson, a team leader, hid bundles of old £20 notes in her bra and knickers and hid them in her locker.

She would take the notes from sealed cages which contained up to £2 million, while a colleague distracted other workers. The cage had two keys, one black and the other white. Mrs Gibson only had keys to the white lock, so her accomplice substituted another white lock painted black.

The fraud was only discovered when Mrs. Gibson and her husband paid £100,000 for investment bonds to Reliance Mutual. The couple emptied bundles of notes on to a desk as staff looked on in amazement. Counting took several hours. The police were later called, as a result of routine checks that follow any large cash transaction.

The Gibsons appeared to rely on Mrs Gibson's £15,000 salary. Yet they owned two brand new four-wheel-drive cars, had £32,000 in savings and a mortgage-free home.

The Bank of England has now installed concealed cameras and extra staff. It claims that repetition of the crime is now impossible.

Current Employee Fraud

Employees perform many types of minor criminal behaviour at work. They include:

- *Fake clocking on and off:* The employee 'swipes in' at the start of the day, and then leaves. They arrive shortly before the end of the day to clock off again. But at online giant Amazon, workers have a handheld device directing them to goods. These tags also measure their productivity in real time. If they lag behind, the machine bugs them and they are issued with warnings.
- *False signature:* An employee signs their own worksheet, rather than getting a manager to sign it. This allows them to skip work. This type of fraud is typically carried out by sub-contractors and agency staff sent by temp agencies to companies when they aren't known.
- *Trading products for benefits:* Technology website Gizmodo alleged that an Apple regional manager discounted computers to local plastic surgeons in order to receive a stomach stapling procedure. Other employees made fictitious transactions, swapped old products for new ones (new iPhones for old ones) and gave away a computer to a local bar in return for cheap drinks.

- *Travel and subsistence fraud*: Employees can inflate their expense claims, claim fictitious trips or seek payment for entertaining clients. The author knew an employee who used to holiday in Europe on the pretext of business trips.
- *Using company facilities*: At one firm, the managing director got the maintenance department to build a new mast for his sailing boat. In another, staff mended lawn mowers and sewing machines at work.
- *Taking company goods*: Employees frequently take the firm's goods home with them. This ranges from protective clothing and tools to computers. Policies should be made and adhered to, so that no one is in doubt about the correctness of their actions. In Chapter 18 on People Risks, we also look at departing employees who steal clients.

Next we consider a major area of fraud: CV fraud by potential employees.

Candidate Fraud (CV Fraud, Résumé Fraud or Job Fraud)

When people are desperate to get a job, particularly in times of high unemployment, they are more likely to doctor their CV. This has implications for employers who will set them to work in jobs for which they are unsuitable.

Pre-employment screening firm HireRight says that 36 per cent of financial services professionals securing a new role in EMEA (Europe, the Middle East and Africa) have 'some sort of discrepancy' on their CV. Other recruiters suggest that in the Middle East, this figure is over 40 per cent. In the UK, Powerchex found 17 per cent of 3,876 applicants in the financial sector had discrepancies. CV fraud includes the following:

- fake degrees or licences;
- fictitious former employers;
- fake current employers – this masks unemployment;
- fake employer references – relatives and friends often create false home-based businesses that 'employed' the applicant;
- gaps in employment;
- prizes for academic or career success;
- taking credit for successes, for example an acquisition or a project, when their involvement was slight.

In New Zealand, Personal Verification Ltd highlighted numerous cases of CV fraud, including the following:

- Rito Henry Miki, a convicted child sex offender used a fake CV to teach in six schools.
- Shadrach Darren Mitchell, a practising employment lawyer, failed to mention 39 convictions, including drug possession, burglary, theft, assault, 13 charges of intentional damage, disorderly behaviour, breaching bail, four drink-driving convictions and five months in prison.
- Trevor Esera, an IT manager who created a false CV to get a job as an accountant at Rinnai New Zealand Ltd, and then stole $1.75 million from the company, using fake invoices from non-existent suppliers.

This emphasizes the need to thoroughly check people's CVs and references thoroughly.

Industry-Specific Fraud

Each industry is vulnerable in its own way. Here are a few:

- *Building societies*: Mortgage fraud – fraudsters aim to get mortgage cash, by over-valuing properties, getting mortgages from several companies on one property and by getting mortgages through identity theft.
- *Credit card companies*: Card fraud – paying for goods on credit cards and then absconding.
- *Construction*: Specifications being written to include or exclude certain bidders, bid rigging by contractors, duplicate invoices, using lower-quality materials than specified, improper signing off.
- *Healthcare*: In the USA, small doctor's offices are used for insurance fraud.
- *Retail*: Fraud involves employees stealing stock, reclassifying stock as faulty and giving it to friends, or selling stock brought in by themselves. In addition, buyers are vulnerable to inducements.
- *Universities*: Some academics fake the results of their research, and some plagiarize others' work. When discovered, this results in academic papers having to be withdrawn from journals, causing embarrassment and loss of reputation to their employer. At one university, a senior lecturer was known to be secretive; he told his researchers not to engage with other scientists in the faculty. Later it was found that he had been faking the results. The research involved breaking rats' legs and assessing the healing process. When the results didn't conform to his thesis, he would break more rats' legs. And when all else failed, he simply changed the numbers. At another institution, the academic put in an application to a major grant-making body, and accompanied it with a fake letter of recommendation from his vice chancellor on university notepaper. In both cases, lack of supervision was part of the problem.

By identifying the typical vulnerabilities in your industry, you can reduce the risk of fraud in your organization.

INVESTMENT FRAUDSTERS: THE POINTS TO WATCH

Since companies and their directors deal with large amounts of money, they are often targeted by investment fraudsters. It is curious that rich and famous people – and often those who are known for their business acumen – regularly fall for the fraudsters' schemes.

Bernie Madoff swindled £40 billion out of many well-known people including actor Kevin Bacon, CNN's Larry King and Lady Victoria de Rothschild. He was jailed for 150 years.

Allen Stanford, the cricketing fraudster, was jailed for 100 years for a £4.5 billion swindle. His fame including landing a helicopter at Lord's and fondling the wives of the English cricket team.

Kautilya Nanden Pruthi was jailed for 14 years for a fraud of £115 million. His 800 victims included the cricketer Darren Gough, actress Frances de la Tour and actor Jerome Flynn. Pruthi claimed to be providing bridging loans to export businesses that had run into cash flow problems, and offered investors a remarkable return of 13 per cent a month.

Roger Levitt relieved Frederick Forsyth of £2.2 million, among others, and went bust owing £34 million. Levitt recommended 20 shares to Forsyth and suggested that to 'save him the bother' of writing 20 cheques, Forsyth should just write one to Levitt. 'I never saw the money again,' said Forsyth.

Nicholas Levene amassed a £16 million fortune. He had a chauffeur-driven Bentley, organized £10,000-a-day pheasant shoots, and spent £588,000 on his son's bar mitzvah party. He was eventually jailed for 13 years for swindling investors out of £32 million. He had spent the money on spread bets. He swindled Stagecoach bosses Brian Souter and Ann Gloag out of £17 million, while London's Ivy and Caprice restaurants lost £5 million.

Greed, the thought of making exceptional money, lies at the heart of these losses. But the fraudsters are persuasive.

According to Louise Brittain, partner at Deloitte, who unwound Levene's estate after he was declared bankrupt, the fraudsters 'are all much the same – very strong personalities, charismatic, able to emotionally manipulate people. They start small and build trust, so they get a ring of people around them happy to recommend them'.

Fraudsters of this sort wear expensive suits, dine at pricey restaurants and have flash offices in expensive parts of town. Initial returns come from newer investors' money, rather than from genuine profits.

Fraudsters usually promote a little-known investment, offering exceptional returns. It is usually an investment that's 'too good to be true', sometimes in an unregulated market. The victim is often recommended by a friend, and wants to be welcomed into an exclusive club of rich and famous people.

Preventing Fraud

The most unlikely people are often found to have committed fraud. So the prevention of fraud must seek to minimize the opportunity for fraud and to implement proper controls. Fraud prevention starts with a written fraud policy.

INTRODUCE A CORPORATE FRAUD POLICY

The fraud policy should include the following:

> *The organization's attitude towards fraud.* Make it plain that fraud is a form of theft, and as such it is a criminal offence. The company should not condone other activities such as giving or taking bribes.
> *The corporate policy towards giving and receiving inducements.* Spell out the policy on giving and receiving entertainment, gifts and the payment of commission. Clarify the difference between commissions and bribes.
> *Methods for controlling and investigating fraud.* The company should have systems for preventing fraud and procedures for checking against fraud. This should include the reporting of fraud to the statutory authorities, and the recovery of losses. The details of confidential methods of investigation should not be included in the policy.
> *Responsibilities for fraud control* should be defined. The ultimate responsibility should rest with the chief executive. This is because many frauds are carried out by senior staff, and because the executive responsible for managing fraud must have sufficient authority.
> *Resources must be allocated to fraud detection.* This is because fraud does not come to light in normal audits. You need to carry out a vulnerability analysis, and conduct audits. This needs people and money.
> *Channels for reporting fraud.* Employees should be told how to report suspected cases of fraud. You may need physical suggestion boxes. Whistleblowers must feel sure they will be safe and listened to.
> *Policy of dealing with wrong-doers.* This will include a policy on dismissing and prosecuting wrong doers, reporting to the police, and references for dismissed employees.

How to Minimize Fraud

As we have seen, the company should assess where it is vulnerable to fraud, and introduce a fraud policy. Then it should take a series of measures to reduce the likelihood of fraud. They comprise the following points, which we consider in more detail below:

- reducing the likelihood of employing fraudulent staff;
- reducing the fraudster's motive;
- reducing the number of assets worth stealing;
- minimizing the opportunity to steal;
- increasing the level of supervision;
- improving financial controls;
- improved detection.

REDUCE THE LIKELIHOOD OF EMPLOYING FRAUDULENT STAFF

The company can reduce the employees' wish to commit fraud by taking care to recruit honest people. This involves properly investigating applicants' CVs and taking up references.

REDUCE THE MOTIVE

The company should also be seen to operate in a fair and honest manner. What reason is there for an employee to behave honourably if his company condones bribery in overseas markets, allows office politics to flourish or pays its top executives unduly high salaries.

REDUCE THE NUMBER OF ASSETS WORTH STEALING

The company may be able to sub-contract certain types of work, or to operate a JIT system which prevents valuable raw materials from being available. The company could also make the assets more difficult to sell, for example by marking them as corporate property.

MINIMIZE THE OPPORTUNITY TO STEAL

The company can minimize the opportunities in various ways, such as by having secure warehouses or perimeter fences. This is discussed further in Chapter 11 on Security.

INCREASE THE LEVEL OF SUPERVISION

Increased supervision is only necessary in areas of potential fraud. It includes such simple things as requiring all executives to submit expense claims to a superior, or by requiring cheques to be signed by two directors.

Supervision may also include video surveillance and employee searches (where this is necessary, acceptable and legal).

Be alert for groups of employees who seem above the law, or untouchable. A group of 'clubby' managers can work together to commit fraud. Ask them for files or other information, and be wary if they don't hand them over. Ask yourself: are there documents or printouts the business never gets to see?

IMPROVE FINANCIAL CONTROLS AND MANAGEMENT SYSTEMS

Improved financial controls will ensure that procedures are in place and are properly followed. For example, records should not be written up later, and mail might need to be opened by a team of employees.

- Auditing procedures should check for adjustments, management over-rides and procedural breaches.
- Prevent sole responsibility for complete financial transactions from occurring.
- Ensure that corporate purchases are subject to formal tenders. This includes making sure that sufficient tenders are received, and that they are from genuine and independent companies.
- Ensure a complete audit trail, so that money paid by the company can be traced to the goods or services it bought.
- Introduce mystery shoppers who check that staff issue receipts, hand over corporate goods (rather than their own) and that procedures are observed (such as closing the till after every transaction).
- Make use of business ratios, such as net profit to sales. If profit is slipping, it may be that sales are not being recorded, or that costs are being inflated. Compare your results with the industry average. It is worth noting that the Inland Revenue uses this technique to check that it is receiving enough tax from each business. The company should also make comparisons between branches.
- Ensure that computer programs are verified and their integrity maintained. Do not allow changes to be made. Critical areas where fraud could take place should be examined.

DETECTING FRAUD IN EMAILS

Ernst and Young (ey.com) has a program that searches emails for phrases used by fraudsters. The software looks for 3,000 common phrases that indicate fraud. Developed in collaboration with the FBI, the phrases include 'cover up', 'nobody will find out', 'off the books', 'grey area' and 'failed investment'.

The software also checks for phrases that suggest the individual is worried about being overheard, such as 'call my mobile' or 'come by my office'.

Fears of getting caught are found in phrases such as 'no inspection' and do not volunteer information'.

When fraudsters are under pressure, they use comments like 'not comfortable', want no part of this', don't leave a trail' and 'make the number'.

Opportunities for new fraud are indicated by words such as 'not hurting anyone', 'off balance sheet transactions' and 'pull earnings forward'.

In bribery cases, people use phrases like 'special fees' and 'friendly payments'.

The benefit of the program, according to E&Y's Rashmi Joshi, is that emails are usually only seized after the damage has been done. Because the data in emails is unstructured, consisting of just words, email is often not analysed; yet email is the common means of communication among employees, clients and third parties.

The program is a proactive search for areas where fraud is about to be undertaken. E&Y believes it could save companies from losing a lot of money, before the fraud grows in scale.

The program, which initially anonymizes the data, can detect code words and can spot changes of tone. Apart from fraud, it also looks for breaches of information and employee misconduct. The software is especially suitable for monitoring the financial services industry, where the opportunity for fraud is high.

Conceptually, the program is straightforward, so any competent IT person should be able to replicate it by carrying out simple search inside Outlook or Gmail accounts, or by setting up a routine search of all emails.

IMPROVE YOUR CHANCES OF DETECTION

- Staff should be made regularly aware of auditors in the business, with special audits being undertaken in areas of high vulnerability. This will discourage employees from considering a fraud.
- Ensure that all members of staff take at least two weeks' consecutive holiday. This will ensure that other members of staff carry out their vacationing colleague's work, which can lead to detection.
- The organization should have a clearly communicated policy on whistleblowing, and ensure that whistleblowers are protected. Fear of retribution often discourages staff from coming forward.

IMPROVE YOUR RECORD KEEPING

- Matching of documents should be required to trigger a payment or the despatch of goods. For example, the warehouse might not be allowed to release goods without first receiving the order form.
- Sequential numbering of forms helps to stop employees removing items to cover up theft or fraud. Prominently numbered forms make gaps or additions obvious.
- Using coloured pre-printed forms and changing their colour from time to time will also prevent the criminal from photocopying an old form and changing it.
- A signature should be required on forms wherever possible, though this is not a very effective deterrent.
- Accounting systems should be explicitly defined in a manual, and you should educate staff in adopting the appropriate procedures. This will stop staff from claiming that they were ignorant of methods. It will also speed the auditing process, and help auditors assess whether the system is operating as it should.

CASE STUDY: LACK OF MANAGEMENT SYSTEMS COSTS RBS £290 MILLION

As it slapped a massive $325 million (£208 million) fine on the Royal Bank of Scotland (RBS) for illegally fixing interest rates, the US Commodity Futures and Trading Commission (CTFC) said the bank 'lacked internal controls, procedures and policies'.

This is the language of ISO 9001. As we've seen before in Chapter 5, it's a system for ensuring that operations go as planned:

- the business decides how things should work;
- it tells staff how things should be done;
- it carries out audits to see whether staff actually did what they were supposed to do.

The UK's Financial Services Authority (FSA) was no less scathing when it fined the bank an additional £87 million. It said, '[A year ago] RBS attested to the FSA that its systems and controls were adequate. The attestation was inaccurate.'

The beauty of ISO 9001 is that as long as you implement the system properly, it's impossible for such problems to happen.

If a business wilfully or through ignorance fails to spot obvious opportunities for fraud, or fails to implement simple checks, no system in the world will detect illegal behaviour. Organizations need a certain minimum level of insight into human nature for a management system to work. If individual executives or traders are a law unto themselves, it's a crisis waiting to happen.

How Fraud is Uncovered

According to the survey by KPMG quoted earlier in this chapter, only one in four cases were detected by a management review. Most fraud cases were discovered following an employee blowing the whistle or a tip-off from an anonymous or external third party.

The organization should make it easy for people to report fraud.

What To Do on Discovering Fraud

Fraud, once discovered, may be distressing or difficult to handle since it may involve a senior or longstanding employee. That is why the company should have written procedures and should scrupulously follow them. There should be procedures to:

1. preserve the evidence;
2. freeze misappropriated assets;
3. report fraud to the authorities and shareholders;
4. implement a public relations plan if the company's image is at stake.

In most businesses, fraud is gross misconduct, and gross misconduct means dismissal. However, some businesses are inclined to be lenient to the perpetrator, who often has extenuating circumstances. This means giving them a written warning and making arrangements to repay the money. This can be a mistake because fraudsters are often repeat offenders. In one company, management wanted not to sack someone who had committed fraud because she was a single parent and the loss of her job would have been traumatic. So she was kept on. Three years later the business discovered she was committing fraud again.

What Companies Do to Prevent and Report Fraud

According to an Ernst & Young survey, 52 per cent of companies have a formal fraud prevention policy, and more than half the surveyed companies had trained their staff in fraud awareness in the past year.

The three best ways of discovering fraud (according to the respondents) are, in order, internal controls, whistleblowers and internal audit. However, external audits were thought to be less effective in detecting fraud than 'by accident'.

This indicates that internal company control mechanisms and audits should be a common and well-managed part of every business. To reduce the cost and improve their effectiveness, the audits should be targeted on suspected areas of fraud. Further, there should be a way for suspicious employees to convey their concerns in confidence, working with tandem with modern legislation which protects whistleblowers.

Many frauds are never reported. This is because of the time that it would take, the embarrassment that would be caused to the business and the fact that the money has been recovered. It is estimated that 40 per cent of frauds are detected but not investigated.

Risk Assessment

By answering the questions below, you can check to see how vulnerable your business is to fraud.

Topic	Question	✓
Assets	Does the organization have assets worth stealing?	
	Does it have commercial secrets that a competitor would pay for?	
Staff	Are any employees responsible for finance or assets unsupervised?	
	Is any employee addicted to gambling or drink, or have heavy financial commitments?	
	Do any employees have close links with suppliers or customers?	
	Does any employee display wealth greater than their income would permit?	
	Does any accounts employee never take a holiday?	
	Is pre-employment vetting weak?	
Systems	Is record-keeping weak in some areas of the business?	
	Is there a lack of written procedures in parts of the business where fraud might occur?	
	Does the organization fail to conduct fraud audits?	
Total points scored:		

Score: 0–3 points: low risk. 4–6 points: moderate risk. 7–11 points: high risk.

The Appendix has a summary of the checklists. By entering the results of this one, you can compare the risk of fraud against other categories of risk.

14
Staying Financially Healthy

'Does that mean you're getting rid of jobs?' says one reporter, sticking a microphone under Phil's nose.

Phil ignores him and goes on reading from a prepared statement. 'The business has been impacted by a mix of soft demand for our services in a sluggish globalized economy plus competition in the sunrise economies.' Phil always talks in management jargon.

'Conditions have deteriorated more than expected,' he went on. 'People will understand that with the cost reductions we're making, some of them will come through labour number reductions. And now I'm going to pass you over to my colleague who will talk you through some of the details.' And with that he hands you the microphone, and disappears. The press surge forward expectantly.

Financial Risk Made Simple

Financial risk is basically about profitability. This in turn comes from the interaction between revenue and margins on the one hand, and costs and debt on the other. We show this in Figure 14.1.

In a few cases, financial problems are caused by financial mismanagement, such as failure to manage debtors. But financial problems are mostly created through business risks that are caused by managers or staff in other departments.

Thus the finance department has a strategic role to play in controlling others' desire to spend and their unwillingness to cut costs. And line departments need to be aware of the impact their decisions can have on the future of the organization.

We start this chapter by looking at the external environment as it affects the organization's finances – the economic cycle. This affects most kinds of organizations, even government bodies, because purse strings get tightened in a recession and are loosened when the economy is growing. In Chapter 20, on Marketing, we also look at some other macro economic factors.

Managing the Economic Cycle

Whenever we think we have the business under control, along comes something unexpected. And downturns are something that most people don't see until it's upon them, perhaps because we don't want to think about the pain it will bring.

But it's vital to see where you are in the business cycle. More specifically it's important to know how far away you are from the next recession. That way, it won't take you by surprise, and you can plan for it. While that may sound pessimistic, companies that are realistic about the future are less likely to suffer.

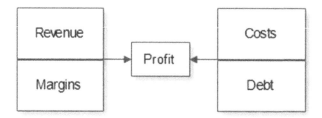

Figure 14.1   Financial risks

Knowing when the next growth period will be is also important, because it is a time for investing in new plant and more call centre staff. But except where very long time scales are involved, that is a nicer, less fraught task.

Let's remind ourselves first how the business cycle works. It involves a growth or boom period, which reaches a peak. This is followed by a downturn, which deepens into a recession. Finally, there is a recovery period (Figure 14.2).

At the start of the cycle, demand begins to grow, fuelled by delayed purchases, optimism and government investment. This leads to a plateau or peak, with most firms doing well. Towards the end of the cycle, competition grows stronger as more businesses enter the market. Inflation then grows as it begins to match borrowing which in turn is getting out of hand. The economy overheats.

Assets, such as property, get over-valued at this stage. Buyers begin to lose confidence and disappear. Sales that came easily are no longer there. House sales slow down.

Finally, there is a period of recession: firms go out of business as they find revenues no longer meet their costs, while creditors grow tired of waiting and put them into receivership.

The government seeks to reduce inflation by seeking to reduce borrowing. The economy stagnates. Consumer demand falls, businesses fail and asset values slump.

This leads to a clearing away of failed businesses through insolvencies, the so-called 'creative destruction' effect. At that point, the cycle is at its trough.

Subsequently, there is a recovery period, as demand begins to grow again. Consumers begin to feel more confident, and go out to make purchases they have postponed. House sales start to rise, leading to a growth in construction. And so another cycle begins.

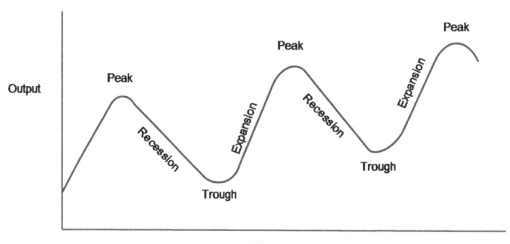

Figure 14.2   The business cycle

The cycle provides special risks for those businesses whose revenues come from discretionary spending or government funding.

Durables like cars and appliances suffer in the downturn, but benefit in the upturn. Construction and real estate agency suffers in recessions, usually with a big decrease in numbers employed.

Addictive substances like cigarettes and the heroin and cocaine trade are relatively inelastic, that is to say they are unaffected by changes in the economy, because their users must have them.

Utilities companies are relatively unaffected because people and businesses need to heat and light their property.

The cycle is unpredictable, but being able to recognize its stages allows the business to respond better.

Many industries are affected by the business cycle; but some firms manage it better than others, thanks to effective planning and responsive systems.

In Chapter 20, we point out that most businesses need to accept that sales will decline in a downturn and recession. Yet few are able to admit this. And those who do usually resort to planning on seizing market share. It is sensible, however, to forecast a decline in sales.

The business can then reduce staff and cut overheads before it is forced to, allowing it to do things in an orderly fashion. And that way, it can get through the recession unscathed.

It is common, but risky, to cut sales and marketing expenditure in a recession. Businesses need every order they can get, and businesses that reach out to customers in the downturn come out the other end in better shape.

Identifying Stages of the Business Cycle

Economic indicators tell us where the economy stands, and these are best seen over time on a graph. Leading indicators suggest what the future may hold. The Conference Board (www. conference-board.org) has an index of leading indicators for many countries and the major economic areas. For the UK, they comprise the following:

*   Order book volume (source: Confederation of British Industry).
*   Volume of expected output (source: Confederation of British Industry).
*   Consumer confidence indicator (source: European Commission).
*   FTSE All-Share Index (source: FTSE Group).
*   Yield spread (source: Bank of England).
*   Productivity, whole economy (Office for National Statistics).
*   Total gross operating surplus of corporations (Office for National Statistics).

Going back to 1970, this index has successfully signalled turning points in the UK business cycles.

Lagging indicators, meanwhile, confirm what has happened. They include the average length of unemployment, the consumer price index and commercial lending activity.

Borrowing and Investment

Most businesses need to borrow, to invest in the future. As we saw above, this is particularly true in the recovery period of the economic cycle.

But borrowing has four effects that increase risk:

1. *It adds to the burden of overheads.* Some revenue goes to repay loans rather than meet costs or be invested in new products.

2. *The bank or shareholders may lose confidence* in the company's ability to repay the loan. Shareholders may sell their stock, while the bank may press for repayment.
3. *The company may reach its overdraft ceiling* and be refused further borrowing. At this point, it will be unable to pay its debts. If it cannot find new sources of credit, it either finds a buyer or goes bust.
4. *The bank may decide to call in its overdraft.* It's been a long time since the local bank was a reliable friend to any business. You never know if a decision made far away will result in a letter giving you 30 days notice that your borrowing facility is being cut. Becoming reliant on an overdraft is a sign that the business is not making headway, and it puts the business at risk from unpredictable bankers.

A highly geared company (one that is borrowing a lot) is where debt is more than half the shareholders' capital (a debt: equity ratio of 50 per cent). When the debt exceeds shareholder's capital, the company is heavily in debt.

For the four reasons quoted above, excessive debt should be reduced. This can be done quickly, by selling assets; or slowly, by making more profit.

Much of this chapter is about improving profit. We look at selling assets towards the end of the chapter. But first we look at how companies come to labour under excessive debt.

BANKS AND FINANCIAL RISK

For a bank or building society, the capital ratio is an important measure of risk.

The ratio relates to the bank's reserves, the amount of cash it has to pay people who want to withdraw their money. This is measured by the leverage ratio: the money it owns as a percentage of the loans it has made. The standards vary by country and by definition; but regulators keep pushing to raise the amount of cash their banks hold. Bad debt, losses and drops in demand can reduce this. The international standard is 3 per cent, meaning the bank can loan 33 times what it has in the bank. If the bank lends too much money, loans unwisely and suffers losses, it may run out of money and go bust, taking depositors money with it.

Excessive Debt

Excessive borrowing is usually caused by:

* expanding too fast;
* failed acquisitions;
* being the victim of private equity deals;
* long-term lack of profit.

Excessive debt is sometimes turned into a crisis by an increase in interest rates. But the underlying problem is the debt itself. Let's look at the four prime reasons mentioned above.

EXPANDING TOO FAST

Rapid expansion is the cause of many companies' failure. New businesses and those who are expanding into new products or markets need more capital than they imagine. Profit always lags far behind investment; and there is a saying: 'Work out how much it will cost, and how long it will take. Then double the cost and triple the time.' There is a certain truth to this. Consumers are often wedded to the brands they know and are often unaware of new developments in the market. And businesses often find they lack the cash to survive.

Sephora started as a chain of 54 loss-making French beauty products stores owned by Boots. It was bought by the luxury brand LMVH, the champagne to luggage company. As Sephora expanded into Italy, the USA and elsewhere, it reached 461 stores in 12 countries, and in doing so lost $50 million, mainly in Japan. It looked like LMVH was about to sell it. But Sephora got a change of management, introduced low-price own label products, and started to achieve success. It was a close run thing.

FAILED ACQUISITIONS

As we see in Chapter 21, sixty per cent of acquisitions fail. It costs money to buy a business, and this typically comes from additional borrowing. This loads debt on to the business, making it harder to break even, harder to make a profit, and harder to invest in core activities.

It's not that acquisition is a bad thing, but management has to be aware of the consequences of failure.

BEING THE VICTIM OF PRIVATE EQUITY DEALS

This is the converse of failed acquisitions. Here, the acquired company is carrying debt obligations to the venture capital company that bought it. Again, it's harder to make profit when the owners need to take money out of the business.

The buyout expert Jon Moulton has said 'a third of mid-sized firms subject to leveraged buyouts will fail or require restructuring', the latter meaning cut backs following losses.

LONG-TERM LACK OF PROFIT

No matter what kind of organization you're running, an iron law operates.

As shown in Figure 14.3, costs have to consistently be smaller than revenues, leading to profit.

While that might sound obvious, it is at the heart of every failed business. Here are some examples:

- New projects get investment but they fail to win the customer's heart. This is typical of many internet startups.
- An organization relies for too long on old products or services, which gradually go out of fashion, and revenues slip below the level of costs. Every old-fashioned menswear shop that appeals only to old men falls into this category.
- The business fails to achieve a sufficiently high price for its services, either because price competition is too severe, or because it hasn't costed its services properly.
- A sudden downturn or loss of business sweeps away a business that had only marginal profits.

Few people in an organization have an understanding of the costs and revenues. This isn't surprising: salaries are a big secret, and every profit and loss sheet has embarrassing lines that the finance director would rather not reveal to employees. Meanwhile the average employee comes to work every day, in the fond belief that their job is permanent and that their pay cheque will always arrive. Only the finance director worries about making the payroll.

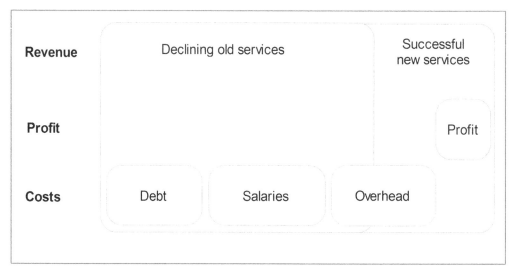

Figure 14.3   The gap between revenue and costs

WARNING SIGNS OF AN UNPROFITABLE BUSINESS

Management makes regular requests to the bank for a short-term increase in the overdraft to cover this month's wages, pending the arrival of a big customer's cheque.

The business pays bills in cash, that is, by credit card, to get additional stock. The credit card gives then an extra 30 days credit.

The owners pay company bills with their personal credit cards.

The company stops making its payments to HMRC or the tax office, since those payments don't seem urgent. But the taxman puts more companies into receivership than most, something that many troubled businesses overlook.

Management discounts its products heavily to generate cash.

Letters come from the bank pointing out that the business has exceeded its overdraft limit.

The business requests an increase in borrowing from its bank or a supplier.

Management pays only part of each creditor's bill.

The company gets repeated calls from suppliers chasing payment of their bills.

Management spends its time trying to survive rather than running the business.

The Need for Healthy Margins

A strong margin is the cornerstone for a healthy business. There is little point in growing a big empire which cannot pay its bills.

Some businesses are investing in the future or engaging in a 'land grab' mode, so that for a period of time their investors accept losses or operating at breakeven. An example of this is Amazon which for 15 years made virtually no profit (it's making just 1 per cent at the time of writing).

Another example is new corporate projects which can take time to come to fruition. But these are the exceptions. Mainstream ongoing businesses are a better yardstick. Healthy margins are a combination of the following:

* adequately high prices;
* sufficient volume;
* low costs.

It is hard to get the first two right – high prices and volume – especially in changing markets. But the third element, low costs, is very much within the company's control. Next we consider the two main elements of cost – production (or operational) costs, followed by overheads.

Reducing Production or Operating Costs

A high production cost demands a high selling price, which can lead to a loss of business. Alternatively, it can cut margins, which moves the company closer to the danger zone.

Benchmarking, market research and feedback from customers will tell the company whether its production costs are excessive.

In Chapter 5 we examined several factors (such as design) which affect production risk and cost. In this chapter, we start by looking at ways of reducing variable costs; through efficiency, automation and order size.

Increased productivity – getting more output for the same cost – is an often neglected way of looking at reducing risk and improving margin. Increased productivity means

- achieving more output with no increase in costs; or
- achieving the same output for less cost.

We look at these next.

ACHIEVING MORE OUTPUT

Improved efficiency reduces costs. This may entail better planning so that materials are at hand when needed. It can also involve reorganizing work or payment systems so that the workforce produces more products or fewer rejects. It may also entail removing bottlenecks and reducing manning levels.

ACHIEVING THE SAME OUTPUT FOR LESS COST

Reducing variable costs often involves increased automation: replacing people with machines. Companies evaluate the value of the investment by comparing the cost of the machine against the savings in labour and material.

Small orders are often expensive, because the profit is often outweighed by the cost of fulfilling the order. Such business is sometimes managed better by a wholesaler or retailer. Many businesses are moving towards 'batch of one' production (where every product is different). While this is a laudable marketing aim, it adds cost and therefore risk.

Cutting Overheads

Every maturing business tends to add fixed costs (staff, property and equipment) to help it meet demand. An increase in profit lets the company invest in better IT or more advertising. And managers want more support staff.

But if sales start to fall, these overheads become a burden, and some of them may be pruned without affecting sales.

Costs are easy to add and, once in place, are difficult to cut. High costs put the company at a disadvantage because they raise the breakeven point. When sales slide, a low-cost producer can survive longer, whereas the high-cost producer reaches a loss situation earlier, and is therefore less well-equipped to survive.

Reducing total costs and unfixing fixed costs allows the company to make profits on a slimmer turnover. In a time of crisis, the better-run company will survive longer.

GETTING RID OF FLAB

Many parts of a business under-perform. If a company gets its sales force to call on ten customers a day rather than their usual eight, profits could rise by the same proportion, that is, 25 per cent.

Some companies' flab can be seen in excessive overheads. Marble walls will not disguise the fact that the company cannot pay its suppliers. A directors' dining room or a fountain at the front of the building will impress no one if the production equipment is held together with string.

The company should avoid having too many layers of management, and many companies have cut out tiers of middle management. Staff should be encouraged to manage themselves, and senior management should run the business rather than managing day-to-day problems.

Introducing a voluntary severance package allows the business to reduce its headcount in areas where it is over-manned. It causes less resentment and fear than compulsory redundancies. This strategy has the potential to slim costs quickly. Unions prefer them to compulsory redundancies, so they have a particular advantage in unionized workforces. However, you can't guarantee that right people will volunteer to leave. And if you don't accept an offer, you may demotivate the member of staff.

UNFIXING FIXED ASSETS

A production line, complete with manufacturing staff, is virtually a fixed overhead. It will continue to produce products unless drastic action is taken. It is not easy to lay off staff, nor recruit skilled workers when business improves. Every time a company commissions a large factory, it is adding a substantial amount of extra risk. Many factories have huge appetites, requiring a large output every week to break even.

However, some parts of the process can be sub-contracted. This means that the company only pays for what it needs. This is a 'make or buy' type of decision. Computer firms often re-badge other companies' printers or disc drives as their own. This extends their product range without requiring the company to invest in production facilities. Many small and little-known electronics firms are kept busily at work assembling satellite TV receivers and DVD players for famous Japanese companies. Large companies sub-contract their manufacturing because of the brief lifecycle of such products and because they recognize that their expertise now lies in designing and marketing their products.

You can encourage service functions to become separate profit centres, by seeking work from outside the business. An extension of this is to 'privatize' them, whereby they become a separate business and sell their services back to the company.

SALARIES

'Overhead walks on two legs' is a phrase attributed to the late Felix Dennis, millionaire publisher and previously a defendant in the *Oz* underground magazine trial. It reminds us that all the people walking around the office are a cost. And people are the biggest cost in many organizations, accounting for 70 per cent of operating costs in many cases.

Human Capital Management Institute (hcminst.com) suggests taking the humble organization chart and adding people costs to it. A digital organization chart enables the business to identify areas of high cost, and allows you to compare the cost of individuals and departments against performance and usefulness to the business. In other words, you may want to keep expensive engineers because they are an asset. Equally, you might find you can increase the average number of people reporting to each manager, and thus reduce headcount.

When a company is performing badly, it should probably be the staff furthest from the customer and the production line who should be made redundant quickest. That often means R&D people, designers, health and safety officers, personnel assistants, training managers and environmental managers.

However, organizations often seem to get rid of sales staff. This risks losing contact with customers; but you can mitigate the impact by concentrating on the most important customers and either abandoning small ones or leaving them to be serviced by agents or wholesalers.

As we will see below, apart from getting rid of people the business has other choices. This includes outsourcing, as well as making salaries more variable by relating them to corporate performance.

Making salaries more variable

Employees' salaries often represent one of the largest costs to the company. Traditionally, the only way to reduce salaries was by making redundancies. In other cases, management has asked for a wage reduction from staff.

Employees can also be given *fixed-term contracts* of employment for one to five years rather than permanent contracts. This reduces management's commitment to the employee, but it also reduces the employee's loyalty because they know their contract might not be renewed. It doesn't reduce costs in the short term, and sometimes leads to the organization making payouts of up to one year's salary to redundant employees rather than the three to six months pay it might have cost them to get rid of the formerly permanent staff.

Some firms use *part-time or seasonal staff* to meet peak sales or production periods. In slack periods the firm can reduce its use of part-timers without risking the core staff.

Companies can make part of their employees' salaries dependant on corporate profit or turnover, known as *performance-related pay*. This allows employees to earn more in times of profit than they could if their salary was fixed (and less when the company is making losses). This evens out the peaks and troughs of the business cycle, and thus reduces the amount of hiring and firing. They are unlikely to be popular in times of poor sales, and may lead to employee dissatisfaction.

Zero hours contracts (ZHC) give the employer absolute flexibility over salaries, because the business doesn't have to guarantee any salary at all. At time of writing, 90 per cent of employees at McDonald's and the retailer Sports Direct are on zero hours, as are 80 per cent of employees at JD Weatherspoon, the pub group. Fifty-five per cent of all domiciliary care workers are on ZHC, while 19 per cent of hotels and restaurants and use ZHC. In many cases shifts are cancelled at short notice; and the employee may have to seek permission from the employer to work for someone else.

ZHC has been criticized as unjust, and those using it risk attracting negative publicity. Moreover in times of low unemployment staff will leave. A better solution might be to offer a minimum guaranteed hours' contract. Chartered Institute of Personnel and Development (CIPD) research shows that 38 per cent of those employed on ZHC are employed full time, working 30 hours or more a week. Companies therefore need to weigh up the undoubted flexibility and cost savings of ZHC against the advantages of a loyal and settled workforce and no risk of media criticism.

SALES SUPPORT COSTS

Companies sometimes put sales staff on *commission-only* payment. This has its drawbacks. It can encourage sales people to go for short-term sales, and to sell products to people who don't need them. It also discourages those activities which don't lead to immediate sales, such as client servicing. And this leads to a lack of team work.

Other firms convert their sales people to *self-employment*, and allow them to sell the products of non-competing firms. The company is then relieved of salary costs, paying commission costs instead. This has the disadvantage of reduced control, reduced loyalty and poorer customer servicing.

A similar solution is to transfer the sales function to an *agency*. This reduces fixed costs. On the other hand, an agency sales force is selling many companies' products, and therefore some firms will lose out.

A better solution may be to focus sales effort on *key accounts*, and to service smaller customers through telesales, which is less expensive. This reduces cost while often improving effectiveness. It also matches the trend in most markets towards a concentration of buying points.

HOME WORKING

Organizations can reduce their costs by arranging for staff to work from home. The first stage is to disband regional offices and for regionally-based staff to work from home.

Reducing the number of staff who need desk space allows the company to move to smaller premises. At some businesses, staff book a desk space when they arrive at work (known as hot desking), and all phone calls are automatically routed to the telephone number associated with that desk.

Apart from IT staff and sales people, home working is suitable for management consultants, architects and other high-grade professionals. But companies should take care when considering home working. Its disadvantages include lack of supervision, loss of face-to-face contact and the danger of reduced productivity.

OUTSOURCING

Companies have for a long time outsourced any functions which are not seen as core skills, or which can be better provided by a specialist business. They include:

- computer services;
- premises management;
- vehicle fleet;
- catering;
- maintenance;
- customer support.

Using outsourcing relieves the firm of capital costs, which are built into the service contract. It can also reduce variable costs, because a specialist firm can often carry out the process more efficiently.

Jobs that are most suitable for outsourcing are the ones a supplier could do more cheaply and effectively, and where the supplier specializes in that field.

With 1,000 cars and 500 light commercial vehicles operating from 70 sites, one industrial group estimated that by outsourcing the acquisition, day-to-day management and disposal of its fleet it would save £500,000 a year.

In their turn, fleet car companies like Avis Fleet Services outsource activities like public relations, advertising and catering.

However, management's view of what constitutes a core competency can change over time. JPMorgan Chase, the financial services group, first outsourced its IT services to IBM for $5 billion (£2.7 billion), and then reversed the deal two years later. JPMorgan said IBM had performed well and had missed no milestones. But the company now regarded technology as a source of competitive advantage. For some firms, dealing with third-party suppliers can't match the ease of working with colleagues in the same organization.

Some managers who try outsourcing regret it. A worldwide survey by Gallup for Proudfoot Consulting (AlexanderProudfoot.com) showed that one-third of managers said outsourcing had either delivered less than expected or had been a complete failure. And a study by PA Consulting (paconsulting.co.uk) showed that two-thirds of companies were disappointed. But Phil Morris of outsourcing advisors Morgan Chambers (now Burnt Oak Partners) said when outsourcing deals fail it is usually because they were badly handled from the start. He said the complexity of bringing it all back in-house when it goes wrong and transferring staff again means that most companies opt to simply try a different supplier instead.

'There are a substantially greater number of companies that solve failing outsourcing by going to a different outsourcer and opting for a different sourcing and delivery model. We see more re-letting than un-outsourcing,' he said.

Senior management often doesn't understand its processes clearly, especially technical ones like IT. This means they sign an outsourcing contract without really understanding its implications. Companies sometimes manage better the second time around, after they have come to understand the limitations and difficulties that outsourcing brings.

For governments, outsourcing should mean the work is done cheaper or better than in-house. But the Project on Government Oversight (pogo.org), a watchdog group, told a US Senate hearing on intelligence contractors that the Government had paid contractors 1.6 times what it would have cost if Government employees had performed the work. As for doing better work, having 500,000 private sector employees with top secret clearances makes security breaches more likely, according to The New York Times.

In the construction industry, there is substantial use of sub-contractors, who in turn sometime use temporary labour. This has advantages in terms of getting flexibility in the workforce, but there can be quality problems, with corners being cut and a fatigued or under-skilled labour force.

OFFSHORING

At one time, it was blue-collar jobs that were offshored, but there has also been a growth in moving clerical and then professional jobs abroad.

As a result of globalization and the internet, many jobs once carried out in the West can be done profitably in low-cost countries such as China and India.

The jobs that are most easily offshored are those that are not customer-facing, and that involve routine work such as data processing.

If a job can be done via the internet or on a PC, or with the use of scanned documents, then it can be offshored. This includes mortgage applications and much insurance work.

Organizations have also been able to offshore other white-collar jobs such as accountancy. Gartner (gartner.com) reckons that 25 per cent of high-technology jobs in the West could be offshored.

The jobs that are more difficult to offshore involve:

- interacting with others – for example a teacher, social worker or security;
- physical jobs such as veterinary surgeon, shelf stacking, cooking or drain cleaning;
- on-site jobs such as firefighter, drain cleaner or receptionist;
- 'core competencies' such as engineering or proposal writing – however, some competencies which might be considered core, such as design, are often offshored;
- strategic 'head office' jobs such as planning or marketing.

Companies can even turn this work into a profit centre. Having set up a business process outsourcing (BPO) firm in India to do its back-office work, General Electric subsequently sold a majority share in it, and raised $500 million. Gecis, now Genpact (genpact.com), and its 16,000 employees were intended to provide support services for GE, including training financial analysis and administration. Following the sale, Gecis was then able to accept work from other companies, and become a revenue earner rather than an overhead. And GE, in turn, was able to use the $500 million to find new growth areas such as security technology.

Offshoring risks

When it comes to offshoring call centres, 28 per cent of people who sought technical support for desktop PCs reported some kind of communication problems, according to a survey in Consumer Reports. Six out of ten complainants said that the support staff's English was limited or hard to understand. Dell stopped sending US corporate support calls to India, amid customer complaints. But many of their home users are routed to an overseas call centre, according to USA Today.

An alternative to manufacturing in the developing world is to cut costs in Western manufacturing plants. This has been seen in Germany, where Volkswagen aimed to cut personnel costs by 30 per cent, by introducing a number of cost-cutting measures. This included a two-year wage freeze and more flexible work rules. In return the company pledged to manufacture certain models in Germany and thus secure jobs at the plants that made them.

BUILDINGS

Leased buildings can be sub-let or the lease re-negotiated. Sometimes the company can downsize into fewer buildings, selling off the surplus properties. When a building is jettisoned, many other costs go with it: insurance, repairs, canteen staff, telephone bills, receptionists, gardeners, rates and water rates.

Buildings are often millstones, as Athena found when its high-cost shop leases bankrupted the firm; every business should be careful about signing long property leases in uncertain times. You should avoid the kind of leases that commit the company for 20 years, have upward only rent reviews and have no facility for sub-letting. It is better to seek a shorter lease which allows the business to vacate the property or pay less in a case of high interest rates. Such contracts are more easily found when the market is at its lowest.

Breaks should be written into the lease, allowing the business to get out from the building if things don't turn out well.

EQUIPMENT

Depreciating assets like equipment should normally be leased rather than bought. This will spread payment in line with the income that they produce. Even existing plant can be sold to a leasing company on a sale-and-leaseback agreement.

Some production-led companies dote on their machines, and often buy shiny equipment which is then under-used. Sometimes equipment is leased before a downturn in demand. WPM Engineers leased three £80,000 machines, only to find that the demand for its products was falling. The company considered allowing the machines to be repossessed, but feared that they would be sold at a very low price. Instead the company has itself sold one machine, and used the money to pay off the outstanding debt. Note that selling leased equipment without the owner's consent may be a criminal offence.

Care should be taken to avoid contracts which commit the company to extended or onerous payments.

STOCKS OF RAW MATERIALS AND FINISHED GOODS

Excessive stock ties up capital. It also risks being written-off if it becomes out of date. This in turn leads to losses. Tight control of stocks can only be achieved if the company has a good production planning system, uses quality suppliers and has a close relationship with them.

JIT is well established in the car industry, supermarkets and elsewhere, allowing businesses to reduce the cost of their stock holdings. Garages hold few stocks, relying instead on next day delivery for the parts they need.

TRANSPORT AND DELIVERY COSTS

If your goods are moved by a contract carrier you are relieved of the problems of leasing and managing a fleet of vehicles, and paying the drivers. They avoid most of the fixed costs of distribution.

Distribution companies have the advantage of pooling many customers' deliveries and can therefore share the costs between them. They also have specialist knowledge in that field.

An exception is where the organization has only short-run local deliveries. Having your own vehicles might be more cost effective in this case. Wine retailer Majestic gets its store staff to drive as few as six bottles of wine to the customer, rather than having dedicated drivers.

PROMOTIONAL COSTS

The cost of promotion can be up to 10 per cent of revenue in some fast-moving consumer goods companies and even higher among online businesses.

Companies which are anxious to cut back on promotional spending are often those which cannot see a benefit from their expenditure. Sometimes advertising becomes a symbol of virility, with competitors comparing the size of their annual spends. Research suggests that companies which cut back on advertising during a recession do less well when the economy picks up – but this is usually research funded by the advertising industry.

The costs of pay per click internet advertising are much more responsive to a downturn in demand than traditional print or television advertising. When parents are less interested in buying toys in January, toy retailers will notice a drop in their online advertising costs.

IT

Effective IT can cut the time taken to do jobs, and therefore the number of people required. After installing an automated order entry system, one of the author's businesses reduced the time taken to process an order from 45 minutes to less than ten. It also removed any clerical errors.

On a different scale, Hewlett Packard said it was going to get rid of 9,000 staff due to productivity gains from data centres and automation.

More organizations are outsourcing their IT, including their email systems. This gives them professional support, avoids investment in hardware and software, and removes the problem of managing IT people. The bill for outsourcing is not cheap, however. And it can lead to frustration and inflexibility when the workforce discovers that the outsourced staff will not make the changes that they are accustomed to. The outsourcing company will require all such changes to be paid for, something that the clients rarely anticipate.

USING IT TO SPEED UP RECRUITMENT AND CUT COSTS

The HR department of a university with 6,000 staff introduced new software that requires all job applicants to apply online.

In addition, departmental managers now post vacancies on the corporate jobs site themselves. This has got rid of delays caused by each department having to submit its vacancies to the HR department, and then wait for HR to advertise the posts.

The departmental managers can now see applications as soon as they arrive, without waiting for HR to pass them over.

This has cut the time to view applications from 30 days to one day, and the new system has empowered and pleased departmental staff.

The new recruitment system got rid of two full-time posts whose holders simply photocopied applications all day long. And there are now fewer, more highly-qualified HR staff, whose jobs now involve advising departmental managers rather than shuffling paper.

Unprofitable Prices

'Low prices' are attractive words to the consumer, but they can endanger the company's survival. Companies underprice their products for two possible reasons:

1. they lack adequate information about costs;
2. they are (or believe themselves to be) in an overly competitive market.

You overcome this by having adequate knowledge of your costs, and by refusing to sell products and services without a reasonable margin.

> *Management accounting*: Getting management information about direct costs is vital, even if the allocation of overheads to different products can provoke debate.
> *Customer profitability*: Unilever companies, which make a range of supermarket brands from Persil to Flora, assess the profitability of each major customer. This is useful in cases where the customer (for example, a retail buyer) is demanding more discounts. From the annual revenue, the analysis deducts the cost of production, cost of discounts, promotional offers, sales support and advertising contributions. This shows whether the price charged is profitable. It can help the company withstand further demands for price reductions.

In bidding situations, such as construction or tendering for government services, it is tempting to make a very low bid, simply to cover direct costs and to stay in business. This inevitably leads to loss-making contracts, especially as unforeseen costs usually arise. That in turn leads to the cutting of corners, which adds more risk. Bidding in such industries is therefore a major risk and job pricing needs to be carefully managed.

If you have accurate information about your costs, it is perfectly acceptable to win business that merely covers the variable costs and makes a contribution to the overheads, if it leads to the company staying in business. On the other hand, competitors often respond by cutting their prices, leading to a downward spiral.

Risks in the Cash Flow Cycle

The most common reason to go bust is lack of cash, rather than lack of sales. Many finance directors spend their time managing the cash flow cycle, which looks like Figure 14.4.

In Figure 14.4, the business buys stock, either paying cash or using credit. It converts the stock into sales, leading either to cash back in the bank, or a group of debtors – people who have bought the goods or services, but not yet paid for them.

The goal of any finance director is to speed up the payments into the business, while slowing down payments out of it. Keeping the value of stock down is another priority.

The dotted lines are ones to be paid with reluctance, while the continuous lines should be pursued with zeal. Stock and debtors are shown 'under water', because they aren't bringing in cash. This is the cash flow cycle.

It should be said that in a balance sheet, debtors, stock and cash would be viewed as assets. But in terms of the all-important cash flow, it isn't always like that.

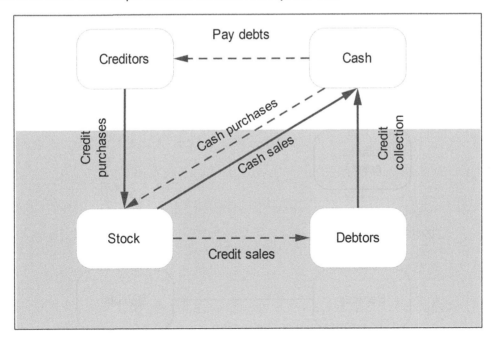

Figure 14.4   The cash flow cycle

SERVICE BUSINESSES CASH FLOW

Not all businesses wrestle with cash flow, and stock doesn't play a role in many organizations. Service organizations' finances tend to look like Figure 14.5.

The finance director spends less time thinking about the costs of stock, and more about whether sales will cover the cost of salaries. The finance director also worries whether investment in new projects will ultimately produce revenue.

In a service organization, the business uses investors' cash and creditors to pay salaries and invest in assets. These convert into services. The services produce debtors, which eventually turn into cash.

So while many people focus on the time between selling a service and getting paid (say, 60 days), the *real* time period is longer, and the outflows larger.

- The cash conversion process includes the time taken from making the goods to getting paid. This starts well before the cash collection process. It means that your money is outstanding for much longer than people realize.
- That extra time involves you spending money on operations and marketing salaries, and the advertising and selling costs that are needed to deliver a customer.

This is particularly true for startup businesses. It takes time to create a range of services, and for the sales funnel to fill with prospects. More established organizations have a bank of clients and services. But when your services begin to age, or your market dries up, it takes time to create new services and explore new markets.

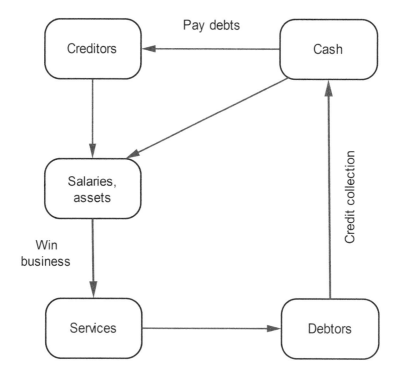

Figure 14.5   Service businesses

Companies with a good cash flow (getting money into the bank before paying out to suppliers) can easily become complacent. Similarly, many organizations only start applying strict cash controls after cash flow becomes an issue. It is better to apply it during the good times, which will ensure the survival of the organization when times get harder.

Bad Debt

Unpaid invoices cost UK business more than £60 billion per year, this being the amount of bad debt chased by members of the Credit Services Association (csa-uk.com).

And according to Intrum Justitia (intrum.com), around 40 per cent of all EU company invoices are paid late. This costs companies 0.7 per cent of turnover in Denmark and Sweden to 3.2 per cent in Spain.

As invoices get older, the chances of collecting the money declines. 60 per cent of three-month-old invoices are paid, in contrast to only 20 per cent of 12-month-old ones.

And if your business has a 10 per cent profit margin, a bad debt of £1,000 will require £10,000 worth of new business to cover the loss. If your margin is 5 per cent, you will need £20,000 of revenue to replace the £1,000 bad debt.

Late payment and bad debts threaten the survival of many otherwise successful companies. There are two elements to controlling debt: prevention and collection, which we turn to next.

DEBT PREVENTION

It is better to prevent bad debts from occurring than spending time chasing them once they have occurred. Prevention is always better than cure, and sound credit management procedures can reduce bad debts by 90 per cent according to some experts.

For new customers, the company should take up references from credit reference agencies and other traders. Your staff should be aware of the need to get references from genuine suppliers.

In a growing economy, companies often feel they need to join the stampede, for fear of being left behind by their more daring competitors. This applies especially to financial services. Corporate policies are overturned, and a new spirit of adventure pervades the firm. This is a risky period, and companies which have sound financial policies tend to survive better than those that don't. It is better to forego risky business, than have it go turn into a bad debt later.

Companies should state payment terms clearly, and charge interest for late payment. All debts should be chased, first by letter and then by phone. There is a true saying: 'He who shouts loudest gets paid first.'

Management should get regular reports on outstanding debts, and never let debts rise beyond a predetermined limit. When they do, it is time to collect them, a topic we turn to next.

COLLECTING DEBTS

Insensitive debt collection can cause a company to lose customers. That is why handing the problem to a debt collection agency is not a simple solution. It is often better to get a member of management or the sales force to visit or phone the late payer, discuss the situation and tactfully seek payment. Other ways to prevent bad debts include factoring and credit insurance.

In factoring, the company sells its invoice to a third-party organization called a factor when issuing it, and gets up to 80 per cent of the value straightaway. The balance is paid when the debt falls due (usually 30 days).

Similarly, credit insurance ensures that the company will get up to 80 per cent of a debt if a customer goes bust. The insurer will usually set overall credit limits for major companies, while new accounts can be opened providing they are less than a specific sum.

BOOKKEEPING, BUDGETING AND FORECASTS

As we've said earlier, access to timely and accurate information is essential. Accounts always arrive after events have happened, which is why budgeting is important. It lets organizations see whether they are veering away from what was planned. It's another way to reduce risk.

Pension Peril

At the time of writing, the world's largest 100 companies had a pension deficit of €290 billion, while the FTSE 100 companies deficits stood at £41 billion.

For many organizations, pensions are a growing problem because each month payments to pensioners are larger than the fund's income, leading to ever-growing deficits. This is serious because such large liabilities could make businesses insolvent.

The pension shortfall has come about because:

- In previous decades, businesses offered generous staff pensions.
- Employees started living longer. Retirees who originally lasted seven to eight years are now surviving for 25 years.
- As stocks rose in the 1990s, businesses stopped paying into their pension funds, a so-called 'pension holiday' that amounted to a cut of £18 billion. For seven years Unilever took pension holidays that saved it millions of pounds. Then in 1999 it also removed the fund's £270 million 'surplus', adding it to the company's profits and subsequently given to shareholders.
- In recent years, returns in equities and bonds have fallen.
- City fund managers have done a poor job of selecting the right investments.
- Businesses have sought to reduce risk by moving their investments from equities to bonds, but bond yields have also fallen.

And as a final woe for pensioners, rather than their employers, low inflation has led to low interest rates, which in turn produce a smaller monthly income.

To reduce their pension liability, organizations are adopting various strategies, involving paying down the debt, paying out less and reducing the future risk. They range from the simple to the abstruse.

- *Put more cash into the pension fund.* Of £21 billion paid recently by UK companies into their pensions, £11 billion was to reduce the deficit rather than to increase benefits to members. BT alone paid an extra £2 billion into its pension fund in a single year to reduce its shortfall.
- *Pay out less.* Move employees from a 'defined' or guaranteed pension to a 'money purchase scheme', one that provides benefits based upon the amount of money that is in the employee's own pension 'pot' when benefits are due to be paid.

- *Transfers.* The fund can offer members the opportunity to transfer out of the scheme through enhanced transfers.
- *Annuity buyout deals.* The trustees pay an insurer to assure the future payment of members' pensions. While this removes the risk, it is an expensive option.
- *Longevity swaps.* These remove the risk of members living longer, with the pension trustees paying a fixed regular premium. However, this places a further burden on the pension fund.
- *Contingency assets.* You can hand over assets such as the organization's IP rights, to the pension fund, sometimes in place of company contributions. The assets will produce cash if the company goes bust or the scheme needs funding.
- *Hedging.* The fund can hedge against inflation or interest rates.

However, these increasingly racy schemes, with their trend to securitizing rather abstract assets, mirror the CDO financial products of the late 1990s which eventually unravelled and led to the sub-prime mortgage crisis.

Companies still see their pension a source of savings in troubled times. Trinity Mirror cut annual contributions to its pension fund from £33 million to only £10 million a year. The three-year cut was in order to obtain a refinancing, which in turn let it pay off £168 million to US creditors. The company, which publishes the *Daily Mirror*, had suffered a 40 per cent slump in pre-tax profits, due to falling advertising revenues and increased newsprint costs. Meanwhile, Trinity Mirror's pension deficit rose £55 million in that year to reach a total funding deficit of £172 million.

Similarly, Premier Foods, maker of Mr Kipling's cakes, deferred its deficit contribution payments into its pension fund for two years. This helped it get a refinancing package, under which 28 lenders extended the term of its £1.2 billion of debt.

Tax Risks

Organizations seek to minimize their tax payments by a variety of means. Richard Murphy of Taxresearch.org.uk includes the following:

- Setting up intermediate holding companies in low-tax regimes:
- Setting up subsidiaries, ones that can charge other parts of the organization, in low-tax regimes
- Setting transfer prices between different companies in the organization, to the benefit of those in low-tax regimes. In some countries, however, abuse of transfer prices is illegal.
- Specifying in which country sales take place. If a product that originates in a high-tax regime work can be improved or modified in a low-tax one, the added value may be claimed in the latter. Or the product can be sold to a central marketing operation which then makes the profit. Where the product is a download, as in the case of software, it is harder for a nation state to claim that the download took place from their jurisdiction.
- Loading costs in high-tax countries. This reduces the profits attributable to that country's exchequer.
- Locating assets in countries that give tax relief on them.
- Siting staff in low-tax regimes. This applies particularly to mobile managers who operate around the world.
- Getting loan capital from a part of the business located in a regime that taxes interest at a low rate. This helps business reduce profits by paying high interest to the subsidiary, rather than pay taxable dividends on share capital.
- Making charges for IP rights. Virgin, based in the low-tax British Virgin Isles, claims royalties from its businesses operating in the UK and elsewhere.

- Negotiating grants or tax holidays in return for investment in a location. Countries are usually desperate for companies to set up manufacturing or offices, and companies can play one country off against another.

However, as Murphy says, there are disadvantages about excessive ploys to avoid tax.

Tax havens reduce a company's transparency, weaken its governance, and can lead to an increase in fraud.

And as we see in Chapter 19 on Ethics, there is growing criticism of companies that practise tax avoidance, such as Starbucks and Amazon. And this could damage goodwill and sales.

Although many companies claim they have a duty to maximize profit, this is not part of shareholder law in many countries. Even in the UK, directors have latitude in how they run the business. This claim of 'duty to shareholders' is either misguided or a useful cover for maximizing profit.

Overseas Investment

Chapter 12 examined the security risks of overseas investment and travel. The financial risks of international activities are also a hazard.

The management consultancy A.T. Kearney (atkearney.com) surveys the executives of leading global firms to ascertain which markets they find most attractive for investment.

Asia dominates the top ten positions on the Global Services Location Index, with India (1), China (2) and Malaysia (3), as well as Indonesia (5), Thailand (7), Vietnam (8) and the Philippines (9). The strengths of these countries vary: India has a wide skill base while Vietnam is the cheapest country.

The Middle East and North Africa are increasingly attractive because of their proximity to Europe and vast talent pool.

Generally, the survey shows a preference for countries with strong growth prospects and emerging markets.

However, before making any overseas investment, you should consider the potential consequences. Different companies and different industries will have quite different requirements – a business which is looking to benefit from offshore call centres would look for a large, educated and yet cheap labour force, whereas a manufacturer would want to consider transport links, supplier availability and potential local demand for the product.

When NEC (now Renesas, www.renesas.com) was looking for a location for its new $1 billion microchip plant, it revealed that California was 'a very hostile environment' to set up manufacturing facilities. The company was put off by stringent environmental regulations and labour laws which made staff costs very expensive.

The company eventually sited its new plant in Scotland. A strong factor was the productivity of the company's existing Scottish plant, which was 10–20 per cent higher than Japanese factories. Regional grants worth 5–10 per cent of buildings' costs were another factor, as was the growing European demand for chips.

INTERNATIONAL FINANCE RISKS

Exporters sometimes lose money because, by the time they get paid, the exchange rate has changed for the worse. Toyota's profits fell by almost 10 per cent in the space of three months when a weak dollar made imported Japanese cars less competitive in price.

The author discovered that his South African profits were zero because the value of the rand against sterling had fallen by half in four years, while our prices had remained the same. We resolved the problem by taking out costs, which restored the business to profit.

HEDGING

Exporters (and companies dealing in commodities) often insure against a change in exchange rates by the use of derivatives (futures and options contracts). Some companies only take out forward contracts. This lets them buy a currency at a set rate in advance. It is thought to be cheaper and less risky than buying options, which gives the company the option to buy a currency at a set rate at a set time. But unlike forward contracts, buying an option doesn't lock the company into a potentially unfavourable exchange rate.

Used as simple insurance, derivatives are straightforward and risk-free. The problem occurs when companies use them to speculate. In the UK, local authorities lost £600 million by investing in interest rate swaps.

The most famous case was Nick Leeson of Barings Bank. Leeson gambled on the future direction of the Nikkei 225 and was responsible for 10 per cent of the bank's profits. However, by allowing Leeson to essentially be his own manager, the bank made it easy for him to disguise enormous losses. Leeson disappeared from his offices and left behind £1.3 billion of liabilities. Although Leeson was later found and prosecuted, the losses led to the collapse of the bank.

Despite this, derivatives trade has grown sharply in the past decade, now estimated at $1,200 trillion, which is 20 times larger than the global economy. This has concerned many experts who point to the role that derivatives played in the last recession.

Speculating in derivatives is attractive because of the huge sums that can be made if the market moves the right way. The average company should recognize its inexperience and the scale of the risk, and restrict itself to insuring foreign revenue on the forward markets.

COMMODITY PRICES

A business that processes a commodity (such as a coffee manufacturer) is at risk from commodity prices. The company can reduce this risk by buying commodity futures. This need not be expensive, but such futures cannot only produce a temporary cushion. Eventually, rising prices must be paid for by the consumer, and excessive price rises will ultimately cause them to switch into substitute products (such as tea or cola, in the case of coffee).

The growth of China has had a major impact on raw materials such as iron and steel. Over a ten-year period, Chinese demand for iron ore grew by 16 per cent each year on average. But with the rebalancing of the Chinese economy, and the expansion in supply around the world to meet China's demands, some experts believe that the price of such raw materials could stabilize or even fall.

MANAGING CURRENCIES AND RAW MATERIALS

Hedging will smooth the organization's costs in the short term. But in the medium term you must pass rising costs on to the customer in the form of higher prices.

This means factoring currency and commodity risk into your pricing strategy, by planning future price rises.

And when the currency or commodity goes in your favour, you can choose to either take the extra margin or go for more volume by dropping your prices.

If your organization is subject to fluctuations in currency or raw materials, you should have a quarterly review of prices.

You should also buy and sell locally as far as possible, using local staff and regional business units. This will keep your costs and revenues in the same arena.

SEVEN WAYS TO REDUCE FINANCIAL RISK

1. *Maintain a healthy margin.* All things being equal, a higher margin gives a company more room for manoeuvre, and more time to sort out problems. It allows the business to build up greater reserves, to invest and to acquire other businesses. The nearer the margin gets to zero, the closer it gets to making losses, and the more likely the business will fail. Margins are simply the difference between revenue and costs, which highlights the need to examine both.

2. *Build financial reserves.* Financial reserves will see a business through a recession, or help it escape from an unprofitable new venture.

3. *Have saleable assets.* This includes profitable divisions which other businesses would want to acquire. Many companies survive a trauma by selling some of their subsidiaries.

4. *Avoid financial adventures that 'bet the farm.* This includes acquisitions whose impact on the business could be ruinous. If the outcome could be catastrophic to the business, the opportunity should be forgone, or the operation sold. This could be used as an argument against all investments, but it harks back to Table 1.9 'Attitudes to risk' and Figure 2.14 'Risk severity and probability'.

5. *Bullet-proof the business.* This can include reducing its reliance on a few big customers, heavily cyclical markets or risky processes.

6. *Have a clear oversight of the organization's finances*, with forecasts, good control measures in place and effective audits.

7. *Keep costs low.* The lower the costs, the easier it is to stay in profit. As we have seen, it is important to control fixed costs, something we discuss later in this chapter.

Risk Management for Financial Services

For financial services, good risk management includes:

Better transparency – if errors are spotted and reported early, there is less risk that they will get hidden and multiplied, as happened with Nick Leeson at Barings Bank.

A risk management culture – people need to be aware of the risks that face their department. This requires training.

Better supervision – the financial services industry pays high salaries to high flyers who often work alone. As with transparency above, better supervision might have saved Barings.

Operational risk management – the organization needs controls in place to prevent overly risky decisions being taken.

INSIDER TRADING RISKS

People in the financial services industry routinely have temptation placed in their way when they are privy to insider knowledge. And the courts have been handing down increasingly heavy punishments.

James Sanders and his wife Miranda were each sentenced to ten months in jail after pleading guilty to charges of insider trading. Miranda's sister, Annabel McClellan, wife of a Deloitte's M&A division in San Francisco, had passed her information about forthcoming deals.

London City investment banker Christian Littlewood was jailed for three years; and he and his wife Angie were ordered to repay more than £1.5 million following a confiscation order made by the FSA. He had tipped off his wife to buy shares on learning that their value was set to increase. The court ruled that if the money was not paid within six months, the jail term would be increased by a further three years.

Garbage In, Garbage Out (GIGO)

The phrase 'garbage in, garbage out' (GIGO) tells us that information is only as sound as the original numbers. If you make decisions based on wrong data, those decisions will be faulty. Therefore you need to be on your guard against simply accepting numbers as truthful or accurate. 'Garbage in' takes two forms:

> *Error:* A utility company found that misplaced parentheses in a spreadsheet caused projected gains to fall from $200 million to $25 million.
>
> *Fraud:* Employees can provide deliberately erroneous numbers to mislead readers or hide problems. A healthcare group presented auditors with a spreadsheet containing false numbers that inflated their assets. This falsely boosted the company's value by $3.5 billion, according to Deloitte.

Spreadsheets are especially risky. It is hard to detect errors hidden among dozens of rows and columns. And yet spreadsheets are very common. In a Deloitte survey, 42 per cent of companies said they used spreadsheets as part of their risk management.

CASE STUDY: ALLFIRST BANK SPREADSHEET FRAUD

At AllFirst Bank, the US subsidiary of Allied Irish Banks, rogue trader John Rusnak lost $691 million through unauthorized currency trading. Rusnak got away with his losses by changing numbers in his spreadsheet which the bank used to assess Value at Risk (VaR), a measure of likely losses.

Raymond Butler, of the spreadsheets risk group EuSpRIG (eusprig.org), says the lessons for audit and control of spreadsheets are:

- Be alert for the possibility of deliberate manipulation of spreadsheets, as well as innocent mistakes.
- Assess and test all important spreadsheets.
- Check key data items and constants against their source and critically review them. The calculations in the AllFirst VaR model were correct; Rusnak interfered with the data.
- Use software such as SpACE (Spreadsheet Audit, Compliance and Examination) to check spreadsheets.
- When potential weaknesses or risks in a system are detected, follow them up.

Reporting Adverse Results

The City expects to see a continued growth in profits, year after year, and any forecast of a fall-off sends tremors through the investors.

Reporting poor annual results can cause shares in a public company to slide. If the market loses confidence, and the slide continues, the company could be taken over. (Typical targets also include small companies which are doing well, as well as larger companies which are doing badly.)

The markets should always be warned about poor results. Keeping in touch with city analysts and the financial media is essential for companies which want to keep a steady share price.

All public companies can and should invest in financial PR. If the company keeps in touch with investors, if it explains what it is doing and why, it will be in a much better position in the event of a takeover bid. While the City is notoriously lacking in loyalty or gratitude, investors are more likely to trust a management team with whom they are familiar and whose strategy they understand.

Responding to a Takeover Bid

Strictly speaking, a hostile bid is mainly a risk for the people on the receiving end, rather than the business itself. Takeovers are merely a way for the capitalist system to reward profitable businesses, and allocate assets efficiently.

In the event of a hostile takeover bid, the business should assess the effect of the takeover. If it decides to reject the bid, it should:

- Gather experts in finance, banking, the law, and city PR.
- Tell shareholders why they should reject the offer. Explain the disadvantages of the offer and the benefits of maintaining the status quo. The arguments should be based on finance not emotion.
- Demonstrate conviction and resolution. Shareholders and the takeover firm can sense signs of weakness or vacillation.
- Consider alternatives to outright rejection. This could include mounting a bid for the other company, or making life difficult for it by adopting liabilities. You could create a shareholder rights plan or 'poison pill'. Typically this allows stockholders to buy shares at a discount if one shareholder acquires 20 per cent of the shares. This makes it more expensive for an acquirer to buy the stock. If the plan can only be revoked by the directors, an acquirer will be forced to get approval from the Board, and this puts power in the hands of the directors. It also gives the Board time to find another buyer or 'white knight'. News International introduced a mechanism that allowed existing shareholders to buy stock at a 50 per cent discount if any new investor acquired 15 per cent of the business. However, there are also disadvantages. Such a move would reduce the value of each share. It also protects weak management, and offers directors the opportunity to enrich themselves as the price of accepting the takeover. It deters institutional investors from buying stock. And regulators may deem the poison pill illegal. The UK's Takeover Panel forbids them.

There are numerous variants of the poison pill. PeopleSoft guaranteed its customers that if it was acquired and product support were then reduced, its customers would get a refund of between two and five times the fees they had paid for their software licences. Sports teams seeking to acquire a star player sometimes make offers containing poison pills that make it impossible for the player's existing team to match.

Reporting Risk in the Financial Accounts

Annual reports often convey minimal useful information about the real trading position of the business. To combat this, and to protect shareholders, accounting bodies are increasingly urging companies to be more open in their reporting.

Companies should identify in their annual reports the risks the business faces. This would include an analysis of:

- the main factors affecting a company's performance and market position;

- the risks and uncertainties it faces (including sensitivity to interest rate rises and currency fluctuations);
- how risks are being managed;
- in quantitative terms, the potential impact on results.

According to the UK's Accounting Standards Board, factors liable to affect future results include:

- scarcity of raw materials;
- reliance on major suppliers or customers;
- self-insurance (whereby an organization sets aside money to cover risks rather than paying it to an insurance company);
- skill shortages;
- environmental costs

These should be reported, says the Board, irrespective of whether they were significant in the reported year.

What to Do if a Cash Crisis Happens

If a cash crisis occurs, you have to develop a survival plan and implement it swiftly.

1. Immediate
   - Stop cash flowing out of the business. This will include the following actions:
     - stop all purchases which are not essential to the short-term survival of the business;
     - put all capital projects on hold;
     - reduce costs, especially staff and non-productive sites;
     - collect outstanding debts;
     - slow the rate of payment to creditors;
     - communicate with the bank and other creditors. Renegotiate loans or credit terms.
2. Medium Term
   - sell divisions or assets to raise cash;
   - close loss-making operations.

How Likely is the Business to Go Bankrupt?

There are several models that predict business bankruptcy. Springate is one such model. It uses four out of 19 popular financial ratios that best distinguish between sound businesses and those that failed. The Springate model takes the following form:

$$Z = 1.03A + 3.07B + 0.66C + 0.4D$$

Where
A = Working capital/total assets
B = Net profit before interest and taxes/total assets
C = Net profit before taxes/current liabilities
D = Sales/total assets

The higher the score, the more financially sound is the company; and the lower the score, the greater the danger of the company becoming insolvent. Where Z is less than 0.862; the firm is classified as 'failed'.

At root, this model simply looks for insufficient liquidity, excess debt, insufficient sales and lack of profit.

According to Bankruptcy Action (bankruptcyaction.com), this model achieved an accuracy rate of 92 per cent using the 40 companies tested by Springate. When a different 50 companies with an average asset size of $2.5 million were tested by another researcher, the model achieved 88 per cent accuracy. A third test on 24 companies with an average asset size of $63 million found an accuracy rate of 83 per cent.

You can also use this kind of predictor to assess new suppliers. Dun and Bradstreet offer a service based on this kind of statistics to tell you how reliable a supplier will be.

Risk Assessment – Finance

By answering the questions below, you can assess the company's vulnerability to financial risk. Score one point for every 'yes' answer.

Topic	Question	✓
Business cycle	Is the organization dependent on the business cycle?	
Financial position	Have sales fallen for two years or more?	
	Is any major part of the business making losses?	
	Does the organization have excessive borrowings?	
	Does the company have a large amount of bad debt?	
	Does the business have substantial overheads?	
	Does the company lack a fast and informative management information system?	
Takeover	Is the company vulnerable to a takeover?	
Markets	Is the business in a declining or low-profit market?	
	Is the company dependent on commodities, derivatives or other high-risk areas?	
Total points scored:		

Score: 0–3 points: low risk. 4–6 points: moderate risk. 7–10 points: high risk.

The Appendix has a summary of the checklists. By entering the results of this one, you can compare the scale of financial risk against other categories of risk.

15
Avoiding IT Disaster

The line from Bahrain is a bit faint. 'It's all backed up, I think. So the data should be OK.' Helen is trying to sound optimistic. She's had her laptop stolen. 'I was working at a table in the hotel restaurant,' she says. 'One minute it was there, the next minute it was gone. I didn't see anything.'

'Was anything very sensitive on it?' you ask. 'Well,' she says, 'only the usual stuff. Um, some customer lists, Oh, and that new project we're working on. But chances are it was just an opportunist thief. They've probably wiped everything and sold it by now.'

You put the phone down and chat to IT. 'I don't suppose you have any way of wiping the data off the laptop remotely?' you ask. Terry the bearded IT guy shakes his head slowly.

How Prevalent is Computer Failure?

Is computer failure common? And if so, what causes it, and how serious are the losses? We start by setting out the facts.

In the last 12 months, according to the UK Government's Information Security Breaches Survey (pwc.co.uk), two-thirds of large businesses suffered an incident where they had to restore significant data from backup (for example, a systems failure or data corruption), as shown in Figure 15.1.

The survey was self-selecting, which could mean that problems are over-reported (because only the people who have suffered feel they have anything to say). Nevertheless, it gives us some insight into the problems affecting IT.

Other businesses in the survey suffered different problems. 59 per cent were infected with viruses. 53 per cent suffered theft or fraud involving computers and 73 per cent suffered from hacking.

Over half of all larger businesses (57 per cent) who answered the survey said they had suffered a 'serious' to 'extremely serious' event.

Overall, this indicates that computer failure is clearly common. Only 11 per cent of large businesses suffered an outage of one to seven days while 3 per cent were down for seven to 30 days. Nevertheless that represents a major risk to business, with big implications for business reputation, additional costs and loss of data.

With the social networking sites like Facebook, phishing by fraudsters, the growth of cloud services, malevolent viruses, the risks from tablets and personal smartphones and the continued hacking by outsiders, the job of protecting the organization's IT is becoming ever more difficult.

Other surveys, ones that look just at the causes on internal data loss, come up with slightly different figures. Lacie (lacie.com) puts *hardware failure* at 44 per cent of all problems, and *human error* at 32 per cent (Figure 15.2).

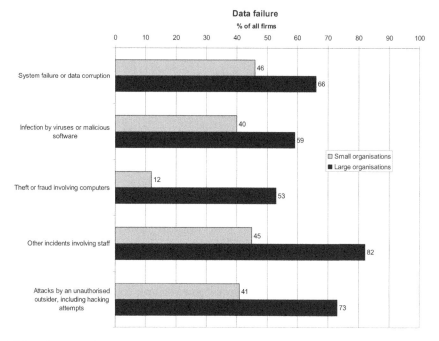

Figure 15.1   Causes of data loss by organization size

Source: PwC

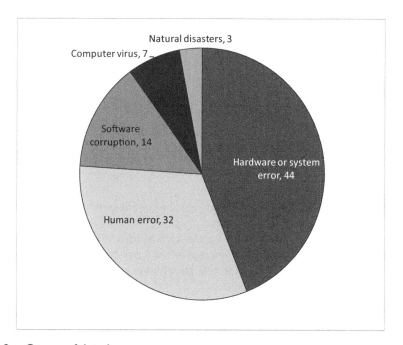

Figure 15.2   Causes of data loss

Source: Lacie

These two causes dwarf all other causes. In other words, there are two main ways to stop data losses: be ready for your hardware to fail, by taking regular backups; and train staff how to avoid losing data.

How Serious is Data Loss, and How Much Disruption is Caused?

87 per cent of businesses are significantly dependent on electronic data, according to the same survey quoted above.

In a study of Canadian CEOs and CIOs by Ernst & Young, 34 per cent of respondents said computer system failure was the most significant risk to business continuity, ahead of other threats such as recession, commodity prices and exchange rates, disasters and terrorism.

Table 15.1 shows that 44 per cent of businesses that suffered disruption. But 14 per cent suffered a major or very major disruption for more than a day.

Table 15.1   Extent of disruption

	No disruption	Less than 1 day	1–7 days	7–30 days	More than 30 days
Very major disruption	44 per cent	1 per cent	2 per cent	1 per cent	0 per cent
Major disruption		8 per cent	9 per cent	2 per cent	0 per cent
Minor disruption		13 per cent	6 per cent	1 per cent	1 per cent
Insignificant disruption		10 per cent	2 per cent	0 per cent	0 per cent

Source: UK Information Security Breaches Report

What Causes Computer Failure and Data Loss?

When asked what the most serious kinds of IT incidents they have encountered are (Figure 15.3), the most common (25 per cent) was systems failure or data corruption. This is the heart-stopping moment when the business finds its computers are no longer working.

Almost as important is infection by malicious software (23 per cent), according to the survey.

Two-thirds of large businesses and a half of small organizations reported systems failure in the last 12 months. This is an extraordinarily high figure. It means that the organization loses its ability to process data or provide services to its customers. At that point, the business stops operating, and staff sit around twiddling their thumbs.

Typically this is caused by a hard drive failure, following by the realization that the backup has failed.

How Easy is Recovery?

In the Ernst & Young survey quoted above, 65 per cent of those who said it was critical to restore their systems within 24 hours admitted they could not do it.

So it is clear that many businesses would have trouble recovering from a systems-wide computer failure before it caused a disruption in their business,

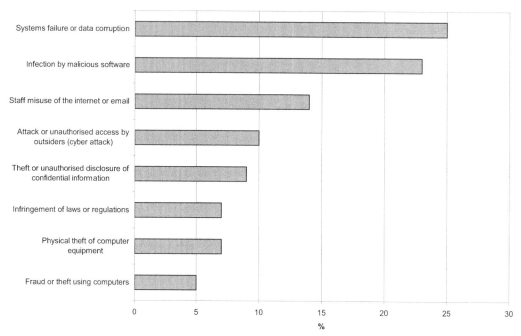

Figure 15.3   Most serious incidents

Source: UK Information Security Breaches Report

The same survey showed that 26 per cent do not have a business continuity plan; 25 per cent have no computer disaster recovery plan and 41 per cent have no overall crisis management plan.

This reinforces the feeling that data is not easily recovered, and that companies are not taking sufficient steps to ensure survival.

HOW MUCH AT RISK IS BUSINESS?

A survey by DataFort (datafort.com) showed that only 18 per cent of businesses have a comprehensive and secure back up plan. Forty per cent of companies are in the company's 'extreme risk' category, with either no backups or backups that would not protect them adequately.

If these companies were to suffer data loss as a result of a virus, technical failure, fire, flood or theft or other incident, the future of the business would be in jeopardy, claims the company.

DataFort surveyed 150 IT-dependent businesses in various markets over a two-month period to monitor backing up habits. They found that an astonishing 83 per cent of companies have an inadequate backup strategy, and that 40 per cent of companies are at 'extreme risk' of losing data.

Half of those who backed up with tape did not have any way to verify whether the backup had been successful. The survey also found that, once the information had been backed up, 19 per cent of respondents stored it close to their computer systems rather than storing it at another location – an essential precaution against hazards such as an office fire.

DataFort commented: 'Given the recent publicity surrounding back up and disaster recovery, you would think that there would be more people taking this issue seriously.'

Protecting the Business from 14 Types of IT Risk

There are 14 main computing risks which cause problems. They are shown in Table 15.2.

Table 15.2   Computing risks

Hardware and software problems	Problems caused exclusively by outsiders	Problems caused by outsiders and/or staff	Problems caused exclusively by staff
1. Systems failure or data corruption	4. Denial of Service' Attack	6. Infection by malicious software (viruses)	10. Data protection and loss of corporate information
2. Hardware and software changes and upgrades	5. Attack or unauthorized access by outsiders (cyber attack)	7. Theft of IT equipment	11. Data loss due to staff error
3. Piracy and licensing problems		8. Fraud or theft using computers	12. Staff misuse of the internet or email
		9. Outsourcing of IT	13. Staff BYOD (bring your own device)
			14. IT Project failure

Next we consider how to prevent these problems from happening.

1. SYSTEMS FAILURE OR DATA CORRUPTION

With a hard drive spinning at 7,200 revolutions per minute, the outer tracks will be travelling at 67mph. The read/write head accelerates at 550g (by comparison, a car crashing into a brick wall at 62mph has a force of 100g). The head flies at a fraction of a millimetre above the disk (0.07mm or less), which is like flying a jumbo jet a few feet off the ground. The room for error is tiny.

If you handle the computer roughly or let it drop, the discs could be damaged, causing them to no longer fly straight, which results in disc failure.

Even if you treat the computer with care, a small speck of dirt on a platter could smack the head, and scratch off the disk's magnet coating, causing it to fail. And even a dust-free drive is working frantically most of the time.

If the power fails (or surges) when a file is open, the user can lose what they were working on, and the file can get corrupted. You can prevent this by using an uninterruptible power supply (UPS). Saving files frequently (or ensuring that automatic saves are switched on) also reduces mishaps.

So you should treat every drive as though it's going to fail tomorrow, and have iron-clad backups. It isn't a matter of *whether*, but a case of *when* it will fail. In the author's personal experience, you can expect total failure every five years, with minor failures each year.

The SMART monitoring system (http://en.wikipedia.org/wiki/S.M.A.R.T.) can predict failure, while if you spot any of the symptoms shown in Figure 15.4, you need to check the system's integrity.

Companies need to develop in-house expertise in data management and data recovery. The more that users understand how their computer can fail, the less likely they are to have a problem. This means training staff to avoid problems. Not enough companies teach staff about the need for backups, not pulling leads out of computers while they're working, not moving computers while they are writing data and not exiting from a program by switching the 'off' button.

Computers should be stored where they cannot be damaged by coffee, accidental movement or floods (so don't put your servers in the basement).

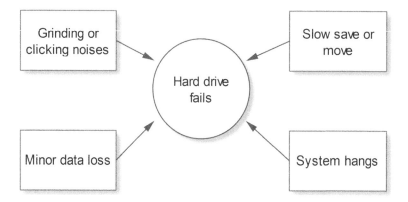

Figure 15.4   Signs of computer failure

MINIMIZING THE LIKELIHOOD OF LOSING DATA

Use Raid: A Raid drive writes data to two separate hard drives, rather than relying on the usual single drive. This means if one hard drive fails, your data won't be lost. However, if the computer was stolen or burnt, both drives would be lost.

Backup daily: All data should be backed up daily, and this should be automatic. The data should be backed up to a separate storage system, so that a failure of the hard drive or mainframe will not affect the data. In other words, if you backup to another directory on the same computer, you will lose the data if the drive crashes.

Keep backup data off site: Same computer backups are not proper backups. When a fire engulfed the premises of Forgeville, an engine filter company, the records kept in supposedly fire-proof cabinets, were reduced to cinders. The company lost catalogue part numbers, invoices, bank details, VAT receipts and 20 years of records. Outsiders were unsympathetic: the VAT inspector said the fire was no excuse for handing returns in late.

Automate the process: It is unwise to assume that staff will carry out backups. Software will perform the work automatically, but IT staff should check that the backups are being carried out, and are readable. It is not uncommon for IT staff to switch off or remove backup software, for example when a server is replaced or upgraded.

As Figure 15.5 shows, there are varying scales of backup, ranging from weak to best.

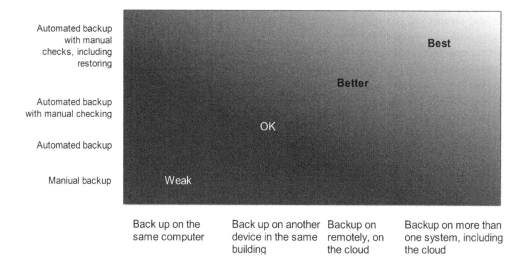

Figure 15.5   Backup model

Even when a hard drive fails, or data is accidentally deleted, the data is still in the machine. Companies can sometimes recover lost data using a remote recovery service, for example Ontrack (krollontrack.com).

CASE STUDY: THE DAY THE DATA DIED

With just 20,000 residents, the affluent Washington state city of Mukilteo is ranked by Money magazine as the ninth best small US town to live in. But disaster struck when the city council's hard drives crashed. The city lost its planning documents, financial data and police records. Initially the town managed to recover only 30 per cent of its data, and staff spent 100 hours re-entering just one recent month's data.

According to *The Herald* newspaper, the city got lucky: it eventually got 95 per cent of its data back, three months later, at a cost of $36,000. Nine months before the incident, the cooling system in the room containing the city's file servers had failed and the room overheated. But the hard drives continued to work for the next nine months. A few weeks before the crash, the city's network engineer had left for another job, leaving only the town with one IT specialist. The system had two backups: a digital copying system and a tape backup. The tape system had broken down four months previously, and wasn't repaired earlier because officials were relying on the digital copying system. The second backup system also failed during the crash. 'We never thought all these systems would fail,' said city administrator Joe Hannan.

Councilwoman Jennifer Gregerson said requests for better equipment and more staff have come from the administration but never were acted on by the City Council. 'For a long time we've understaffed and underfunded the IT department,' she said. 'I'm disturbed that the decision to cut corners and try to do it cheaper, that those choices were made.' After the incident, the city sent its hard drives to an outside company that, after three weeks and a $5,000 payment from the city, said it could not recover the information. Subsequently a second firm recovered 95 per cent of the data. The Mukilteo failure is regularly replicated in other small organizations. There are several lessons to be learnt from this story:

- Backup systems can lull you into a false sense of security.
- Backup systems can fail, even if you have two systems in place.
- If staff are overloaded with work, backup systems can get neglected.
- A problem, in this case overheating, can have consequences much later (in this case nine months subsequently).
- If an organization is too small to afford enough internal IT expertise, and if its data is critical (as it is for a public organization), it should seriously consider outsourcing its IT.

---

2. HARDWARE AND SOFTWARE CHANGES AND UPGRADES

Changing hardware or software always brings upheaval, uncertainty and temporary loss of productivity. Good changes can bring improved speed, efficiency and information; while unnecessary changes produce few benefits.

Many companies are reluctant to upgrade their operating system to the latest version of Windows or move to a newer edition of Microsoft Office. They are often impelled either by Microsoft dropping its support for older versions, or because new equipment arrives with the software installed. IT projects are notoriously difficult, a topic we address in Chapter 16.

Failure is not uncommon during software upgrades. When NatWest updated its software, customers were left without access to their accounts for four days. A programmer managed to delete or corrupt files used to update transactions such as ATM withdrawals to the bank's master copy. So for three days, the nightly updates to 10 million accounts didn't happen.

The Unite union blamed the recent outsourcing of this work to India, where staff didn't have the years of experience that local staff had. Others pointed out that updates should be tested before going live.

Banks and other businesses have systems that are added to over the years. The original programmers have long since left, and old 'legacy' software lies in the program, some of it dormant, and other parts active. No one understands the programs in their entirety. An update in Knight Capital's software caused old code to come to life, causing the company to inadvertently buy billions of dollars worth of shares in 148 stocks on the New York Stock Exchange. The company only narrowly escaped bankruptcy after being bought in a rescue package by a group of investors.

Programmers say financial institutions have failed to invest properly in IT. They also say systems are patched together badly (or 'kludged'). Meanwhile, managers say that programmers love to play with code, and simplicity is not their goal. In the US, state variations cause further complexity. There is no one easy solution, but commitment to, and an understanding of, good IT is the starting point.

CONFIGURE OR CUSTOMIZE?

Modifying third-party software such as SAP or Oracle to suit the needs of the business can also bring problems. You can either Configure or Customize third party software, but the latter carries more risk.

- Configuration means simply adding fields or modifying drop-down lists, but staying within the software's boundaries.
- Customization involves changing the code to create new functionality. When you upgrade, you may lose that customization. There are ways to preserve it, but that involves modifications after the upgrade, and you risk forgetting how the customization was done.

So while third-party software has numerous advantages, there are downsides, too. Specifically you cannot easily modify your program.

---

3. SOFTWARE PIRACY AND LICENSING

Owning unregistered software may not seem like a big risk, but using copied discs or illegal downloads exposes the company to the wrath of Business Software Alliance (bsa.org), an organization comprised of leading software companies. It reckons half of small and medium-sized businesses are using illegal software.

The BSA pays up to £20,000 to people reporting the use of pirated software, and much of this is thought to come from disgruntled employees or contractors. When individuals are laid off, they sometimes 'shop' their former employer as an act of revenge.

Tullis Russell, a papermaker based in Glenrothes, was taken to court by the BSA, for the illegal use of software. Tullis Russell eventually paid the BSA £18,000 in an out-of-court settlement.

The exercise started after received a call on its confidential hotline. Following enquiries, Tullis Russell was found to be using 154 unlicensed products, including Attachmate, Visio, Symantec and Microsoft software.

Twenty-four per cent of BSA's settlements come from organizations within the IT sector, while construction/engineering and architecture/design together represent 27 per cent of settlements; educational organizations make up 8 per cent.

Since the BSA compares records with leading software companies like Microsoft, it isn't hard for them to determine whether the software is legal. But the BSA says it target 'egregious' use of pirated software rather than poor companies that suffer from poor record keeping.

The average fine is in the region of £25,000, though the BSA sometimes settles out of court. Solicitors advise that companies which receive a questionnaire from the BSA should answer it. Unanswered questionnaires are followed by a BSA visit.

The BSA website helps companies collate information about their licenses.

Software licensing as a business issue

Software licensing is a thorny issue for larger organizations. The perils include fines for non-compliance, over-spending on unnecessary applications and badly-negotiated support deals.

Many companies find themselves paying for licenses for already licensed software. Others complain about the byzantine nature of licensing – 95 per cent of the SAP User Group say SAP's software licensing policy is overly complicated, according to Concorde Software (ConcordeSoftware.com). *CIO Journal* says that businesses dislike the overhead involved in keeping on top of licensing, and that software migrating to the cloud causes additional headaches.

Organizations should use effective Enterprise Software Advisors (ESAs) to ensure they only install what they need, especially where they have signed a multi-year Enterprise Agreement (EA). They can also use tools to establish software usage, and hire independent specialist auditors (Software Licensing Advisors) who can negotiate with Microsoft.

Alternatively, some organizations may decide to move to open source software, such as the Linux operating system and Open Office, or to Google apps.

4. 'DENIAL OF SERVICE' ATTACK

A hacker may launch a 'distributed denial of service' (DDoS) attack on your server or website. Hackers can send your server thousands of messages a minute – too many for your computer to respond to. It is like having thousands of commuters all trying to push into a stadium at the same time, causing a blockage at the barriers. Everything on your system slows down or stops working. Visitors to your website will get a message telling them the site isn't available. Staff won't be able to use the intranet. To outsiders it can look as though the organization is defunct. That has major repercussions on your reputation and image.

Any organization can suffer DDoS. The more visible you are, the more likely you are to be attacked. And you don't even have to be a controversial business.

The business needs systems that warn of system failure, and a process for managing such an attack. You will need the active support of your ISP, which means having the relevant telephone numbers and passwords on hand, and not locked in the office. You can also simulate an attack to see how effective your response is.

If you suffer a DDoS attack, the measures you can adopt. These include:

- throttling the incoming traffic;
- route all traffic to larger servers;
- rejecting requests that have a common fingerprint that identify them as malicious (for example the use of specific ports);
- have redundant servers on standby.

5. ATTACK OR UNAUTHORIZED ACCESS

Unauthorized access into the system can take two forms.

1. A member of staff may try to get to confidential data, either out of curiosity or malice. This can result in a loss of company secrets.
2. An intruder, normally a hacker, may try to get access to your data. The results can be a loss of company information, a financial loss (if the company bank accounts are infiltrated), or the loss of data (if hackers put viruses into the system).

There are several ways to avoid unauthorized access or theft of data by staff or hackers:

- *Restrict access to parts of the computer network* which contain confidential data (for example, through password protection). The back end should not be accessible from remote devices. For example, your staff may be able to remotely access and amend their own data in your HR database, but not access anyone elses nor amend the system settings.
- *Use a firewall*, both at the perimeter and desktop level. This will block all computer ports other than those intentionally left open by the IT administrator. Firewalls can provide an effective defence against many hackers but should not be considered a complete solution.
- *Restrict the number of computers with writeable DVD drives.*
- *Block USB ports* using a physical or password lock.
- *Keep some computers off the network* (for example, those in Board members' offices).
- *Adopt closed user grouping*, which allows a company to send confidential information down standard telephone lines.
- *Use strong passwords* for wifi, network connections and important documents. They should be unique to the individual. The system should not allow dictionary words (for example carboatplane) which can be cracked in seconds, nor a phrase, nor substitute numbers for letters (such as 4exampl4), a birth date or a telephone number.

A password that uses dictionary words will be cracked in seconds, and cracker programs add new passwords to their database whenever they locate one. A program like Cloud Cracker (cloudcracker.com) can check 300 million words in 20 minutes, for just $17. And according to ArsTechnica (arstechnica.com) a PC running a single AMD Radeon HD7970 GPU chip can try 8.2 billion password combinations a second.

Once inside the network, hackers can destroy files and damage the company. Businesses should therefore monitor the number of failed password attempts to check whether your computer is being targeted by a hacker. The number of tries should be limited, this being a powerful way of protecting the system. Medusa (foofus.net) and other dictionary software tools can audit an organization's password vulnerabilities. Many are free, such as oclHashcat-plus (hashcat.net).

Any kind of logic that humans like is easier to crack. The programs know that you're likely to start with a capital letter, add four or more letters, and end with a few numbers (such as Hello45). They understand our way of thinking. And if hackers can't crack a password by intelligent guesses, they can use 'brute force' software, which checks text from AAAA0000 to ZZZZ9999 until it finds the right combination.

Programs like LastPass (LastPass.com), Roboform (roboform.com) and 1Password (agilebits. com) can generate strong random passwords, and store them in a vault locked by a master key.

Server passwords should be 'salted', a process that adds characters to the password, which in turn makes the cracking process much longer.

If a hacker is discovered, they should be locked out by severing external communication links. A hacker who discovers they are under surveillance may instantly take drastic steps to cover their tracks, such as wiping all the corporate data.

But it's just as likely that an employee, working from home, will connect to the company's network using remote access or a VPN (Virtual Private Network). If the home PC is infected, this gives viruses and worms access to the corporate PCs and servers. This means ensuring that each work PC is equipped with virus checkers. Home workers should be given this software too.

6. INFECTION BY MALICIOUS SOFTWARE, AND HOW TO PROTECT AGAINST IT

Many viruses find their way into the corporate system when employees open email attachments or click on email links. It also happens when they download programs from the net, such as toolbars or games, or when they bring their own USB flash memory device to work. Other viruses come from hackers that find their way past firewalls.

SCAMS, PHISHING AND PHARMING

Users often receive an email purporting to come from their bank, Ebay or Paypal, asking them to update their credit card details, to continue using their online services.

The email actually comes from criminals who harvest the information and run up a bill on the user's credit card. While the sophisticated reader might be amused to think that anyone would be stupid enough to fall for this trick, known as 'phishing', for the fraudsters it's a numbers' game. If they send the email to enough people, a percentage will be fooled. Pharming is similar to phishing, except that the user is sent to a fake website, misleading the user into believing it is their bank's site, for example.

You have to imagine the hazard posed by your least able employee. Mostly this kind of scam might get only the user's personal details, but there are reports of phone calls from 'the IT department', asking employees to confirm their password to the server. This is a quick way to gain access to company records, and would be invaluable for industrial espionage.

The US Government has openly accused the Chinese military of mounting attacks on US Government systems and defence contractors. This means that any organization involved in the world of defence, infrastructure, transport, research or energy could become the victim of an attack from a hostile government.

- *You must have up to date anti-virus scanning software* such as Avast or AVG. Anti-virus activity used to focus on the perimeter – at the point where emails and internet came into the system; now it must be done at each desktop. And since the anti-virus needs to be updated daily, this must be done by the IT department. It can't be left up to individual users.
- *Give the anti-virus program to mobile employees* for their home PCs.
- *Be wary of replying completely on anti-virus software.* When Imperva (imperva.com) tested 82 new computer viruses in 40 anti-virus products, the initial detection rate was less than 5 per cent. On average the anti-virus companies took a month to identify the new viruses and remove them.
- *Protect access to important work databases by toughened passwords.* Users typically employ the same soft password on multiple sites. Hackers can then use these passwords on other sites. LinkedIn, the business networking portal, suffered a security breach that led to the theft of 6.5 million user passwords.
- *Protect programs from being altered;* you can do this through software that stops the computer's registry from being changed. Staff should not be able to alter programs.
- *Stop staff using pirated or 'borrowed' discs,* or discs from suspect or unknown origins. A policy, which we discuss later, should be introduced, widely disseminated and actioned.
- *You may also decide to remove CD Rom drives or USB ports* from vulnerable PCs.
- *Preventing access to the internet* prevents users from downloading infected games and other files. Using a stand-alone PC for important work prevents it from being infected by other machines in the company.

- *Education is vital.* The organization must educate its staff about the hazards of viruses. To cut 80 per cent of virus attacks, staff need to be told just two rules, according to Tony Dyhouse, cyber-security director at the Knowledge Transfer Network:
  - don't open email attachments;
  - don't click on email links.

Semantic has provided short 30-minute workshops at a global bank, according to *The Guardian* newspaper. The workshops avoid overly technical details, and include the risks from phishing and social networks from a personal point of view, not the bank's. The bank encourages staff to attend the security awareness sessions by offering attendees a new mobile gadget as an incentive; and this has been very successful.

THE DAMAGE DONE BY VIRUSES AND OTHER PROGRAMS

The first computer virus was called Brain. Invented in 1987 by two programmers from Pakistan, it was intended not to cause harm but merely to self-replicate.

The Christmas virus was also developed in the same year. It affected 250,000 IBM users around the world, and caused the company's email system to fail for two hours. Started as a prank by a German law student, the virus drew a picture of a Christmas tree, and then duplicated itself to all the names and addresses on users' distribution lists.

The lost time caused by the Christmas virus among the first 50,000 users is estimated to have cost $2 million. The virus did not destroy data, but shows how rapidly a virus can spread.

Modern viruses are not so benign. The security firm mi2g (*mi2g.com*) estimates the cost of the Netsky.B virus to be at least US$3.12 billion worldwide. This was calculated on the basis of 'helpdesk support costs, overtime payments, contingency outsourcing, loss of business, bandwidth clogging, productivity erosion, management time reallocation, cost of recovery, and software upgrades'.

There are various species of damaging program, not all of which are strictly speaking viruses. *Logic bombs* are designed to lie dormant in a computer, awaiting a particular date. They then 'explode' by deleting files. Unlike viruses, bombs act only once, though their tasks might be to release a virus.

Worms sit in memory, reproducing themselves. Because they reproduce exponentially, they quickly clog up a computer.

Trojan Horses are a species of virus which only start after the user has interacted with them. In the case of the Christmas virus, users were asked to type the word 'Christmas' to get a picture of a Christmas tree.

Some computers are even infected in the factory: Microsoft found that counterfeit software was being installed on newly manufactured PCs in the factory. These contained viruses that would take control of the computer when it was switched on.

Half of all Fortune 500 computers are infected with at least one piece of malware, the DNSChanger virus, according to security firm IID (internetidentity.com).

Nor are Apple Macs immune any longer. More than half a million Apple computers have been infected with the Flashback Trojan, according to a Dr. Web, a Russian anti-virus firm. The virus allows the computer to be hijacked and used at as 'botnet'. More than half were based in the US.

---

7. THEFT OF IT EQUIPMENT

Computers are routinely stolen. A UK Home Office study showed that computer theft made up 18 per cent of non-residential burglary, with the average cost being £2,616. Since thieves know that computers are quickly replaced, they often return to steal the replacement equipment. Twenty-five per cent of all computer crimes from commercial properties in the Home Office survey were 'repeats'.

It is extraordinary that such expensive, useful and portable equipment can be left loose on executives' desks.

- Computers should be locked to the desk: there are several systems on the market for doing this, including laptop locks. The insistent thief can cut through a padlock and wire, but it will deter the casual thief.
- Computers and printers should be security marked, and their serial number recorded. Labels which cannot be removed will make the computer less saleable.
- Other devices which make burglary less attractive are stickers warning the thief that the computer will not load from the floppy drive, and that the operating system is encrypted.
- The thief can also be warned that the computer contains a traceable radio bleep, or that it is registered on a national database. Computers can also be alarmed, so that when moved they set off a siren.
- Walk-out theft by staff can be minimized by the procedures suggested above, and by requiring staff to exit the building past a security desk.

Twelve hours after the author moved into new offices, and before the CCTV and alarms were installed, thieves broke in and stole every single PC in the building including the server. However they overlooked a small black box sitting on a shelf. It contained the backups, which meant the business could continue with relatively little disruption. We subsequently posted large '24-hour video surveillance' notices in prominent office windows.

8. FRAUD OR THEFT USING COMPUTERS

We have covered computer-related fraud in Chapter 13. A typical example is where a purchasing manager provides fake supplier invoices, and then pays the money into his own bank account.

When changing its payroll system, and migrating its staff's details into the new database, one large employer found that many staff had not been allocated to a specific point on the 'pay spine'. This meant that when pay increases were awarded or adjustments had to be made, the payroll staff had to make the changes manually, allowing for fraud to creep in.

9. OUTSOURCING IT

Outsourcing isn't a problem like a virus attack would be, but it represents a business risk if not handled properly. We looked at outsourcing as a general issue in Chapter 14, but IT outsourcing carries its own special hazards.

Given the technical nature of IT, its need for 100 per cent uptime, its costs and the changing technology, outsourcing is an attractive option.

Another rarely voiced benefit is a reduction in the threat of industrial action. On one-day strikes at a university, the network frequently goes down. This may be because IT staff are providing guerrilla support for the industrial action. Alternatively, it could be that network problems are common and that on days of action they go unfixed, either because the IT people are absent or because they're supporting the industrial action. Either way, outsourcing IT reduces the IT staff's industrial muscle.

A Gartner survey (Gartner.com) showed that 83 per cent of IT costs were for upgrading and operating existing applications; while a mere 17 per cent is spent on innovation. This suggests that the bulk of the costs are routine and therefore low value added, and might therefore be outsourced without loss of core skills. Thus a company that uses Oracle might get a certified outsourcing firm to maintain its database and add new functions.

Another factor is potential for cost reduction. In India and China, the cost of writing software can be as little as 10 to 40 per cent of the cost of doing it in the West. So a lot of IT work has been transferred there; and more could follow. However, cut-price distant providers may cause quality problems.

UTILITY OR GRID COMPUTING

'Utility computing' gives responsibility to another company to supply all your computing needs, including servers and storage. Thus the supplier will provide extra computing power when the business needs it. For example, some businesses have greater computing needs once a month for accounts or payroll; while banks need more computing power at Christmas.

Also known as grid computing, on-demand computing and other names, it works in the same way that a utility firm provides whatever electricity you need, when you need it.

The advantages lie in getting rid of redundant backup servers. Since the supplier will be looking after many clients, computing power can be shared. It should also make costs easier to control, and allows the company to respond to unpredictable demands from customers without investing in technology that may never be needed.

In addition, it frees the company from seeing IT as an area constrained by budgets, and therefore lets it become more flexible.

Outsourcing companies point out that businesses may use only 60 per cent of a server's processing power, and none of the backup server attached to it. Therefore there can be a lot of unnecessary capital tied up in IT. According to one estimate, companies use 20 per cent of PCs, 20 per cent of Intel-type servers, less than 40 per cent of Unix servers and 40 per cent of storage space. If this is shared among many companies, there will be a cost saving.

The companies most suited to utility computing are enterprise-sized organizations, and especially retailers who suffer peaks and troughs in sales – because a utility supplier can provide computing power as needed. Suppliers point out that you don't have to switch the whole business over to this method; you can start by moving specific departments.

One such supplier is Amazon Web Services (aws.amazon.com), an offshoot of Amazon, which can provide both data storage and a retail platform.

However, several large retailers have brought their IT back in-house. They include Target, the US discount retailer, and Marks and Spencer.

Outsourcing email

Email is another area that normally requires in-house technical support. Businesses are finding that Gmail (mail.google.com), Microsoft 365 (office365.microsoft.com) and similar services give them flexible computing power and substantially reduce the need for tech support.

Decide, however, whether you need your outsourced email supplier to maintain its servers in a specified location such as Europe, due to EU legislation on international transfers of internal data.

Consider also how long you need your supplier to commit to maintaining your service and your data. You may want a contract requiring them to support you for ten years, and not all will do that.

The dangers in IT outsourcing

However you may get locked into a service contract, lose expertise in this area and become reliant on the supplier. It may thus be hard to break out of the contract and retake control of your IT.

Moreover, the supplier may want the business to integrate its systems through all departments, and use a standard operating system and storage. This could entail a learning curve as employees get to grips with mew methods and applications. Equally, the organization may end up being more lean and effective as a result of this.

According to the *Journal of Global Information Management*, (igi-global.com), the greatest concern in IT outsourcing is the unexpected costs of transition. In other words, outsourcing IT is a messy business with unexpected outcomes.

Table 15.3 shows some of the main risks associated with IT outsourcing. They can be combated by taking the steps listed below.

Table 15.3   Outsourcing risks

IT Outsourcing Risks	
Initial risks	In-house staff resistance
	Unexpected transition costs
	Delays in setup
Service risks	Service deteriorates or never reaches in-house standards
	Poor quality and reliability
Ongoing costs	Costly contractual amendments
	Cost escalation, hidden service costs
Strategic issues	Loss of in-house competencies
Security issues	Damages due to security breach
	Disaster and recovery costs
People risks	Poor communication
	Mistrust
Future	Vendor lock-in
	Switching costs (repatriation or transfer)
	Disputes and litigation

Ten ways to reduce risk in IT outsourcing

1. *Get in-house staff buy-in.* There has to be benefits where possible, including voluntary redundancy, and greater responsibility and job enhancement for those who remain.
2. *Manage vendor selection effectively.* Ensure the widest possible longlist. Provide a detailed brief. Talk to clients of the shortlisted suppliers.
3. *Test outsourcing with a small project initially.* Only move to full-scale outsourcing if it works.
4. *Outsource only non-core activities.* Outsource what is repetitive and low value added, leaving in-house staff to work on higher-value projects.
5. *Get a clear understanding of costs.* These include initial transition costs, the costs of amendments and the costs of repatriating the service. Run 'what if' scenarios.
6. *Set clear, measurable goals.* This must include a Service Level Agreement. Agree penalties including an out-clause for non-performance.
7. *Have a written contingency plan.* This should include disaster recovery, and how to terminate the relationship.
8. *Measure key performance indicators.* Share the data.

9. *Have members of the outsourcing supplier on site*, if the scale of the project allows for it. They need to be embedded, to resolve problems
10. *Have procedures to actively manage the supplier.* Effective communications are vital. Have an escalation process, to ensure problems are resolved.

10. DATA PROTECTION AND LOSS OF CORPORATE INFORMATION

Embarrassing discoveries and news stories involving the loss or publicizing of customer information come from many sources:

- Old PCs that are sold or dumped, and are later found to contain confidential information. Brighton and Sussex University Hospitals NHS Trust was fined £325,000 by the Information Commissioner's Office (www.ico.gov.uk), after sensitive patient information was found on hard drives sold on eBay. The data also included staff home addresses, their criminal convictions and suspected offences. The Trust had engaged an individual to destroy 1,000 old hard drives, but he secretly sold at least 252 of them on the internet.
- Staff losing laptops (or having them stolen), which contain confidential information. The same applies to memory sticks
- Hackers getting access to an organization's servers. Sony was fined £250,000 after its PlayStation Network database was penetrated. The hackers got details of 77 million accounts, including credit card information.
- Staff send emails or letters to wrong people. Plymouth City Council was fined £60,000 when a social worker sent details of a child neglect case to the mother of another child. This was due to a mix-up when the pages were sent to the wrong printer at the Council's offices.

These risks can be combated by:

- setting clear procedures for managing confidential information;
- auditing the processes regularly to ensure no breaches are taking place.

The company that generated and owns the data (known in the UK as the 'data controller') remains liable for it even if it passes it to a third party to manage it or dispose of it. Scottish Borders Council got a £250,000 fine after a contractor who digitized its records then put the paper records in a paper recycling bank in a supermarket car park, where they were spotted by a member of the public and reported to police.

Similarly, if you store confidential records in the cloud, it is likely to be your responsibility if the cloud storage is broken into. Online encryption is essential, and you need to know whether and how the data might be decrypted by a criminal.

11. LOSS OF DATA DUE TO STAFF ERROR

Staff often delete files accidentally, lose data by misfiling it and fail to save files to the corporate server.

Mobile devices mean that more files are stored on executive's laptops and tablets. This is particularly problematic in walled systems such as Android and Apple's IOS, where software is less likely to Microsoft's, and the backup systems (if the user has implemented them) may go to a proprietary system such as Apple's iCloud. When this happens executives' files may not be synced to the server.

The business needs to ensure that all mobile devices are synced with Microsoft Exchange, if that is what the company uses. And staff need to be educated about the need to save or backup to the corporate system.

As we discussed earlier, saving files frequently (or ensuring that automatic saves are switched on) also reduces mishaps. The program should be set to 'always create a backup'; and 'Save Autorecover' will recover files a user was working on if power is lost.

At one workplace, random directories kept going missing, only to be discovered inside another, unrelated one. The author suspected that it was due to the arthritic hands of an old but valued employee accidentally dragging the files in Windows Explorer. The case was never proven.

Staff need to be shown how the corporate IT filing system works. And they need to be discouraged from keeping files in their own private corner of the system, one that is often not included in the backups.

IT systems should have screens that ask 'Are you sure you want to delete this file?' before data can be deleted. And the organization needs to regularly test whether it can retrieve files from the backup system.

12. STAFF MISUSE OF THE INTERNET OR EMAIL

Staff misuse their online access in a host of ways. They include the following activities:

- Wasting time surfing the net. This includes looking at porn and visiting social networking sites.
- Sending offensive material such as porn by email. This is an HR matter, which we examine in the Chapter 18 on HR.
- Sending confidential information by email to outsiders. This is a security matter, which we examine further in Chapter 11 on Security.
- Bullying or harassment of peers or subordinates by emails. This is an HR matter, best controlled through HR policies.

Time wasting

The internet is a huge drain on staff productivity. Without monitoring systems in place or the ability to constantly look over an employee's shoulder, it is hard to know whether an employee is hard at work in front of a screen, or posting on Facebook.

Staff Monitoring Solutions (staffmonitoring.com) has some appalling statistics about cyber slacking:

- non-work-related internet surfing results in up to a 40 per cent loss in productivity each year at American businesses;
- 70 per cent of all internet porn traffic occurs during the 9am–5pm work hours;
- 70 per cent of companies have had sex sites accessed using their network;
- 30–40 per cent of lost productivity is accounted for by cyber-slacking;
- 25 per cent of corporate internet traffic is considered to be 'unrelated to work';
- 32 per cent of workers surf the net with no specific objective – men are twice as likely as women;
- 37 per cent of workers say they surf the net constantly at work;
- 27 per cent of Fortune 500 organizations have defended themselves against claims of sexual harassment stemming from inappropriate email.

According to a survey by International Data Corp (IDC.com), 30–40 per cent of internet access is spent on non-work-related browsing, and a staggering 60 per cent of all online purchases are made during working hours. At the author's distance learning business, very few sales take place at the weekends. Most enrolments for self-employment courses occur during weekday work hours, implying that many people are misusing their employer's time.

WAYS TO REDUCE STAFF MISUSE OF THE INTERNET

1. Have a written policy on internet use, and inform employees about it. The policy should cover the following:
   - non-work use of the internet, use of social networking sites, porn;
   - the sending of offensive messages, pictures or jokes to other employees;;
   - using the internet for private work;
   - abuse of subordinates by email;
   - the need to report infringements to management;
   - penalties for infringing the policy;
   - corporate internet monitoring activities (for example use of monitoring software).
2. Connect computers to the internet only by wireless. This means that only those employees who need the internet will be connected.
3. Install software that blocks access to specific sites, such as social networking, porn and gambling sites. Inet Protect is one such program (blumentals.net/inetprot).
4. Install software that tracks online use. Such programs include Spytech (spytech-web.com), or WebWatcher (webwatchersoftware.com). According to Staff Monitoring Solutions, 63 per cent of major US companies monitor workers' internet connections and 47 per cent store and review employee email.

At one company, a male staff washroom window looked out on to the company's offices. While using the urinal one day, one manager saw a member of staff perusing porn at his desk and reported it to HR. The HR department sent its only male member of staff into the toilets to confirm the sighting. The member of staff, who was gay, agreed that the other member of staff was indeed looking at porn; but wryly declared himself traumatized because the porn featured lesbian sex.

13. STAFF BYOD (BRING YOUR OWN DEVICE)

The smartphone and tablet revolution took many IT departments by surprise. All those carefully developed plans whereby a member of staff had an official company laptop and possibly a BlackBerry came to naught when senior staff demanded to be allowed to use their iPhone. Directors insisted that they should be able to access the company intranet using their iPad.

This isn't surprising. Employees become more productive if they carry their corporate diary on their smartphone, if they are able to update it at will, and if other members of their team can instantly see those changes.

In the face of this demand, companies have given way, and every business is now awash with Android and Apple devices. In addition, employees are also storing data in the cloud, on DropBox, SkyDrive and Google Drive. They are using Google Docs to save data and transfer it to other devices. This is known as BYOS (bring your own software). Again, there are risks in the use of software that is not controlled by the business. Sophos (sophos.com) suggests the organization therefore needs to know:

1. *Can the business enforce its policies on BYOD devices?* If the employee owns the device, how much control can the company exercise over it? Can the business audit whether policies are being adhered to, by checking individual tablets and smartphones? This is particularly relevant if the organization operates in a regulated field such as financial services.
2. *To what extent will employees comply with corporate policies?* If the business can't check whether, say, an employee is saving files to a private DropBox account, is the policy worth adopting? How will the business get devices back from ex-employees?
3. *How important is the data?* Is any information confidential? Is it encrypted? And could it be subpoenaed?

BYOD security measures

The business should require the following security measures:

1. *Devices should be protected with a passcode.* Each iPhone or Android tablet should have a four-number password.
2. *Devices should have anti-virus protection.* Apple machines, not formerly the target of hackers, are now more at risk due to their popularity. Apple claims its devices are impregnable, a comment *Forbes* magazine saw as 'complacent', especially since devices can be 'jailbroken'. It brings to mind the last words of John Sedgwick, the US Union Army general who strode around in the open, saying: 'They couldn't hit an elephant at this distance.' Seconds later, he was shot in the head by a confederate sharpshooter and died.

    Various third-party companies provide software that provide anti-virus information for Apple devices, but at time of writing, none can scan an Apple device. If implemented, a policy requiring anti-virus protection would prevent an employee from using Apple devices, something that is clearly impractical at present.
3. *Only use cloud storage that encrypts data.* Many cloud services don't encrypt the data they receive.
4. *Confidential data on removable media should be encrypted.* After giving a presentation at a national conference, a company director left her memory stick in the conference laptop. The organizers later posted it back to her. Such sticks often contain confidential information in addition to the presentation itself.

    Another executive found a memory stick on the floor of a toilet in a train, possibly having fallen out of a trouser pocket. He decided that the simplest solution was to put the stick deep inside in a waste bin, rather than try to identify the owner from the data. Criminals would not be so civic minded.
5. *Wipe data.* Users must be able to wipe the data off a portable device if it is lost or stolen. Apple users need to set up 'Find my phone' before the device is lost. The company should have a strategy for wiping retired devices.
6. *Support a limited number of mobile devices.* Employees should be willing to accept certain restrictions on their freedom, in order to ensure good data security.
7. *Apply Application Control.* Corporate computers can check for specific activities and stop them. This prevents viruses from moving from portable devices to the company's servers. Semantic (semantic.com) gives the following examples:
    - *Block attacks from removable drives:* (Network worms often penetrate via USB devices). Application control will still allow the USB stick to function normally.
    - *Prevent PDF attacks:* Web-based attacks can hide inside a PDF file. An Application Control rule can stop Acrobat and Acrobat Reader from writing code to the server.

- *Prevent new browser helper objects being installed.* Browser Helper Objects (BHOs) are used by browsers to provide additional toolbars. They can report on the user's browsing behaviour. Application Control can block unwanted BHOs, such as the Alexa Toolbar.

Implementing a BYOD security plan

A BYOD security plan could involve the following actions:

- engage stakeholders;
- gather information about usage;
- assess impact on the corporate network;
- implement a trial plan among a small group of users;
- roll out the policy among all employees;
- audit compliance;
- update the policy in light of new technology and changing circumstances.

14. IT PROJECT FAILURE

IT projects are notoriously prone to failure. This isn't surprising given that organizations set out to create something out of nothing. Unlike a construction project, where the materials and methods have been known for hundreds, even thousands of years, IT projects are usually a step in the dark. This is a problem we discuss in Chapter 16.

Setting a Security Policy

Just over half of large companies have a security policy and contingency plans in place. But in some cases the security policies are not strictly enforced.

The company should have a written policy on computer security, and this should be endorsed by senior management. The policy should cover:

- lines of responsibility for the IT system;
- password complexity;
- encryption;
- data loss prevention; backup and recovery systems;
- firewalls, anti-virus and malware procedures;
- use of mobile devices, including smartphones and tablets;
- access to internal data by staff;
- use of cloud services;
- use of the internet by staff, porn;
- policy on personal emails, cyber bullying;
- retiring old equipment;
- collection of equipment from employees leaving the business.

This policy should be supported by written procedures, specifying the action to be taken to protect the data. To ensure that the procedures are implemented, the company should regularly carry out audits.

Contingency Planning

A Society for Human Resource Management (SHRM.com) survey found that 78 per cent of randomly polled members said they had a contingency plan for computer viruses.

However, it depends on how viable that plan really is. IBM (IBM.com) believes that companies pay insufficient attention to contingency planning for their computers. Its research shows that four out of five companies don't have a viable contingency plan for computer disaster.

All companies which rely on their computer should have a contingency plan, and the plan should be monitored all the time.

This means assuming that the computer system has failed. The company should check what would happen if its programs became faulty, its data was lost or its computers were stolen.

We discuss contingency planning more fully in Chapter 23.

MINISTRY OF DEFENCE 'LACKS CONTINGENCY PLANS'

The UK Commons Defence Select Committee criticized the Ministry of Defence (MOD) for having 'no proven back up'. The committee said: 'The government needs to put in place – as it has not yet done – mechanisms, people, education, skills, thinking and policies which take into account both the opportunities and vulnerabilities that cyber [attack] presents. It is time the government approached this subject with vigour.'

The committee said: 'The evidence we received leaves us concerned that with the armed forces now so dependent on technology, should such systems suffer a sustained cyber attack, their ability to operate could be fatally compromised.'

The MPs worried that the MoD's suppliers and organizations that work with the military were vulnerable to cyber attacks, and that enemies might find a 'backdoor' route in to the MoD.

The MoD said it has a range of contingency plans in place, 'although for reasons of security, we would not discuss these in detail'. The UK Government has told its departments to put cyber defence at the top of their agendas, and set aside £650 million to boost the UK's cyber defences.

Managing IT Risk through ISO 27001

Companies looking for a comprehensive way to manage their IT security can use ISO 27001. This is an Information Security Management System (ISMS), whose scope is shown in Figure 15.6. It helps organizations do the following:

- You identify the risks facing your information, and introduce controls to protect it.
- Personal information is kept secure. Organizations with ISO 27001 are unlikely to suffer the scandal of confidential records turning up in rubbish dumps. So the business will be meeting its legal obligations, especially those operating in the regulated sphere.
- Trading partners will know that the organization protects and controls their information, and will therefore be ready to exchange information.

ISO 27001 involves the following actions:

1. Define the organization's information security policy. This must demonstrate management's commitment.
2. Define the scope of the system (what will be included?). This depends on the business requirement, the assets to be protected, the location and the technology.
3. Assess the risk: Identify the threats and vulnerabilities to assets and the impacts to the organization. Identify the areas of risk to be managed.

4. Illustrate the business processes and flow of information, to indicate the role of IT in the business.
5. Have an up-to-date list of IT assets.
6. Select the controls that will be used.
7. Document the selected controls. Decide what measurements will be taken to check them.
8. Document the procedures that must be followed. Show how they are monitored.
9. Get management and employees committed to the system.
10. Define responsibilities and authority within the organization. Evaluate competency and undertake staff training.
11. Allocate sufficient resources (money and people) to the system to ensure it operates properly.
12. Audit the system periodically, and get management to review its findings and act on them.

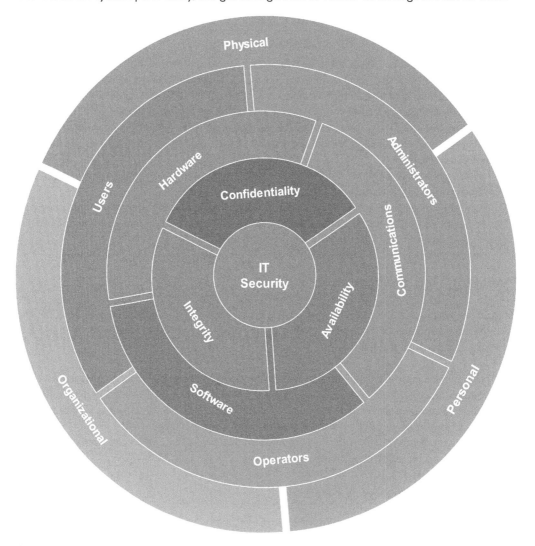

Figure 15.6   ISO 27001 issues

To get the full benefits of the standard, the company should gain certification from an accredited body, namely one that is accredited with the national standards agency, such as ANAB (USA), UKAS (UK), SCC (Canada) NABCB (India), INAB (Ireland) and so on.

Risk Assessment

By answering the questions below, you can assess your vulnerability to computer disaster. Tick all applicable boxes.

Topic	Question	✓
Computer use	Do IT systems involve real-time applications (such as mail order or financial services)?	
Impact on the company?	Would the loss of computers for two days disrupt the business?	
Contingency planning	Does the organization lack an IT contingency plan, one that has recently been tested?	
Finance	If computer data was lost, would the business know who owed it money?	
Accessibility	Can data be accessed from PCs without passwords?	
Backups	Are backups not taken, or (if they are made) are they left in the same building?	
Projects	Are you undertaking a major IT project?	
Staff	Do many managers lack training in the use of IT security?	
	Do staff store data on their own laptops and tablets?	
Outsourcing	Do you outsource any of your IT?	
Total points scored:		

Score: 0–3 points: low risk. 4–6 points: moderate risk. 7–10 points: high risk.

The Appendix has a summary of the checklists. By entering the results of this one, you can compare the scale of computer risk against other categories of risk.

16
Minimizing Project Risk

Ann has big visions. 'No more paper,' she says. 'Everything will go into one system. Clients will enter their own data.'

Do you have a view on costs and timescale?' you ask. 'Outsource it,' she says, abruptly. 'Freelancers could do it in four months.'

'But there are integration issues, training issues…' You run out of words.

Ann pauses, her hand on the door knob. 'I need it doing, not excuses,' she says firmly. 'I need a plan by the end of next week.' And with that she's off.

You stare out of the window for a while. Is she right? Or will it be an expensive fiasco? Only time will tell. You sigh.

Projects: Inherently Risky Activities

Projects take many forms – from road building to databases, from new products to overseas factories, and from corporate acquisition to drug research.

But they also have much in common: a big investment, a complex mix of people and assets, an uncertain outcome and a high risk of failure.

Many businesses are wholly project-based, whether management consultants or film studios. So their entire existence is routinely at risk. And failed projects have a devastating effect on corporate success.

Whatever format they take, projects routinely over-run on cost and time, and don't work as they should. In many cases, the project is shelved.

Ford spent four years and millions of dollars developing an Oracle ebusiness procurement system called Everest before abandoning it and reverting to older technology. Sources blamed complexity, poor performance and suppliers' unwillingness to use Everest to compete online for work.

Common to many major projects is the cost. A semiconductor plant costs up to $1 billion. This level of investment makes even the largest and most profitable firms nervous.

PROJECT LIFECYCLE

Each project has four phases: inception, planning, execution and close. There are risks at each of these phases, as shown in Figure 16.1. As the project develops, changes become ever more expensive; therefore the two early stages (Inception and Planning) are especially important, and are often the ones to which insufficient attention is paid.

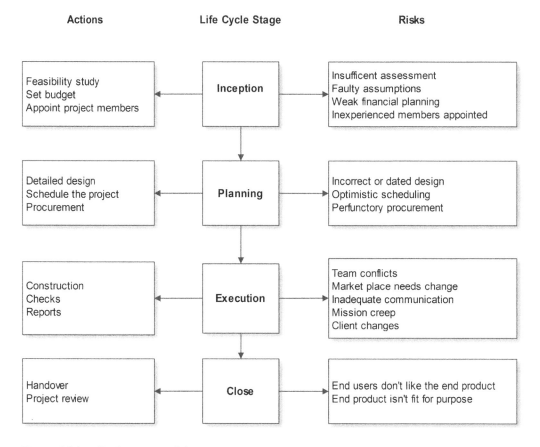

Figure 16.1   Project stage risks

THE PROJECT TRIANGLE

The so-called project triangle or iron triangle (Figure 16.2) shows that any project has three constraints. These are time, cost and scope. The three elements lie at the heart of the project's risks. You can't change one without either affecting the others or producing an inferior end product (shown in the figure as 'quality').

Quality could mean software that does the job or a new product that succeeds in the marketplace.

So, if a project has to be delivered fast, there is a risk of higher costs and the project may need a more limited scope. If you balk at the cost, you may have to suffer accept a longer delivery time and put up with a more limited scope. A wider scope, meanwhile, will increase costs and take longer to deliver.

Delivering a project on time is hard, and a delayed project is often due to insufficient thought having been given to the project's complexity or scope.

Keeping within budget is a failing often seen in government projects (if only because they are more open to the public gaze). Projects that go over budget can be caused by wishful thinking, failure in estimating, project creep or weak project management. We examine these and other problems in this chapter.

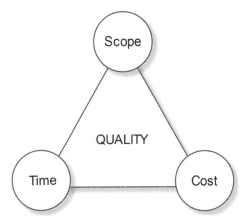

Figure 16.2   The project triangle

DEPENDENCIES

A dependency is where one task is dependent on another one being completed before it can start or be completed. In a construction project, the decorators can't start until the plasterers have finished. The plasterers can't start until the electricians have finished. And the electricians can't start until the carpenters have finished. A delay by the carpenters sets the whole project back.

You need to focus on the important dependencies, and monitor those, rather than get bogged down in every possible dependency.

The 11 Main Risks in Projects

There are many ways in which projects fail. Often it's a combination of factors. Below, we list 11 types of risk that cause failure.

1. failure to define the scope;
2. scope creep;
3. wishful thinking and budgeting failure;
4. management and governance failure;
5. project team staffing problems;
6. underestimating the complexity of the task;
7. optimism in timing;
8. failure to engage stakeholders and users;
9. partner delays and failings;
10. the end product under-performs;
11. handover failure.

We examine these in more detail in the section that follow.

1. FAILURE TO DEFINE THE SCOPE

Management and clients are often vague, and fail to put in writing what they want. This may be because they lack technical knowledge or a 'big picture' people.

Hence some projects start out without the client having properly defined their scope. By 'scope', we mean the end result or deliverable. Scope should explain what the client expects to get.

Failing to define the scope gives rise to two problems: firstly the various stakeholders may have different results in mind, and so differences of opinion will emerge as the project develops, leading to conflict and delays. Secondly, the project can grow in size because there is no way to reject the client's requests, no way of saying 'But you didn't say you wanted that'.

By properly defining the scope, clients will receive what they asked for.

2. SCOPE CREEP

As the project develops, the team often finds that it has omitted some functionality. The team finds new features to add to the project, or the customer wants the project to incorporate new developments occurring in the outside world. This results in delays and cost over-runs. It can be avoided by careful change control, whereby any request for a change has to be formally reviewed.

3. WISHFUL THINKING AND BUDGETING FAILURES

Government mega projects always seem to end up costing much more than estimated. Some managers or politicians lack planning skills, are insufficiently numerate, and rush into a project without thinking too much about it. They can be carried away by their enthusiasm.

Sometimes, it is said, government projects go over budget because the project would never get started if the population was told what the true final cost would be.

That implies some degree of cunning on the part of the civil service. An alternative view is that managers are optimistic when it comes to price, do not foresee problems and meet unexpected delays (for example if an archaeological dig is required). All these add to the cost.

Even when budgeting is carried out, failure to correctly estimate costs, or to include every cost, will add to the final over-run.

The project manager should carefully evaluate estimates from partners and suppliers, based on experience. Getting estimates from several suppliers will allow the team to judge what the costs are really likely to be.

4. MANAGEMENT AND GOVERNANCE FAILURE

A £160 million UK Department for Transport (DfT) shared services system was criticized by MPs after savings of £57 million were converted into a loss of £81 million. This came about largely due to management failure.

The DfT failed to put the contract out to tender. Instead it used an existing deal with IBM for a similar system. It then failed to identify the requirements. When DfT realized that its project assumptions were incorrect, only two months in, it failed to change the original plan. It also saw a number of key performance indicators as unimportant. According to the MPs, the DfT cut corners in an attempt to meet a deadline. It cut testing time and, when the system went live for seven other agencies to use, it was not working properly.

5. STAFFING PROBLEMS

Management can overburden project staff with additional tasks, which gives them less time to concentrate on the project in hand. This inevitably leads to delay. Management loses interest in tasks once they have been started, and flits to the next shiny object.

This has the added problem that managers fail to stay involved with the project. You have to be alert to projects that lack that no one is actively controlling.

6. UNDERESTIMATING THE COMPLEXITY OF THE TASK

Britain dumped a new computer system and went back to paper CVs for junior doctors after its online job application system failed to work properly. The £6.3 million Medical Training Application Service (MTAS) system didn't grade applicants well, and as a result unqualified applicants were invited to interview.

Thousands of junior doctors were left without posts when the computer wrongly assumed they were immigrant workers. It was said that the system was rushed through, when more time should have been taken. At one point, it also left applicants' personal details, including their sexual orientation, visible to everyone.

The lesson is clear: it's vital to test the project thoroughly before making it live. And it is dangerous to set deadlines that prevent enough time for the project to be completed.

7. OPTIMISM IN TIMING

Humans are hopelessly optimistic, and no one wants to be seen to seen to be a pessimist when asked to do a job. This is why people under-estimate how long a project will take.

When projects are delivered late, it can be due to many things mentioned in this section of the chapter, including supplier delays, scope creep and underestimating the complexity of the task. You can reduce the likelihood of delay by:

- setting out in detail the tasks to be undertaken;
- defining an accurate critical path;
- using experienced and capable partners;
- assessing the risks and reducing them (for example reducing the scale of the project).

8. FAILURE TO ENGAGE STAKEHOLDERS AND USERS

Listening to end users will help to keep you safe. Staff at the sharp end have years of accumulated knowledge yet are often ignored. It's an old management consultant's trick to discover the cause of problems by asking the operational staff. It's known as 'borrowing your watch to tell you the time'.

Sometimes project managers think communication is just about telling people what they are doing. But it's more about listening and probing.

Working on a parts management system, a Quantas project team wouldn't talk to the engineers on the ground. 'We wouldn't ask the engineers what their views on our software systems were,' said Qantas's CFO. 'We'll put in what we think is the appropriate for us.'

The $40 million system was so hard to use that engineers threatened industrial action. Eventually Quantas dumped it and started the project all over again.

9. PARTNER DELAYS AND FAILINGS

Sub-contractors often fail to deliver what they have promised, for all kinds of reasons (or excuses). This can mean essential parts not being delivered, or some element of construction being held up.

This can sometimes be because the contractor is inefficient, or was optimistic about the number of staff needed to deliver the project. Equally, it can result from unexpected problems that occur.

If you chose the lowest bid, or beat the contractor's price down, you may find this results in low-quality work.

Partnership sourcing involves treating a contractor as a member of the team, and engaging with them early. We review this in Chapter 7 on Procurement.

10. THE END PRODUCT UNDER-PERFORMS

It's essential to know in advance how powerful the end product needs to be, or what its essential functions should be. In a computer system this might be measured in transactions per minute, or the ability to compare specific kinds of information.

This failure comes about when the project team doesn't understand the needs of end users. When the project is delivered, it simply doesn't meet customers' needs. Of course, all end users complain when they first start to use a new computer system, but we have to distinguish between the tweaks that are easily made, and fundamental problems that mean the product has to be torn apart and re-done.

A $370 million cashless card system that would allow people to use Sydney's buses, ferries and trains was dumped after seven years in development when it was found to be 'riddled with software faults'. RailCorp, the state transit authority (STA), said that the system was full of bugs. ERG, the contractor, said it was forced to incorporate buggy legacy software, and that the STA changed its point of sale software and its internal accounting software without telling ERG, so that ERG's software 'was then inconsistent or incompatible with the upgraded RailCorp software'.

11. HANDOVER FAILURE

The project team can finish the project properly, but then it fails when it gets put into practice. When the fifth terminal at Heathrow Airport went into operation, it managed to lose 28,000 bags, and hundreds of flights were cancelled. A backlog of passengers started to back up in the terminal, and British Airways had to stop accepting hold luggage.

The problem was blamed on staff training and poor morale meant that other staff wouldn't help out. Staff didn't know what they were supposed to do, and management didn't listen to those who warned there would be problems when the terminal opened its doors (see http://news.bbc.co.uk/1/hi/uk/7322453.stm).

The Project Risk Log

You can put the risks into a log, allocating each an impact factor, and listing an action, an owner and a contingency plan. This is shown in Table 16.1.

Table 16.1   **Project risk log**

Date	Risk	Impact	Action	By	Contingency plan
01.06	Old equipment at client site may render development work unusable	H	Clarify whether upgrade of equipment is likely, or downgrade project specs	MR	Revise project schedule
14.06	Lack of engagement from client	H	Get JR to involve DA	FL	Set out requests in writing for client to approve before moving forward
17.0	Urgency is causing shortcuts	M	Memo to DA seeking decision	FL	Employ temporary staff

Major Types of Project

In the following sections, we examine the main types of project and see where the main risks lie. Specifically, we look at:

- construction, mining, petrochemical and civil engineering projects;
- aerospace, automotive and manufacturing;
- organizational change;
- risk in event projects;
- new product development risks;
- risk in IT projects.

RISK IN CONSTRUCTION, MINING, PETROCHEMICAL AND CIVIL ENGINEERING PROJECTS

The main risks in this area are: weather, distance, finance and the use of many contractors. We look at each of these next.

Weather: Building roads, new homes, factories and chemical plants involves outdoor working, where unexpected weather can halt work for long periods of time. With climate change, this is becoming more pronounced.

Distance: These projects usually take place far from the company's base, so there are additional problems of communication, especially when they take place in the more remote or hostile parts of the world.

Finance: The projects are often a major investment. Hence, if the project doesn't run according to plan, it can be financially ruinous.

Multi-contractor: frequently there are many specialists on site, often working for different companies. This poses the risk of conflict and misunderstanding and lack of team work.

On the other hand, such projects usually involve reasonably well-known methods of assembly or construction.

RISK IN AEROSPACE, AUTOMOTIVE AND MANUFACTURING

Typically, such projects involve building a new manufacturing line. They vary in complexity. Installing an established printing press at a new site and making sure it runs properly is relatively risk-free, whereas it is much more risky to build a new design of aircraft where the wings and engines are made in different countries and by different companies.

Complexity is added where the final product has to be used by a client, or many customers, which involves training and trouble shooting.

RISK IN ORGANIZATIONAL CHANGE

We will look at organizational change in Chapter 18. Organizational change is distinguished by having employees rather than equipment at its heart. Examples include the restructuring of a workforce, or changing a process.

Organizational change often meets project management when a new IT system is introduced, which requires staff to work in different ways. It is sometimes seen as the soft side of project planning, but is often a critical success factor.

Staff can be unwilling to adopt new ways of working. They are more comfortable with the old structures and systems. They sometimes resist change to the extent of sabotaging it. When one organization set up a new payroll system, it opted to run both the old and the new alongside each other. This parallel running was to check that the new system was accurate, and to iron out any bugs. The payroll team caused major delays to the new system because they kept prioritizing the old system and failed to enter the same data in the new one. This put the launch back by several months.

Staff in these circumstances will be quick to blame the new system, partly because they are unfamiliar with using it and because they would prefer to go back to the old system. Assuming that the project has a future, management has to push past these objections, which will decline as time goes by. But in the meantime you must be ready for the system to be vilified.

RISK IN EVENT PROJECTS

Events include sporting activities as varied as marathons and Olympics, a product launch, gala dinner or a theatrical performance. An office move also comes into this category.

An event is typically a one-time only project. So whereas a computer project can be put back a few weeks, or a vehicle's introduction delayed, the event has to work right first time, especially if it is a live, non-repeatable one.

Live events are often outdoors, so the weather is a risk. And everything has to work perfectly and on time, including lighting, performers, sound, front of house and speeches.

Risks include fire, health and safety (with people tripping and falling and being trampled), food hygiene, sanitation, drug use, criminal activity including theft, the risk of protest or terrorism. This calls for the crafting of a carefully thought out risk assessment.

For commercial events, there is the added risk of insufficient ticket sales causing financial losses. A friend of the author used to stage an exhibition that coincided with the Cannes film festival. It was profitable until the year that followed the Twin Towers disaster: the US visitors stayed away in droves. He ended up closing the business and had to sell off assets.

Similarly, a small business that ran affordable art fairs in Ireland did well until recession hit, and people suddenly stopped buying art. The business closed.

You can buy insurance against cancellation, postponement or abandonment of the event, as well as public and employee liability.

NEW PRODUCT DEVELOPMENT RISKS

Ninety per cent of new products are said to fail. The problem lies in the uncertainty of knowing whether customers will want the product. You can stop consumers on the street, or get them into a focus group, and ask them whether they would buy a new product. But it is hard for people to predict their future buying behaviour, and people like to be kind to the interviewer.

When it comes to pharmaceutical products, even the most promising ones fail when near to launch when the side effects (from headaches and nausea to increased heart attacks) are found to outweigh the benefits. And often they fail to outperform existing drugs.

Unlike other projects, we need to accept failure in new product development. That means having enough projects going through the pipeline, and learning from each failure.

It also calls for input from creative minds and designers, and an efficient system of getting close to the end user.

Managers will always call for changes as the business environment alters; so change control – ensuring that any calls for change are examined and approved before being accepted – is vital.

RISK IN IT PROJECTS

Installing a new computer system or database is often expensive, with major systems costing millions of pounds. Management may have no experience of introducing IT systems, which is why so many go wrong. According to the Standish Group's 'Chaos' research, a staggering 31 per cent of IT projects will be cancelled before they get completed, and 53 per cent of IT projects will cost over 189 per cent of their original estimates.

The public sector is well known for its failed IT projects. This may be because of civil servants' lack of IT skills. But other factors are at work too. The nature and size of public sector activities means that its IT projects are more complex and larger than those of the private sector. Moreover, private sector IT failures can be quietly buried, which isn't the case with public money.

- Failings in a £456 million computer system designed by EDS for the UK's Child Support Agency meant that one in eight single parents awaiting government payments were receiving them. At that time, single parents were owed £750 million and were waiting four to five months for maintenance payments.

  More than half of cases entered into the new computer system were delayed by 'major problems', though this was subsequently reduced to 10 per cent. Staff were 'breaking down in tears of frustration' as they were reduced to calculating cases on paper. Glitches in the system meant they had no way of correcting on-screen errors and had to start files again from scratch every time a mistake was made.

- The London Stock Exchange spent five years and £300 million on a paper-free settlement system called Taurus. The system did not work and had to be abandoned.
- The Performing Rights Society (PRS), which collects artistes' royalties, lost £8 million, one-third of its annual income, on a computer which failed to work. A report blamed the prevailing attitude of staff and managers in the PRS, whose attitude, it said: '...was not conducive to the successful conduct of large-scale computer projects. This requires the managers and staff to be open about mistakes, to learn from them, to expose their work to review by peers, to welcome criticisms, and to accept responsibility.'
- Wessex Regional Health Authority lost £20 million on its computer system.

There is no shortage of horror stories where computers are involved. The UK public sector spends some £14–£21 billion each year on software. But only around 16 per cent of its IT projects can be considered truly successful, according to a report by the Royal Academy of Engineering and the British Computer Society. The Office for National Statistics puts the success rate at 30 per cent.

A typical private sector failure is the management information system (MIS). Finance directors regularly demand one because it should, in theory, bring together data from production, sales and purchasing. The goal is to show, on neat interlinked screenfuls of information, how the company is doing.

In practice, MISs often fail because the design is wrong. They don't show the information that they should, or they are too complicated to use. Eventually, managers give up and start to rely on the old information methods. The computer department, meanwhile, grumbles that management doesn't know what it wants.

CASE STUDY: SAP IS SUED FOR $100 MILLION

America's biggest refuse collector, Waste Management, announced it was suing SAP for $100 million as a result of the failed implementation of SAP's ERP package. ERP integrates information from across the organization, including production planning, purchasing and sales.

Waste Management said that the Waste and Recycling Software was 'undeveloped, untested, and defective'. The damages sought by the company included its expenses in implementing the package plus punitive damages.

One of the biggest reasons for an ERP failure is people unwilling to change existing business processes (as we saw earlier in 'Risk in Organizational Change'). Sometimes people have a vested interest in their existing processes and are reluctant to change.

While it is tempting to scrap a failed implementation rather than throw good money after bad, companies should see if they can fix the problem first.

Most systems have terrible failings when first unveiled, but tweaking can often make the system work much better. Companies should define their business processes, have a methodical tendering process, allow time to get it right and avoid feature creep.

Finally they should also be aware of the limitations of generic systems which don't always fit their industry. Sometimes bespoke solutions are best. In one of the author's businesses, it took three years for the database to evolve, including a failed first attempt. Now the current software does everything but make the tea.

THE MAIN PROBLEMS IN IT FAILURE

Enough prominent IT failures have been reported to indicate the main problems to be avoided.

Lack of involvement by line managers. IT, like accounting, is a subject which managers often don't understand and leave to experts. When the project finishes, the managers are given a system which they don't want or understand.

Consultants who don't understand the client's business. Consultants sometimes win business on the strength of contacts with the chairperson or chief executive. They may not take the trouble to understand how the business works. Understanding the business process is central to the success of the project. This is where the analysts' time must be invested.

Box-shifting suppliers. Some firms are uninterested in whether their systems meet the client's needs.

Over-ambitious projects. It is dangerous to try to build a Tower of Babel. Any project which is 'state of the art' is risky. The consultants may be learning on the job, because they have never worked on this type or scale of project before.

HOW TO AVOID PROBLEMS WHEN LAUNCHING A NEW IT PROJECT

Though IT project share similarities with other types of project, they also have their own special problems. Below we discuss how to overcome them.

Ask whether you need a bespoke solution. If a standard software package will do the job, this is a safer option.

Formally assess the risks. This should, at the very least, ensure that the risks are openly discussed and forecast. Responsibility will then be shared throughout the organization, rather than being laid at the door of a single IT manager, when perhaps it was caused by a Board decision.

Involve line management in drawing up the brief, and in approving the project's stages. The project must include detailed conversations with users, to understand how their departments work, the information that is needed and any other criteria. Most end users want to switch on a machine and find the information on the screen. They don't want to memorize computer protocols, or get involved with technical details.

Do not become too dependent on consultants. Gain knowledge talking to vendors and consultants and by reading about the subject.

Be aware that vendors are not independent. Vendors want to sell their system.

Don't underestimate the costs of the project. As the Standish research (above) shows, you should double any cost estimate, and perhaps the time too.

Beware of selecting a system that is too advanced. Leading-edge technology is more likely to fail. A tried and tested system with many users will have had its bugs removed. Many computer consultants are in love with new technology; anything a few years old is boring to them.

Break big products into manageable steps. This will make the project easier to control.

Send tenders for any major project to several competing consultants. The same applies to hardware and software specifications.

Choose consultants who demonstrate an understanding of your business. The consultants must have implemented a similar system elsewhere. Let someone else pay for their learning curve. Get references, and take them up. Remember that consultants' fee rates vary enormously; the bigger the practice, the bigger the day-rate.

Consider using a software house that carries certification. This will ensure that the company has systems in place for managing the project professionally.

Undertake extensive trials before the system goes live. The old and new system should continue to operate in parallel until the new system works flawlessly. Do not set tight deadlines, and don't allow your business to become dependent on the new system until it is working properly. Allow time for resolving the problems in the new system.

Software can be delivered on time, on budget and to the right specification – but then requires more effort to make it work properly. It is rare for a software project to work first time.

Give managers and users formal training in the system. Put money into the budget for this purpose. Ensure that users are trained. Some experts recommend spending a third of the budget on hardware, a third on software and a third on training.

Have a disaster recovery plan. Do not bin the old system until the new one is working properly.

Be aware of changes that could alter the outcome of the project. If change occurs, check to see if it alters the risk.

Build in flexibility. Ensure that the system is upgradeable. Consider using an 'open system' that uses industry-standard hardware and software. Don't become dependent on a single supplier's proprietary system.

Hold regular reviews. Don't sit back and wait for the contractor to deliver a complete solution.

CASE STUDY: PROJECT RISK – FROM THE CONTRACTOR'S VIEWPOINT

One electronics business (let's call it ABC) admitted to suffering many difficulties over a recent project. ABC had wanted to break into the food-processing market. It aimed to prove its credentials to the client and to the other companies in the industry. Determined to win the project, ABC bid a low price.

The project required ABC to install manufacturing controls in food factories around the world. Disadvantages included unfavourable payment terms; a lack of familiarity with the client's industry; and having to sub-contract some of the engineering work to an overseas subsidiary.

When the project was 20 per cent complete, ABC found that the costs were much higher than expected.

Originally, it had hoped to gain additional work from the client, but this is now a poisoned chalice. The client is now familiar with what needs to be done, and this makes it difficult for the contractor to raise the price of future contracts. And so ABC may be destined to carry out more projects for the client at a marginal profit, in order to recoup some of its losses. Equally, the contractor can't cut corners because of the risks that this would entail.

This study highlights the risks of pitching for work with several unknowns – in this case working in an unfamiliar market, in overseas operations, at low prices and dealing at arms lengths with a sub-contractor.

Classifying the Project to Determine the Risk

Being able to classify the project helps you identify the level of risk. This is shown in Table 16.2. It ranges from the project's technical complexity to the effect on the organization's reputation.

If you find this classification helpful, you could take it one step further and evaluate the riskiness of your project by ascribing a 1–5 score for each factor, and then adding up the total score.

Table 16.2   Nine factors affecting a project's risk

Project characteristic		Score 1=low, 5=high
1.	Technical complexity	If the project involves cutting edge technology, it will be much more liable to fail than one using tried and tested solutions.
2.	Geographical location	Will the project take place in the company's own workspace, or in a river basin thousands of miles away? Will there be one single location, or is the project implemented in 300 retail stores? The simple the geography, the less risk.
3.	Industry or profession	Is the aim to design a new tablet for consumers, or running shoes for athletes? In both these cases, it will involve designers, drawings, prototypes and an overseas manufacturing plant. But if the project is for a new children's toy, which is a fickle market, the risk of failure may be higher.
4.	Clarity of the end product	Some products are easily seen – such as a building on an architect's plan, or a vehicle on a blueprint. Others are invisible, such as software code which has to be written. If the product can't be seen, it is the most risky of all projects.
5.	Duration	The longer the project takes, the bigger it is likely to be. McKinsey found that project duration wasn't a factor in project success or failure, as long as milestone meetings were held.
6.	Cost	A major oil rig is a big cost, while refitting a washroom is less significant. Normally, the bigger the cost, the higher the risk; but see 'Effect on reputation' below.
7.	Number of workers involved	The number of people as it affects risk really relates to the people on the team (for example the number of building labourers). The quantity of people using the end product (for example, army boots) is much less significant.
8.	Urgency	Disaster relief charities deal with urgent projects. They have become used to dealing with urgency, and they can reduce the variables by storing the necessary materials (such as tents and blankets, or having suppliers who can produce at short notice). But even so, they find it difficult to be effective when they arrive at a disaster area, where transport is in short supply and visas haven't had time to be arranged. For anyone else, urgency spells high risk. Taking longer time allows for better planning, and less chance of things going wrong.
9.	Effect on Reputation	Sometimes a small project can have a big impact on company reputation. A relatively minor update to NatWest's computer system meant customers' bills weren't paid for several days, which caused the company embarrassment.
Total score		

Score: 1–18 points: low risk. 19–27 points: moderate risk. 28–45 points: high risk.

Ten Steps to Minimizing Project Risk

You can minimize risk in major projects by adopting the nine steps shown in Table 16.3. We examine each in turn.

Table 16.3   Ten steps for managing risk in projects

Planning for risk	
1.	Obtain adequate information
2.	Examine all the options
3.	Carry out a risk assessment
4.	Allocate experienced staff
5.	Create a project plan
Controlling the Risk	
6.	Invest one step at a time
7.	Build in flexibility
8.	Review progress regularly, review external information
9.	Spread the risk
10.	Manage the identified risks

1. OBTAIN ADEQUATE INFORMATION

You need to undertake sufficient research and obtain technical data, including information about similar projects undertaken by other organizations.

This is not always possible because much data is a commercial secret. You should quantify the information, showing how much money is at stake at different stages of the project. And you need a sensitivity analysis, showing the results of different outcomes.

Decisions often have to be made about building a new factory, depot, store or call centre. This may be because the current plant is operating at maximum capacity, or because management wants to invest in a new opportunity. You need to know what the total demand in the market is, what share of the market it has and what is likely to happen to demand over the life of the plant.

If several companies decide to build new plants at the same time, there will be excess capacity in the market. On the other hand, if you do not invest in the new plant, you may be at a disadvantage and perhaps be left with higher production costs in the old plant.

Information is vital at all stages of the project, especially information about costs and timing. Regular review meetings should ensure that data is shared.

But beware of putting too much faith in complex models such as Monte Carlo simulations. Their results often prove only what their authors want to show.

2. EXAMINE ALL OPTIONS

It is rash to undertake a major project without considering the options. Rather than launching new stores abroad, you might buy an existing chain. Rather than investing in your own IT system, you might outsource your computing facilities. Rather than launching a new product nationally, you could opt for a test market or regional launch. Rather than spending $2 million on a new plant, you could make the existing machines produce more products.

3. CARRY OUT A RISK ASSESSMENT

Your risk assessment should examine the project's sensitivity to various factors. This might include a 5 per cent rise in raw material costs, a 10 per cent downturn in demand, or a catastrophe (such as fire or explosion).

Companies often fail to identify the threats to the project. As a result, they are ill-equipped to fight back if these threats appear on the horizon. Businesses should be ready to expect problems so they don't turn up as surprises. With a little forethought, many could have been foreseen.

To identify the risks, management can set up a brain-storming session or involve experts from different disciplines. The risks should then be logged in a risk register. They should be reviewed at intervals, to see whether new risks have emerged while others are no longer present. A project is dynamic and therefore the risk register must match that.

The business should also consider whether it is financially and managerially equipped to carry out the project.

The risk measurement should be done by the people who will be held accountable for the project.

4. ALLOCATE EXPERIENCED STAFF

Projects often go wrong because the inexperience of the project manager. The company should appoint a manager who has previously carried out a similar project. If consultants are used, they should have worked on similar activities in the past. IT people complain that project managers know how to use a GANTT chart but are clueless about IT.

CASE STUDY: THE CONSULTANT WHO FLED

One business appointed an experienced project manager to install a new HR computer system. It involved seeking bids from software suppliers, choosing the best bid and working with the supplier and local staff to implement the program.

Some months in, the project manager started to complain that key staff were uncooperative. The HR director held meetings with the staff, and asked them to be supportive. They replied that the consultant was remote and didn't engage with them. Three months before the project was due to go live, the project manager resigned, saying she had a new job to go to. Senior management entreated her to stay on until the project was complete, but she refused.

The company pushed back the launch date and hired a new project manager who gained the trust of staff, and the program was launched successfully.

In hindsight, senior managers believe the consultant lacked confidence in handling internal staff who, in turn, may have felt sidelined by the outsider's appointment. She had a good CV and track record, and everything indicated that she was resourceful and effective. But more thorough taking up of references might have highlighted that weakness.

---

5. CREATE A PROJECT PLAN

Having undertaken the first four steps (getting information, examining the options, doing a risk assessment and allocating staff) you can now create a project plan.

This will contain all the necessary tasks (the work breakdown structure or WBS), costs, milestones and other data needed. For a project of any size, it is important to use project management software, involving Programme Evaluation and Review Technique (PERT) and critical path measurement (CPM).

Software that allows different members of the project to input and review the information is especially useful.

At this point, you can modify the design, the plan (people, resources, time and money) and the contract, according to the likely risks.

6. INVEST ONE STEP AT A TIME

It is often possible to avoiding committing the company to the entire costs of investment. For an engineering project, the phases of development might be as follows:

- market research;
- engineering design;
- site preparation;
- order equipment;
- install and commission equipment.

At any one of these investment points, a decision can be made to go ahead or stop the project. The decision to proceed will only be given if the current stage is successful. This is common in oil and gas companies, where extraction comes after a long process of geological surveys and exploratory wells.

Once equipment has been ordered, most companies have passed the point of no return. However, capital projects with long lead times may be cancelled (as is the case of defence contracts for aeroplanes).

7. BUILD IN FLEXIBILITY

If creating a new plant, vehicle or plane, you can build in flexibility so that the plant can produce more than one kind of product. Then, if the market forecasts are incorrect, the plant won't become a white elephant. While adding flexibility can increase costs (because it may involve more expensive equipment), you may decide that the reduction in risk will be worth it.

Building a smaller plant which can be expanded is an option for some companies. Retro-fitting is often unsatisfactory; but where output can be increased by, for example, adding more injection-moulding machines, this is a sensible option.

8. REVIEW PROGRESS REGULARLY

You need to continually reassess whether the facts that prompted you to develop the project remain true. This will allow you to either alter the project or halt it. You should review the 'business case' – the reasons behind the project – to see whether it is still valid.

You should also compare the milestones – progress against target – and identify where problems may be occurring.

9. SPREAD THE RISK

You can share the cost and risks of the project with other partners. Although this reduces the company's profit, it also means that the potential losses will be equally reduced. Moreover, the other partners may have skills or assets which you do not possess, whether in political contacts, experience of similar projects or marketing skills.

Another way of spreading the cost is to lease the equipment. This helps to delay payment and minimizes the effect on cash flow. Where you can buy insurance, this is a good idea.

10. MANAGE THE IDENTIFIED RISKS

Where risks have been identified, there are four ways to manage them, as we have seen elsewhere in this book. They merit repeating here:

Avoid. You may be able to avoid the risk. You may decide not to proceed with the project. Or you might avoid using suppliers that you have no experience of.

Mitigate. You can choose to reduce the risk. For example, you might reduce the impact of bad weather by undertaking the project in the summer months. Or you might split the work into smaller sections, to avoid any one piece of work causing a major problem.

Transfer. You can transfer the risk to someone else. This might involve buying insurance. Or you might get an outside expert, for example, a consultancy, to undertake the work out of house.

Accept. In some cases you might deem the risk to be acceptable, and be prepared to live with the consequences.

RISK/REWARD CONTRACTS

Risk/Reward contracts give the supplier a stake in the successful outcome of the project. Some conventional contracts have penalty clauses for failure, though most don't. The risk/reward contract offers the supplier an incentive to meet the objectives. These include providing a workable solution, on budget and within the agreed timeframe.

For example, an IT project might be worth £1 mllion. The client can offer to pay 90 per cent of this as normal, but will pay an additional 20 per cent if the supplier meets the objectives.

Thus the supplier stands to earn 110 per cent of the contract if they deliver on time.

Companies often put 10–40 per cent of the value into the risk/reward element of the contract.

Such contracts are harder to manage than normal ones, and are at risk of dispute. For example, what happens if the client changes the specifications half way through the project, as so often happens, to meet changing circumstances?

If you find it hard to get suppliers to agree to a risk/reward contract, it means they are unsure of meeting your requirements. This should signal a warning, and you should consider changing the parameters of the contract until suppliers are willing to shoulder some of the risk and reward.

In projects designed to achieve savings, and where the project contains uncertainties, you might share both the costs and the savings with the supplier. For example, doctors may receive incentive payments if they inoculate a specified number of older patients against flu, or if they conduct a certain number of 'well woman' interviews.

In one case, a health authority wanted to incentivise pharmacists to reduce the cost of treating diabetic patients, based on the cost per patient in the previous year. However, this risks a conflict of interest with the patient's needs, an issue discussed in Chapter 7 on Procurement. In other words, it might encourage pharmacists to offer patients cheaper but less effective treatments.

Methodologies and Management Standards

There are two established major management standards, PMBoK Guide and Prince2. ISO 10006 has also entered the fray, and is based largely on the PMBoK Guide.

PMBOK GUIDE

Created by the Project Management Institute (PMI.org), the PMBoK Guide (Project Management Body of Knowledge Guide) divides a project up into its five stages: initiating, planning, executing, monitoring and controlling and closing.

It also has ten knowledge areas, such as the project's time, scope, people and cost, not to mention its risks. The PMBoK Guide has several processes, each of which needs to be controlled for the project to succeed:

* develop the project's charter;
* develop the project management plan;
* direct and manage the project's work;
* monitor and control the project's work;
* perform integrated change control;
* close the project or phase.

PRINCE2

PRINCE2 (standing for PRojects IN Controlled Environments, version 2) is a project management methodology. It was originally developed by the UK Government and is therefore more widely used in the UK, especially in Government projects.

PRINCE2 is based on seven processes, including initiating the project, managing the product delivery, managing the stage boundaries and closing the project. It also has 'themes', such as business case, organization. There are also 'principles', such as continued business justification, learning from experience, and having defined roles and responsibilities.

ISO 21500 AND 10006

ISO 21500 provides guidance on project management. It can be used by any kind of organization, whether public, private or community based; and for any type of project, irrespective of complexity, size and duration.

It describes the concepts and processes that are considered to form good practice in project management.

ISO 21500 helps people understand how project management fits into a business environment, and provides a basic guide on project management.

Finally, there is also ISO 10006, which gives guidance on how to apply quality management in projects. It is a guidance document, and is not used for certification or registration purposes.

AGILE AND SCRUM

Agile methods specifically relate to IT projects. Agile is based on the principle of producing working software in incremental builds. The team produces a continuous series of software in short cycles called sprints. The team assesses each build, identifies problems, and reviews and prioritizes the backlog of tasks. The team then schedules the most important tasks for the next sprint. This regularly gives the client something to see, rather than waiting until the project is finished.

Scrum is a widely used Agile method. It involves daily scrum meetings at which the previous day's work is reviewed, and revisions to the workload made. Scrum involves teamwork, and active regular feedback on what is being done and what needs to be changed. The client is actively involved.

THE LEAN STARTUP

As the name implies, The Lean Startup is aimed at new businesses, especially software ones. However its principles are equally suited to projects in some (but not all) other areas. The principles of the Lean Startup involve the following:

1. *Create a minimum viable project (MVP).* This is a basic working model, without any unnecessary refinements. Take this to the market. Only by asking people to buy your product will you discover whether it is needed. If no one wants it, you have saved hours of development time. If customers buy it, you can improve it.
2. *Learn what the customer thinks.* Use actionable metrics to discover more about the product, such as the repeat purchase rate. This will tell you whether the product is viable in the longer term.
3. *Carry out split testing.* If your product is online, offer an alternative price or design to every other.
4. *Pivot.* If the product is slightly wrong, be prepared to amend it. Many successful products have emerged out of ideas that the market didn't want.

Risk Assessment

You can assess your vulnerability to project risk by answering the questions below.

Question	✓
Is your main project simple, rather than complex?	
Have you evaluated alternative options?	
Are milestones planned and project meetings well attended?	
Is senior management leadership engaged?	
Has risk assessment been carried out?	
Is the project run by an experienced member of staff?	
Have project plans been created?	
Does the project allow for flexibility?	
Is the risk shared with other partners, through leasing, sub contractors or insurance?	
Has the organization committed itself to taking only one step at a time, rather than being committed to the entire project?	
Total points scored:	

Score: 0–3 points: high risk. 4–6 points: moderate risk. 7–10 points: low risk.

The Appendix has a summary of the checklists. By entering the results of this one, you can compare operations and production risk against other categories of risk.

17
Corporate Governance: How to Comply

It's the day of the shareholder's meeting. The central London hotel conference room is stuffy and the coffee is sour. But the Chair is cheerful. He looks relaxed, chatting briefly to fellow directors as the meeting comes to order. As he gets up, he takes the shareholders through the results of another reasonably good year.

But a few voices of dissent emerge. Why are there no women on the Board? Why is the CEO also the Chair? And are the newspaper reports about his private dealings true? The Chair brushes off these criticisms with a breezy manner.

Your eye is caught by the sight of some rather youthful and slightly scruffy shareholders. Surely they won't try anything on? Has anyone thought to organize some security, you wonder?

Corporate Governance Risks

Corporate governance is the management of the organization at the highest level. It is designed to ensure that top management behaves with transparency and integrity, and is held accountable for its decisions. And because it relates to strategic management, usually of large organizations, major risks are involved.

Governance has become a hot topic in recent years, largely due to the scandals where senior management have behaved irresponsibly or engaged in criminal activity, such as:

- manipulating their company's share price;
- misleading investors and the public;
- making false statements about the company's profitability;
- hiding their losses;
- paying excessive salaries to top staff despite poor performance;
- treating company money as a personal expense account.

At the same time, Boards have rubber stamped the CEO's proposals, didn't question rash plans and waved through large pay rises for directors.

The regulators were often found to be toothless or inactive. The principle of 'light touch regulation' meant that regulators didn't take action, not wanting to impede growth and entrepreneurialism.

The result was a parade of business failures and crises, including Lehman Brothers and Enron. Warren Buffet summed it up by saying to his Berkshire Hathaway shareholders:

After sitting on 19 boards in the past 40 years ... too often I was silent when management made proposals that I judged to be counter to the interests of shareholders. In those cases, collegiality trumped independence ... A certain social atmosphere presides in boardrooms where it becomes impolitic to challenge the chief executive.

Regulation Has Been Strengthened

As a result of these failures, governments around the world have imposed greater controls on the way businesses are run, especially publicly quoted ones, in order to prevent the scandals of the past and reduce the risks to shareholders. This goes under the name of corporate governance.

Corporate governance is about the principles and processes used to manage a company and to report on its progress. It's about ensuring that the company is run with integrity and transparency, and for the benefit of shareholders – and increasingly other stakeholders.

In the US, the move to stronger regulation started with the Sarbanes–Oxley Act, while in the UK the Government introduced the Corporate Governance Code and the Companies Act.

Around the world, this kind of legislation has different names, but the principles are similar. Governments and regulators are trying to ensure that directors and their advisers don't misuse their considerable power. As the world of business becomes more transparent, companies have to ensure their businesses are run properly, and they must be able to demonstrate that.

The penalties for non-compliance have been slight to non-existent in the past, but have now become more severe.

Let us look at some specific legislation, starting with the UK's Companies Act, to see what it requires and what the risks for non-compliance are.

The UK Companies Act

Running a company used to mean maximizing shareholder returns. The UK's Companies Act 2006 added wider provisions, including the following.

Directors must act in a way that benefits the shareholders as a whole, but also take into account:

- the long-term consequences of decisions;
- the interests of employees;
- the need to foster the company's business relationships with suppliers, customers and others;
- the impact on the community and the environment.

Directors must 'avoid conflicts of interest, and must not accept benefits from third parties'.

Companies listed on the UK's stock exchange have to publish a business review. These should include information on:

1. factors likely to affect the future of the business;
2. environmental, employee and social issues;
3. contractual and other arrangements essential to the company's business.

This is designed to give shareholders better information about quoted companies, and give directors a discipline for stating the business' strengths and weaknesses. The review looks not only at the past but also the future, and therefore goes beyond the traditional backward-looking annual report.

Most of the penalties for non-compliance with the Companies Act are trivial, typically a £1,000 fine. However, a few clauses provide for serious penalties:

- If the company wrongly acquires its own shares: up to two years in prison (Section 658).
- If you help someone buy the business by giving them money: up to two years imprisonment (Section 678 and S679).
- Fraudulent trading: up to ten years imprisonment (Section 993).

The Legal Aid, Sentencing and Punishment of Offenders Act

The Legal Aid, Sentencing and Punishment of Offenders Act extended the penalties in the Companies Act. It allows for potentially unlimited fines in the magistrates courts for corporations, directors and wealthy individuals.

Whereas in the past any case brought in a magistrates' court would usually cause minimal financial impact, unlimited fines mean that is no longer the case. This could have an impact on claims for injury compensation, and on health, safety and environmental prosecutions.

In turn, companies and directors may seek to have their case heard in the crown court, where a legally qualified judge will preside, rather than a lay magistrate.

The UK Corporate Governance Code

The UK Corporate Governance Code came about from the Financial Services and Markets Act 2000, and applies to companies listed on the London Stock Exchange.

It came in the wake of scandals such as Polly Peck (where Asil Nadir fled to Cyprus after channelling the company's money there), Robert Maxwell (who robbed the pension fund) and BCCI (which become insolvent, having funded terrorism and dictators).

In such cases, the CEOs were hardly accountable to shareholders, and were able to conceal illegal activity.

The Code requires companies to disclose how they have complied with its tenets, and explain where they have not applied the Code. This is what the Code refers to as 'comply or explain', which we discuss later in this chapter.

There are five sections, on leadership, effectiveness, accountability, remuneration and relations with shareholders. What follows is a summary.

LEADERSHIP

- Every company should have an effective Board which is responsible for the long-term success of the company.
- There should be a clear division of responsibilities between the Chair who runs the Board and the CEO who runs the company's business.
- No one individual should have unfettered powers of decision.
- Non-executive directors should challenge and help develop proposals on strategy.

The separation of the Chair and the CEO's role is seen as a litmus test of good corporate behaviour, because it moderates the exercise of an individual's power. By contrast, the $33 billion News Corporation has had Rupert Murdoch as both Chair and CEO, and is controlled by him and his family. News Corp has come under fire for many reasons. Journalists from one of its titles, *The News of the World*, colluded with police, and hacked into the phone of dead soldiers and a murdered schoolgirl, Milly Dowler, giving her parents hope she was still alive. As a result, a British Parliament committee said that Murdoch was 'not a fit person to exercise the stewardship of a major international company'.

EFFECTIVENESS

- The Board and its committees should have the right balance of skills, experience, independence and knowledge of the company.

- There should be a formal, rigorous and transparent procedure for appointing new directors to the Board.
- All directors should be able to allocate sufficient time to the company to discharge their responsibilities effectively.
- All directors should receive induction on joining the Board and should regularly update their skills and knowledge.
- The Board should receive information in a timely manner, and in a form and of a quality that enables it to discharge its duties.
- Each year the Board should evaluate its own performance and that of individual directors.
- All directors should be submitted for re-election at regular intervals.

Companies also need to consider the Board's diversity, especially gender. This is because a male-dominated Board means a lack of diverse thinking, and could lead to 'groupthink'. The National Employment Savings Trust (NEST), which is set to become one of the UK's biggest fund managers, aims to vote against the director who chairs a Board nominations committee if that committee doesn't have a gender target.

CASE STUDY: THE CEO GIVES HIMSELF A PERSONAL PROFIT FROM HIS COMPANY'S SALES

The Board of the $11.6 billionn Chesapeake Energy company, a leader in natural gas, said they had not reviewed or approved a controversial $1.1 billion mortgage raised by the company's founder, Chairman and CEO, Aubrey McClendon.

The mortgage was raised on his stake in the company's oil wells, and was used by him to gain a 2.5 per cent interest in every well drilled by the business. Analysts said this created a conflict of interest and violated corporate ethics, something the company rejected.

The Board also approved a $112 million pay deal for McClendon, making him the highest paid CEO, in a deal which was criticized by shareholders and the subject of lawsuits. McClendon also ran a hedge fund, Heritage Management Company, which traded in oil and natural gas, a role that was not disclosed to shareholders, and led to accusations of a conflict of interest. *Forbes* magazine had named McClendon one of America's best performing executives.

ACCOUNTABILITY

- The Board should present a balanced and understandable assessment of the company's position and prospects.
- The Board must assess any significant corporate risks.
- The Board should maintain internal control systems.
- The Board should establish formal and transparent arrangements for maintaining an appropriate relationship with the company's auditor.

The JP Morgan scandal involving losses of £4.1 million was a textbook failure of a Board's inability to maintain internal controls. The loss was caused by a London-based JP Morgan trader, Bruno Iksil, whose bet on derivatives went the wrong way. At a US Senate hearing, Ina Drew, head of JP Morgan's trading operation, accused Iksil and members of the London team of hiding the bad news from her. She said Iksil failed to value his accounts properly, hid his losses and failed to report the projected risks.

But members of the US Senate sub-committee who investigated the event, said that shortly before JP Morgan disclosed the massive loss, the company had changed the model it used to measure risk. The change gave the appearance of halving the loss.

The committee also heard that Jamie Dimon, JP Morgan's Chief Executive, had withheld key data from the regulator, the Office of the Comptroller of the Currency. JP Morgan, meanwhile, blamed Drew for not being in control of the traders.

It was noticed that Jamie Dimon was Chairman, President and CEO of JPMorgan, again placing a lot of power in one individual, and making it hard for others to stand up against him.

However, the investment industry is generally plagued with such scandals, because massive profits can be made from taking bets on future share, commodity or currency prices. And when the bet fails, as it must sometimes do, the natural urge is to hide the loss and seek to reduce it by making further bets. It is unclear why the investment industry doesn't have systems in place to identify growing losses.

REMUNERATION

- Remuneration should be sufficient to attract and retain directors of the quality required to run the company successfully.
- But a company should avoid paying more than is necessary for this purpose.
- A significant proportion of executive directors' remuneration should be structured so as to link rewards to corporate and individual performance.
- There should be a formal and transparent procedure for developing policy on executive remuneration. No director should be involved in deciding their own remuneration.

Fifty-nine per cent of Aviva's shareholders voted down the executives' pay deal, in an AGM that was punctuated by slow handclaps and calls for Board members to resign. Shareholders objected to the £2.7 million pay deal for CEO Andrew Moss when the company's shares had fallen 61 per cent in five years, and analysts said the insurance giant had lost its sense of direction.

Many countries have initiated 'Say on pay' legislation, which requires companies to present executive pay deals to shareholders for their approval. In Australia, if more than a quarter of shareholders vote against the directors' remuneration package at two consecutive meetings, the directors have to stand for election again in 90 days. In the USA, most quoted companies have non-binding annual shareholder votes on compensation.

RELATIONS WITH SHAREHOLDERS

- The Board should ensure a satisfactory dialogue with shareholders.
- The Board should use the AGM to communicate with investors and to encourage their participation.

When investment analysts were asked about the quality of information contained in companies' financial statements, 53 per cent rated it as 'fair' or 'poor' (Table 17.1).

Yet this information is clearly important. In the same survey, 78 per cent of investment professionals said they frequently analysed companies' top hazards.

In a test by PwC, one group of analysts were given an abridged set of accounts relating to Danish firm Coloplast, while another group was given the full report complete with non-financial information. According to PwC, the results were startling. Those with the full report were overwhelmingly in favour of buying the stock, whereas those with the financial data all decided to sell.

Table 17.1   Evaluating the quality of information on potential hazards in financial statements

Quality of information	%
Excellent	2
Good	45
Fair	31
Poor	22
Total	100

Source: Global FM

Improving Shareholder Communications

Given that Boards are weak at communicating with shareholders, with this job largely falling to the CEO, there are numerous ways in which it can improve:

Shareholder–Board days. Companies can set up 'Meet the Board' days, where investors can meet investors.

Board visits to major shareholders. Boards have been known to do a road show, meeting major investors in different cities.

Investor meetings with Board committees. There is no reason why an audit committee should not meet with investors and discuss current topics.

Virtual shareholder meetings (VSM). A VSM is an interactive online meeting that allows for dialogue, and can permit voting by verified shareholders.

Shareholder relations committee. A set of directors could be tasked with regularly engaging with shareholders.

REPORTING RISK

Reporting risk to shareholders has come to the fore, because regulators want businesses to demonstrate that shareholders can make investment decisions based on full knowledge. Publicly quoted organizations must therefore report on material future risks, including:

- supplies and suppliers;
- environmental;
- legal;
- financial, including debt;
- IP, including expiring patents;
- regulation;
- competitors.

As well as future risks, shareholders also need to know about activities in the past year that might give rise to risks, including:

- industrial action;
- adverse weather impacts;
- technology problems;
- loss of customers or markets.

Organizations should also comment on how they assess risks and control them, and provide an outline of their risk management structure. Where uncertainties about risks exist, the report should identify them.

Comply or Explain

The Corporate Governance Code is entirely voluntary. As we saw earlier, it uses the device of 'Comply or explain'. This is the requirement for a company to either confirm that it complies with the code or explain why it does not. For example, a section of the Code might not apply to the firm.

'Comply and explain' is also used in many other jurisdictions. Australians tend to call it, 'If not, why not?'

The idea is that investors will shun companies that fail to comply, and this will encourage companies to conform.

Most companies comply, rather than explain, But where they explain, the comments tend to be thin, with phrases such as 'We believe this to be in the best interests of the company and its shareholders', implying a reluctance to engage in a dialogue about company strategy.

'Comply and explain' has improved corporate transparency and governance. But it was born in the era of light tough regulation and reflects that approach. As such, it doesn't stop those who lack integrity or have criminal intent. And because it relies on investors engaging with Boards, it is irrelevant to short-term investors who focus on immediate profit rather than governance. Any future corporate failures will serve to question how appropriate a voluntary code is; and the bad therefore risks driving out the good.

Five Board Weaknesses

Despite the Code, five weaknesses remain:

1. *Executive pay.* The code has failed to stop the growth in executive pay. Disclosure has merely served to increase CEOs' pay, with each company attempting to show that their chief was worthy of being in the top quartile of earnings.
2. *Overbearing CEO.* An overly strong CEO has the power to enforce his or her decisions. The Barclays LIBOR scandal was thought to be symptomatic of that.
3. *Board appraisal.* Companies have been unwilling to engage in reviewing the skills and performance of Board members.
4. *Audits.* External auditors have sometimes either been captured by their client, or else put their income above the needs of shareholder information. Either way, too many audits have been a whitewash, leading sometimes to the failure of their client and the loss of the auditor's credibility. The UK's FSA fined PwC £1.4 million and severely reprimanded it for failing to spot that billions of dollars of client money had not been properly ringfenced at JPMorgan Chase.
5. *Board diversity.* Women comprise only 12 per cent of FTSE 100 Board members, and 16 per cent of non-executive directorships. Yet companies with three or more women Board directors achieve return on equity 45 per cent higher than the average company (30percentclub.org. uk).

   Higgs & Tyson ('Balancing Boards', opportunitynow.bitc.org.uk) found that almost half of the directors they surveyed had been recruited through personal friendships and contacts. Only 4 per cent had a formal interview, and only 1 per cent had obtained the role through answering an advertisement.

Such weaknesses imply lower standards in these areas. Low standards can create risks, and risks can cause failure.

In 'Comply or Explain: Market Discipline and Non-Compliance with the Combined Code' MacNeil and Li found a strong link between share price and non-compliance with the code. In other words, companies that do not comply perform worse, though cause and effect are hard to establish.

So, although there is no legal penalty for failure to comply, unwillingness to conform and failure to rectify the five issues above point to corporate vulnerability.

Shareholder Activism and Revolt

Shareholders seem to becoming more activist. Novartis, the Swiss pharmaceutical group had to abandon the award of a $78 million 'golden gag' non-compete payment to its outgoing chairman, Daniel Vasella. This was aimed at preventing him from advising competitors after he stepped down from the pharmaceutical giant's Board. The agreement was cancelled after shareholders issued a criminal complaint against Vasella and the company's compensation committee.

Companies can only prevent such actions by ensuring that all proposals are reasonable.

Best Practice in Governance

There are symptoms of best practice which, if instituted, reduce financial risk. They comprise the following:

- *Board structure*: There is a properly structured Board in place, capable of taking independent and objective decisions. It is of the right size. Chair and CEO are separated. Members are selected through competition. The Board contains an adequate number of non-executive and independent directors who take care of the interests of all stakeholders.
- *Board activity and performance*: The Board remains in effective control of the affairs of the company at all times. It meets regularly, has access to independent advisors, and its performance is periodically reviewed. Papers are provided in advance. The Board is regularly briefed on strategic issues such as business development, technology and regulatory trends. Non-executives visit company sites unaccompanied by senior management.
- *Shareholder and regulatory communication*: The Board keeps the shareholders informed of relevant developments impacting the company. Appropriate disclosures are made. Applicable accounting standards are followed, subject to any major departures that are explained in the Financial Statements.
- *Risk management*: There are internal audits for all relevant risk areas. There are comprehensive financial controls and procedures. Procedures are in place to prevent and detect fraud.
- *Corporate social responsibility (CSR)*: Conflicts of interest are transparent. Potential spark points are audited, such as health and safety, political contributions, employee relations, bribery and whistleblowing.
- *Strategy, planning and monitoring*: There is a strategic plan. Financial planning takes place, with annual budgets for income, expenditure and capital. The Board approves the overall budget. The Board receives regular reviews of financial results, involving reports on variances and updates on forecast out-turns. Management succession is planned.
- *Management responsibilities*: The company's management has clearly defined responsibilities and authority. Management performance is objectively measured. Remuneration policy is transparent, based on performance and not managed by the recipients.

- *Board committees*: At a minimum, there are remuneration and audit committees. They are independent and objective, and meet sufficiently often.

Legal Risks in Buying and Selling Shares and Businesses

So far we have looked at the governance facing issues facing conventional quoted companies. But the financial services industry is in a league of its own when it comes to governance risks.

The UK's Financial Services and Markets Act 2000 introduced controls into the banking and financial service industry. These were designed to stop insider trading and ensure that investors were not misled. To a large extent the act was about corporate governance of the financial services industry, specifically its investment side.

The Act allowed for unlimited financial penalties. For various offences, the regulator may 'impose on him [the transgressor] a penalty of such amount as it considers appropriate'.

The Act created a regulator, the FSA, to police the industry. For a long time the markets didn't take the FSA (now the Financial Conduct Authority) seriously, but in time it began to impose heavier fines.

- The FSA imposed a fine of £7.2 million on David Einhorn, a US hedge fund manager who sold a big holding in Punch Taverns, Britain's largest pub owner, just before a rights issue diluted the value of the shares. The FSA accepted this was a 'non-deliberate transgression'.
- After Cattles, a doorstep lender valued at over £1 billion, went bust, the FSA fined three former senior managers £700,000 for issuing misleading statements about the company, and banned them from working in financial services.
- Ian Hannam, the Head of Capital Markets at JP Morgan Cazenove, was fined £450,000 and subsequently resigned after sending an investor emails containing news about a JP Morgan client – insider information that could have given the recipient an unfair advantage.

Regulators are equally active in the US and elsewhere. The Securities and Exchange Commission (SEC) fined SAC Capital, a US hedge fund, a record $600 million (£397 million) for selling shares based on inside information. A doctor had tipped off the hedge fund that Bapineuzamab, an arthritis drug, was about to report poor results in clinical trials. The company then sold $690 million worth of shares in its maker Elan and Wyeth. At the time it was seen as the most lucrative insider trading scheme ever. SAC Capital allegedly made a profit of $276 million on selling the shares. Mathew Martoma, the hedge find's manager, received a bonus of $9.3 million after the deal, and at time of writing was facing criminal charges.

In summary, directors and brokers must take great care not to mislead the market, nor to provide or seek inside information. Casual emails can lead to future embarrassment and fines.

The US Experience: Sarbanes–Oxley, Jobs Act, Frank Dodd

Sarbanes-Oxley (known as SOX) is a controversial US law passed in the wake of corporate and financial scandals, notably Enron. Named after the two politicians who sponsored it, the law aimed to restore investors' trust in companies' accounts, by placing more controls on annual reports and directors' behaviour. Similar laws have been passed in many other countries.

The law is aimed at accountants and the directors and managers of public companies, and makes them personally responsible. CEOs and CFOs must certify the accuracy of their company accounts, and this means top officers can no longer plead ignorance – they are personally responsible for the integrity of an audit. They can be fined up to $5 million and imprisoned for 20 years for failing to obey the law.

The law also applies to all companies whose subsidiaries are listed on a US stock exchange. This means a European company with US interests may have to introduce SOX controls into the business.

The requirements of SOX include the following:

- *Loans to Directors.* Companies may not make loans to directors and executive officers. This ban may even preclude personal loans and the use of company credit cards and the cash-less exercise of share options. This means that directors can no longer treat the company as their personal piggy bank.

  Such controls might have prevented the scandal at Hollinger International, the newspaper company that accused its chief executive, Lord Black, of plundering its assets. Some of the minutiae included jogging attire for Lady Black (his wife) $140; exercise equipment for Lady Black $2,083; a leather briefcase for Lady Black, $2,057; opera tickets for Lord and Lady Black, $2,785; and summer drinks, $24,950.
- *Audit committee.* Companies must have an independent audit committee which will appoint, compensate and oversee the work of the auditing firm, and to which the auditing firm must report directly. Members of the committee may not do other work in the firm.
- *Auditors.* The accountancy practice must rotate the partners in charge of a corporate client every five years. This will prevent the relationship from getting too cosy.
- *Non-audit services.* The law bans accounting firms from providing a number of non-audit services. These services include bookkeeping; design and implementation of information systems; appraisal or valuation services; actuarial services; management and HR services; investment advisor or investment banking services; legal services and other expert services. This means the accounting firm can't advise on acquisitions, and then approve them in the annual report.
- *Internal controls.* Section 404 of the law requires the company's annual report to contain an 'internal control report', which assesses the effectiveness of the company's internal financial controls. This requires stringent internal auditing. And, according to a survey by Financial Executives International (FEI), it greatly increases the company's costs, including a 40 per cent increase in the fees charged by external auditors.

Some US companies have been deterred from going public, due to the high costs of regulation. This includes biotechnology companies that would otherwise raise money for trials through floatation.

Some UK and European companies have chosen to avoid floating US subsidiaries on grounds of the increased complexity. Other companies have complained about the added burden that SOX imposes. However, this is now thought to be less than 0.036 per cent of revenue and falling as companies have bedded in the systems. And if a business wants US funds, there may simply be no alternative.

The JOBS Act (Jumpstart Our Business Startups Act) relieved some of the regulatory pressure on smaller and emerging businesses. For example, the time allowed for compliance among new public companies has gone from two to five years. And the bill increased the number of shareholders a company is allowed before it is required to register with the SEC.

The Frank Dodds Act was aimed at the US financial services industry, and designed to ward off the kind of financial meltdown that was seen in the 2008 financial crisis. It increased the regulation of the financial services industry, including greater transparency of derivatives trading. Non-US firms with interests in the US have been affected, and the Act meant that such companies have had to get to grips with new rules and different regulatory agencies, as have overseas investment businesses. As ever, companies simply have to manage the change.

Risk Assessment

By answering the questions below, you can check to see how vulnerable your business is to governance problems.

Question	✓
Is the CEO able to act without restraint?	
Are the powers of CEO and Chair combined?	
Does the business have a cosy relationship with its auditor?	
Are non-executive members merely there to rubber stamp Board decisions?	
Do women constitute fewer than 20 per cent of Board members?	
Has the company been criticized for its executive pay?	
Has the Board performance not been appraised in the last two years?	
Are new Board members appointed without a proper procedure?	
Do Board members get insufficient information, delivered too late?	
Does the Board lack a fully range of skills?	
Total points scored:	

Score: 0–3 points: low risk. 4–6 points: moderate risk. 7–10 points: high risk.

The Appendix has a summary of the checklists. By entering the results of this one, you can compare your governance risk against other categories of risk.

18
Risk is All About People

Simon, a senior IT manager, is taking you to an employment tribunal – because the Board hasn't automatically given him the departing IT director's job. You shake your head in disbelief. 'The job is being advertised on the open market, and no one is stopping them from applying,' you point out.

You ring your lawyers to brief them, only to find that the partner who handles your work has been poached by one of their competitors. She'll doubtless be looking to take on your work. But will her contract allow it, you wonder?

It's time for that union meeting – they say your performance management programme is tantamount to bullying. From where you stand, every risk centres around people.

The HR Department and Risk

HR people are often somewhat unengaged with risk management. That's perhaps not surprising since risk management tends to focus on hard issues: regulations, structures, procedures and audits.

There's also the alienating language. 'Risk people can talk Klingon,' admitted Alex Hindson, head of group risk at insurance firm Amline in an interview with *HR magazine* (hrmagazine.co.uk). 'And the thing about HR professionals is that they are often in the job because they like dealing with people, not filling in forms.'

Yet whenever there is a crisis or catastrophe, it's usually down to someone doing something wrong. And when the problem is later analyzed, it was often due to a bad culture: a CEO who's out of control or staff who don't care. At other times it's a lack of skills or a herd mentality.

So if the business wants to get people aware of risk, HR is the department that knows how to put a message across and engage people.

HR has a lot of data at its fingertips but, as Mary Young of Canada's Conference Board points out. 'HR can be woolly and not quantitative, so the risk guys will say it's not worth the paper. But if you frame people issues in risk, you automatically get people's attention. Talk risk, and there will be visibility right to the top of the board.'

HR risk is changing. While it can sometimes seem as though the business is beset with risk, it isn't all one-sided. Global competition for jobs and investment has meant that staff wages have been static or falling in many countries, zero hours contracts (ZHC) are common in some industries, and employment legislation is unlikely to get tougher. In addition, corporate taxation has been dropping, and companies have unprecedented freedom to move assets around the world. Trade union membership is low, and legislation has made trade unions considerably less powerful.

On the other hand, the difficulties of attracting talent have grown, and young employees' attitudes and behaviour is different from those HR has been used to. Figure 18.1 shows the main risks that HR can have an impact on.

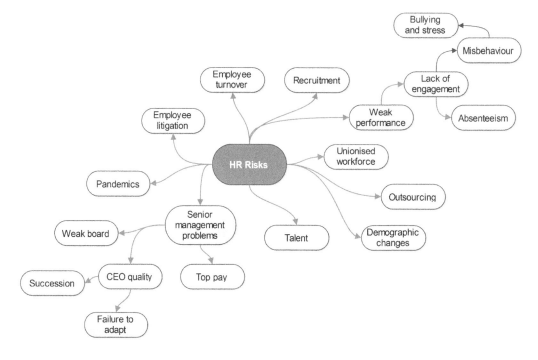

Figure 18.1   HR risks

Risk is About People

It isn't business risks that bring down a company, but the management which fails to manage them. And any organization is at serious risk if the management and workforce are weak. The main actors – people and posts that are crucial in HR risk management – are shown in Figure 18.2. The risks include:

- a Board that brings bad PR to the business by agreeing unmerited pay rises or excessive perks to directors;
- a chief executive who lacks leadership or common sense, or who is weak in a crucial area such as marketing or finance;
- having a weak finance director, or even a non-existent one in smaller firms;
- managers who base their decisions on hunch rather than on fact. Some Boards never see or call for research reports, financial data or sales figures;
- an organization that is slow to respond quickly to changes, such as new technology, falling sales or increased debt;
- lack of corporate structure, especially in smaller firms which fail to allocate responsibility for key functions – often the chief executive oversees sales or finance on a part-time basis;
- organizations that are demoralized or demotivated, or where there is internal strife and politics.

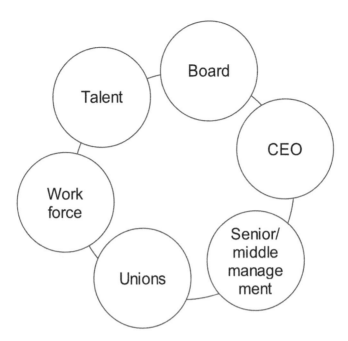

Figure 18.2   Actors in HR risk

The Leadership Challenge

It's difficult not to feel some sympathy for CEOs upon whose shoulders rests the herculean task of leading large organizations at a time of unparalleled change.

Most large businesses are more at risk than ever before from left-field upstarts who can appear from nowhere.

Such uncertainty can lead to misguided swerves in corporate strategy, such as when Marconi changed from a reliable telecoms business to a risk-laden, highly-leveraged outfit that went bust.

This can even influence a business such as Enron which saw energy generation as old fashioned.

But it is vital to distinguish between thrill seeking and taking carefully judged risks. We need to know that the CEO is not a cynical, mistrustful individual who lacks concern for others. We need to be sure they aren't moody or unwilling to admit mistakes.

Equally, it's vital to find a CEO who is visionary, capable of making decisions and of leading a team.

In recruiting senior management, especially CEOs, you may therefore want to employ an assessment tool such as the Hogan Development Survey which aims to identify whether or not the individual suffers from psychopathic tendencies. This and other failings are shown in Table 18.1.

FAILURE TO ADAPT

One of the biggest management challenges is failure to adapt. Success brings ossification. Over the years, a successful organization finds it ever harder to adapt, because that involves changing the methods that brought it success. Organizations that were founded before the arrival of computers, still less the internet, are at risk of being wedded to old systems. Indicators of the hidebound organization are shown in Table 18.2.

Table 18.1   Top 12 CEO failings

Top 12 CEO failings		
1.	Greedy for personal gain	8. Is weak in a core area, such as finance, IT or marketing
2.	Doesn't listen	
3.	Focused on short-term issues	9. Avoids meeting customers or staff
4.	Fails to make decisions	10. Overly controlling
5.	Displays sociopathic behaviour	11. Bullies
6.	Appoints weak subordinates	12. Lacks vision
7.	Is narcissistic	

Table 18.2   11 indicators that reveal a hidebound organization

11 indicators that reveal a hidebound organization
1. Doing it the way we've always done it: sticking to the existing production or operational process. Resistant to change.
2. Manual, paper-based systems.
3. Maintaining the standard industry business model; failure to see alternatives.
4. Belief that customers should respect the ways that the organization works, rather than the other way around. This is 'management for the benefit of the workforce' rather than the customer.
5. Maintaining hierarchic or bureaucratic structures; many organizational levels.
6. Promotion by longevity or seniority rather than merit.
7. Nepotism.
8. A bias against women.
9. Reliance on old sales channels.
10. The 'canteen culture': casual racism and sexism is the norm, and senior management is unable to impose new values.
11. An attitude of 'business as usual'. Complacency and inertia.

Sometimes the boss can see what needs to be done. At other times the organization knows that change is needed, because nothing has changed in 20 years. Here are six methodologies for implementing a change programme.

- *Benchmarking exercise* – looking at what competitors are doing. Establishing best practice. The limitation of this is that if your competitors are not doing what the customer needs it can lead you into a false sense of security.
- *Review of other industries* – for example airports have looked at how theme parks have managed crowds and queuing.
- *Market research* – identifying customers' attitudes. The drawback to this is that customers can never tell you what they want; only what they think of the current offerings. They can only tell you what is wrong with the current offering, which can provide pointers.
- *Using consultants* or futurologists to identify necessary areas for change.
- *Gap analysis* – define where you are, where you need to be, what the gap is, and how you will bridge it.

CEO'S PAY AND PERKS

Directors' pay awards and perks have been controversial for some time. Excessive boardroom pay continues to be an issue:

- In the USA, the average CEO of a company on the S&P 500 Index earns 380 times the average American worker's wage.
- At time of writing, the highest-paid CEO was Apple's Timothy Cook, whose total compensation was nearly $378 million. The 100th highest-paid CEO, Heinz's W.R. Johnson, had total compensation of more than $18 million, 543 times the average worker's income. Twenty years ago, according to the US Internal Revenue Service, CEOs earned only 40 times more than the factory floor worker.
- Average FTSE 100 CEO total pay is around 145 times the average salary for UK workers.

On many levels this wouldn't matter, were it not that many academic studies have shown that within-firm pay inequality is associated with lower firm performance. A study comparing compensation data on 1,500 US public companies using data from Standard and Poor's found that business productivity is correlated with pay disparity between top executives and lower level employees.

Treating the business like a personal fiefdom has reduced, though older examples stand out for their brazenness:

- Ross Johnson, CEO of RJR Nabisco, once sent the company jet to transport his dog, Rocco, from Palm Springs, California to New York. He named it 'G. Shepherd' on the passenger list.
- The SEC accused Tyco CEO Dennis Kozlowski with looting the company of $600 million. This included a $2 million Roman-themed birthday party for his wife. At the party was an ice sculpture of Michelangelo's *David* whose penis streamed Russian Stolichnaya vodka into crystal glasses.
- In the course of a divorce between Jack Welch, former boss of General Electric, the US's largest company, and his wife, it emerged that despite having retired, he was entitled to unlimited use of the company's Boeing 737 (value $291,000 a month) and a Central Park apartment, as well as a $9 million pension.

Undue perks damage a company's profits. David Yermack, an economist at New York University, looked at 200 large American companies, comparing those that let their CEO use company jets for personal purposes with those that did not. Even after accounting for other factors, Yermack found that the long-term stock-market performance of perk-rich companies was dramatically worse than that of their peers. This cost shareholders hundreds of millions of dollars a year.

While unpopular in Anglo-Saxon business cultures, worker representation on remuneration committees is also something to be considered, as a way of moderating undue top management pay. In Germany, worker representatives comprise half the members of remuneration committees at most of the largest businesses.

THE WEAK BOARD

All of this points to the need for a strong Board. A weak Board is dangerous because it allows the firm to drift, missing opportunities, creating inertia, stifling initiative and causing good staff to leave. Some major failings are shown in Table 18.3.

The Chair and non-executive directors should ensure that the boardroom seats are occupied by people of true merit, and should seek to remove those who fail to demonstrate results (while avoiding short-termism). It is important to distinguish between results that stem from adverse external conditions, and those which arise from poor management. More than one chief executive has been removed while battling in difficult trading conditions.

Appointing non-executive directors (and thus strengthening the number of outsiders on the Board), and setting up a remuneration committee, may give the business an independent outlook that could save it from copying some of the damaging decisions reported in the media in recent years. We examined this in more detail in Chapter 17 on governance.

Similarly, there is a need for high-quality evidence-based decision making at Board level. Boards sometimes suffer from the herd instinct. Herd mentality happens when everyone in the industry is doing the same thing, and senior people are loathe to miss out. The Tulip Mania (1637), the South Sea bubble (1711), the Wall Street Crash (1929), the dotcom bubble (1997–2000), the housing bubble (2007–2009), and every other bubble happened when top people wanted to avoid missing out on what looked like the right thing to do.

Table 18.3   Top 10 Board failings

Top 10 Board failings			
1.	Fails to make evidence-based judgements	6.	Lacks leadership
2.	Lacks strong non-executive members	7.	Members are ill prepared
3.	Prefers to discuss trivia	8.	Supine, fails to challenge the CEO
4.	Poor attendance	9.	Members are hostile to each other
5.	Lacks strategic vision	10.	Takes decisions for departmental benefit

LACK OF SUCCESSION

Some companies face strategic risks through the lack of corporate management or succession. A family-owned business sometimes suffers discord among family members as to the future direction the company should take. This affected Clarks, the shoe company, when some family members wanted to sell the business. It also affected the 264-year-old Pedro Domecq, which eventually sold out to Allied Lyons (now Allied-Domecq) for $739 million.

The situation is also likely to affect the public sector. In a survey of federal US agencies and public sector organizations, only 28 per cent had, or planned to have, a succession management programme, despite 56 per cent of the same respondents indicating they believed their organization was seriously short of leaders to meet emerging changes in their organizations.

Institutional Investor magazine estimated that more than $600 billion is currently managed by hedge funds whose founders will reach at least their 60s within the next decade. Many of these managers are stars whose names are closely linked with their firms, and often part of the brass door plate.

The problem affects all types of business. According to the Family Business Institute (FamilyBusinessInstitute.com), only 13 per cent of US family-owned businesses survive with a third-generation owner in charge, in part perhaps due to a lack of succession planning by the founder.

How to design a succession plan

To prevent a vacuum developing at the top, leading to a stagnation and future crisis, the business should adopt the following steps:

- get management buy-in for succession planning;
- identify the future leadership needs of the organization;
- conduct staff reviews to get feedback, identify problems and opportunities, and identify competencies;
- identify high-potential talent;
- recruit external future leaders, where appropriate (for example, graduate management trainees);
- give staff experience of different departments and roles;
- create opportunities for people to grow within the company;
- implement training, mentoring and coaching programmes, both formal and informal.

And in the case of a family business, you should take the following steps:

- put in place a clear management structure to identify roles and responsibilities, rather than support an extended family;
- ensure effective governance consisting of an experienced Board that includes outsiders;
- hold performance reviews of all employees, including family members;
- require family members to work outside the business, and acquire adequate qualifications if they aspire to director-level positions;
- ensure the younger generation acquire knowledge of the business and the motivation to maintain it.

CHANGE MANAGEMENT FAILURE

For many staff, change seems never-ending. They live in a perfect storm of technology advances, heightened expectations among customers, and even a rapid turnover at CEO level and ownership.

Management knows that change is essential. Any organization that hasn't changed in five years is probably at risk of being outpaced by competitors or facing criticism from regulators.

Hence senior managers are desperate for staff to show initiative, multi-task, be more productive and smile at customers. On top of that, staff are more closely observed and measured. And that's before we get to the possibility of redundancies.

So it's no wonder that any change is greeted with suspicion or even despair. Employees have often seen previous rounds of change, many of which didn't work. Many reckon they simply have to wait it out and things will go back to normal.

Only three in ten change programmes succeed, according to consultants McKinsey (mckinsey.com).

Employees employ various strategies to prevent organizational change. They can:

- find fault with the detail ('That'll never work because we won't have enough time…');
- develop an 'us and them' mentality, with anything suggested by management being intrinsically bad;

- fail to engage (I'll just carry on as normal...');
- sabotage ('If we don't fill in the forms, they can't see what's going on...');
- be hostile to colleagues. When job losses will take place, employees may denigrate their colleagues, to improve their own chances of survival).

On top of this, managers often create the conditions for failure by failing to communicate, understand staff motivations and show visible leadership.

Seven Factors That Permit Successful Change

To succeed in organizational change, managers have to:

1. *Recognize employees' anxiety* and the loss of their familiar and reassuring environment. Let people have an honest discussion. Employees must be able to express their concerns openly. Psychologist Kurt Lewin said that most people prefer to operate within zones of safety. For that reason, they actively resist change. Management therefore has to help people 'unfreeze' so they can transition to a new style of working. Eventually, they will 're-freeze' into the new mould.
2. *Understand employees' motivations.* Corporate efficiency isn't a high priority in the minds of many, and retaining market leadership isn't that important to them. McKinseys say that employees at all levels, from director to cleaner, are tugged five ways when it comes to loyalty. Those five motivations are society, the customer, the company, the working team and 'me' personally. Each of them has an equal claim on the employee's heart and mind (see Figure 18.3).

    Hence managers frequently ignore 80 per cent of employees' concerns. To succeed, a change management programme should explain why the programme will benefit all five motivators. It is useful to remember the phrase 'What's in it for me' (WIIFM) when managing people.

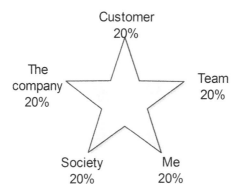

Figure 18.3   All employees have five equal concerns

3. *Provide clear, unambiguous communication and leadership.* Do not let staff believe that management has a hidden, unspoken agenda.
4. *Engage staff in finding solutions.* You must consult about the need for change and the best way to achieve it, and encourage initiative. Managers must give staff time to see the need for change. Management has often lived with an issue for weeks or months: staff need time to get Board and take ownership.
5. *Give employees some respite from constant change,* by making changes in a planned way. This will reduce the level of 'change fatigue'.

6. *Give staff a positive takeaway.* When seeking saving and greater efficiency, something that is seen by staff as 'demolition', add an element of 'rebuilding' and development into the process.
7. *Ensure that training includes active on-the-job practice and coaching,* to embed new procedures and values. Restricting training to classroom learning risks people going back to their old ways once they return to their desk. Set quantitative goals, with group incentives for success.

Talent Risks

Many businesses depend on key workers – senior managers and Board members responsible for creating change, managing businesses and keeping customers. If they move to a competitor, the firm may be at risk from their knowledge and contacts. Former employees can poach customers or key members of staff, set up in opposition or pre-empt the company's strategic plans.

Advertising agencies are often dependent on a well-known creative director. If that person leaves, the agency may lose existing and potential clients. If the individual's name is on the agency door, this poses added problems.

To manage that risk, organizations can adopt both defensive and aggressive solutions. Defensive solutions include contracts stopping ex-employees from working with the agency's clients for a period of time. But this is, at best, a short-term solution.

A better plan is to give key executives rewards that encourage them to stay. This will include:

- financial incentives (salary and profit sharing);
- managerial responsibility, or freedom from this kind of work. Companies often promote their best sales person, only to find that person is no good at, and doesn't like, managing people;
- lifestyle benefits (giving them the respect and credit their egos require, or the lifestyle they like).

TALENT SHORTAGE

Recruiting the best people is essential for the organization's long-term prosperity. Yes recruitment is often badly managed, with managers hiring people in their own likeness or recruiting 'safe' candidates. All too often the organization fails to check the successful candidate's background, leading to unwelcome surprises later on.

Some companies also face a critical skills shortage. This can stop them growing. HR needs to reflect on where the gaps are growing, and plan to plug them. To make the case, you have to identify the skills that are needed, collect the data and present an analysis.

Talent shortage can be reduced (or even worsened) by the quality of the recruitment process, which we look at below. In addition, you can instigate Workforce Planning, a methodology that involves assessing the organization's current skills, analysing the skills it needs for the future and deciding how to bridge the gap.

THE CREATIVE DEFICIT

In a survey reported by *Wired* magazine (Wired.com), 1,500 chief executives from all over the world said the most important skill they seek in employees is creativity.

Creativity is the one attribute that allows a company to stand out from the competitors. It gives both giant advances and small improvements. It produces effective new products and leads to improved ways of working,

In the twenty-first century, ideas rather than muscle will win. Nimble businesses that produce a string of better products and services usually rise to the top, overtaking the lumbering market leaders. An organization that lacks creativity, or downplays its importance, is therefore at risk.

But creativity needs to be carefully managed, for creative employees are unruly and don't abide by the rules.

Moreover, it's easy to ignore and disparage such employees, for an organization is threatened by people who aren't team players, and who disrupt the status quo. Managers are happiest when their subordinates do what is expected of them. And so someone who questions a manager or suggests alternative solutions may be seen as a threat.

Managers need to be helped to recognize, tolerate and support creative people. This isn't easy for your managers, because they're judged on the smooth running of their departments or divisions. And those who get promotion often do so because of their ability to manage in a disciplined way; so the uncertainty and disorganization that ideas foster can be inimical to them.

In addition, creativity needs to be supported by structures that ensure things get done. Creative people are likely to be immersed in unworkable ideas and surrounded by half-completed projects.

But creativity need not be a skill possessed by only a few. It is a skill that can be learnt. Like a muscle, it gets better with practice. To foster creativity, you need the following:

1. Implement management training to understand and develop creativity.
2. Take staff away from their desks and tasks to discuss strategic or business issues.
3. Discourage people from working from their inbox.
4. Allow people respite from the continuing pressure of work. It will give them time to think.
5. Encourage debate and discussion.
6. Foster collaboration. If people work with other disciplines or outside organizations, it produces cross-fertilization.

Research suggests that brainstorming isn't very effective, apparently because participants are usually banned from criticism. Lack of debate leads to weaker thinking.

DRUG MISUSE

Some observers have blamed the 2008 crash on cocaine use, with bankers and traders regularly high on the white powder, leading to what Alan Greenspan, former US Federal Reserve chief, called irrational exuberance.

Quoted in *The Guardian* newspaper, Dr Charles Luke, an A&E specialist at Cork University Hospital, who has studied the effects of cocaine on bankers, says 'prominent figures in financial and political circles made irrational decisions as a result of megalomania brought on by cocaine usage'. It was reported that Jimmy Cayne, former CEO of Bear Stearns, had an antacid medication bottle filled with cocaine.

You will want to watch out for signs of drug misuse, since it can lead to aberrant behaviour and possibly the downfall of the business.

Workforce Risks

So far in this chapter we have examined the HR risks relating to the Board, the CEO and the talent.
Now we examine the risks that relate to all members of staff, not just the high fliers.

In the complex twenty-first century world there is a multitude of HR risks that affect the organization. Below we highlight the major ones, and then discuss each in turn.

> *The demographic challenge:* An older workforce that is resistant to change, and a lack of fresh talent, will pose problems for many organizations.
>
> *Recruitment* is still not as effective as it should be. Managers still make irrational choices, and the process is flawed.
>
> *Lack of engagement:* Alienation and *bad industrial relations* all lead to poor performance.
>
> *Absenteeism* is rife in some organizations, for which organizations pay a heavy price.
>
> *Weak performance:* Over time, this causes customers to defect because it will be increasingly apparent that competitors' staff offer a better service. As roles increasingly require complex problem solving rather than routine procedures, the training and performance requirements will rise.
>
> *Bullying and stress:* A culture of bullying is often a proxy for remote management and/or ingrained workforce behaviour that's hard to change.
>
> *Legal threats:* Driven by no-win no-fee solicitors and the loss of deference, more individuals are suing their employers for real or imagined harm.
>
> *Outsourcing:* Hospitals use contract nurses, while builders use sub-contract labour to build houses. Who is in control of the twenty-first century workforce?
>
> *Stealing clients, data or other assets:* The business can be at risk of losing clients to departing employees, as well as databases.
>
> *Pandemics:* They're a 'Black Swan' event. That's to say, they are never expected to happen — until they do.

DEMOGRAPHIC CHANGES

Over the next decade, demographic changes will make it more difficult for companies to recruit staff. In many parts of the West there will be a decline in the number of young people and a gradually ageing population, as the bulging 'baby boom' post-war population ages.

According to PwC, Japan's median age is 45, and in the UK it is 40 (and is predicted to rise to 42 by 2035). Affected by its one-child policy, China will suffer a big decline in its working age population by 2050, as will Russia.

With low population growth rates, the number of people of working age will drop. This will lead to wage pressure, skill shortages and more flexible working patterns.

The *removal of the retirement age* means that people will stay in the workforce longer, with attendant issues over reluctance to adopt new technology and adapt to changing circumstances. Meanwhile these older workers will occupy posts that would have gone to young new recruits, with the result that the organization will fail to hire vigorous young people with fresh ideas. This can be partly overcome with:

- reverse mentoring, whereby younger staff help older ones adapt;
- healthcare support for older employees;
- support for phased retirement.

INEFFECTUAL RECRUITMENT

Recruiting a poor performer adds greatly to the risks to which a company is exposed.

Yet recruitment is still poorly managed in many firms. This is partly because judging personality and effectiveness is notoriously difficult. Simply measuring IQ or assessing personality at an interview will not identify whether the candidate will perform well. In Europe, many firms have adopted graphology, yet research shows this is equally flawed. Companies can minimize their risk by being methodical. This includes:

- preparing a job specification;
- creating a personal specification, outlining the individual's required experience, abilities and qualifications;
- undertaking the right method of recruitment, whether through head hunting or through advertising;
- training managers to be effective interviewers;
- adopting practical activities to identify how the candidate actually performs;
- checking CVs and taking up references.

Some companies hire MBAs and while this adds rigour to corporate thought there is a risk that company thinking becomes overly structured and planned, with less importance placed on intuition, creativity and risk-taking. *The Scotsman* newspaper unkindly described today's MBAs as 'corporate civil servants'. 100,000 MBAs graduate in the USA each year, as do 10,000 annually in the UK.

There are plenty of myths in recruitment. An outgoing personality is often thought to be essential for a successful sales person. But according to research by IBM Kenexa (kenexa.com), the most important attribute for sales success is emotional courage, the ability to keep going after knock-backs.

Google has moved away from judging candidates on college entrance exam scores and grades, according to *The New York Times*. These factors did not predict success, according to Prasad Setty, the company's vice president for people and analytics. Google found that the most innovative workers, who are also the happiest, are those with a strong sense of mission about their work and who feel they have personal autonomy. We examine engagement below.

Other firms are turning to Big Data. Recruiters can identify effective programmers by seeing whether their code is well regarded by other coders, and whether it gets re-used. By examining a programmer's involvement on community sites, a business can identify skilful coders – and it may not be those with the best CV. Some are young or complete outsiders.

ENGAGEMENT PROBLEMS

The larger the organization, the harder it is to maintain employee motivation and engagement.

This is particularly difficult in a period of organizational change where employees are demotivated.

Organizations can assess engagement through surveys, exit interviews and retention statistics. To achieve engagement, you need the following:

- *Recognition* – People must feel that they are people are recognized and appreciated. Recognition programmes are thought to be a useful tool for improving engagement.
- *A sense of justice* – Staff must see that good people are promoted and the rule of law prevails. Management must not be arbitrary, quixotic or angry.
- *Ethical* – Employees must believe the organization performs a valuable service or produces a worthwhile product. This is hard to achieve if you make cigarettes or adopt questionable ethics (see Chapter 19 on Ethics).

EMPLOYEE TURNOVER

The average UK and US employee turnover figure is 15 per cent, but this varies enormously by industry (as shown in Table 18.4). For example, the telemarketing industry expects 26 employees to leave each year for every 100 employees.

The main risk is excessive turnover. One business in the pub sector reported a turnover level of 120 per cent, meaning that for every ten employees, they had to recruit an extra 12 each year. This has big implications for staff skills and customer satisfaction.

However, employee turnover can also be too low. If there isn't enough turnover, organizations can stultify due to lack of fresh blood.

Recording high levels of satisfaction, as measured by staff surveys, may encourage the organization to believe it's doing a good job. But as Edward E. Lawler has observed in *Forbes* magazine (Forbes.com), staff satisfaction doesn't necessarily mean that staff will perform well. It merely indicates they don't feel the need to look for another job. And satisfied employees tend to RIP (retire in position). Hence the need to concentrate on performance, something we examine next.

Table 18.4   Average employee turnover, USA

Average Annual Turnover	
Services – accommodation, food and drinking places	35%
Arts, entertainment and recreation	27%
Retail/wholesale trade	22%
All industries	15%
High-tech	11%
Government/public – state/local	9%
Association – professional/trade	8%
Utilities	8%

Source: SHRM Human Capital

WEAK PERFORMANCE

A survey of 170 organizations by XpertHR (xperthr.co.uk) showed that 90 per cent of employers were concerned about under-performance. The two most common poor-performance issues cited were Capability (74 per cent) – that is to say, employees' skills, ability and aptitude; and Attendance (61 per cent).

Attendance is a much bigger issues in public-sector businesses – 87 per cent of respondents in this sector said that absence is a common issue compared with 44 per cent of manufacturing companies and 60 per cent of other private-sector employers. Other performance problems included:

- unacceptable attitude/behaviour towards other colleagues;
- poor attitude towards customers or clients;
- failure to meet deadlines;
- general misconduct.

Issues that are more likely to hit manufacturing companies are:

- failure to meet set work objectives;

- lower than expected output;
- accidents or mistakes with equipment.

This is an 80:20 issue. Managers spend 80 per cent of their time dealing with the 20 per cent of employees who are problem cases. Most employees do a reasonably good job, and employers spend little time helping the majority of employees to perform better.

Problem employees are often either canny and/or lacking in self-awareness. They are often two steps ahead of the line manger in terms of legal knowledge, and in a unionized workplace they frequently complain of bullying.

How to manage poor performance

As ACAS (acas.org.uk) points out, performance management has its limits, if the wrong people have been recruited in the first place.

Companies find it hard to manage poor performance. In the XpertHR survey, 28 per cent said their organization's efforts have been ineffective. To manage performance, line managers need to do the following:

- determine the necessary skills and behaviours for each job;
- agree the individuals' goals, competencies and development needs;
- agree a personal development plan, involving coaching and training;
- give regular feedback;
- manage under-performance;
- hold an annual appraisal review with the job holder, possibly supplemented by interim half-yearly reviews.

Of these, the most difficult issue is managing under-performance. This can be done by:

- self-appraisal – asking the employee to identify their strengths and weakness and training needs;
- regular dialogue and feedback – helping to identify outputs and behaviours that don't meet the standard;
- coaching and training;
- capability procedure or performance improvement plan; for people who are 'incapable' rather than unwilling to perform, possibly followed by performance meetings;
- disciplinary process – if all else fails, the organization needs to pursue disciplinary procedures.

It is vital to train managers to manage under-performance, and to have a regular dialogue with their staff. The issue must be dealt with early and not allowed to drag on.

ABSENTEEISM

Absenteeism causes problems in organizations, because many teams or departments can't operate properly if a key member is absent. And even where someone can cover for the absent person, there is a loss of output, an increase in stress and, frequently, a reduction in customer satisfaction.

Significant numbers of employees around the world admit to calling in sick to work when they were not actually sick, as shown in Table 18.5.

Table 18.5   Employees playing hooky

	Percentage of employees admitting to calling in sick when not unwell
China	71%
India	62%
Australia	58%
Canada	52%
USA	52%
UK	43%
Mexico	38%
France	16%

Source: Kronos Incorporated, Global Absence Survey

The Unscheduled Absence Survey, by US employment law firm CCH (cch.co.uk), found that 66 per cent of last-minute absences were due to employees dealing with personal or family issues, while only 34 per cent were due to 'personal illness'. The 66 per cent broke down as follows:

Family Issues 22 per cent
Personal Needs 18 per cent
Entitlement Mentality per cent
Stress 13 per cent

More than two-thirds said there is a pattern in unscheduled absences, with 37 per cent of organizations reporting the most noticeable pattern is people calling in sick on Mondays and Fridays. This was followed by 17 per cent recognizing the most noticeable pattern occurring around holidays such as Christmas.

Absenteeism is also an indicator of an unhappy workforce. In Britain's NHS Trust hospitals, four out of ten employees have fallen ill or felt unwell from work-related stress, according to staff surveys. This was due to their workload and was exacerbated by poor management, with two fifths of respondents noting that they had too many conflicting demands on their time.

As CCH pointed out, people today are juggling the demands of their busy personal and work lives. Organizations need to recognize that employees have a life outside work, and recognize family obligations. Accepting the reality of family life makes for a greater probability that employees won't have to dissemble.

Steps that companies can take to reduce absenteeism include:

- introducing flexitime, job sharing, compressed work weeks, telecommuting;
- greater acceptance about taking unpaid time off;
- better job design and stress reduction activities such as counseling;
- more healthcare support, such as flu injections and healthy workplace programmes;
- improved communication between supervisors and their staff.

THE ROLE OF ANNUAL APPRAISALS

According to a study by the Society for Human Resource Management (SHRM, shrm.org), nine out of ten employees say that annual appraisals don't work.

Apart from the fear that an appraisal inspires, annual reviews are often ineffective because:

- managers lack the courage to say what should be said;
- many employees are simply mediocre – but not bad enough to be sacked. In such cases, managers find it easier to accept the status quo and say nothing.

In a survey of 770 HR professionals by the SHRM, nearly half of HR managers think annual performance reviews are not a good gauge of the individuals' performance.

Some organizations use 360 degree reviews, or crowdsourced feedback, seeking the views of peers as well as the individual and their manager. Customer complaints and feedback are another option.

In his book *Good to Great*, Jim Collins identified the following tactics as essential for constructive feedback of all types:

- lead with questions, not answers;
- engage in dialogue and debate, not coercion;
- conduct autopsies, without blame.

CASE STUDY: PROBLEM EMPLOYEES

Problem employees' are especially difficult to deal with because they are rarely stupid enough to engage in tangible misconduct which might lead to disciplinary action and dismissal.

For several years the author put up with an employee (let's call her Moira) who lowered the mood of her department by controlling what her colleagues' opinions should be, and isolating anyone she didn't like. The staff were simply scared of her. Moreover, Moira played a vital role in the organization, being the only person who really knew how the company's despatch operation worked.

The issue was solved when, in an annual appraisal, another employee alleged she was being bullied. This gave the author the opportunity to challenge Moira who declared that she was resigning, and walked out. Moira asked for her job back the following day, an offer we respectfully declined; and you could almost sense the change in the mood of the department the following day.

Another employee fostered an 'us and them', 'management versus workforce' outlook. The author offered her six months' pay to leave the organization, on condition of signing a Settlement Agreement, formerly known as a Compromise Agreement. The alternative would be a managed process to support her to change in her behaviour, and that if the programme failed, it could lead to disciplinary action.

Shocked, she accepted my offer to pay for advice with a solicitor of her choosing, and she subsequently accepted the offer. Again, the organization became a happier place almost overnight.

WORKFORCE SCANDALS AND STAFF MISBEHAVIOUR

Staffing scandals occur on a regular basis at nursing homes or hospitals.

- At Stafford Hospital a member of the nursing staff taped a dummy to a baby's face to keep it quiet. Previously, the hospital had paid out over £1 million compensation to families of patients who had suffered 'inhumane and degrading' treatment. At the same hospital, Reni Biju, a nurse, surfed the net, rather than carrying out vital hourly checks on a patient with chronic heart problems who was then found dead in her bed.
- At the Cygnet Hospital in Beckton, East London, a nurse, Victor Ogbeide, bombarded a vulnerable teenager with 600 text messages, and was banned from working with patients

under the age of 60. Ogbeide made dozens of calls to the 19-year-old woman who had a history of self-harm.

- Martha Sekete, a nurse at Heath House in Birmingham, left an elderly patient with her hands on a hot radiator and dragged another resident across the floor on a bed sheet. She also snatched a sandwich from one man, and threw it in the bin before shouting at him and grabbed a woman by her hair and neck to move her up the bed. Sekete has been thrown out of the profession.

These scandals imply a disregard for patient dignity and a failure to understand how to behave.

To combat the risk of staff misbehaviour, the organization has to instil a shared culture and set of values. Putting the patient or client first must be paramount.

- Staff need to feel valued. All too often healthcare staff believe no one listens to them.
- Staffing levels also need to meet the needs of the situation, so there is no need to cut corners.
- Management needs to be visible, and set the tone.
- Staff need to be taught about how to conduct themselves, and how to respond to situations where conflicts would arise, how to deal with aggressive or irritating patients, how to exercise self-control and how to manage their anger.
- Staff must be able to report those who misbehave. Whistleblowers must be protected.

On a less dramatic but just as important level, the same applies to all customer-facing staff, especially where staff are facing large numbers of the public, such as in public transport and at unemployment offices.

EMAIL USE

Since the business is responsible for the employee's actions at work, it is liable for any email or internet use. Sexist emails at KPMG, where female staff were rated on a one to ten scale for attractiveness, caused embarrassment for the business.

Whether a business is allowed to monitor an employee's internet use or emails will vary from jurisdiction. In Australia, three employees sacked for sending pornographic emails were reinstated by a tribunal on the grounds that no training on email policy had been given and that the staff had not been given time to digest the allegations. This serves to remind us that any disciplinary action needs to be carried out within the legal framework.

There are risks of various kinds with staff emails and social media posting. Apart from bullying and sexism, they include breaching customer confidentiality, leaking company secrets and bad mouthing the business. We review the problem in Marketing (Chapter 20) and Legal Risks (Chapter 22).

THEFT OF CLIENTS AND OTHER ASSETS

Many organizations are structured around providing services to clients, and such businesses are always at risk of losing them to departing staff. This includes law firms, financial service businesses, advertising and PR agencies, and anyone with a sales force.

It is common practice to put a restrictive clause into the employment contract, banning staff from poaching clients for six or 12 months. This non-solicitation clause gives the company time to get clients used to a new account handler. And it can be boosted by a letter from the company's solicitors, warning the ex-employee against any such action. However, enforcing a restrictive clause is difficult and expensive, with County Court action starting at £15,000.

When a key employee announces they are leaving, it is good practice to put them on 'gardening leave', and to prevent them from having access to company databases or clients. Many companies have the employee summarily escorted to the front door by security.

But there are more proactive solutions as well. It is good to rotate clients around staff, every 12 to 18 months, to reduce them becoming dependent on any one individual. You can also ensure that other members of the team are actively engaged with the client, to demonstrate that the company is more than just the individual.

Since many staff plan to set up a business before they leave, you can also check for changes in behaviour, such as a fall in lead conversions, high levels of unallocated time on timesheets or unusual absences. The author recalls his CEO arriving unannounced to congratulate him on winning a new client. While he talked I had to nonchalantly cover up all the papers that were strewn over my desk – papers that indicated I was setting up my own business. The CEO only had to glance at the papers to see that they were nothing to do with the company. But he didn't have that most important of skills: the ability to read papers upside down.

The same risks attach to people in other roles. If a university development director, the person responsible for getting donations from rich donors and alumni, moves to another university, he or she will take with them the knowledge of donors' preferences and attitudes, and could use this to benefit their new employer. And a key accounts person who goes to a competitor would inevitably encourage a buyer to stock her new employer's products. You can adopt the same processes mentioned above: rotating employees, exposing the client or buyer to more members of staff, and being alert to changes in behaviour.

PANDEMICS

For business, a pandemic is represented by a sudden and huge loss of staff due to illness. Unlike the Black Death and earlier pandemics, most cause thankfully fewer deaths these days, but they decimate staff numbers. Their absence leads to a loss of output and a failure to meet customers' needs.

When it strikes at suppliers, it leads to a loss of stock and raw materials being delivered; while your corporate customers will cancel orders, delay projects and fail to pay invoices on time.

Pandemics are often caused by a new and highly infectious strain of virus such as the H1N1 or 'swine flu' virus that killed 17,000 across the globe in one year alone. The viruses usually subside and normal patterns return. But while the pandemic is active, the business can be brought to a standstill.

You need a continuity plan that sets out how you will manage a pandemic. We examine this in Chapter 23.

THE OUTSOURCED AND OFFSHORED WORKFORCE

The very reasonable fear of being stuck with unnecessary overheads means that an increasing proportion of the workforce these days is contract labour.

And the attraction of low-cost manufacturing and services puts even more staff on the other side of the globe. HR traditionally has had little control over this, not least because you have no responsibility for the recruitment, retention and discipline of those outsourced employees.

The fact that we have covered outsourcing and offshoring in Chapter 7 (on Procurement) and Chapter 14 (on Finance) demonstrates where control of this type of staff lies.

Nevertheless, when looked at holistically, the HR function needs to reflect on the risks attendant on a far-flung staff who owe allegiance to another employer.

HR and line management can reduce some of the risk by ensuring that it has some measure of involvement, and that communication and engagement with outsourced staff is no different than for internal employees.

Working Conditions

With these factors in mind, the organization needs to adopt good working conditions for its staff. This will include:

- An open and equitable management style;
- A culture that values teamwork and excellence, and conditions that assist female (as well as male) employees, such as flexitime, childcare facilities and part-time working;
- Training and re-training, to develop a workforce capable of producing products and services in a rapidly changing world. The willingness of the workforce to accept rapid change will be higher in well-managed firms which have the trust of the labour force.

A company that cares for its workforce will be typified by its *willingness to listen* and to learn, and to implement changes that improve the workplace conditions. Sharing corporate information with the workforce, including sales and cost figures, is a prerequisite to becoming world class.

> *Involving the workforce in decision making* will be a hallmark of the successful company; and this will include long-term decisions concerning corporate strategy and staffing levels.
> *Welfare problems* frequently cause crises. They include perks, maternity arrangements, withdrawal of canteen facilities, staff anger over unfair dismissal and unsafe working conditions. Many of these topics will seem unimportant to management, but they quickly spark a dispute.

STRUCTURING FOR MAXIMUM EFFECTIVENESS

Many organizations have reduced entrepreneurial risk by ensuring that as many of their staff as possible meet the customer. Some organizations are merging their sales and marketing departments, and re-structuring to provide 'category management'. In this way, companies are recognizing that old boundaries may no longer be the most effective way of managing the business. In Chapter 20 on Marketing, we consider the relationship between consumer marketing, selling and customer service.

WORKPLACE BULLYING RISKS

Six in ten workers across the UK have been bullied or witnessed bullying over the past six months, according to a survey by the trade union Unison (unison.org.uk). This represents a big rise on bully statistics in the past. According to Dave Prentis, the union's general secretary, bullying has doubled in the past decade.

It looks as though as managers are increasingly required to achieve results, and in turn put pressure on employees to perform.

Fraser Younson, head of employment at the law firm Berwin Leighton Paisner (blplaw.com), told *The Guardian* newspaper: 'Managers are chasing things up, being more critical. If they are not trained to deal with increased levels of stress, then we are seeing them do this in a way that makes staff feel bullied.'

Others say that managers take advantage of a downturn to become more aggressive to subordinates.

Contrary to myths, bullies can attack popular and effective employees as well as the more obvious shy or retiring victims. Nor is it exclusively bosses bullying subordinates. Clients can bully professionals, juniors can bully superiors and one colleague can bully another.

Sexual harassment and demeaning comments about the victim's race, gender, disability or age are also forms of bullying. Each can have potentially explosive results through resignations and compensation claims. Bullying can take many forms:

- physical touching or sexual gestures;
- isolation at work or exclusion from social activities;
- gossip, jokes and offensive language;
- the release of confidential information;
- online comments or writing on walls;
- stalking or pestering;
- personal comments.

Other kinds of bullying include verbal assault, shouting and offensive language. So too is belittling the victim's opinions, constant criticism, overwork and jokes about the victim's appearance.

Bullying can often happen in organizations where there is a large, long-established, manual and often male workforce. However, financial services is the main area where employees take legal action about bullying.

> *Deutsche Bank* had to pay ex-employee Helen Green £850,000 damages for personal injury and consequential loss. She said that she had suffered psychiatric injury caused by harassment and bullying by her fellow employees.
>
> She said that her colleagues ignored her, laughed at her, made raspberry noises with each step she took and told her 'you stink'. The High Court said that the Bank's management was 'weak and ineffectual'.
>
> *Cantor Fitzgerald*, the US broker, had to pay a £912,000 damages award, later reduced by £116,000, to a former employee who claimed he was subjected to a 'culture of bullying and abuse'.
>
> *Credit Suisse Bank* settled another alleged bullying case for £200,000 when a Pakistani-born trader claimed he was treated like a slave by colleagues, and warned off dating white women.

CASE STUDY: BULLYING AND ITS AFTERMATH AT ROYAL MAIL

A Royal Mail internal inquiry concluded that supervisors drove one of its employees Jermaine Lee to kill himself at the Birmingham Mail Centre in Newtown, Birmingham.

Royal Mail set up two action groups in the wake of the eight-month investigation described as 'one of the biggest ever conducted in our 350-year history'.

The Royal Mail created a 'shop-a-bully' hot-line. It also introduced its 'dignity at work' policy, and within 12 months had trained almost a quarter of its staff in issues relating to diversity. 'It was common knowledge that bullying and harassment was widespread in Royal Mail,' says Satya Kartaria, the company's Director of

Diversity and Inclusion. 'We wanted to turn that round, make it a great place to work. It was a very macho culture and we knew that in the past no challenges had been made to unacceptable behaviour. We recognized that training managers and all employees would be crucial in challenging the existing culture.'

To prevent bullying requires a change of culture. Royal Mail now has an active training programme, and has appointed 20 anti-bullying investigators.

As a result of the inquiry, Royal Mail sacked a senior executive, and suspended six managers and supervisors; and an area manager resigned.

Five years after Jermaine Lee's death, Royal Mail's Chief Executive said the company had 'still not beaten the bully boys, with cases of alleged harassment still a cause for concern'. But he said great strides had been taken to stop the bullies and the group was determined to stop workplace harassment.

LEGAL ACTION OVER BULLYING

Claims for bullying have to show the following:

- the claimant has suffered a psychiatric or physical injury;
- the behaviour of the perpetrator amounted to bullying or harassment;
- the perpetrator knew or ought to have known, that that their actions would cause harm;
- the perpetrators could have taken steps to avoid the injury, by exercising reasonable care.

This demonstrates the need to have systems in place in order to avert claims for bullying.

HOW TO PREVENT WORKPLACE BULLYING

- Implement and promulgate an anti-bullying policy. Specify that bullying is unacceptable.
- Instigate training for all staff on what constitutes acceptable behaviour at work.
- Establish systems for investigating and dealing with conflict. This should start with counselling and mentoring, leading to conciliation and arbitration if unresolved, and finally disciplinary action. Inform staff what actions they can take, and ensure no victimization takes place when bullying is reported.
- Investigate complaints quickly. Maintain discretion and confidentiality, and protect the rights of all individuals involved. To make an effective case, you need detailed written records over time.
- Appoint a discrimination and bullying advisor.

Stress

Stress occurs when an employee is unable to cope due to undue pressure at work. As we've seen in Chapter 6 on Health and Safety, stress accounts for 40 per cent of all work-related illness.

Stress at work accounts for 13.5 million lost working days in Britain according to the Health and Safety Executive. One in five workers rated their jobs as 'very' or 'extremely' stressful. Stress is now recognized as a major cause of heart disease, which in turn produces 21 per cent of male absences from work.

However, HR specialists also say that stress occurs when an employee has a vulnerable personality and is therefore unable to deal with stress, or where they are incompetent to perform their job adequately. Large organizations have many employees who are not competent to do their job or who are not resilient enough to cope with the demands of a brisk working environment.

Some employees find it hard to juggle conflicting demands – with the phone ringing, deadlines looming and competing requests from colleagues, bosses and customers.

While stress can be lowered through training or job design, most jobs involve a certain amount of uncertainty and conflict, and some may be unable to cope. Therefore organizations may need to move some employees to a less demanding job or even manage them out.

THE RISK OF LEGAL ACTION OVER STRESS

Few claims about made ever make it to court, though the ones that do are often high profile. To make a legal claim for stress, an employee has to show:

- They have suffered a psychiatric or physical injury.
- The injury was clearly related to stress at work. Japan Tobacco Inc. paid out £246,000 to the bereaved family of Saburo Sanada, a 54-year-old manager who suffered *karoshi* – sudden death by overwork. Saburo had been working 400 hours a month, supervising the construction of a hotel. *Karoshi* is a major issue in Japanese business (there were 150–200 reported deaths a year)
- The harm was reasonably foreseeable. The employer knew or ought to have known about the risk.
- The employer has failed in their duty of care to take reasonable steps to prevent or reduce the risk of harm to the employee.
- The harm was due to the employer's breach of duty of care. In other words, the employer failed to take action. Despite several complaints to the employers, from both herself and her union, Jan Howell was left in sole charge of a riotous class of largely non-English speaking pupils, 11 of whom had special needs. Despite suffering one breakdown, nothing was done to ease the difficult situation and Mrs Howell suffered a second breakdown, resulting in her retirement. The Local Education Authority was forced to admit liability and pay £250,000 in compensation.

These cases demonstrate the need for organizations to have systems in place that are capable of recognizing a stressful environment and taking steps to alleviate it. Table 18.6 lists the main signs of stress.

Table 18.6   Signs of stress

Headache	Increased use of alcohol, tobacco, drugs or
Tiredness	sleeping pills
Eczema	Depression
Muscle tics	Feeling powerless
Stomach problems, diarrhoea, constipation	Irritability with customers or co-workers
Anger, frustration, violence, aggression	Problems at work, such as forgetfulness
Anxiety	Absenteeism

Employee ill-health endangers not only the employee but also the business itself and other people. The Clapham rail crash which killed 35 people was allegedly caused by faulty work done by a technician who had taken only one day off in the previous 13 weeks.

HOW TO REDUCE WORKPLACE STRESS

Employers now need to manage the whole working environment, not just equipment or the type of work. And this starts with a concern for people. Solutions for reducing stress and absenteeism (as well as ill health and industrial disputes) include the following:

- undertake an audit of employee attitudes and stress levels;
- provide clear job descriptions and lines of reporting;
- ensure regular upward and downward communication, and ensure employees' opinions are heard;
- make work more fulfilling (for example, cell production which lets a group of employees complete an entire task rather than just part of it);
- increase the amount of control that employees have over their work (for example, overtime and pace of work);
- ensure that workloads are managed and balanced;
- increase employees' technical work skills;
- give employees better coping strategies, such as improved diet or exercise, or strategies for dealing with abusive customers;
- improve support and supervision;
- enhance working conditions (noise, breaks, fumes and so on);
- implement a fair reward system, perhaps avoiding piece work;
- increase job security and career development;
- increase flexibility in working arrangements;
- introduce stress management training, as well as substance abuse training;
- improve ergonomics, for example for employees working at computer screens.

Industrial Relations and Disputes

The hazards associated with industrial disputes are clear: lost output, bitter industrial relations and loss of reputation among customers, the city and other important stakeholders.

Most disputes can be foreseen, as relations between management and unions gradually deteriorate. Grievances can grow over many years, and a workforce that believes it has been treated unfairly is more likely to take industrial action.

The company should have mechanisms to ensure that grievances are heard and treated seriously. Management must make efforts to communicate the reasons for changes, and preferably gain the acceptance of the workforce, before they are implemented.

The company should assess the probability of a strike, consider what damage it could cause and analyse how it could be pre-empted. The best solutions are to ensure that the company is seen to act fairly and honestly.

RESPONDING TO A LABOUR DISPUTE

In responding to a dispute, the organization should take the following steps:

- Assess the needs of the union or workers. A demand for a pay rise could be masking another grievance (such as bad industrial relations).
- Maintain a dialogue with the union throughout the dispute, irrespective of how impossible a solution may seem.
- Identify the outcome of different results, in terms of impact on profitability or future negotiations.
- Avoid giving any undue benefit without gaining a corresponding concession from the union.
- Be consistent in your dealings with the union. Demonstrate commitment to your course of action at all times.
- Provide corporate financial data to the unions, because you should have nothing to hide. The wages bill is frequently the company's biggest cost. You should also give information on competitors' costs, if it demonstrates that yielding to demands would make the company uncompetitive.
- Give regular briefings to the media, emphasizing the risks that agreeing to union demands would bring (for example, job losses).
- Communicate directly with staff, for example through notice boards or public meetings, or even letters to the workers' home address (though this can be seen as threatening and invasive).

The business should also require staff to obey the law. In some countries legislation prevents walkouts or wildcat strikes. The company should respond to lawbreaking by using whatever remedies are available (for example, seeking a restraining order or damages for lost revenue).

It is wise to avoid actions that may produce short-term solutions but will store up trouble in the future, such as strike-busting through bussing in non-union workers. The company will have to continue operating after the dispute.

To bring the dispute to an end, the company may need to identify face-saving concessions for the union, which will enable both sides to leave the dispute with honour. However, if a dispute seems inevitable, the business may need to:

- build buffer stocks;
- move production or operations to other sites;
- outsource or sub-contract operations or production

Management also needs to act smart, and stay one step ahead. One hospital administrator found there was always uproar if he wanted to close a ward. But people couldn't object to redecoration. So he would put ladders and pots of paint in the ward, and say it was closed for renovation.

Risks of Being Sued by Employees

The chances of being sued by an employee seem to be growing, as is the cost of meeting a claim. One US survey suggested that 25 per cent of private companies have been sued by a current or former employee, and that 44 per cent said they expected to face a claim in the future.

Massachusetts attorney Jay Shepherd reckons that a business with 100 employees in his state stands a 28 per cent chance of being sued by a member of staff each year. An employer with 250 employees faces a 57 per cent chance, while a 500-worker employer has a 81 per cent likelihood.

That means if you employ 1,000 employees, you have only a 3.5 per cent chance of *not* getting sued this year.

A unionized workforce increases the chances of being sued, while the industry is also important: restaurants, hotels and hospitals are more likely, while professional service firms are less so. And the jurisdiction also matters: US states have varying degrees of employer or employee friendliness.

This is likely to get worse. More employees are likely to regard themselves as disabled, in particular those who find themselves depressed and taking anti-depressants.

Employees are more likely to talk to a lawyer these days, and to go to court, including vexatious claimants. However, the cost for employees to go to court has risen, which has deterred many would-be claimants.

There is less stigma about mental health and more information available about it. Legislation is more supportive of the individual, including the UK's Equality Act, which allows for hard to diagnose impairments such as ME, fibromyalgia and chronic fatigue syndrome. For example, an alcoholic could require special consideration on the grounds of their liver damage.

And many companies seem to be less knowledgeable about the law than they should be. This is evidently an area of risk, and one that receives less attention than it needs. Common claims are for the following:

- equal pay;
- unfair dismissal;
- redundancy pay;
- discrimination (sex, race, disability and so on);
- breach of contract;
- working hours;
- unauthorized deductions from wages;
- failure to consult in a redundancy.

Organizations can protect themselves against being sued by the following measures:

- have written policies on major staff issues, backed up by effective employment contracts;
- implement regular staff appraisals and performance reviews;
- have graduated and consistent disciplinary measures for employee failings;
- instigate a robust grievance policy that ensures problems are heard in good time and don't get out of hand;
- maintain written notes of all relevant issues;
- implement staff surveys to identify problem areas, and then acting upon them (because knowing about a problem and not taking action will strengthen the employee's case;
- allow for mediation which can be a faster, cheaper and more private route compared with being sued in open court.

Risk Assessment

You can assess your vulnerability to HR risk by answering the questions below. Score one point for each box ticked.

Question	✓
Does the CEO get criticized for undue rewards?	
Does the CEO lack leadership skills?	
Are any major roles lacking a post holder (such as IT, marketing or finance)?	
Is the organization slow to respond to change?	
Does the organization lack an organization chart?	
Are staff demotivated?	
Does the organization fail to hire or promote talent?	
Do many people under-perform?	
Are people disengaged?	
Is absenteeism well above the industry average?	
Is there a culture of bullying?	
Is the organization likely to suffer an industrial dispute in the next 12 months?	
Doers the organization lack an effective anti-stress programme?	
Has any employee sought to sue the organization in the last 12 months?	
Total points scored:	

Score: 0–4 points: low risk. 5–9 points: moderate risk. 10–14 points: high risk.

The Appendix has a summary of the checklists. By entering the results of this one, you can compare your HR risk against other categories of risk.

19
The Ethical Dilemma for Organizations

'Everyone does it. It's standard practice,' says Liam. He wants to run a sales promotion that involves giving holiday vouchers to retailers who buy sufficient product during the month. 'Yes,' you say. 'But it's basically bribery.'

'Oh, grow up,' says Liam. 'It's not illegal. There is no rule saying we can't do that'. You try to reason with him: 'But we're encouraging people to get a benefit for themselves rather than for the good of their company.'

'You should be in social work,' says Liam, witheringly. 'I'll talk to Frank. He'll agree this promotion. Your views don't count.' And with that he's off.

The Tug Between Profit and Ethics

Every organization wants to minimize its costs and maximize its income. If taken to extremes, this means providing the cheapest product or service for the highest price.

The enticements are greater where competition is imperfect or absent, or where the customer is weak or lacking in knowledge. 'Cost-plus' defence contracts have resulted in poor value for money (because the contractor makes more profit by increasing the costs of the project), and pharmaceutical companies have overcharged for drugs. Schering-Plough agreed to pay $346 million in fines and damages to settle charges that it overcharged for drugs sold through the US Medicaid, the US Government's health programme for the poor.

Other factors encourage companies to neglect ethics. The need to win contracts can make businesses bend the rules. A feeling of 'us against the world' can lead managers to believe that 'the end justifies the means', with a resulting loss of honesty. According to a survey in *Le Monde* (LeMonde.fr), 64 per cent of French company CEOs believe that corruption is endemic in business.

But ethics is increasingly recognized as a major business risk. Damaging newspaper reports and court cases involving bribery and other forms of dishonesty have serious consequences for the corporate reputation and future profits.

In this internet age, an activist in their bedroom with a laptop has as much clout as the biggest company in the world. All of which goes to show that businesses have to be ethically sound – and must be able to prove it. This is generally known as Corporate Social Responsibility (CSR), although CSR has unfortunately become known in some companies as just corporate PR.

Changing the Corporate Culture

Companies often profess to act honourably but behave unethically. Many corporate cultures have 'winning' as their most important value. This means that ethical behaviour comes to be seen as an impediment.

Managers who preach the importance of good ethics may be seen as naive, negative or obstructive. They may be told they're standing in the way of progress, or suffering from divided loyalties. And they may be passed over for promotion, or be made redundant when the company is downsizing. This is particularly true when a company is not making profits, or when the economy is in recession.

Until the corporate culture changes, the company cannot set about developing an ethical position. If top management regards itself as ethically neutral, and for ethics not to be a boardroom issue, the company is at risk from unethical corporate behaviour.

If senior management regards ethics as an academic or political issue, rather than something fundamental to the business, the firm risks being damaged by corporate malpractice.

The Main Areas of Ethical Risk

If discovered, unethical activity can do irreparable damage to a company. Customers may cease doing business with that company. In a market with few customers (such as aerospace), this could be a major problem. Or the company may be entangled in embarrassing or expensive lawsuits. It may end up paying heavy fines.

Illegal activity also undermines the company itself. It infects an otherwise honest organization, leads to the creation of secret accounts, and to people turning a blind eye to illegal behaviour. It lets individuals in the company claim exemption from internal investigation due to their work for 'special' clients. Staff will not be loyal to a dishonest company.

It is also difficult to keep unethical activity secret because of the existence of forces opposed to it; in particular:

*   'whistleblowers' – employees who inform on the company;
*   the media, which likes nothing more than a corruption story.

However, some behaviour is more borderline, and since it does not constitute a criminal offence, it is more difficult to manage or stand up to. Below are listed the main categories of ethical failings, following which we examine how the business can forestall them.

There are nine main areas of ethical issues. They are shown in Table 19.1. We discuss each in more detail below.

Table 19.1   Nine main areas of ethical risk

1.	Relations with government
2.	Attacks on competitors
3.	Unethical alliances with competitors
4.	Product and service quality
5.	Managing staff
6.	Relations with suppliers
7.	The environment
8.	Dealings with customers
9.	Bribery

1. RELATIONS WITH GOVERNMENT

Examples of corrupt relations with governments include:

- tax evasion or undue avoidance;
- excessive payments to political parties;
- selling to tyrannical regimes.

Tax evasion and avoidance

As discussed in Chapter 14 on Financial Risk, there are many opportunities for companies to reduce their tax burden. These range from the innocuous, such as using capital equipment allowances, to extreme offshoring.

Within each industry, different companies make their own tax choices. In recent years, the courier company UPS has paid a US tax rate of 24 per cent, while its competitor FedEx pays less than 1 per cent.

Meanwhile California-based Apple not only pretends it is headquartered in Ireland, but the Irish company is in turn 'owned' by a business called Baldwin Holdings, located in the Virgin Islands without even an address or a phone number. This keeps the taxes paid by Apple, the world's most profitable tech company, down to 2.2 per cent, and reduces its payments to the USA Government by £2.4 billion.

Google has a group profit of 33 per cent, but declares losses in the UK. Austin Mitchell MP told the company: 'Either you're running the business badly or there is a fiddle going on.'

And each year the author's accountants offer him ever more intricate ruses that involve offshore islands. When their offer is politely declined they say it is their duty to advise me of such opportunities, adding that a surprising number of businesses also reject these blandishments.

Critics argue that it is wrong for a corporation to expect employees to benefit from roads, schools, colleges and healthcare without contributing to those costs. It is also argued that in lieu of tax from corporations, the tax burden falls more heavily on private citizens and small businesses.

JK Galbraith used to talk about 'private wealth and public squalor', where corporations and a few individuals earn large sums of money, amid crumbling roads, schools and public hospitals. However, this contrasts with the view held by many that:

- governments waste money;
- it is better husbanded by private companies;
- companies provide employment and thus tax revenue.

The central question is whether corporations are paying an adequate compensation for the benefits they get from operating in a particular jurisdiction.

But the debate surrounding tax avoidance has been growing louder, and there is likely to be growing stigma against such activity. Starbucks has become the target of activist attacks by UK Uncut for claiming losses in all but one of the 15 years it had operated in the UK, while its rival Costa Coffee paid 25 per cent tax.

The company's reputation also fell. According to polling firm YouGov, Starbucks' 'buzz' score, a measure of the number of negative and positive comments consumers have heard about a brand, fell from +0.7 to -13.9. Starbucks' reputation score also dipped: from +4.6 to -3.9 following the tax revelations.

A YouGov spokeswoman said: 'To say this story has been a disaster for the Starbucks brand would be a bit of an understatement. [But] it's still too early to say what the long-term impact of this is going to be.'

As long as a company provides a unique and desirable product or service, such as Amazon or Apple, customers are likely to shrug off the criticisms of tax avoidance. However where the market is more competitive, for example as high street coffee chains multiply, or where values such as friendliness are an integral part of a company's brand identity, the business may lose its customers' loyalty.

CASE STUDY: STORM OVER ABF'S TAX PAYMENTS IN ZAMBIA

ABF, the firm that includes Ryvita, Primark and Kingsmill bread, was accused of exploiting Zambia, one of the world's poorest countries, when Action Aid issued a report saying that AF's Zambia Sugar subsidiary paid only 0.5 per cent tax during the last five years.

One-third of the company's profits were paid out to sister companies based in tax havens. And the company paid $3 million to a one-man company in Mauritius for 'trade contacts with customers in the European sugar market, transportation of sugar to Europe, foreign currency management and the availability of effective credit terms'.

After various tax reducing measures, the eventual tax payable on leaving Zambia was only 5 per cent because the owner of Zambia Sugar is a Dutch co-operative, whose owners in turn are based in Mauritius and Jersey. They are therefore classed as members not shareholders, and so their income is not classified as taxable dividends.

ABF pointed out that Zambia Sugar had invested £150 million over the last five years, and capital allowances therefore resulted in no corporate tax being payable.

The storm resulted in articles in the UK's *Observer* newspaper, *Lusaka Times* and elsewhere. Supporters of ABF pointed to the tax relief on its investment, and the jobs created. Detractors highlighted the use of tax havens, while supporters said ABF had the right to use perfectly legal devices to reduce their tax liability.

Political contributions

Excessive political contributions are wrong because they can undermine the democratic process and corrupt the politicians. Companies can be accused of giving donations in order to win public contracts rather than assisting democracy.

Although it is difficult to prove a link between political donations and favours for business, the scandal that surrounds suspicious cases is particularly harmful. The damage is done not only to the business involved but also the political party.

- SAB Miller, the world's second largest brewer, gave more than £400,000 in campaign donations for one set of general elections in South Africa. The money went in proportion to the share of votes won in the previous elections. Cyril Ramaphosa, deputy president of the country's biggest part, the ANC, is a non-executive director of the company. These donations and the capture of a senior figure can hardly fail to affect the country's leaders when it comes to making decisions that affect the company, and may give rise to corruption.
- The SAB Miller case was openly disclosed. Not so was the payment in Alberta, Canada, by billionaire Daryl Katz, owner of the Edmonton Oilers ice hockey team. He allegedly donated $430,000 to the province's Progressive Conservatives, 14 times the limit for political contributions. According to the *Times Colonist* newspaper, Katz wanted the Government to spend $100 million on a new arena for his team, and wrote a single cheque. The political party then divided the contribution into smaller amounts and said they came from Katz's family members and business associates.

Some companies have recognized the potential risks associated with political involvement and withdrawn from the practice. BP now has a no-donations policy. It says it will 'make no political contributions, whether in cash or in kind, anywhere in the world'. However, *The Washington Post* says it has donated $4.8 million in recent years to political causes, largely to Republican-aligned political groups that were working to defeat state ballot initiatives in California and Colorado that might have increased taxes on the oil and gas industry.

Selling to tyrannical regimes

The video surveillance industry has come under attack from human rights campaigners. Honeywell, IBM, United Technologies and General Electric have been accused of selling surveillance technology to the Chinese Government. The companies say their products are used for controlling crime, while their opponents say they are used for political repression. Some cameras are placed inside Tibetan monasteries and were used to detain nearly 200 monks who participated in a protest at a monastery in Gansu Province.

As a response to the 1989 Tiananmen Square massacre, US corporations are banned from selling crime control products like fingerprint equipment, photo ID card material, or night vision technology. However, the regulations appear to be much flouted. At a Beijing Trade show, Motorola was selling police radios and wireless video surveillance systems.

2. ATTACKS ON COMPETITORS

Examples of unethical attacks on competitors include:

- illegally obtaining information about competitors (for example, by luring away or bribing their staff);
- making false allegations about competitors, for example through the sales force.

Companies can post false reviews on Amazon about competitors' products, and add negative comments to forums. Just a few hostile statements can sow the seeds of doubt in customers' minds, and seriously undermine a product's sales. 'Sock puppet' accounts are those set up on Facebook, Twitter, Amazon or a forum by someone to give fake reviews under a false name.

3. UNETHICAL ALLIANCES WITH COMPETITORS

Cartels are against the interest of customers and are everywhere banned. Suppliers are prone to indulge in price fixing, and are increasingly being found out.

As shown in Table 19.2, the European Commission fined three producers of cathode ray tubes (CRT) glass used in televisions and computer screens. Japanese firms Asahi Glass, Nippon Electric Glass and Germany's Schott AG were fined a total of €128 million for operating a cartel. Samsung Corning Precision Materials was given immunity for giving information about the cartel. The group held meetings to co-ordinate prices.

The EU fined four companies, ACC, Danfoss, Embraco and Panasonic, a total of €162 million for holding meetings to maintain market shares and co-ordinate prices for their refrigeration compressors, as used in fridges and freezers. Tecumseh was not fined as it revealed the existence of the cartel to the Commission.

Table 19.2   Largest fines imposed by the EU on individual firms

Firm	Industry	Fine (€ million)
Saint Gobain	Car glass	896
EON	Gas	553
GDF Suez	Gas	553
ThyssenKrupp	Lifts and escalators	479
Hoffmann-La Roche	Vitamins	462
Siemens AG	Switchgear	396
Pilkington	Car glass	370
Sasol Limited	Paraffin wax (used in candles, car tyres, chewing gum and so on)	318
ENI	Synthetic rubber (for tyres and so on)	272
Lafarge	Plasterboard	249

But the largest fine, of €1.3 billion went to members of a replacement car glass cartel. Asahi (again), Pilkington, Saint-Gobain and Soliver controlled 90 per cent of the annual €2 billion market, and met to discuss prices and customers. The EU started the investigation after a tip off from an anonymous source. St Gobain's fines were increased by 60 per cent because it was a repeat offender. Asahi's was reduced by 50 per cent for helping expose the cartel.

The Commission can fine companies up to 10 per cent of their annual worldwide turnover, while offering immunity to whistleblowing companies. The fines are paid into the Community budget. The fines thus help to finance the EU and reduce the tax burden.

Individual governments are no less stringent. The Australian Federal Court imposed fines of AUS$58 million on airlines operating an airfreight cartel. They included Qantas, British Airways, Japan Airlines and Korean Airlines. A Qantas executive was jailed for six months in the USA.

The same court also fined Singaporean and Indonesian paper companies $8 million for fixing the price of photocopy paper. The secret meetings were held in south-east Asian countries, especially those that had no anti-cartel laws. But the court ruled that the cartel had been implemented in the Australian market and harmed local consumers

The UK levied fines of £225 million on two tobacco companies, Gallagher and Imperial Tobacco, and ten supermarkets for fixing the price of cigarettes. Among the retailers were Asda, the Co-operative Group, Morrisons, Safeway, Sainsbury's and Shell.

Similarly, the UK's Office of Fair Trading (oft.gov.uk) fined Mercedes Benz and three commercial vehicle dealers £2.6 million for co-ordinating prices, rigging the market and exchanging commercially sensitive information. This serves to remind us how easy it is for sales people to try and co-ordinate markets, not necessarily realizing that their activities are illegal.

4. PRODUCT AND SERVICE QUALITY

Ethical product failures include the following:

- producing products which are environmentally unsound (for example, surface mining, unnecessarily disposable products, or those that produce toxic waste);
- testing non-medical products on animals;

- selling products or services which are poor value for money, or which could harm people (for example, cigarettes or weapons).

We discuss some of these issues in Chapter 8 on the Environment.

UK banks offered PPI (payment protection insurance) with loans to people who for whom PPI would be inappropriate. PPI would only pay out to people in permanent full-time employment, and many policies excluded mental or back problems. So serious was the FSA's concern that the regulator required the banks to write to 12 million customers to tell them they may have been miss-sold. Lloyds Bank upheld 81 per cent of customer's complaints in the customers' favour. The banks set aside £12 billion to reimburse customers.

The banks complained that 25 per cent of claims were fraudulent, egged on by the no-win, no-fee legal firms, but observers saw such results as nemesis or poetic justice – the banks brought bad luck upon themselves by their own actions.

The banks are regularly held up for ethical failures. Previous banking scandals have included pensions mis-selling, endowments mis-selling, money laundering and manipulating the Libor rate.

So it is clear that short-term profits gained from unethical products often have repercussions in future years, in terms of financial losses and damage to the corporate image. For years the banks continued to support the golden goose that was PPI, ignoring the rise in criticisms, and their top executives earned big bonuses from the products. It is hard for individuals and businesses to wean themselves off lucrative products.

The reverse is also true. A life salesman said, 'Honesty is my best weapon. When people know I'm being honest with them, it gives me a big advantage, because customers are desperate to know that they aren't being cheated.'

Companies that sell harmful products (such as the cigarette firms) operate in a difficult sphere. Sooner or later a disgruntled customer or a regulatory authority is going to win a law case that could mean the end of the business.

The many ethical investment trusts have reached a broad consensus on what contravenes ethics. Their policies preclude them from investing in companies which:

- trade extensively in countries which have repressive regimes;
- sell tobacco, armaments or gambling. Some also exclude alcohol;
- exploit animals (though medical research is often deemed important).

The ethical investment trusts, which are set to grow, also seek companies that have a good record in labour relations and in environmental protection.

5. MANAGING STAFF

Excessive payments to members of the Board, especially those who are sacked, is a topic we discussed in Chapter 17 on Governance.

When it comes to other staff, health and safety is a crucial area for ethics. The company should avoid taking short-cuts which might produce short-term benefits but endanger the workforce.

For example, safety procedures should be followed while plant is running. Down-time will be increased if a machine has to be electrically isolated before cleaning begins, but it also avoids the risks of injuring an employee.

Wages for the lowest-paid workers can become a source of embarrassment, and therefore risk. In the past, many organizations outsourced cleaning and catering. This often resulted in cost savings to the business, but the corollary was that the workers sometimes got reduced wages or conditions. There is pressure from the trades unions and activists for management to recognize the lowest paid, and ensure they get a living wage (universallivingwage.org and livingwage.org.uk). The campaigns are particularly prominent in high-cost cities such as London, and in not-for-profit and government organizations.

Another ethical issue is zero hour contracts (ZHC), a topic we looked at in Chapter 14 on Finance.

Overseas staff are an important topic, especially those in the developing world. Companies that relocate production to third world countries are often viewed suspiciously by pressure groups, trade unions and the public. Low pay and bad working conditions are seen as exploitation and the cost of any resulting public relations crisis can damage the brand image. This sort of risk has affected sportswear manufacturers over the past few years, particularly Nike and Adidas.

Activitists re-tagged Adidas merchandise in stores with '34p' ones, indicating the minimal hourly wage rate paid to the Indonesian workers who make Adidas goods.

The campaign group 'Labour Behind the Label' claimed that the basic pay of Indonesian Adidas workers is only £10 a week.

William Anderson, Head of Social and Environmental Affairs for Adidas' Asia Pacific region, pointed out that workers also got performance bonuses and overtime payments. He also said that meant that people in the Pacific Rim need to earn only 20 per cent of American wages to purchase the same goods and services as Americans, the concept of Purchasing Power Parity. And he cited an ILO Global Wage Report (ilo.org)showing that an $85 weekly wage in Vietnam was equivalent to $295 in Thailand and $379 in the Philippines – in other words the cost of living is different in different countries.

6. RELATIONS WITH SUPPLIERS

Corporate customers always seek to drive down suppliers' prices. And while this can sometimes be achieved through productivity gains, such gains will eventually peter out.

In some markets the problem lies with the drive for lower prices, which results in cheaper supplies. Ethical companies have to face this problem, and recognize that being the cheapest sometimes means being the least ethical.

Acting unfairly towards suppliers causes its own problems. A company that extracts large discounts from a supplier will find that it gets poor quality products or service. Eventually, if it cannot make a profit, the supplier will go bust.

CASE STUDY: DIAMONDS AND CIVIL WAR

Diamond mining has paid for wars in Africa and funds human rights abuses. For this reason the United Nations has a programme to clean up the industry, and there is a voluntary pledge for retailers that guarantees the origin of all diamonds sold. But retailers are 'largely unable to provide consumers with meaningful assurances that their diamonds are conflict-free,' says a report from Amnesty International.

Forty-eight companies, including Asprey, Debenhams, Kmart and TK Maxx failed to respond to the survey. And the vast majority that did respond failed to provide details of how their policy works in practice. Random visits to 579 stores in the USA and UK found that fewer than half have a policy on conflict diamonds; while only a tiny minority provide a warranty confirming that the diamonds come from legitimate sources.

Global Witness director Charmian Gooch said: 'The sad truth is that most consumers still cannot be sure where their diamonds come from, nor whether they are financing armed violence or abusive regimes.'

If it doesn't go bust, the supplier will look for more profitable customers, and cease doing business with its original customer. In either case, the firm has to continually look for new suppliers. Other examples include:

* exploiting suppliers in less-developed countries;
* using child labour in less-developed countries (an issue discussed in Chapter 7 on Procurement);
* entering into illegal agreements with suppliers' staff to dishonestly obtain goods.

7. THE ENVIRONMENT

Example of ethical failure include failing to take precautions against damaging the environment; or knowingly causing damage (for example, by illegal disposal of waste).

As discussed in Chapter 8, environmental protection has only become a management issue in recent years; and many companies have been left a legacy of dirty plants and unsatisfactory methods of waste disposal.

Some businesses are in a dilemma because their products are both essential to modern living and cause environmental damage. For example, paper making has a record of environmental damage. But the industry has taken steps to minimize the damage, with the result that some companies and some papers are less polluting than others. An organization must seek to reduce its carbon footprint, minimize waste and reduce the resources it uses.

8. DEALINGS WITH CUSTOMERS

Unethical behaviour towards customers involves three main actions:

* giving misleading information or advice to customers which will encourage them to place orders;
* selling products to inappropriate customers;
* giving bribes to customers in return for business

We examine the first two issues next. Because of its pervasiveness, bribery is covered in the subsequent section.

Giving misleading information

Prudential was forced to set up a £1.1 billion compensation fund for victims of the personal pensions scandal. Along with other providers, Prudential mis-sold pensions to millions of customers who would have been better off if they had remained with their employer's pension schemes.

Withholding information

Few sales people voluntarily proffer information that could lose them a sale. But in a society that increasingly requires full disclosure in sensitive markets, this is what companies have to do more and more. Financial service companies have to show their client how much money their intermediary is making from the sale of a service like life assurance.

GlaxoSmithKline agreed to pay the state of New York $2.5 million to settle its lawsuit alleging that it concealed problems of efficacy and safety in its drug Paxil. It also agreed to publish summaries of all its trials in a registry. Drug manufacturers have been criticized for seeking to publish in medical journals only favourable information about their products.

Selling products to inappropriate customers

Nestle has long been criticized for selling baby formula as milk, especially in less-developed nations, where contaminated water can lead to gastrointestinal illness and even death. *The Guardian* newspaper said that Danone's Indonesian subsidiary Sari Husada, was incentivizing Indonesian midwives to promote sales of formula milk. In Indonesia, where 45 per cent of the population has no access to clean water, the use of baby milk is life threatening as well as expensive: it can cost 400,000 rupiah (£26, $40 US) a month, equivalent to half the family income. Because of the cost, mothers dilute the milk, leaving the baby malnourished.

This raises an ethical dilemma. How does a business encourage its staff to perform well, which often means selling enough product, but at the same time discourage inappropriate selling? What happens when a product that is perfectly suitable for Western mothers who can't breastfeed is at risk of being mis-sold in the developing world?

Strict guidelines must be put in place, and then the policy should be monitored to ensure that zealous sales people don't overstep the mark.

The same issue occurs, though less dramatically, when a bank tries to sell an inappropriate pension, loan or insurance policy. Governments have put controls in place to stop this kind of mis-selling, but the practices continue, usually with new products that the regulators or courts haven't yet caught up with.

In such cases, the senior management must show that such behaviour is wrong and unacceptable.

9. BRIBERY

Bribery is endemic in many countries, especially where state employees are poorly paid. While most bribes are secret, a few come to the public attention. Here are just four examples, in oil, civil engineering, pharmaceuticals and printing.

> *Oil*: The Norwegian oil company Statoil was found guilty of paying $15.2 million in illegal 'consultancy fees' in return for securing contracts in Iran's lucrative oil industry. The company was fined $3 million dollars, despite there being no evidence that the bribe was to any effect – many experts believe that the level of corruption in the oil industry is so high that $15 million would not amount to much influence at all.
>
> *Civil engineering*: In Lesotho's Highlands Water Project (LHWP). Masupha Sole was the chief executive responsible for overseeing the construction of two dams, both funded by international development agencies such as the World Bank and the EU. Sole received at least 18 million South African Rand (ZAR) in bribes from four companies involved in the construction. Lahmeyer International, the largest engineering consultancy firm in Germany, was fined 12 million ZAR for its crimes, Acres International, a Canadian engineering firm was fined 13 million ZAR and Schneider Electric, a French electrical company was fined 10 million ZAR. For his part, Masupha Sole was jailed for 15 years.
>
> *Healthcare*: 'I bribed and corrupted doctors to prescribe Glaxo drugs,' said a former Glaxo sales representative, according to the Sunday Times. The report accused the company of giving GPs

gifts and entertainment. It was on the understanding that the doctors would prescribe Glaxo's drugs to patients.

This is not the only occasion on which Glaxo was said to be involved with bribes. Flavio Maffeis, President of Glaxo SpA, the company's Italian subsidiary, was named in a corruption scandal, in which the company was alleged to have paid £100,000 in bribes and paintings to a senior government health official in return for favours.

Printing: Managers at bank note firms owned by Australia's Reserve Bank of Australia (RBA) are said to have paid sales agents to bribe public officials in Malaysia, Indonesia and Vietnam in order to secure banknote contracts. This was noteworthy, because of the damage it did to Australia's standing and to that of the RBA. It is said they paid millions of dollars across Asia. An OECD report on bribery found that 75 per cent of Australia's top 100 companies operate in a high-risk sector, a high-risk country, or both.

When bribes are seen as necessary

In some countries, bribery is seen as an essential pre-requisite to doing business. Transparency International (TI) has a map showing the scale of corruption by country (Figure 19.1).

An Ernst and Young survey (EY.com) found 15 per cent of companies are willing to make cash payments to win or retain business.

Andrew Kakabadse, a Professor of International Management Development at Cranfield University (cranfield.ac.uk), said, 'In two-thirds of the world, one can't do business unless a bribe is given … It is called bribery in the West; in other countries it is called a transactional cost.'

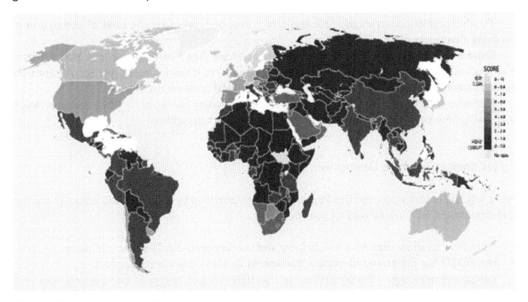

Figure 19.1   Corruption by country
Source: Transparency International

Kakabadse says that 80 per cent of deals outside Europe are affected by attempts to extract money. These are sought by various individuals:

* ruling elites;

- middlemen;
- police officers;
- politicians;
- civil servants (planning officers, local government staff and customs officers).

All of these demand cash and favours to:

- win a new contract;
- get invoices paid;
- release impounded goods;
- protect existing contracts.

Alternatively, they require 'facilitation payments' to get:

- electricity switched on in offices;
- a contract signed.

Richard Williams, Group Managing Director of engineering firm Mott MacDonald says firms must be wary of 'jumping into new export markets, blind to the potential for corrupt practices'. He says poverty wages for civil servants in central Europe and elsewhere exacerbate the problem. 'It is generally state organisations that put a lake of treacle in front of you that must be crossed to win a contract,' he said 'And even when you win, there is the possibility that you have become the victim of a corrupt practice that is going on without your knowledge. In some countries we have restricted ourselves to working with the private sector.'

TI, which ranks countries by their levels of corruption, says that the level of corruption is declining after years of campaigning.

Others would argue that companies cannot condone illegal behaviour in one part of their business, and not expect the corruption to infect other parts. It cannot adopt one set of standards in, for example, developing countries, and a different moral code elsewhere.

Moreover, as the examples above show, it is the purchasing power of rich Western companies that make these payments. They are thus the source of the corruption.

Export Credits Guarantee Department (ECGD)

In the UK, exporters who used the insurance service provided by the Export Credits Guarantee Department (ECGD) are subject to controls:

- firms have to show they have anti-bribery and corruption (ABAC) systems in place;
- the ECGD has the power to inspect documents to check payments to agents;
- companies must sign a document stating that they will not engage in corrupt practices, and will take action against employees who transgress;
- the ECGD will refer evidence of malpractice to the appropriate authorities.

Foreign Corrupt Practices Act (FCPA)

Bribery of foreign officials to secure benefits, whether a contract or (in the case of Chiquita, bribing the President of Honduras to lower taxes) is a common temptation when large sums of money are at stake.

The US FCPA applies to any organization, including foreign firms, which create a corrupt payment while in the US. In recent years, Smith & Nephew, Biomet, Marubeni Corporation (a Japanese firm), Siemens AG (a German business) and BizJet have all paid millions of dollars of fines to the Department of Justice under the FCPA. Siemens alone paid $450 million.

Staff Behaviour

So far in this chapter we have looked at corporate unethical behaviour. But staff also do bad things of their own accord.

The Huffington Post reported that staff in an Apple store allegedly wiped the hard drive of any customers they didn't like. If a customer complained, they would point out the customer had signed a form waiving the company's responsibility for losing data. Another employee rode a customer's hard drive like a skateboard because the customer had said the store smelled.

Dealing with Unethical Competitors

Companies sometimes justify unethical behaviour by saying, 'Everyone else in the market does it. We'll lose out if don't.' Taken to extremes, this is the same argument as selling heroin to children 'because other people are doing it'.

If competitors are acting illegally, the company should report the fact to the authorities. In the long run, unlawful activities will be found out and punished. The ethical company should seek to present moral leadership, not follow the worst example. It should do this because illegal activities will eventually come back to haunt the perpetrator.

The Impact of Unethical Behaviour

As Figure 19.2 shows overleaf, it is inevitable that companies combine illegally or offer bad products. Just as inevitably, this will be uncovered.

Where there are few suppliers, business leaders meet and get to know each other. With common interests, and the need for 'orderly marketing' to forestall price wars, companies soon discover that unethical products or behaviour lead to improved profits and bonuses.

Similarly, when customers are vulnerable, such as needing a loan, or a business is in a dominant position, it can lead to the business selling unethical products.

In both cases, this leads to increased profit. But usually it ends with a whistleblower leaking information, or lawyers instating a class action. The final stage is the denouement in court, and fines being imposed.

Undertaking an Ethics Audit

The organization should carry out an ethics audit, based on the nine categories referred to in this chapter. The audit should find out whether the company undertakes illegal activity. The auditors need to be sufficiently senior to demand honest answers, and they will need to interview the most senior members of staff and those in key positions (such as accountants and purchasing managers). Senior management should put its support behind the audit, to prevent employees from simply denying any immoral activity.

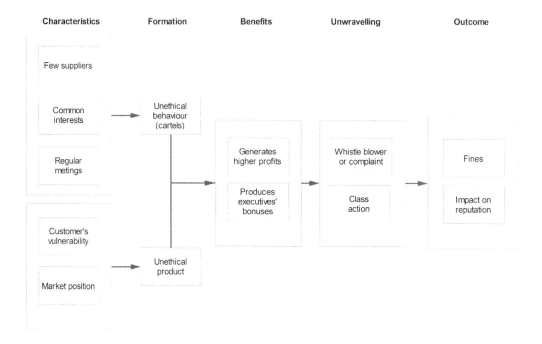

Figure 19.2   The five stages of ethical failure

The audit will represent a 'smoking gun', because any evidence of illegal activity discovered in the audit could make senior management liable. For example, if illegal activities are not reported, senior officers could be guilty of withholding evidence about a crime.

In practice, senior management will already be aware of any major illegal activity, such as bribery. Assuming that the company has halted this activity, if it ever existed, the company should then turn its attention to lesser wrongs, and especially those which, while not necessarily illegal, constitute a grey area. This is the area which, if discovered, could give the company a major PR problem.

Issuing an Ethics Policy

The company should issue a code of practice that will clarify individual responsibility and alert people to malpractice. The code should cover illegal or immoral activity towards customers, competitors, suppliers, other members of staff and the public. It should lay down guidelines for the giving and taking of bribes, incentives, gifts and hospitality. It should also cover the issue of employees working for other organizations. There should be an ethical filter for new product development.

Staff must have a system for reporting ethical failures. This mechanism should ensure that whistleblowers are not victimized. The policy should be rigorously enforced. If it is not, staff will quickly learn that it can be ignored. Enforcing the policy means that staff who transgress it must be disciplined, which for gross breaches means terminating their employment. See also the fraud policy in Chapter 13.

CASE STUDY: GAS PRICE WHISTLEBLOWER SACKED

Seth Freedman was sacked by his employer ICIS Heren, a wholesale gas reporting business, after blowing the whistle on alleged price fixing by energy companies.

Freedman had reported wild movements in the unregulated over-the-counter gas market to his employer, and then raised them with the FSA fearing inaction on the issue.

ICIS Heren said it was forced to remove Freedman because he had lost the trust of price reporting agencies and his colleagues. Reed Business Information, which owns ICIS, said it had offered Freedman three other roles on the same terms and conditions within the business. Freedman said the jobs had no relevance to his experience or skills. He said he was being victimized, and said, 'Traders have made it clear to me that manipulation of gas prices is taking place on a regular basis. They name big companies of trying to rig prices and reap profits.'

At the time of writing, the big six energy companies denied they were engaged in anything illegal. The UK's FSA and the energy regulator Ofgen were expected to continue investigating for several months.

COMMUNICATING THE POLICY

The CEO should launch a policy, perhaps by email or letter to all members of staff. Table 19.3 contains a sample code of conduct.

Table 19.3   Code of ethical conduct

1. Letter from Chief Executive	Presents top management's message of the importance of integrity and ethics to the organization Introduces the code of conduct: its purpose and how to use it
2. Goals and philosophy	Considers the entity's: • culture • business and industry • geographic locations, domestically and internationally • commitment to ethical leadership
3. Conflicts of interest	Addresses conflicts of interest and forms of self-dealing Speaks to personnel and other corporate agents and those activities, investments or interests that reflect on the entity's integrity or reputation
4. Gifts and gratuities	Deals with giving of gifts and gratuities, setting forth the entity's policy, typically going well beyond local law Sets standards and provides guidance regarding gifts and entertainment and their proper reporting
5. Transparency	Includes provisions dealing with the organization's commitment to complete and understandable social, environmental and economic reporting
6. Corporate resources	Includes provisions dealing with corporate resources, including IP and proprietary information – whom these belong to and how they are safeguarded
7. Social responsibility	Includes the entity's role as a corporate citizen, including its commitment to human rights, environmental sustainability, community involvement and environmental and economic issues
8. Additional conduct-related topics	Includes provisions regarding adherence to policies established within specific areas of company activity, for example: • employment issues such as fair labour practices and anti-discrimination • governmental dealings such as contracting, lobbying, and political activity • antitrust and other competitive practices • good faith and fair dealing with customers, competitors and suppliers • confidentiality and security of information • environmental practices • product safety, quality

Source: Coso

The Triple Bottom Line

'People, planet and profit' was a phrase coined by environmental consultant John Elkington. Known as 3P or TBL, the triple bottom line broadens an organization's focus away from simply making money (or delivering government services) to considering the wider effects of its operation.

> *People*: this refers to the social impacts of a business. The organization should treat people fairly, whether employees, customers, or suppliers and their workforces. That means avoiding the use of child labour or indentured labour.
>
> *Planet*: the business must seek to reduce its environmental impacts, as we discussed in Chapter 8 on the environment.
>
> *Profit*: the organization should take the costs and benefits of society into account, rather than simply measuring the return on its own capital. For example, if the opening of a supermarket closes smaller businesses and reduces the total number of jobs in a town, there is an overall in the overall benefit to the population. This requires 'full cost accounting', which covers the organization's environmental and social costs as well at its conventional revenues and costs. This can help unearth costs and failings which the business can then seek to remove.

ISO 26000

ISO 26000 is not a management standard, in that you can't get certification for it. However, it sets out a model for managing CSR.

It covers seven topics, including labour practices, the environment and governance (Table 19.4).

ISO 26000 asks the organization to consider two fundamental elements of social responsibility:

1. Recognize its social responsibility within its sphere of influence.
2. Identify and engage with its stakeholders.

Once the business has grasped these principles, it should identify how it deals with the seven core subjects of the standard.

It then integrates social responsibility into its decisions and activities. This involves practices such as:

- making social responsibility integral to the company's policies, organizational culture, strategies and operations;
- building internal competency for social responsibility;
- undertaking internal and external communication on social responsibility regularly.

Rather than suggest its own methodology, the standard refers the reader to other initiatives such as:

- The Global Reporting Initiative (GRI);
- CSR 360 – Global Partner Network (Business in the Community UK);
- EFQM Framework for CSR & Excellence Model (efqm.org);
- Social Accountability International (SAI – SA8000 standard).

Table 19.4   ISO 26000 core elements

---

1. Organizational governance

2. Human rights
- Due diligence
- Human rights risk situations
- Avoidance of complicity
- Resolving grievances
- Discrimination and vulnerable groups
- Civil and political rights
- Economic, social and cultural rights
- Fundamental principles and rights at work

3. Labour practices
- Employment and employment relationships
- Conditions of work and social protection
- Social dialogue
- Health and safety at work
- Human development and training in the workplace

4. The environment
- Prevention of pollution
- Sustainable resource use
- Climate change mitigation and adaptation
- Protection of the environment, biodiversity and restoration of natural habitats

5. Fair operating practices
- Anti-corruption
- Responsible political involvement
- Fair competition
- Promoting social responsibility in the value chain
- Respect for property rights

6. Consumer issues
- Fair marketing, factual and unbiased information and fair contractual practices
- Protecting consumers' health and safety
- Sustainable consumption
- Consumer service, support, and complaint and dispute resolution
- Consumer data protection and privacy
- Access to essential services
- Education and awareness

7. Community involvement and development
- Community involvement
- Education and culture
- Employment creation and skills development
- Technology development and access
- Wealth and income creation
- Health
- Social investment

Source: ISO

Risk Assessment

You can assess your vulnerability to ethical risk by answering the questions below. Score one point for each box ticked.

Topic	Question	✓
Governance	Does the organization lack an ethics policy?	
Customers	Do you have major contracts to governments and large organizations?	
	Are incentives, gifts or bribery an accepted part of doing business?	
	Does the business work in less-developed countries?	
	Do you have weak or vulnerable customers?	
Competitors	Does the company spread false information about competitors or conduct 'dirty tricks' against them?	
Suppliers	Does the company take advantage of its buying power to force prices down?	
	Does the business use child labour or indentured labour?	
Staff	Do you expose workers to danger?	
	Does the work cause undue stress to managers or staff?	
Total points scored:		

Score: 0–3 points: low risk. 4–6 points: moderate risk. 7–10 points: high risk.

The Appendix has a summary of the checklists. By entering the results of this checklist, you can compare the scale of ethical risk against other categories of risk.

20
How to Succeed in Marketing by Being Aware of Risk

'A JV with Eddie's organization will leverage our skills and their reach,' says Will, spouting his usual marketing jargon. 'Any questions?' he enquires, looking around.

'Don't we risk teaching a potential competitor about our market?' you ask slowly. 'And if they mess up, it'll our reputation on the line.'

'I've known Eddie Gardner for years,' says Will. 'He's solid. Trust me.'

The meeting ends, with everyone agreeing that Will's JV is the start of great things for the organization. You're not so sure.

Marketing is Opportunity Risk

Risk isn't usually associated with marketing. And marketing people don't usually talk about risk. Cautious finance people and CEOs think about risk, while marketing people are more cavalier – or that's the way the world sees it.

Yet there is no shortage of marketing threats. Loss of a major customer, competitor innovation, low-cost imports and erosion of margin by increased competition – it's hard to keep profits and market share buoyant in a rapidly changing world.

With today's 'lightweighting' of the corporation, many businesses have fewer assets. They are less at risk from fire, say, or environmental pollution than in the old days when the corporation was a smoke-breathing satanic mill. Today organizations depend more on branding, design and IP. So marketing risks have assumed a greater importance.

Awareness of marketing risks allows a business to stay clear of the pitfalls, and helps to clarify some strategic issues, such as:

- What markets should we invest in?
- How are our customers changing?

The Sources of Marketing Risk

Marketing risk comes from two categories: outside the business and inside it.

Outside risks are the big strategic issues that are formed on the world stage, such as recession and tariffs. The *inside risks* are down to the company alone. They include failure to innovate and reliance on too few customers.

In this chapter we look at the big external threats, and then the insidious internal risks often associated with complacency, lethargy and bad decision making. They break down as follows:

1. External risks	2. Internal risks
• Government policies and economic change	• Weak product performance
• Impact of the business cycle	• Inadequate promotion
• Changes in the market	• Poor design, weak branding
• Pricing problems	• Failure to innovate
• Counterfeiting and mimicry	• Weak processes
	• Over-reliance on major customers
	• Distribution problems
	• Social networking and PR issues

Many of these are inter-related. We look at each of these categories in turn, starting with external risks.

External Risks

I. GOVERNMENT POLICIES AND ECONOMIC CHANGE

All organizations are, to a greater or lesser extent, vulnerable to change in government policies. There are three inter-related macro-economic factors that can harm an organization: legislation, recession and the business cycle.

Legislation

Changes in legislation can include the banning of a raw material, curbs on waste disposal, or new health and safety regulations. Alcohol companies, for example, face curbs on drink advertising, minimum price legislation and where alcohol can be sold.

The Swiss chemical firm Sandoz (Sandoz.com) came under pressure as governments tried to halt rising healthcare costs by reducing the drugs bill. Governments do this by reducing the level of payment for drugs, by de-listing some drugs and by encouraging the use of generic drugs. Sandoz sought to counteract this squeeze by launching new products, by controlling costs and by more efficient marketing.

Other industries suffer as a result of changing tariffs and taxes. Governments and supranational bodies such as the EU seek to achieve their political and economic aims. We are likely to see more restrictions on products that cause climate change or energy use.

Turf wars also affect business. Sales of Geest's Caribbean bananas suffered when GATT agreements allowed South American bananas to be sold more widely in Europe. Similarly, the EU steel industry was affected when the US Government imposed tariffs of up to 30 per cent on steel imports. The same action is also estimated to have cost the US economy 200,000 jobs in industries which required steel, due to increased steel costs.

Recession – increases in interest rates or inflation, leading to a fall in demand

A rise in interest rates can cause problems to some industries. House sales fall as interest rates rise, so real estate agents suffer.

Most industries will be affected by a recession. Those that escape can include value retailing, food manufacture, funeral parlours, cosmetic surgery, and luxury goods such as watches.

Sometimes high inflation and interest rates are signs of an imbalanced economy that is out of control. At other times they are simply part of the business cycle, something we look at next.

2. IMPACT OF THE BUSINESS CYCLE

In Chapter 14 on Finance, we examined the risks inherent in the business cycle, as the economy moves from growth to boom, then to downturn, recession and stagnation, before a recovery starts again.

Most businesses need to accept that sales will decline in a downturn and recession. Few are able to admit this, and those who do usually resort to planning on seizing extra market share. It might be more sensible, however, to forecast a decline in sales.

That in turn will allow operations or production to lay off staff which, while painful for those involved, will allow the business to get through the recession intact.

Senior management also cuts sales and marketing expenditure in a recession. This, too, is a bad thing. Grabbing every sale is vital in a downturn, and companies that maintain their advertising in a downturn usually are the ones with the biggest market share when the recovery arrives. Large, well-established companies tend to survive a recession, while smaller businesses with fewer reserves and less of a grip on the market are more likely to fail. Your size and position in the market will indicate what strategy you should adopt in recessionary times, and also which divisions need investment or divestment.

Innovation is also important during a downturn, because that will gain market share. However, consumers often prefer tried and tested products during a recession. Hot dog stands will do better than salad bars. The author launched a dog food brand just as a recession was getting underway, and the brand never took off. He ascribes this in part to the loss of small retailers, many of whom ceased trading, leaving multiples and supermarkets with a bigger share of the market. Equally, it might just have been a bad product.

Marketing in an upturn is usually a joy for marketers, because it comes with bigger budget and a greater willingness on senior management's part to agree new activities.

3. CHANGES IN THE MARKET

Changes in the market are due to disruption in:

- channels;
- business models, sometimes based on technological change;
- fads and fashion.

Changes in channels – growth of the internet

The internet has wrought the greatest change in channels ever seen, often bringing suppliers and end users together and displacing the intermediary. This is the process known as disintermediation.

Businesses that once sat in the background, such as insurance companies, have developed direct sales strategies. This has caused some brokers to disappear, such as Norwich Union's Hill House Hammond chain.

Some major retailers like Morrisons were slow to react to internet shopping. Sometimes the retailer felt that online shopping didn't fit their strategy. In other cases they didn't see online as a credible threat, or that they lacked the necessary skills. Many felt that having a store in every high street was necessary to give them adequate coverage.

Companies whose senior management have a weak understanding of IT, or who rely on other skills such as people management, are vulnerable to more alert competitors.

The high street too has changed. At one point the high street had become dominated by service businesses such building societies, but many of these transactions have moved online or disappeared. It's a sign of how swiftly even established businesses can be swept away by change.

IS THERE A FUTURE FOR RETAIL?

In the future, retail will be limited to services and products not easily obtained online, and those people buy for pleasure rather than necessity. This will include the following categories:

- Personal care: nail bars, hairdressing and possibly education/childcare.
- Healthcare: dentists, complementary therapists and alternative.
- Consumption: coffee shops, bars and restaurants.
- Fashion: shoes and clothes, which don't lend themselves to online transactions.
- Niche brands: especially specialist local food and crafts.
- Business services: financial transactions, estate agency, but in more limited numbers; and newer services such as mediation or life coaching.
- Entertainment: leisure activities often involving new technology, sometimes short lived.
- Display showrooms, especially for online brands and prestige businesses

There may be more interactivity, introduced through so-lo-mo (social, local, mobile) activity, involving consumers' phones, with stores offering personalized deals based on specific customer. However, the scale of this remains to be seen. After a decade of supermarket loyalty cards, stores have not been able to make much use of the information.

Overall, there will be a reduced need for retail space, causing problems for commercial landlords. Many secondary shopping streets will be re-zoned for domestic housing or self-employment hubs.

An uncertain future exists for the supermarket chains, squeezed by the discounters, and with too many stores overall. Western countries are increasingly over-provided with supermarkets, resulting in them cannibalizing each others' sales. This will be exacerbated by the increase on online grocery shopping (albeit that online sales may reach no more than 10 per cent of all grocery business). As a result, physical store sales may decline, leading to the closure of less efficient outlets, until a new equilibrium is reached.

Retail concentration and the online aggregators

The growth of online shopping has created a new concentration of buying points. Just as the supermarkets squeezed out small grocers and weakened the food brands, so too the online stores like Amazon have captured sales in markets which were hitherto less affected by retail concentration, such as electrical goods. Amazon has 20 per cent of all online sales, holding not only publishers in thrall, but other industries as well.

The same thing has happened with service providers such as insurance companies who now find themselves at the mercy of online aggregators, such as Confused.com and GoCompare.com. The aggregators take 56 per cent of all new UK car insurance, according to Datamonitor (datamonitor.com).

The consumer hasn't learnt that the 'best buys' may only be those that pay enough money to be represented in the listings, while the insurance companies' margins are being squeezed as they jostle to get into the all-important 'top five' ranking, and risk neglecting customer service. However, each country is different, and only 2 per cent of French consumers buy new insurance online. However, the aggregators are likely to see growth in life and home insurance, critical illness, credit cards and savings accounts.

Meanwhile estate agents complain about the money they have to pay to Right Move, which has a nearly 60 per cent market share of online property sites. On the other hand, the estate agents save money by not taking out large advertisements in local newspapers as they used to. But unlike the insurance world, which has gone self-service, home owners still want estate agents to do the work of pricing and advertising their house, and negotiating its sale, an example of where personal service is still needed.

New business models

New business models can be highly disruptive. New competition can take different forms:

- *Low-price competitor.* Cheap pound shops or dollar stores undermine established mid-range retailers. In Turkey, after the Government deregulated domestic aviation, no airline took advantage of the change. So two years later, Esas Holdings, an investment group, bought Pegasus, an air charter company which had just 14 planes, and launched a low-cost airline. In the past eight years, Pegasus has carried 54 million passengers, of which 40 million flew for less than $50. As a result, it is the country's second largest domestic carrier.
- *Larger scale.* In the US, Walmart thrashed its competitors by offering very large, low-priced stores.
- *Technological advances.* In the music industry, vinyl was affected by cassettes, and then by CDs and mini discs. Subsequently consumers shared files using peer-to-peer technology. Then Apple introduced iTunes and consumers could buy mp3 downloads from Amazon.
- *Convenience.* Cars provided a more flexible and convenient form of travel, which had a disruptive effect on bus and train travel. In 1950, over 80 per cent of UK households did not have a car, the figure has now fallen to around 26 per cent.
- *Freemium.* New models that provide hitherto paid-for services free of charge, by getting their revenue from another source (for example, the supplier) rather than the customer or from a related product. Microsoft has been practising this for years. By making its Outlook email service free, it has frozen competitors out, a ploy that helps it retain 43 per cent of the email market. Freemium suppliers may also offer a paid-for higher value service, such as online games that charge for additional weapons. Because of its importance, we look at freemium in more detail below.
- *Going direct.* As we saw above, some intermediaries such as retailers have lost out where the supplier is now able to deal directly with its customers.
- *Self-service.* When ATMs were first introduced, people believed that customers would prefer to talk to a friendly teller rather than deal with an impersonal machine. How wrong they were. Then with the development of interactive online databases, the internet allowed consumers to book their own airline tickets and even choose their own seats. That swept away ticketing agents and travel agencies. Similarly, patients can book their own appointments with a GP, and departmental heads can advertise and manage their job vacancies without the need for HR to be involved.

The pain of change

Until the arrival of the internet, newspaper reading didn't change for three centuries, and so the owners weren't called upon to change their model. The same is true of bookselling, publishing and retailing, with the latter now under threat from 'showrooming', which is where people check products in a shop and then buy online from eBay or Amazon.

So it isn't surprising that today's directors find it difficult to change a well-established model. It is particularly hard when you have a successful business model that has worked well for decades.

THE GROWTH OF FREEMIUM

Traditional video games companies that were accustomed to selling shoot-em-up products on DVDs for £40 have been threatened by the growth of free online games in Facebook such as Farmville.

In the latter case, users get the game for free but have to pay for extras. Similarly you can buy smartphone and tablet games for as little as 99p. Although the best sellers make a lot of money, the investment in programming is a bigger risk for such a low-priced product.

Industries that want to give away a service, as a way of selling their services in a related market, put existing businesses at risk. Companies that sell information goods, such as directories, are vulnerable to competitors providing the information for free, often by getting internet users to collect the data. Thus a retail consultant might create an online retail directory, as a way of attracting users to its consultancy services. Some believe that all information products will eventually become free. However, some value-added information is likely to remain a paid-for product.

There is also a risk that the growth of free services makes consumers reluctant to pay for them. Equally, many freemium models may not be profitable and will therefore disappear.

You therefore have to decide whether any elements of your business are at risk from new models, including freemium, and whether you need to make a product, or a sub-set of their product, free. You may have to find value-added services or different products.

There are also risks with the freemium model. In many markets, it is hard to get users to cross the barrier from free to paying just one cent or penny. However in other markets, such as online data storage (for example Box.net), users gradually come up against the boundary that requires them to buy the premium service.

Freemium works best when there is an 'ecosystem' rather than a one-off product. Users who install Dropbox get committed to using the service, whereas someone who uses a freemium game may be enticed away to a new game by a different provider.

And on the upside, Nicholas Lovell, author of *The Curve*, suggests that offering a free or low-priced product exposes your business to a huge increase in the number of people. This can lead to an increase in the number of people buying your higher-price product. And the goal, he says, is to make premium-priced products available for your 'superfans', people who will pay large sums of money to be associated with you. Businesses mostly ignore the high-value end of the market, believing that no one would pay double or treble the money for a broadly similar product. So we shouldn't be scared of giving away things for free, and charging massive additional money for special benefits.

Dealing with disruption

Companies lose out to disruptive competitors for one of six reasons:

1. The paradigm shift starts off as too small and niche to worry about.
2. The disruptive technology is too alien, and too difficult to handle.
3. The new technology would require the business to reduce its investment in the familiar technology which it must pursue to maintain its position against current competitors.
4. The disruptive products are initially poor in quality (such as the first digital camera which couldn't offer high-quality images).

5. The new products would cannibalize the company's profitable products.
6. The company doesn't have the zeal for the new disruptive system, preferring the comfort of what it knows.

SIX WAYS TO FACE DISRUPTION

The six solutions for dealing with disruption are as follows:

1. Constantly scan the horizon for possible threats and opportunities.
2. Treat disruption as an opportunity.
3. Be prepared to hire programmers, or set up a test system or production unit to learn more.
4. Check out customers' view of alternative models – but don't ask loyal customers who may be wedded to the existing system.
5. Seek out ways to be your own competitor. Set up competing businesses, using a skunkworks, an independent unit that won't be crippled by corporate management.
6. Ask yourself: 'How could a competitor with a different business model put us out of business?' If the answer is clear, you need to devise your own new model, before someone else does.

CASE STUDY: LOGITECH PIVOTS, ILFORD DIES

Since its formation in 1981, the Swiss company Logitech has dramatically changed course several times. In the space of a few years, its computer mice went from being leading-edge products for engineers to mass-market commodity items; and their retail price crashed from $120 to $10. It now distributes its product in over 100 countries and has strategic partnerships with most IT manufacturers.

From a small plant in Switzerland, Logitech moved its manufacturing to a big factory in California and then to low-cost bases in Taiwan and China. Seeking to escape its reliance on mice, the company launched into new markets such as webcams and speakers. Today, Logitech is the market leader in cordless peripherals. Daniel Borel, the company's co-founder, admitted that the upheavals were emotionally draining.

For photographers, Ilford was a much loved icon. The 125-year-old Ilford business company had a 60 per cent share of the black and white film market. But the switch to colour photography and digital cameras led to a 26 per cent fall in sales of its monochrome film in just six months, and the company called in the receivers.

Most companies expect small changes in the market, but they are rarely prepared for major discontinuities. Often a small firm, an internet company or a foreign business introduces the major change.

Fads and fashion

Markets which are subject to fads or fashion, such as toys or clothing, and those which have a short lifecycle (such as IT products) are difficult ones in which to sustain success. In these markets, companies suffer major sales swings.

In the distance learning market, demand for garden design courses reached a peak when the topic was a popular subject for TV programmes, but declined when the TV stations stopped running these shows.

Competitors are always looking for ways to dislodge the market leader and make some money. That means markets are always unstable, no matter how established a business seems.

Staying in the same market too long

Some companies are in the wrong market. The last 40 years has seen the decline in the West of big smokestack industries, like iron and steel, along with shipbuilding and coal mining. Other industries have risen in their place, like electronics, computers and retailing.

Companies can alter their destiny by forecasting what the future will bring, and altering their corporate strategy accordingly. We discuss scenario planning in Chapters 23 and 27.

Sticking with what you do best is sometimes a better strategy. GEC sold its dull traditional industrial and defence businesses, and dived into the fashionable telecoms and internet world, at the height of the dot-com era. Just seven years later, the business was effectively bust. The 109-year-old company that was once valued at £35 billion was sold for just £1.2 billion.

4. PRICING PROBLEMS

Faced with one problem or another, companies often try to win business tactically by cutting prices or spending more on publicity and sales promotion. While this can help the healthy business return to profit, it may compound a weaker company's difficulties, leading to a downward spiral of declining profit. A price war can break out due to various factors:

*   there is excess production capacity in the industry;
*   there has been little innovation in the market;
*   one business adopts an aggressive marketing campaign based on lower prices;
*   there are few suppliers (an oligopoly). Examples include newspapers, soap, beer, washing powder or cigarettes.

A beer price war in the USA saw profits fall for the two largest US brewers, Anheuser-Busch and Miller Brewing. In response to falling beer consumption (largely due to health awareness) Miller, the smaller of the two companies, reduced its prices at a time of year when Anheuser-Busch, the manufacturer of market leader Budweiser, traditionally increased its prices. It initially saw a 2 per cent increase in market share but Bud soon struck back. A year-long price war saw revenues-per-barrel fall by a significant amount.

In contrast, Coors, a third brewer, refused to enter the price war and saw its revenue-per-barrel increase in the period. It seemed that the price war had served only to cheapen the well-known brands, despite intensive advertising, causing many consumers to trade up, despite the higher prices.

Price wars rarely benefit the business, and are often a substitute for innovative marketing and product development.

Ways out of a price war

The situation resolves itself when:

*   One of the companies innovates its way out of the problem. For example, the company can split its brand in two: a 'defender' brand to fight price competition and a premium brand to defend the brand's values. Or it can 'unbundle' its services, allowing customers to only buy what they need.
*   One or more competitors leave the market.
*   Companies scale down their output. New facilities are sometimes mothballed.
*   Demand starts to grow sufficiently so that there is enough business for all the firms.

Often, several of these factors combine to pull the market out of the doldrums.

WHEN FALLING PRICES ARE A SIGN OF HEALTH

Falling prices don't always signal a problem. When technology or mass production reduces production costs, the consumer price falls correspondingly. This typically happens in technology markets.

Sometimes a price war can benefit the market leader by knocking out weaker competitors. UK supermarkets continue to build new and larger superstores, but the consumer is spending no extra money. This puts pressure on supermarkets to win sales, and they do this by dropping their prices. The smaller supermarkets and those which are carrying debt will fail first, leaving the biggest operators to emerge victorious.

Pricing strategy

You need to estimate what levels of sales and profit would result at a given price level. This knowledge will help you respond strategically to competitors' price moves, rather than follow in a knee-jerk reaction.

Some markets grow when prices fall; others are relatively inelastic (insensitive to price changes). Failure to raise prices in line with inflation often leads to sharp falls in profitability. The company should therefore adopt a confident and consistent pricing strategy. It should make small but regular price increases, these being less visible to the customer. Price increases are often best made after the peak sales period or before holidays, so that the rise is less noticeable.

The price of the product or service should include the risk involved. Insurance companies do this as a matter of course: young drivers pay more. And if you want to pay for one of the author's courses over five months, it will cost you 20 per cent more. Rate of return (RAROC-based) models ensure that each product or service reflects the company's risk.

As any finance director will tell you, healthy margins are essential for the future of your business. Unfortunately, if you are sitting on a high-margin niche, competitors usually get to hear about large margins, and eventually swoop in to get their share. This drives down margins but sometimes increases the market size, making the products just as profitable. So while large margins are desirable, they also carry the seeds of their own destruction. You need to plan for scenarios where new competitors come to take a slice of your market.

Own label

Grocery brand manufacturers such as Unilever and Proctor and Gamble have found it difficult to maintain market share in supermarkets as consumers find cheaper own-label products as effective. This is partly attributable to the rise of the 'hard discounter' stores, such as Lidl (lidl. com), initially in Germany but also found in France and other European countries as well.

As a result, the companies have had to spend more money promoting their products, which reduces profits.

In some categories the consumer finds an emotional benefit which serves to keep them buying the branded product. For example, young men continue to buy Unilever's Lynx deodorant (also known as Axe) because it promises to make them more sexually desirable.

5. COUNTERFEITING AND MIMICRY

Paris's Musée de la Contrefaçon (unifab.com) houses the world's earliest fakes. Dating from 27 BC and found in Arles, southern France, they consist of parts of four wine bottles marked with the stamp 'MC Lassius', a leading wine brand of the time. But three of the stamps are fake. The counterfeiters wanted to present their wine as the real thing, and make more money as a result.

Counterfeiting has been with us for a long time, but globalization has turned this problem into a serious issue. According to the museum, a kilo of fake goods is worth eight times more than a kilo of cannabis, so the rewards are clear.

According to the International Chamber of Commerce (www.iccwbo.org), 7 per cent of the world's trade is in counterfeit goods, valued at $350 billion. Industries affected include manufacturers of software, automobile and aircraft parts, pharmaceuticals and FMCG (Fast Moving Consumer Goods) such as foodstuffs, beverages, tobacco, clothing and personal care products.

Counterfeiting is a particular risk for companies with strong brands that sell in South America, Eastern Europe and South East Asia. But the problem can also arise in Europe and the USA. The latter is a major producer of counterfeit goods.

The victim company risks losing not only its revenue, but also its reputation, for many counterfeit brands are of poor quality.

Software and record companies suffer a similar problem over copying. According to the IFPI (ifpi.org), the organization which represents the world's record companies, music piracy is worth $4.1 billion. One in three discs is a pirate copy, and 1.5 million pirate CDs are created each day.

Software piracy is at least as common, with the Business Software Alliance estimating that the illegal copying of business programs costs the US software industry around $11 billion.

Counterfeiting is particularly prevalent in China and Eastern rim countries. According to China's Development Research Center, a research institution affiliated with the Chinese State Council (cdrf.org.cn/en), counterfeiting in China is a $16 billion industry. Other offenders are Pakistan, the Philippines, Eastern Europe and North Africa.

Nor are only brands and software at risk. When Walt Disney first launched Disneyland in 1955, he had pre-sold all 20,000 tickets for the opening day. Unfortunately, someone had counterfeited an extra 15,000 so the site had nearly twice the number of expected customers. The overcrowding caused many problems. According to Martin Sklar, vice chairman of Walt Disney Imagineering, some of the rides broke down (there were only 20). The over-filled ferryboat sank. And with temperatures exceeding 100 degrees, it was a difficult day for staff and visitors. But Disneyland lived to tell the tale.

Figure 20.1   A Sharpie marker, next to a 'Shoupie' marker

Source: Wikipedia

CASE HISTORY: BEATING THE COUNTERFEITERS

100,000 bottles of Smirnovskaya vodka were destroyed in St Petersburg, following a successful court action by Grand Metropolitan, now Diageo. The company rates counterfeiting as one of its main business risks, and continually takes legal action against counterfeiters which produce vodka bearing the crown, shields and other designs of Smirnoff, and with names like Selikoff and Romanoff. In a single year, the company can take action against 50 pirates.

Johnnie Walker, another Diageo whisky brand, is often imitated by brands calling themselves Johnnie Hawker, Joe Worker and Johnny Black, generally with similar red and black labels.

According to Diageo, the fake Johnnie Walker whisky can be 3 per cent Scotch, with the rest being 'local spirits of dubious origin'.

The *Financial Times* says there is a market for empty liquor bottles that can be filled with illicit spirits and then sold as the real brand. In turn, companies like Brown-Forman, which makes Jack Daniels whiskey, have developed technology that makes bottles more difficult to refill.

Companies also create security holograms, seals and inks that are harder to duplicate and can be verified through the distribution process.

MIMICRY

Counterfeiting (above) is illegal, but in most countries mimicry is not. Some supermarket own-label products mimic the brand leaders. Across Europe, 36 per cent FMCG are own label, according to Just Food (just-food. com). In the USA, 62 per cent of grocery shoppers agree that store brand foods are usually on par with the quality of the name brand, according to Packaged Facts (P*ackagedfacts.com).*

Leading brands complain that supermarkets are parasites which have not invested in creating brands. The supermarkets argue that they are offering the consumer choice and lower prices, and well as powerful own brands. No one, they say, forces the consumer to buy own label.

Yet as *The Grocer* magazine pointed out, some of Tesco's own-label brands clearly mimic leading brands. Walkers objected to the company's Sensations crisp range which contained the same livery and same image cues such as potatoes set against a rural background.

In other markets, an innovation introduced by one firm is soon matched by competitors. Much pharmaceutical R&D is designed to develop drugs which match existing competitors' products.

Brand names need to be jealously guarded. A US appeal court ruled that 'Swiss army' knives can come from any country. This has cost the Swiss manufacturers Wenger (*wenger.ch*) and Victorinox (*victorinox.com*) a lot of revenue. More than half of US consumers who buy a pocket knife with the famous red cross think they are getting a Swiss product. In fact, they are probably buying a poor-quality Chinese knife which costs one-fifth of the price, and outsells the original by three to one.

In the IT market, successful online businesses face competition from clone sites. Successful sites like eBay, Zappos or Fab.com are quickly imitated, even down to the website look, and marketed in Germany and other markets. At Rocket Internet (rocket-internet.de), the Berlin-based Samwer brothers have sold cloned versions of Facebook and YouTube for $112 million and $36 million respectively. One internet entrepreneur said, 'Give me six weeks and six thousand dollars and I can replicate any website.'

Internal Marketing Risks

Many marketing problems occur as a result of the company's own failings. As we saw earlier, the main ones are as follows:

1. Weak product performance
2. Inadequate promotion

3. Poor design, weak branding
4. Failure to innovate
5. Weak processes
6. Over-reliance on major customers
7. Distribution problems
8. Social networking and PR issues.

We examine these problems next, starting with weak performance.

1. WEAK PRODUCT OR SERVICE PERFORMANCE

Customers are rational. They buy the product that seems to offer the most performance at the right price.

Usually, the product with the best performance becomes the market leader. But there are occasionally variations on this. Sometimes the first into a market gets a 'first mover advantage' where its brand name and distribution allows it to stay number one, and overcome any performance weaknesses.

Performance means different things in different markets. It might mean attractiveness, ease of use, tastiness, speed of operation, reliability or customer service.

In luxury markets, such as handbags, perfumes and spirits, low price can be a disadvantage, and the branding is all-important. In some markets the brand identity is the critical issue. Millward Brown (*millwardbrown.com*) *says that brand loyalty comes from five factors in addition to price:*

1. *Rational benefits.* These drive success when those benefits make a brand the most desirable choice.
2. *Emotional benefits.* They are an advantage when the brand or the experience of the brand makes people feel good.
3. *Popularity.* This leads to purchase by signalling that a brand is a safe choice, from either a functional or a social point of view, or both.
4. *A point of difference.* This sets a brand apart and gives people a reason to choose it over alternatives.
5. *Dynamism.* A brand with a sense of dynamism is one that is setting trends or shaking up the status quo.

The point is that poor product performance is a big marketing risk. Once customers recognize that a competitor provides a better price/performance mix, they will desert. Therefore companies need to be aware of how their product performs in the market. But first we have to know exactly what we mean by performance.

Defining comparative product performance

Many consumer markets have three price bands: low price, mid price and premium products, as shown in Figure 20.1. Customers get better performance at the premium end of the market, but not everyone wants premium performance. Many people are content to drive a small Fiat rather than a big Mercedes.

The thick diagonal line shows average market performance. Product A is offering better value for money, as is any other product to the right of the solid line. Product B is offering worse value, like any other product that would be located to the left of that line

Performance improves over time (shown by the dotted line), so that next year's models will be better than this year's, giving the consumer better value for money. It is easy for a company to get left behind, offering the consumer last year's model.

Performance applies equally to service companies and business to business:

- a school may have better exam results than others in the area;
- a local authority may have cleaner streets than its neighbours;
- an airtime provider may have better quality signal in the same area;
- an architect's practice may produce better designs or win more awards than its competitors;
- a legal firm may have better systems for managing clients' cases.

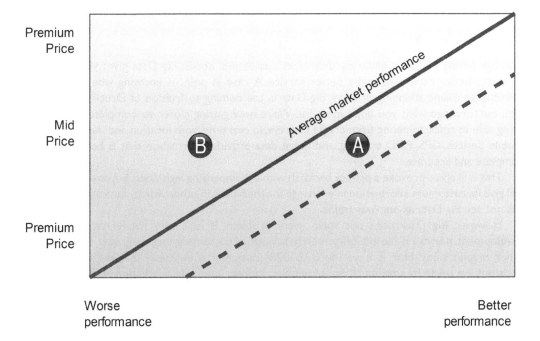

Figure 20.2   Price versus performance

Reducing performance risk by benchmarking

In benchmarking, the company determines, usually through research, what makes a superior product. As we've seen, that could be price, customer service or product design.

The business then decides the best companies to compare itself with, and the best units of measurement. It then gathers the data for its own business and its competitors. A benchmark exercise is shown below in Table 20.1. Once you've done that, you should determine how you can achieve a lead over the competition. You can also determine the product's positioning, whether that's high performance, high fashion, value for money or speed of operation.

Table 20.1   Benchmarking exercise – small electrical goods

Critical success factor	Company to benchmark	To be measured	Method of collecting the data	Action
Customer appeal	Braun	Market share	Published data	Do analysis
Price	Remington	Consumer price, trade discounts	Price lists	Establish profit margins
Distribution	All	Depth and breadth of stocking	Own survey	Brief sales force
Functionality	Black and Decker	Warranty cost, comparative tests	Function testing, Which? reports	Set up programme
Service response time	Philips	Number of days	Survey	Commission research

Learning through Big Data

Through having a clearer understanding of its customers' needs, Big Data gives you get better forecasts, better campaigns and a better service. A case in point is knowing where customers abandon an online shopping process. Big Data is the coming to fruition of Drucker's aphorism, 'You can't manage what you don't measure.' We're now getting closer to 'complete data', that is being able to track someone from cradle to grave, in real time, with location and interaction data. Mobile devices, GPS, RFID tracking, and usage data provide information that is becoming more complete and accurate.

This is likely to provoke a privacy backlash with accompanying legislation, but smart companies will give its customers information in exchange for their data. In other words, successful businesses will not see Big Data as one-way traffic.

However, Big Data does not solve every problem. It isn't very helpful in new product development, nor can it see the off-screen behaviour of customers, including what they buy from other organizations. Nor is it so useful to B2B (business to business) markets where buying decisions are made by small numbers of people at longer intervals. In addition, to provide useful information Big Data needs to be sifted, sorted analysed and interpreted. That requires investment and a dose of creative thinking.

CASE STUDY: WORDPERFECT – HOW A LEADER BECAME A LAGGARD

It is hard to believe that the market leader in word-processing software was once WordPerfect, with over 40 per cent of the market while Microsoft Word had less than 20 per cent.

Yet within two years, their positions were reversed, thanks to Microsoft's introduction of an easy to use icon-based word processor, the now ubiquitous Windows-based format.

Microsoft leapt ahead with Word for Windows by offering users greater ease of use and by automating tasks. People no longer had to memorize arcane lists of keystrokes to format their work – they simply clicked an icon using a mouse. This move was hardly unexpected. Microsoft had borrowed the technique from Apple, which in turn had learnt it from Xerox.

WordPerfect then introduced a buggy WordPerfect for Windows. The company also lacked an Office-type bundled suite. Some commentators also alleged that Microsoft played tricks to beat the competition, such as making non-Microsoft software crash or sending it false error messages.

This reminds us that innovation can upend a market, that the gatekeeper can control the market, and that a product must work properly to retain even loyal users. It also tells us that a determined innovator can wreak havoc among its competitors, by fair means or foul.

2. INADEQUATE PROMOTION

In some markets promotion is everything, and in others it hardly seems to matter. Retailers place more importance on their location, and service companies tend to downplay promotional activity when their staff are fully employed. Marketing takes place when the principals see that long-term projects are ending. This produces 'feast and famine' marketing activity. The same thing happens with engineering companies when they see a decline in orders.

Promotion should be constant. New customers coming into the market need to be made aware of the product, and existing customers only change their purchasing behaviour reluctantly.

The bigger the price tag, the longer it takes to win a customer, and the greater the need for long-term promotion. And where competitors are promoting their products, the business has to maintain an adequate 'share of voice' if it is to prevent customers defecting.

Promotion means communicating properly

Good promotion is *informative*, telling the customer something they didn't know. It should emphasize an *advantage* over other brands. It should communicate a *brand image*, one which appeals to the target market. And it should seek a *response*, especially the urge to buy the product being advertised.

Good promotion is *regular*, because customers are always dipping in and out of the market. It may be *entertaining*, *dramatic*, or *confrontational*, to break the 'glass case of indifference' that surrounds the consumer.

Good promotion does not harangue: it should be the customer's confidant and best friend. It seeks to converse with the customer at different times of day and in different ways, whether conversationally by direct mail or grandiloquently by 48-sheet posters. It talks about the customer's needs and interests, not those of the advertiser.

In the internet age, people want information, not a sales pitch. Traditional advertising such as the TV commercial, is seen as 'interruption marketing'. This contrasts with 'permission marketing', as popularized by Seth Godin (sethgodin.com), where a consumer gives you the right to communicate with them, often after being found online, or where your customer requests information.

EIGHT MAJOR RISKS IN MARKETING COMMUNICATIONS

1. Over-claiming. It leads to disappointment, hostility and possibly legal action.
2. Dumping good advertising executions just when the customer is beginning to recognize them.
3. Failing to be single minded and consistent about the brand's offer.
4. Failure to understand the needs of the customer.
5. Developing ads that win awards rather than customers.
6. Failing to test and measure the advertising, so the company doesn't know whether the campaign is successful, and which media work best.
7. Failure to use all relevant media and tools, including social proof (testimonials), online and viral marketing.
8. Failure to advertise continually.

---

3. POOR DESIGN, AND WEAK BRANDING

Many companies still under-estimate the importance of good design. One exception is corporate logos. Designed by Landor Associates (landor.com), the BP logo cost $211 million, while the Accenture logo, again by Landor, cost £100 million.

Occasionally companies get their logos for very little cost. The Google logo cost nothing, having being designed by Google boss Sergey Brin. The Coca Cola logo was produced by the company's first bookkeeper Frank M. Robinson, and therefore cost zero. And Nike paid graphic design student Carolyn Davidson just $35 for its famous swoosh logo. Later, the company reputedly gave her shares worth $600,000.

Besides logos however, many organizations don't seem to realize that when their product sits next to a competitor's on the shelf, design is what makes a consumer reach for one and not the other. The success of the Innocent smoothie company (innocentdrinks.co.uk) was largely founded on its friendly corporate identity.

Functionality – the way a product operates – is just as important, Apple's success has come from the intuitive and responsive way its products work, as well as their stylish appearance.

But even good design is not enough to make a company immune from marketing problems, as the case history below shows.

CASE STUDY: WHY PEPSI IS NO LONGER NUMBER TWO

Coca Cola remains the number one fizzy drink, despite blind tests that say consumers prefer Pepsi.

Some say that the testing method is flawed: consumers only prefer the sweeter Pepsi for a sip, but not in a full drink. But many people see Coke's supremacy as the result of constant and consistent branding.

Pepsi's ad budget has halved in recent years, and the company has focused on its Frito Lay food business, perhaps accepting the fact that it can never beat Coca Cola.

But Coke has been gaining on Pepsi all the while, so that Pepsi has slipped into third place, after Coca Cola and Diet Coke. Coke now sells $28 billion against Pepsi's $12 billion.

Pepsi's branding and slogans have been inconsistent over the years as it tried and failed to find a message that would resound with consumers.

It nearly succeeded at one point. In the 1970s its clean logo made Coke's look old fashioned, and the Pepsi Challenge made Coke nervous.

Coke replaced its flagship brand with New Coke, rather than offering it as a line extension. This made consumers angry, and led to Pepsi claiming victory in the cola wars. So Coke changed tack, re-introduced the old formula, and sales rose even higher than before.

During three centuries, the vigorous, vibrant Coca Cola script has remained more or less consistent, making it an instantly recognizable icon. Pepsi has regularly amended its clean, modern look, but it never seems to fully satisfy most consumers.

And Coca Cola earns 82 per cent more profit than PepsiCo, despite having 11 per cent less revenue for all its brands.

The message here is about the power of branding: heavy advertising and consistent branding can even overturn taste preferences.

---

4. FAILURE TO INNOVATE

Innovation provides tomorrow's profits. The age of a company's products is often a measure of its vulnerability, and failing companies often have dated products or services.

Even the most conservative markets are constantly changing, as competitors try to get a competitive advantage. Faced with a barrage of new stuff, the customer changes his preferences ('Look, a new shiny!'), and other companies then imitate the innovator. This is the theory of 'competitive rationality'.

It's easy for a firm to fall behind this perpetual motion. Management grows weary of innovating, and prefers to stick to what it knows.

Thus in any market, you can see usually two types of company: Leaders and Laggards (see Table 20.2). The Leader is constantly trying out new ideas, launching new products and looking for new ways to attract the customer. The Laggard usually imitates the leader some months later, but it is easy to see who is driving the market. By its intellectual failure and its lack of vigour, the Laggard, usually slips further behind.

Table 20.2   Leaders and laggards

Leader	Laggard
Innovates	Follows
Makes news	Makes imitative 'me-too' products
Concentrates on strategy	Concentrates on tactics
Manages for the future	Manages for the status quo

A different theory, the 'isomorphism framework' suggests that companies end up behaving like each other as they try to emulate the market leaders, while government regulations, retail buyers and consumer preferences also tend to make them converge. This makes them vulnerable to outsiders who don't know 'the rules of the game'.

The pain of being a leader

Market leaders usually stick to what made them successful. They protect their 'rent stream' from their big-selling products, and find it hard to adopt new technologies and systems that would cannibalize their own products. This makes them easy prey for the innovative new entrant.

Thus a leader can quickly become a laggard. When James Dyson developed a bagless vacuum cleaner, it was the first major design improvement the industry had seen since 1901. The long-standing market leader, Hoover, soon found itself struggling to compete.

Dyson later took Hoover to court for patent infringement relating to his revolutionary design. The case was eventually settled for £4 million, by which time James Dyson had already amassed an estimated £700 million fortune from his vacuum cleaners.

And this points to a painful truth. Marketing failure is often rooted in a company's success. If you achieve a dominant position in the market, it's hard not to become reliant on the products, market and systems that got you there in the first place. It's difficult to stay nimble and to develop new methods of doing things.

This situation is hard to spot, because it applies to apparently invincible companies at the peak of their success.

Kodak invented the original digital camera in 1975, but failed to capitalize on it because the company was wedded to its fat profit margins – up to 70 per cent according to John Naughton in the *Observer* – while digital margins were likely to be 5 per cent.

Product age is usually a good indicator of risk. If more than half of the company's turnover comes from products or services that have not changed in the last five years, change is overdue.

Protecting your brands

Companies can reduce the risk to their products by building new products around them. While the elderly Kellogg's Cornflakes is still a brand leader, it is now surrounded by own-label products, new mueslis and new breakfast foods, many of them made by Kellogg's.

Defender brands: As we saw earlier, these are brands designed to prevent loss of revenue to cheap or 'value' competitors.

Shoulder brands: These are variants of the main brand, offering:
- 'light' or extra strength versions;
- cheap or premium versions of the product;
- slightly different ingredients – Mars has surrounded its Mars Bar with countlines (hand-held bars) each containing different types of biscuit, nut and caramel, for example, Milky Way, Bounty and Snickers.

This reduces the opportunity for a competitor to make similar products that would nibble away at the main brand.

This applies to service businesses just as much as it does to product companies. A company that offers relief from anxiety sells CDs, DVDs, courses, workshops, and a practitioner certification programme. In other words, there is a price point to suit every pocket.

Line extensions: Shoulder brands are slightly different from line extensions, which offer new variants of the same product. Thus shampoo brands launch new varieties (aloe vera, cooling menthol, Amino Collagen Plus, and so on). The biggest shampoo and conditioner brand, Head and Shoulders, has 18 variants. While line extensions add facings and generate launch revenue, they can dilute the brand's core values and spread volume among extra variants.

Unlike most market leaders, Kellogg's has stayed on top. Others are less successful. Large airlines now struggle to compete with budget airlines like Ryanair and EasyJet; and video rental stores have all but disappeared from the high street with the advent of on-demand broadband streaming and online rental services like Netflix.

The new product development process

Every company, especially market leaders, need to plan for a different future. As Figure 20.3 shows, successful new product development stems from generating ideas, testing them, and launching them.

In some businesses the new product development process becomes a bureaucratic quagmire. A new product the author once worked on foundered because the production department wanted the marketing department to pay it for costs that would be incurred in changing the process and for unusable stock.

As we saw in Chapter 16 on project risk, the Lean Startup process, created by Eric Ries (theleanstartup.com) is characterized by creating a 'minimum viable product' and getting it to market fast. Until customers have had a chance to try it, the business can't get adequate feedback. Then if necessary, you 'pivot', in other words find out what element of the service people like, and change to that.

YouTube was originally a dating service, but people liked being able to upload videos. Flicker was an online video game, but customers mostly uploaded photos, so the service changed and become the world's largest photo-sharing site. Twitter, meanwhile, was originally a podcasting service.

The Lean Startup concept is just as applicable to existing businesses that are creating new products as it is to new businesses. In particular, it encourages companies to launch quickly, rather than delay by over-analysing ('analysis paralysis').

That said, the process isn't fully applicable to capital intensive industries. Boeing can't build a rough and ready airplane, and see whether customers like it. But it lets customers alter their planes (MyBoeing) in a wide range of ways, and even has StartUpBoeing, which helps entrepreneurs start airlines.

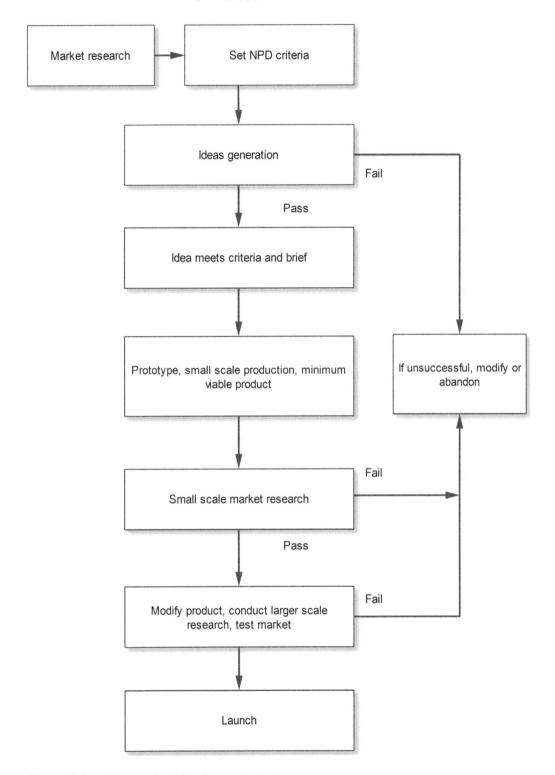

Figure 20.3   New product development strategy

Seeking revenue in unfamiliar markets

Many readers will be familiar with the Ansoff diversification grid (Figure 20.5). It suggests that launching new products in new markets is risky. For example, construction firms sometimes undertake work in high-risk areas they are unqualified for, such as asbestos clearing or working in derelict buildings.

It's a basic truth that the further from your core skills and market, the harder it is to succeed. As we saw in Chapter 1, Richard Branson has sometimes succeeded by teaming up with major businesses and offering a fresh and bright brand image.

It is easy to get bored or frustrated with your existing market, with its petty wrangling and small town feel. Like any prime minister or US president, foreign affairs are always more glamorous. But after being burnt in these forays, companies often come slinking back to their own market and decide to stay there. The author worked for Unilever at a time when its supermarket products were coming under siege by own label. Unilever decided to buy companies in the home decorating industry (Nairn Williamson), car leasing (Ford and Slater) and the builder's merchant (Kennedys). Several years later they had to unload them following continual losses. They were generally bought by smaller businesses in those fields who understood the market better and who were able to add them to their existing portfolio of brands. As Tom Peters said in his book *In Search of Excellence*, you should stick to the knitting.

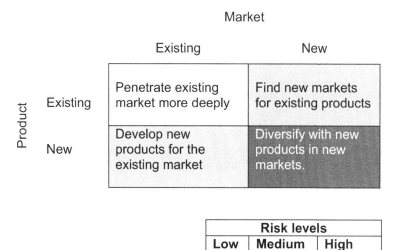

Figure 20.4   Diversification risk

The need to reinvent and rejuvenate

A management consultancy the author once worked for was dependent on one main service: it provided consultancy advice on ISO 9001. Management consultants sell fashionable management ideas, and the demand for any product always declines eventually. As the demand for this service petered out, the company found it had no other big product to sell.

Companies should devote substantial efforts to ensuring that its product remains state-of-the-art. New materials or techniques should never be ignored. Static or mature markets are particularly vulnerable, the exception being brands with miraculous longevity like Coca Cola.

For nearly 20 years Milk Tray chocolates were associated with James Bond imagery. Cadbury then boldly decided to abandon this image, and sales rose by 13 per cent. It demonstrates that taking risks is sometimes the only way to rejuvenate a brand or win new business.

Innovation isn't always easy. For years, Woolworth, one of the world's biggest retailers, tried to find a winning formula to replace the declining 'five-and-ten cent' variety store on which it was founded. As long ago as the 1960s, Woolworth sought to develop speciality stores, eventually succeeding with its Foot Locker sports shoe outlets. In 1997, the company finally closed its remaining 400 US Woolworth department stores, and today the chain now operates 1,900 Footlocker stores around the world, selling sports shoes to its 12–20-year-old customers.

New product development errors

Nine out of ten new products fail. Errors come about because companies develop products which the consumer does not want. This can happen because insufficient research was carried out. An infamous failure was Sir Clive Sinclair's C5 battery powered C5 trikes. Consumers found that the trikes were unsafe for use on the road, being slow and low.

At one time the author was set to develop software for the domestic cleaning industry. There was a clear gap in the market. But small-scale telephone research among cleaning companies revealed that they preferred to keep the information in their head and didn't like using computers. That discovery was critical to avoiding a waste of time and money.

Even major companies suffer. When Unilever added manganese to its Persil washing powder, and called it Persil Power, rumours circulated that it rotted clothes. The Good Housekeeping Institute withdrew its endorsement, and eight independent laboratories agreed that the powder did indeed damage clothes. Lever Brothers reduced the level of manganese content by 80 per cent, but the episode reduced Unilever's profits by £57 million. This was in addition to £200 million spent in development and launch costs. Fortunately, the consumer stayed loyal to Persil.

---

10 WAYS TO FAIL IN NEW PRODUCT DEVELOPMENT

1. Fail to scan the horizon.
2. Launch products without asking the market.
3. Have no new product development projects on the go.
4. Lack an effective new product development methodology or plan.
5. Rest all your new product development hopes on one product.
6. Take too long to get to market.
7. Add too many functions to the new product, and lose clarity of offer.
8. Launch me-too products.
9. Allow emotion to get in the way of reason.
10. Continue to push failed new products.

Reducing risk in innovation

Pay attention to what customers are saying. Encourage users to comment on the product. It doesn't have to be complicated: people can simply email the business. Amazon users wanted parental controls for the Kindle Fire, so they got added.

Go for crowdsourcing. The crowd has more ideas and often knows best. And with today's technology, it's easy to source ideas with third-party businesses such as CrowdSource (crowdsource.com).

Kaggle (kaggle.com) uses competitions among programmers to create statistical models that improve business performance.

Collaborate. Collaborating with competitors, organizations in other countries, or those further upstream or downstream, can help the business achieve traction in new segments or markets. The biggest barriers are fear and lack of trust, for example through losing IP and the risk of having talent poached.

Going beyond the half hearted. Too many businesses fail to do much more than a bit of brainstorming. They also fail to foster a creative environment. This means they fail to see bigger picture, fail to create new business models, and produce only me-too products and simple line extensions.

Fail faster. In a numerate climate, executives are required to produce sound evidence for new product ideas. This means disregarding instinct. If only 10 per cent of new products fail, you need nine failures to get one success. Each failure takes you one step closer to success.

CASE HISTORY: THE TRIUMPHS AND FAILURES OF BARON BICH

Even the most successful innovators sometimes fail. Baron Bich was famous for his disposable Bic lighter, his disposable razors and especially the throwaway Bic ballpoint pen, of which 15 million are sold every day. When he died in 1994, his company had a turnover of £650 million and owned (among other businesses) the largest lingerie business in France and Italy.

But Bich also had his share of failures. He tried and failed to sell inexpensive Bic perfume through supermarkets, newsagents and petrol stations. He eventually withdrew it, after three years of losses totalling £15 million.

His venture into fashion, with the purchase of Guy Laroche clothing, lost £25 million in one year alone. Expansion into windsurfing boards also cost the group profit.

Some products took a long time to come right. Though Bich bought Waterman, the US pen firm, in 1957, the company did not make profits until 1965.

Bich accepted a degree of failure as the price of innovation. He was also unwilling to rest on his laurels, and strived continuously to stay ahead.

The rise of the commodity

Every innovative drug introduced by GlaxoSmithKline (gsk.com) will, one day, become just another low-priced commodity. Drug company patents expire in as little as eight years after a drug is launched.

When the Viagra patent lapsed in New Zealand, the price plummeted from $25 a pill to $8 a pill. Drug companies try to offset the commodity fate by extending a drug's life through the introduction of improved versions. In addition, more people may use the cheaper drug, which will reduce the sales decline.

Sometimes the drug companies face a 'patent cliff', the trough caused by the expiry of a patent before sales of newer big drugs pick up. This requires such companies to carefully manage their new product development plan, so that new drugs arrive as old ones expire.

But that's not easy to achieve: success in new product development can be unpredictable. In the drug industry it can be reliant on accidental breakthroughs (both the hair loss drug minoxidil and Viagra were originally intended for lowering blood pressure) as much as carefully planned execution (a 'pivot' as described earlier in this chapter).

Drug companies are now seeking to develop drugs for the developing world, as a way of combating cutbacks in government healthcare, and the rise of generic drug makers. Fortunately, other markets don't suffer the fate of having their patents lapsing so quickly,

5. WEAK PROCESSES

Service companies rely on efficient and customer-focused processes to survive and prosper. Yet many fail to achieve that. This puts them at risk of losing customers to competitors who make it easy to buy from them.

If you buy food at Tesco or Sainsbury's, the checkout operator will probably smile and look you in the eyes. But at the competing supermarket half a mile away, the staff look bored and avert their eyes, making you feel like you're an intruder in their life.

A leading computer retailer makes customers wait in line to buy their products. At an Apple store, by contrast, the employee has a card reader on their belt, which means you don't have to queue up at a checkout. It's very customer-centric.

The same issues apply to most markets. Once you're a customer, Amazon makes it very easy – almost too easy – to buy goods from them. Other businesses have a time-consuming checkout.

Turnaround time is a key factor these days. The time to process a loan or mortgage application, or to resolve a complaint, can make a big difference to a company's corporate image. Similarly, customers have strong feelings on how effectively a company communicates with them on the progress of their application or complaint. There are wide variations across the financial services industry, in part due to the quality of each company's IT systems.

6. OVER-RELIANCE ON MAJOR CUSTOMERS

Many companies are over-dependent on a few major customers. In some firms 80 per cent of sales come from just 20 per cent of the customers. The UK grocery market is dominated by five major supermarket chains, so food processors are particularly vulnerable to having their brands being de-listed, especially in favour of the supermarket's own-label goods.

Recently a UK advertising agency was earning half its income from one US client. When the client sold its UK business, the new owners decided to spend less on promotion, and the agency nearly went bust.

Many retailers now rely on Amazon for their shop front. However, the company takes between 6 per cent and 25 per cent of the revenue, and companies who depend on Amazon for their visibility are at risk from being de-listed or changed terms of business.

One anonymous retailer quoted in the *Financial Times* said, 'The work we and others have done has paved the way for Amazon to understand what sells and what doesn't. This company aims to stay ahead by launching new designs and finding new manufacturers, saying, "We have to protect ourselves against Amazon, which has perfect knowledge of everything".'

Diane Buzzeo of Ability Commerce, a software business for retailers, told the *Financial Times*, 'They [Amazon] look at your product and then go to the manufacturer and say, "Hey, why don't we sell this direct?"'

In contrast eBay, which also provides a store front for retailers, does not have its own retail business; and that offers some protection.

6. DISTRIBUTION PROBLEMS

Large branded goods companies generally know all the outlets that stock their products. Other companies are sometimes less advanced. Many businesses operate regionally, despite having a product suitable for the country as a whole. Others think nationally, when their product could sell globally.

There are many ways a company can expand: through additional sales people, agents or through exporting. One solution is franchising, which lets a company increase its revenue without adding overheads. The franchisee pays for premises, equipment, vehicles, employees and stock.

The more successful the franchisee, the more royalties he pays, and the more product he buys from the franchisor. The franchisee works harder than an employee because he owns the business. In bad times, the franchisee shoulders the burden of falling sales. But there are also disadvantages. The company loses the management of its distribution, and franchising may be a less profitable way to do business. The same principle applies to alternatives, such as appointing licensees and agents.

Direct sales, channel conflict and company own stores

Many companies have chosen to sell direct, because bypassing the retailer or intermediary allows them to get closer to the customer, and earn higher profits because they cut out the retailers' margin. In many markets this caused hostility among the conventional retail or brokers' channels, but it died away.

A more recent feature has been the vogue for makers of even quite limited ranges to open their own stores. It gives the manufacturer a closer link with its customers, and better insights into preferences.

The main risk for company-owned stores is whether the business has a sufficiently wide product line to justify its own retail outlets. Telephone giants like O2 or even Apple (apple.com) might reasonably be expected to win enough store traffic to make stores viable, but niche players include M&Ms (mymms.com), fabric business Cath Kidston (cathkidston.co.uk), and fashion brands such as Levi (levi.com).

Company-owned stores provide a brand-rich immersive experience for the consumer. But retail is a very different kind of business from manufacturing. And once losses start, they are hard to staunch.

Payment mechanisms

As technology advances, electronic micropayments may increasingly take the place of cash. Methods include an online wallet, a contactless card or a sensor that recognizes a mobile phone. Micropayments push down transaction costs (you don't need a cashier), speed up transactions (no need to type in long card numbers), and allow the purchase of impulse items and low-value purchases.

In the 'offline' world, this can involve paying for a cup of coffee, a newspaper or a bus ticket, while online it could be paying to read an article or download a song. It also benefits the freemium market whereby you might get a racing game on your phone for free, but pay for better cars or more tracks.

For the supplier, there are risks in either failing to enable micropayments, betting on the wrong system, or failing to make a low-cost version of your product (a magazine article rather than the whole magazine).

Finally, there are issues as to whether consumers are willing to engage in micropayment systems (will they prepay into online wallets?), and whether micropayments can be profitable given that transaction costs and the cost of fraud must come out of the micropayment.

7. SOCIAL NETWORKS AND PR ISSUES

Before the internet, consumers tended to be the passive recipients of information from businesses (see Figure 20.5 – Pre-internet). When the internet first arrived, it allowed a two-way conversation between businesses and consumers. But the conversations were still one-to-one. That was Internet 1.0.

But with the development of social networking, consumers began to talk among themselves – internet 2.0. Review sites arrived, online campaigns started, and companies found it harder to control the conversation.

The era of anarchy had begun. Wily or disgruntled consumers have become expert at grabbing media attention for their cause.

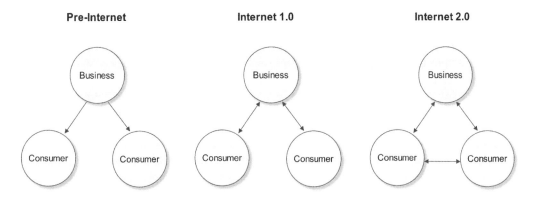

Figure 20.5   The move from simple to more interactive conversation patterns

---

CASE STUDY: UNITED BREAKS GUITARS

When singer-songwriter Dave Carroll saw baggage handlers throwing his beloved guitar around on the airport tarmac, and after failing to get compensation, he wrote a song called 'United Breaks Guitars' and put it on YouTube (http://bit.ly/z2GU5). The song was viewed 1.5 million times within days of going live, and has now been seen over 12 million times.

Four days after the song went on YouTube, United Airlines stock fell 10 per cent, costing shareholders $180 million, a loss due, it is said, to the song.

A year later, United tried to make things right with a $3,000 donation to a music charity, but the gesture was seen as too little and too late. Meanwhile Taylor Guitars got an unexpected boost from the free publicity about the singer's beloved guitar.

This episode demonstrates the importance of being able to respond flexibly to potentially bad PR.

Negative publicity comes from a number of forms:

- campaigns and discussions on Twitter (twitter.com), Facebook (facebook.com) and other social networking sites;
- negative reviews about a product or service on review sites, whether on Angie's List (www.angieslist.com), Yelp (yelp.com) or Tripadvisor (tripadvisor.com);
- 'sucks' sites, where individuals or pressure groups castigate a business.

The mass media then picks up these comments and promotes them. A petition on change.org was signed by 60,000 people, calling for Friends Life to pay out on a life insurance policy for Nic Hughes, a 44-year-old who died from cancer of the gall bladder. The company had turned down his application, citing his failure to mention pins and needles as an existing complaint. The story was picked up by the media and endorsed by celebrities such as Stephen Fry. Eventually the ombudsman told the company to pay out.

Dealing with bad online publicity

- Have a monitoring system for online publicity. This should provide an early warning system for stories that appear to be going viral.
- Appoint someone senior enough to deal with bad publicity. The standard processes for handling complaints will be insufficient. And all too often companies leave the responsibility for social networking in the hands of a young and junior employee, because they are seen as more in touch with that area. But such people probably don't have enough experience or seniority to handle an emerging crisis.
- When HMV (hmv.com) staff were made redundant by the administrators Deloitte (deloitte. co.uk), the employees took over the company's Twitter account to broadcast comments live ('There are over 60 of us being fired at once! Mass execution, of loyal employees who love the brand'). It is vital to control access to social networks, especially at times of upheaval.
- Where a complainant is media savvy, it is often cheaper to pay off the complaint, no matter how ill-founded, than to stick to the company's policies. Many complainants have the time and energy to spend weeks, months and even years pursuing their grievance, as well as the skills to design clever attention-grabbing antics.

CASE STUDY: MARKS & SPENCER IS DRAWN INTO A SUCKS SITE

When negative comments appeared on the popular Mumsnet.com site about Fathers For Justice (F4J, fathers-4-justice.org), a pressure group that seeks better access for estranged fathers to their children, the organization hit back.

F4J set up mumsnet-thenakedtruth.com, which at time of writing ranked fourth for the phrase 'mumsnet'.

Noticing that Mumsnet was sponsored by Marks & Spencer, the pressure group created a spoof Marks & Spencer advertisement, criticizing the store and seeking to persuade it to withdraw its advertising from Mumsnet.

The group's members also stripped naked in the company's head office, to further draw attention to their cause. Three-quarters of Mumsnet's £3 million annual revenue was said to come from advertising, according to *Management Today* (ManagementToday.co.uk).

Dealing with sucks sites

In some cases, complaining to the site's host (ISP) is enough to get negative comments removed. Some organizations attempt to crowd out critical sites by adding additional domains.

Going to court rarely succeeds where the site just offers opinions, and does not defame (by spreading false information), and its use of trademarks are usually deemed 'fair use'. As long as the site is clearly different from the corporate one, people are unlikely to confuse the two, so a claim for passing off is unlikely to succeed.

Going to court can result in the 'Streisand effect'. This is named after a case brought by the singer Barbara Streisand who unsuccessfully sued a photographer for $50 million, in an attempt to get her mansion removed from publicly accessible views of the California coast. Before the case the online photo had been viewed six times, two of which were by her lawyers. In the month following the trial, 420,000 people viewed the image.

Fan sites can be an equal problem because they're unpredictable. Most strategists recommend being supportive of them, to keep them on side. This doesn't always work, however. Fans of one of the author's businesses set up an informal support site. When the site started to promote competitor's products and criticize ours, we asked to be involved, and the domain then turned into a 'sucks site' that attacked us. The domain eventually withered when the ex-fans lost interest; but it occupied management time for many months.

CASE STUDY: THE ETSIO PROTESTS

In addition to sucks sites, most organizations will suffer from criticism on the net. The author launched a training programme called Etsio. It was modelled on the internship model, and because trainees had to pay, it attracted coverage in the national press ranging from thoughtful comment to angry hostility.

The forums were particularly bitter, seeing it as 'paying for a job'; and attempting to engage in debate seemed only to inflame the opposition. Eventually we shut the business due to lack of demand and the comments died away, albeit with the critics claiming victory.

It should be said, however, that many problems connected with social networking are an irritant, rather than a major PR crisis. We cover the latter in Chapters 23 on Continuity and 24 on Crisis Management.

Loss of corporate reputation

When a crisis occurs, one of the first casualties is the corporate reputation. The public believes that many corporate offences are as serious, if not worse than, violent street crimes.

The White Collar Crime survey (crimesurvey.nw3c.org) indicates that 28 per cent of the public believe themselves to be the victim of price misrepresentation (see Figure 20.6). This includes fake discounts in supermarkets or estate agents who over-represent to a seller how much their house will sell for. The public increasingly regards white collar crime as serious.

The book as a whole is dedicated to preventing the company from losing its good reputation. In Chapter 24 we consider in more detail how companies can manage their corporate reputation in times of crisis.

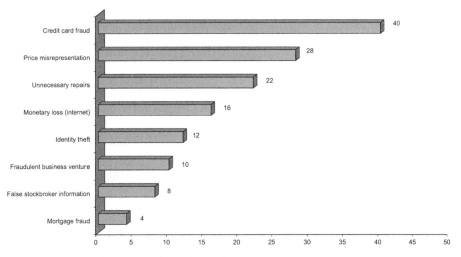

Figure 20.6   Public views on white-collar crime
Source: US national survey on white-collar crime

CASE STUDY: ARTHUR ANDERSEN'S FALL FROM GRACE

A railway director bursts into the small office of Chicago-based Arthur Andersen in 1914, demanding that he approve the company's annual accounts. The company had inflated its profits by failing to properly record day-to-day expenses. Arthur Andersen replies that there isn't enough money in Chicago to make him change his mind. Later the railroad company goes bankrupt, and the firm of Arthur Andersen becomes known for its probity. This reputation for stern integrity continued until the 1980s.

But by the mid 1990s, two-thirds of Andersen's $3.3 billion in US revenue was coming from management consulting. According to the *Chicago Tribune* (chicagotribune.com), the firm's Professional Standards Group that gave advice on tricky ethical and regulatory issues began to be ignored. In 2002 the company was convicted for obstructing a federal investigation into Enron, its leading client. The company surrendered its accountancy licence, sold off its audit division, and became Accenture.

This wasn't Arthur Andersen's only failure. It was fined $7 million by the SEC for 'improper professional conduct'. This included overstating the earnings of their client Waste Management by $1.4 billion. In the same year, Anderson also paid $110 million to Sunbeam shareholders to settle lawsuits stemming from its inflated earnings statements, according to the US Public Broadcast Service (PBS). The firm had also moved its base from Chicago to the tax haven of the Bahamas. The following year it paid Enron investors $40 million to settle claims against its non-US divisions.

Arthur Andersen's role in the Enron scandal led to the US's Sarbanes–Oxley Act, which placed much tighter controls on company accounts and their auditors.

Strategic Marketing Vision

Throughout this chapter we have looked at how external event and internal weaknesses can create marketing failure. This can be defined as declining sales, market share and profit.

Recognizing these threats, and adopting a strategic approach to marketing, will increase the chances that the business will survive and prosper. Figure 20.7 shows the marketing vision the company should adopt.

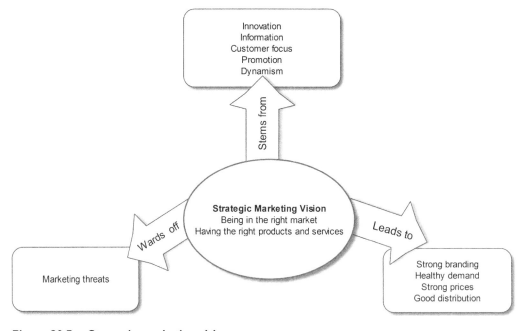

Figure 20.7   Strategic marketing vision

The organization has to have a clear marketing vision. It has to be determined to be in the right market, and to have superior products or service.

Strategic marketing vision goes beyond simple brand maintenance. You need strategic action to realize that vision, whether through acquisition or new product development.

BEING IN THE RIGHT MARKET

Being in the right or wrong market is often a matter of historical accident, and does not guarantee future success. Many of today's companies are in existence because they were offering the right product ten or even 50 years ago. To stay in business, you have to offer products that are right for today's customers.

It is easy but dangerous to focus on the products you sell rather than the needs of your customers. This blurs your understanding of the company's purpose, which is to make profit. The ideal market would have the following characteristics:

- be large enough;
- be growing;
- be profitable;
- have a high cost of entry (to dissuade others);
- require skills not easily imitated;
- not be subject to excessive cycles, or be vulnerable to government policies;
- not be unduly controversial or unpopular (such as the nuclear industry);
- be one you have core skills in.

No perfect market exists, but some are more perfect than others. Within that market, you should:

- be large enough to compete effectively;
- be able to achieve growth (by increasing your market share or market penetration) – if you are already the market leader, your room for growth is limited;
- have core skills or a competitive advantage over your competitors.

If the market or your business fails to match these criteria, you should be seeking new products and markets in which to make money.

Risk Assessment

By answering the questions below, you can assess your vulnerability to marketing risk. Tick all relevant boxes and add up the number of points ticked.

Topic	Question	✓
Macro economics	Is the company vulnerable to changes in government policy or the business cycle?	
The market	Does the organization operate in fickle or fast-moving markets?	
Counterfeiting and mimicry	Do you sell famous brands which can be copied?	
Price	Is a price war likely?	
Performance	Does your product or service perform worse than your competitors?	
Sales	Have sales been static or declined during the last two years?	
Dependency	Are you dependent on a few customers?	
New product development	Is there a lack of new product or service development?	
Promotion	Does the organization fail to consistently promote its service or product?	
The customer	Is the firm weak in fostering good customer relations?	
Total points scored:		

Score: 0–3 points: low risk. 4–6 points: moderate risk. 7–10 points: high risk.

The Appendix has a summary of the checklists. By entering the results of this one, you can compare the scale of marketing risk against other categories of risk.

21
Diversification, Acquisition, Divestment and Partnership

'They've been coasting for years,' says Grant. 'Using our contacts, we could double their sales inside five years. It would pay for itself.' Grant wants to buy a mid-size contractor. 'It's a different kind of market, though,' you say. 'And what happens if sales decline? We're taking on a load of debt.'

'You're such a wet blanket,' is Grant's scornful riposte. 'Get with it. Let's shake things up a bit.'

Stick to the Knitting, or Spread the Risk?

As we saw in the last chapter, a business needs to be in the right markets for long-term success; and achieving this may involve the business in either acquisition or divestment.

The passion for diversification in the 1970s and the rise of the conglomerate was quickly dissipated when many conglomerates got into difficulty. Some companies lacked synergy and the conglomerates, with their wide range of products and markets, were more difficult to manage.

Nevertheless, companies can reduce their risk by operating in more than one market.

There may be strong practical reasons involved: some diversifications are designed to counter seasonal sales troughs. In the 1920s Walls started making ice cream to match the fall in demand for its Walls sausages. And a well-run business can add value to an acquisition through its management skills and capital strength.

Acquisitions: Risks and Rewards

An acquisition can bring many benefits, including:

* economies of scale;
* bigger market share;
* broader product portfolio;
* entry into a new market;
* the acquisition of specialist knowledge;
* preventing competitors from acquiring a useful asset.

When you can't achieve sufficiently fast organic growth, acquisition looks like common sense especially if, as the acquirer, you have a mature business with a pile of cash.

In recent years, successful online businesses like Google have acquired countless smaller businesses, aware of the short life span that tech companies have, and conscious of the need to overcome their reliance on their one 'golden goose' product. Ninety-six per cent of Google's revenue comes from its search advertising fees.

When you have large cash reserves, it's hard to leave it in the bank when investing it would theoretically bring more profit (not to mention additional bonuses for senior executives).

Yet six out of ten mergers and acquisitions (M&A) lose money, according to the M&A Research Centre at London's Cass Business School (cass.city.ac.uk).

Other analysts say the figures are much worse. According to the Hay Group consultancy (haygroup.com), just 9 per cent of 200 major acquisitions it looked at were 'completely successful' in achieving their objectives. For UK acquirers, the 'complete success' rate dropped to an extraordinary 3 per cent.

THREE REASONS FOR ACQUISITIONS FAILURE

Three main causes of failure in M&A failure are as follows:

1. *Failure to integrate the businesses after merger.* Acquirers spend time working on financial and IT integration, but not the culture. AOL bought Time Warner for $164 billion. It then failed to get Time Warner to run its content online; and AOL users didn't get TimeWarner's fast broadband service for five years. There were serious personality clashes between the chino-clad young people at AOL and the 'suit and tie' people at Time Warner. AOL executives took most of the top jobs in the new company, causing resentment, and AOL had no experience of managing a large and diverse business like Time Warner.

2. *Buying a has-been, buying too late:* When AOL bought Bebo for $850 million, the latter had already lost the battle for the social media to Facebook and MySpace. AOL reportedly sold Bebo for just $10 million.

3. *Lack of synergy:* If you buy outside your current market or product portfolio, you may not be able to bring useful support to it, nor it to you.

Excessive Debt – The Private Equity Problem

Some well-known companies have buckled under excessive debt after being bought and sold by private equity firms once too often. This is known as 'extracting value' or 'asset stripping', depending on the speaker. 'Buy it, strip it, flip it' is a phrase commonly heard on Wall Street. Here are just three examples.

- *Biffa:* As waste management company Biffa was being taken over by its banks and other creditors, the GMB union said, 'Biffa is a fundamentally sound business crippled by the greed of private equity owners. The core waste and recycling operation can emerge stronger without the millstone of debt.' After private equity firms had bought it, the firm was saddled with debts of £1.1 billion.
- *The Automobile Association (AA):* After buying the AA for £1.75 billion, the private equity firm CVC cut 3,400 jobs, reduced wages, cut patrols by one-third and reduced training. It took out £400 million in dividends in just three years, and then sold the business to Saga for £1.9 billion. By this time, the AA was loaded with 40 per cent more debt than before. It is little wonder that the company's roadside response time fell from first position to third, as rated by Which.

- *Phase Eight:* The private equity company Towerbrook added £90 million of extra loans onto the loss-making Phase Eight fashion retailer to enable Towerbrook to receive dividends. With 88 stores and 140 concessions in Britain and Ireland, Phase Eight lost £4.6 million on £121 million of sales after having to pay £16.4 million for debt-related interest payments.

BUYING SPREES

Some companies go on buying sprees. Lord Weinstock built up a big cash cushion in his Marconi business. But his successor George Simpson changed the company's policy, and spent a fortune buying businesses including Reltec and Fore Systems, Ferranti, Vickers Shipbuilding, Yarrow Shipbuilders and Plessey.

This activity gave the business massive debts, which forced it into major layoffs. Burdened with debt, and suffering from the end of the dot-com era, the firm needed a £4.7 billion restructuring, but this failed to save it and the company was sold to Ericsson.

Acquisitions

Company acquisitions are a dangerous area because large sums of money are at stake, and the needs of seller and buyer are completely polarized. Leaving aside whether you're getting value for money, and whether the new business is the right fit, you need to know whether there are problems lurking out of sight.

The term 'due diligence' is often associated with self-preservation on the part of business advisors. But for the purchaser, it's vital to ensure that you're getting what you think you are.

The areas to check are as follows:

1. *Assess the financials,* including tax returns, book keeping and debt. Check out:
   - the target company's financial robustness;
   - potential for increasing the target company's profits;
   - sustainability of the target company's profits;
   - level of investment needed post-acquisition;
   - opportunity cost of forgoing other investments and internal business development.
2. *Verify patents and trademarks,* especially if they have financial or marketing value. If patents are about to expire, the company could face new competitors. This is common in the pharmaceutical industry.
3. *Check whether the company is facing lawsuits.* Even if none are on the horizon, the combined assets of the new company make a class action more attractive. You need to ask whether the company operates in a litigious area, such as mortgages, primary mortgage insurance, tobacco, payday loans, medical devices and healthcare products, and automotive products such as tyres and seat belts; or ones that suffer from the outfall, such as insurance. Where the acquired company has been run without proper governance, as for example, run by a single buccaneering entrepreneur, there are higher risks.
4. *Talk to customers* about their view of the company.
5. *Conduct background checks* on the company's key executives.

Fraud is difficult to spot. However, Chapter 13 on Fraud suggests the best places to look, such as finance people who are responsible for both accounts receivable and payable.

While some of this work should be left to forensic accountants, it is risky to leave it all to professionals. Personal involvement often reveals information that cannot be gleaned from a balance sheet.

The price tag will reflect the company's value at its current level of sales. The trick, therefore, is to work out how to increase sales. Otherwise, you will face a lean period of many years while the company's profits simply earn back the price you paid. Some businesses are sold by retirees, and these can be good value, especially if the business is solid but has coasted for a while. A new business sold by a young person is more likely to have problems, so discover why they are selling.

EIGHT WAYS TO MINIMIZE RISK IN ACQUISITIONS

1. Don't rely on revenue growth.
    It is risky to assume you will make the numbers work by gaining more revenue. If the increase in revenue doesn't take place, you will be left stranded. Many people think that the previous owner is an idiot, and that the buyer's superior marketing skills will add an extra 20 per cent on to revenue.
    Yet all too often the hoped-for rise in sales doesn't transpire. It is best to assume sales will continue the same as at present. That way, any extra revenue will be icing on the cake.
2. Build risk mitigation into the contract.
    Retain the seller: You can add terms into the contract that keep the seller engaged in the business for several years, as an employee or consultant or though equity. This ensures that sales don't drop as soon as the buyer walks out the door the day after the contract is signed.
    You can also tie the purchase price to the company's future performance. Earn-outs are less satisfactory to the seller, but some are willing to accept that.
    Liability clauses also help to remove the risk. This holds back some part of the price, or requires the seller to repay it, if unexpected liabilities such as contaminated land or lawsuits are discovered or take place within a certain time period.
3. Know your limit in advance.
    Like buying at auction, it is best to set a private ceiling on what you will pay. Otherwise, the price can get wound up as buyers compete. Even if you lose out, this can sometimes mean that the successful buyer has paid too much.
4. Recognize the deadly allure of the single prize.
    If there is only one company you can buy, it becomes all the more desirable. Then management has to acquire the business at any cost. This is particularly true if competitors also want to buy the business.
    When Kraft bought Cadbury, there were no other confectionery brands with such a heritage brand loyalty and market share. Similarly, Mars wanted to acquire Wrigley because it was the outstanding player in its field. That's not to say either company paid over the odds in this case, but it demonstrates that occasionally a proposition seems overwhelmingly attractive. And the possibility of it falling into the hands of a competitor seems unacceptable. At that point, the acquirer is at risk of paying whatever is required to gain the prize.
5. Identify cost savings – but make them realistic.
    Savings are often found by amalgamating service departments. For example, the buyer can amalgamate the sales force, accounts, purchasing, and even operations or manufacturing. But prior to the acquisition, the buyer's departmental heads need to take responsibility for the cost savings they expect to achieve, and what the costs of amalgamation will be. Otherwise, the savings may prove to be illusory.
    For example, the purchasing director needs to identify exactly how much will be

saved by combining the purchasing function. Knowing they will be held accountable makes people come up with more realistic, even conservative, numbers than optimistic corporate development staff might claim.

6. Stress-test the numbers.

You should put a value on three outcomes: optimistic, neutral and pessimistic. This will help you to identify the 'what-of' issues. What if sales fall? What if raw material costs rise?

The $64,000 question then is: Are we financially able to stand the losses incurred in the pessimistic scenario? Are we willing to take that risk?

All too often the pessimistic scenario turns out to be the right one, because the other two scenarios are inflated by wishful thinking.

7. Plan alterative scenarios.

What if a price war breaks out? What new technology is on the horizon? What if the government imposes a ban on the chemicals used? What if the acquired company loses its biggest contract?

These questions should be used as a way of creating alterative scenarios. This helps identify the company's vulnerabilities, but it also forces you to think what steps you could take if any of the outcomes happened. Are any of these outcomes terminal? Can the business get round these obstacles?

8. Be aware of project fatigue.

Acquiring a business can take a long time, with endless investigations and stressful negotiations. It is common for management to decide on one candidate company just to get the process over with.

Acquirers can prevent project fatigue by allocating enough resources to the project, and by having a plan that identifies what stage the search has reached at any one time.

ACQUISITION ISSUES CHECKLIST

- To what extent does the acquisition meet a strategic need for diversification?
- How familiar are we with the new market?
- Are there any skills (marketing, financial or production) we can use to gain an advantage?
- Have we researched the market? Is demand set to grow, decline or remain stable?
- Have we spoken to the target company's customers and potential customers? What do they think of the company?
- To what extent do the acquired company's profits and revenue depend on key employees, key customers or ephemeral products?
- To what extent is the company's profit due to extraordinary items, such as revaluation of assets or sale of premises?
- What liabilities might we be acquiring, such as liability for cleaning up contaminated land?
- Are property and asset values reasonably stated?
- Are there any parts of the new business which might be sold off, to pay for the acquisition?
- Are we fully aware of the HR issues: TUPE, redundancy payments, potential demotivation of staff and possibly employment tribunals?

How to Make it Work, Post-Acquisition

Post-merger integration (PMI) is where most acquisitions fail. A survey by *Business Week* among experienced US M&A managers found that nearly 40 per cent of them blamed the PMI process for the transaction's failure. Only 27 per cent faulted the price of the acquisition and only 15 per cent cited the particular company that they selected.

In other words, it's the post-acquisition process that leads to failure. In the following section, we review seven ways to make succeed after the merger:

SEVEN WAYS TO SUCCEED AFTER THE ACQUISITION

1. Get clarity on your aim.

   Is the acquisition more about growth or cost reduction? The latter is more painful, and will lead to fear and resentment. This calls for speedy action.

2. Have a detailed plan.

   Signing a contract is only the start. Yet management often thinks their job is done at that point. The acquirer must create detailed action plans for integration; and these need to be written before the local management has time to invent ways to obfuscate and delay them.

3. Get the employees together.

   Employees should be helped to meet at an early stage and to discover points of shared interest. This might involve workshops, town-hall meetings or outdoor bonding sessions.

   Cheerleaders should be appointed in local offices to communicate a vision of a larger, more successful company with access to more skills and wider markets, and with more secure jobs – albeit with some pain in the interim.

4. Understand the critical parts that need to be integrated or removed.

   By the time the ink dries, you should know which parts of the new business need to be integrated. If you can identify the elements that will continue to stand alone, and are not relevant to cost cutting, that will simplify the job to be undertaken.

5. Understand cultural differences.

   If the acquired company is more relaxed, independent, younger or more dynamic than your own, this needs to be taken into account. An increase in apparent bureaucracy and restrictions on previously acceptable practices can lead to the best staff leaving. Sometimes, the acquired company can teach the purchaser more entrepreneurial ways of working.

6. Keep the culture.

   If you buy a business for its innovation, you will want to preserve that spirit. Divya Gugnani, CEO of Send the Trend, which was bought by the home shopping channel QVC, says, 'Our office is the exact same office we had before we got acquired. Same Ikea desks, same team lunchroom, same birthday traditions.'

7. Manage the customers.

   Customers often find that their sales person has left, or that no one can make decisions any more, or that the culture is not as friendly. This can result in customers moving elsewhere, not least because the competitors will be waiting to pick up disgruntled clients. It's vital to pay extra attention to major customers, to understand their drivers, and to make them feel special and wanted.

Joint Ventures

Diversification doesn't have to mean huge investment and big risks. Joint venture (JV) is one way to gain access to new markets or win new projects without risking too much. A Western company that wants to break into China may find it convenient to do a JV with a Chinese company that will have local contacts and an insider's knowledge of local practices.

Where there is no need for local knowledge, a JV is ideal for projects that are too big for either party on their own and lets them share capital and liability

A third type of JV is where partners have different skills. The Virgin Group has been able to expand rapidly into new markets by doing a JV with a major business in a target industry. Virgin lends its brand name and consumer nous, while the partner usually invests capital and operational skills.

There are many risks in entering a JV. Usually there are few problems with successful JVs. But it is when JVs don't go well and relationships sour that CEOs discover more time should have been spent in the assessment phase, and on the terms of the contract.

Both partners need to have compatible goals and visions and operating standards. If one partner finds the other has lax performance standards or unethical practices, the JV can unravel. It is vital therefore to get to know the other business, not just its financial strength but also its outlook and work style.

If you are thinking of doing a JV you need the active engagement of a specialist lawyer. Major areas to be ironed out are:

- investment
- risks
- profit sharing, including its timing
- participation
- control
- taxes
- liability
- conflict resolution
- internal pricing
- transparency of data; auditing.

The new company should have a well-defined operating agreement. This will include strategies for good communication, and the clear assignment of responsibilities.

Selling or Closing Part of the Business

Companies need to look at the businesses they own with a critical eye, and ask whether they should keep them. Unsuitable businesses can drain the whole company of resources; and if their problems grow they can put the whole company at risk.

Selling a division also puts cash into the business, either to reduce debt or allow for other acquisitions.

If you can see problems on the horizon before they become serious, and can sell the business while it's in good shape, that is a clear advantage. It's hard trying to sell a weak business in a declining market.

The checklist in Table 21.1 shows the kinds of businesses which should be sold or kept.

Selling part of the business helps the firm to get working capital and reduce interest payments. It is one of the first things a turnaround expert or 'company doctor' will do when trying to help an ailing company survive. It also solves the problem of a loss-making division.

The company can sell a business unit to its management (as a management buy-out), to a competitor or to a new entrant to the market.

Losses occurring in any part of the business should be stopped before they drag down the profitable parts of the firm. One firm made good profits from making toilet cubicles, but lost money when installing them for clients. As a result it went bust, despite having a full order book.

Closure is logical when no buyer can be found for the loss-making business. There will be one-off redundancy costs, but at least the losses will be stopped.

Table 21.1   Divestment strategy

Divisions to keep	Divisions to sell
Occupies little of management's time	Occupies a lot of management's time
Geographically close to the rest of the business	Geographically distant from the rest of the business
Has synergy with the rest of the business	Lacks synergy with the rest of the business
High barriers to entry	Low barriers to entry
Is part of the core business	Is not part of the core business
Is profitable	Contributes little profit, or makes a loss
Uses new technology	Uses old technology
Its market is growing	Its market is static or declining
Is a brand leader	Is No. 2 or 3 in the market.
Good industrial relations	Poor industrial relations
Competent management	Weak management

ABANDONING MARGINAL PRODUCTS

Companies can often improve profitability by shrinking the business. This involves abandoning marginal products and consolidating their operations to fewer sites.

It is worth checking how many people are employed on different product lines. Sometimes unprofitable products employ many people. Abandoning such products can have a substantial effect on profitability.

As with the unprofitable businesses above, the loss-making products are often those associated with the company founder. Sometimes they were the company's original product. That makes it harder to abandon them.

CASE STUDY: CLOSURE IS RARELY EASY

Management usually hesitates and prevaricates over closing part of the business. Siren voices will tell you that sales will increase, or that a new strategy will turn things around. And most directors or company owners are only too aware of the impact that closure will have on loyal employees.

You may feel protective towards the ailing part of the business. It is often a part that is close to your heart. Closure implies admitting defeat and accepting failure.

As a result, management can delay closure by 12 months while it agonizes. And as each month goes by, the losses drain the good parts of the business, and put the whole edifice at risk.

It is important to respond fast to developing losses. Optimism is dangerous under these circumstances. Too many businesses fail to take action. They hope that next month will bring better results.

The author delayed closing a business for nine months. Eventually the situation was resolved by setting clear financial targets and deadlines, namely that the business had to achieve breakeven averaged over the following three consecutive months. This put a halt to the debates about sales tactics among senior managers, and provided much needed clarity and simplicity.

And when the author finally put the company into voluntary receivership, he felt a sense of relief. You hand responsibility over to an insolvency practitioner, and are no longer allowed to make any decisions about it, or even speak for it. And the main company bank account begins to fill again with revenue from the good businesses, no longer burdened with outflows to the bad one.

At that point, you realize you should have taken that step a long time ago. The numbers that were wrong a year ago rarely get better.

Risk Assessment

By answering the questions below, you can assess your vulnerability to acquisition and divestment problems. Tick all relevant boxes and add up the number of points ticked.

Issue	✓
Has the business made an acquisition in the last two years?	
Has the company a poor record of post-merger integration?	
Are you currently engaged in seeking to buy companies?	
Does the success of the acquisition depend on revenue growth?	
Has risk mitigation been omitted from the contract?	
Have you failed to plan for alternative scenarios, post-acquisition?	
Have you omitted to create a detailed post-acquisition plan?	
Is the target company substantially different in culture?	
Does the target company operate in a substantially different market?	
Have you failed to close down or sell weak divisions?	
Total points scored:	

Score: 0–3 points: low risk. 4–6 points: moderate risk. 7–10 points: high risk.

The Appendix has a summary of the checklists. By entering the results of this one, you can compare the scale of acquisition and divestment risk against other categories of risk.

22

Legal Risks, Intellectual Property and Minimizing the Chances of Getting Sued

The work is, frankly, shoddy. And the contractor has stopped taking your calls. Tim, your lawyer, says you can get another company in to rectify it. 'And then what?' you ask. 'Presumably you kept back some of the money?' asks Tim.

'It was stage payments,' you reply, 'and so we've paid all but 10 per cent. And the cost of remediation could be half the original bill.'

'And was no one watching the work?' asks Tim. 'Well we aren't specialists in suspended ceilings and lighting circuits,' you say, crossly.

'I guess we've got to put this down to experience,' says Tim. 'Come and have a chat next time you want major works done.' You leave his office, pondering where to put the departmental staff while the work is started all over again. What a shambles.

Are You at Risk of Being Sued?

A survey by the Association of British Insurers showed a fall in the number of employers' liability claims in the 2000s; and similar results were seen in the US and France.

However, for some more high-profile organizations the risks of large-scale liabilities lawsuits are growing. This includes action by regulators and action groups, who have previously been more passive. And with a reduction in deference, more people are likely to take to the courts just because they can.

High-profile cases apart, large employers are likely to have a constant stream of low-value but time-intensive disputes with individual employees.

But the fear of a major lawsuit is greater than the reality, and more businesses will be closed down by mundane risks such as fire or lack of sales, rather than litigation.

So on the one hand the risks of litigation are often over-emphasized. On the other hand, we must be aware of major threats, and manage them accordingly.

TRENDS IN LITIGATION

Third-party litigation funding and 'no-win no-fee' lawyers have increased the likelihood of being sued,

Accountancy practice Moore Stephens was sued for £90 million for alleged negligence as an auditor. The case was backed by a litigation funding firm; and although the suit was unsuccessful, it could have bankrupted the practice.

These third-party funded cases are likely to increase. They will be mostly directed against large-scale commercial firms because such cases will generate the necessary returns. It is possible that smaller-scale third-party funding will also develop.

Class actions may also grow. Shareholders successfully sued Shell for $450 million for allegedly misrepresenting its proven oil and gas reserves. Companies need to ensure that their insurance covers likely class action.

Another development is Forum shopping – with the claimant choosing a jurisdiction in which they have a higher chance of success or a bigger payout. This is also growing. Some US states generally award higher payouts than others. This means that litigation can occur in unexpected jurisdictions, especially if the defendant operates in many countries.

SOURCES OF LITIGATION

It is wrong to assume that litigation will come from only one source. Companies can get sued for specific reasons, by specific stakeholders. They tend to conform to specific areas, as shown in Table 22.1.

Litigation is often the outcome of poor management in a specific area of the business. Let's look at who might sue your organization:

1. *Employees* may sue for wrongful dismissal, harassment or failure to be promoted. They might also suffer from health and safety problems. We examine these risks in Chapter 18 on HR risk.
2. *The consumer* could sue for product or service failings, or professional failure. These risks are reviewed in Chapter 5 on operational risk.
3. *Business customers* might claim for failure to meet the requirements of a contract. As with point 2 above, we consider these risks in Chapter 5.
4. *Suppliers* are usually on the receiving end of litigation. It is usually when they fail to deliver on time or provide a poor-quality service or product. So you as a client would be litigating against them. Whereas if you fail to pay a supplier, due to cash flow problems, you would be the object of litigation. We discuss the latter in Chapter 14 on financial risk.
5. *Pressure groups*: You can be attacked for failing to adhere to your CSR, for example if your suppliers in the less-developed world treat their employees badly. We examine these kinds of risk in chapters 7, 8 and 19.
6. *Regulatory authorities*: Failure to comply typically produces a statutory fine rather than litigation, but the impact (costs and loss of reputation) is similar. Some risks relate to health and safety incidents (examined in Chapter 6. Or they may be concerned with some aspect of service delivery, such as restaurant hygiene, the kind of risk covered in Chapter 5). Alternatively they may relate to corporate governance, which we discuss in Chapter 17.

Leaving those risks to the chapters mentioned above, we're left with only one main stakeholder, namely:

7. *Competitors* might litigate for defamation (a marketing issue), unfair practice (an ethical issue), or for infringing their IP, such as a trademark. We examine the issue of IP below.

Table 22.1   Source of legal risk by stakeholder and area

	Employees	Consumers	Business customers	Competitors	Pressure groups	Regulator
Product or service		●	●			
Health and Safety	●					
Fire	●					
Pollution					●	●
Security	●					
Ethical failure			●		●	●
Marketing claims and statements				●	●	●
IP				●		
Continuity		●	●			●

Legal Risk Assessment

If you have undertaken a risk assessment as suggested in Chapter 2, you will have identified the main risks to your organization.

You will then have taken steps to control those risks, as discussed in Chapter 4.

As we saw above, the succeeding chapters have covered almost all the risks you are likely to encounter, and legal risk mostly stems from failures in specific departments (for example employees behaving badly, as covered in the Chapter 18 on people risk).

It is nevertheless worth having a legal expert conducting a legal risk assessment. This involves examining the organization's processes, and identifying the severity and probability of legal risk.

The individual who carries out the legal risk assessment should be equipped with the risk assessments carried out in other parts of the business, to see whether they are comprehensive and sound.

The legal risk assessment should examine applicable legislation, to see whether the organization is compliant. Some areas are particularly fraught, such as:

* governance;
* filing of legal information;
* privacy;
* disclosure;
* packaging and waste disposal;
* emissions to air and water;
* employment issues such as equality, diversity and right to work;
* health and safety;
* whistleblowing;
* ethical issues such as price fixing;

- contracts with suppliers, landlords, agencies and other business partners;
- data protection and information security;
- IP.

Legal risk comes to the fore in:

- acquisition;
- JVs;
- purchase of major assets such as land and buildings;
- doing business abroad.

It is important to see whether employees understand their legal obligations. In some organizations staff are unaware of corporate policies.

If the assessment encounters any area of legal risk, you can put measures in place to reduce their likelihood and severity.

INTELLECTUAL PROPERTY (IP)

As we've said before, we live in an age of weightless companies, whose manufacturing is often outsourced and the important assets are brand names and the company's knowledge. An early example is Nike, whose products are made by sub-contractors and whose strengths lie in design, branding and marketing.

The ownership of IP, such as trademarks and patents is important for three reasons:

1. IP can be used to gain a competitive advantage and to keep out competitors.
2. A company's value often hinges on its IP. Where a company is valued much higher than its asset value, the difference is often accounted for by its IP.
3. Globalization means that an unprotected asset can be copied and sold in other countries, leading to a loss of revenue to the business.

CASE STUDY: BT COULD HAVE GOT A ROYALTY FOR EVERY INTERNET LINK CLICKED

Using IP actively and wisely is essential. Buried among its 15,000 global patents, BT discovered one that it reckoned would allow it to own the hyperlink. Filed in 1977 but not issued until 1989, the patent was designed for use in text-based information services such as Prestel and Viewdata.

Success in the courts would have meant every software and computer company in the world would have to pay rights to BT.

In the end, a US judge threw out the application on the basis that it related to dumb terminals (1970s technology) rather than between computers (the structure of the internet).

All of the areas below require proper assessment and active management.

- *Patent:* gives an inventor ownership of an invention. IBM has made more than 5,000 patent applications a year, as has Samsung (4,500), Microsoft (3,000), Canon (2,500) and Panasonic (2,500). When Apple sued Samsung for infringing its 'pinch to zoom' patent, the US courts slapped a $1 billion fine on Samsung. Owning such rights can keep competitors out of a marketplace.

- *Trademark or service mark*: a symbol or brand owned by the person who created it (for example the famous Coca Cola trademark is owned by the Coca Cola company).
- *Internet domains*: not just part of the organization's branding, but for many an integral part of its marketing and selling and distribution operation.
- *Geographical indication*: products whose production is limited to a specific place. Roquefort cheese is protected, but cheddar is not.
- *Copyright*: the exclusive rights to written, musical, photographic and other artistic material. Films, records and books are protected by copyright.
- *Trade secrets*: these can be protected through employment contracts and confidentiality or NDAs, as well as injunctions and litigation over their breach.

CASE STUDY: FRENCH WINE COMPANY LOSES A CASE TO A COPYCAT CHINESE BUSINESS

The French wine company Castel had to pay a Chinese business $5 million over trademark infringement. Castel was ordered to pay the Shanghai-based Shanghai Banti Wine the money after a court upheld a ruling which deemed it had unlawfully used the trademark 'Ka Si Te'. Castel had been using the words *Ka Si Te*, a Chinese transliteration of Castel, for its marketing in China. But the French company failed to register it as a trademark. This allowed the Chinese wine company, which had spotted its popularity, to register the mark for itself.

On discovering it was being used by the Chinese company, Castel attempted to get it revoked. But the Chinese business sued for unlawful use and, after a ruling which deemed the Shanghai business the rightful owner, Castel was ordered to pay the fine.

The case study shows the importance of registering trademarks.

Patent trolls

The term 'patent troll' describes an organization that sets out to get money from organizations which, the troll says, are infringing a patent.

The trolls often buy up patents from bankrupt IT businesses, and then seek out companies that are vulnerable to the threat of injunction. Many companies pay up in order to settle early and ward off litigation. Patent trolls cost organizations $29 billion a year in the US alone. Defences include the following:

Preventative measures:

- *Monitor new patents.* You should monitor new patents for potential infringement.
- *Conduct a clearance search.* Before developing a product, you should check for patents or pending applications that involve important features of the product.
- *Buy insurance.* You can buy insurance against infringing someone else's patents.
- *Actively buy patents.* Rather than let trolls buy up the patent, you could buy them for the business, thus pre-empting them.

After a patent troll contacts you:

- *Change the design.* You can create a different design and cease using the patent.
- *Start opposition proceedings.* You can oppose the trolls' patent, and have it re-examined.
- *Litigate.* You can defend your position by demonstrating the product was in operation prior to the patent. Or you can challenge whether you are actually using the technology in such as way as to infringe the patent.

- *Settle early.* You might choose to settle early, as a way of reducing costs. This is what trolls rely on.

INDUSTRIES FOR WHOM INTELLECTUAL PROPERTY (IP) MAY NOT SEEM A PRIORITY

In some industries, such as retailing, IP appears to be less relevant. Retailers exist to split bulk materials into smaller quantities suitable for end users. And so their strengths and risks lie in achieving the right stock, store size and store location. Their core skills also include logistics – getting the product on to the shelves in a timely manner – managing stocks and low prices.

But retailers who are merely efficient lack a competitive advantage. And that means the consumer has little to distinguish them from competitors' stores. The answer lies in developing their own brands. Since these are available only in their stores, own brands create additional store loyalty.

In addition the brand identity can include your logo, livery, advertising catchphrase, uniforms, staff behaviour, special services or methods of delivering the service.

Other service industries, such as banks, insurance companies and solicitors, might also claim that IP isn't important. They often see client servicing and location as their critical success factors.

But there are important areas of IP that service companies sometimes overlook. For example, software that might be developed through a software firm, which might in turn start selling it to competitors.

To give themselves a competitive edge, service companies develop exclusive products and services, and business processes. If competitors copy and use them, you lose that advantage. Therefore you need to investigate the extent to which you can protect such processes and services from use by others.

And there are financial issues at stake. After the city of Munich decided to convert its 14,000 computers from Windows to Linux, its hand was stayed over patent issues relating to Linux and open source software. It feared that companies holding software patents could issue a 'cease and desist' order to Munich's city hall, which would shut down the city's computer systems or force it to pay licence fees.

However, litigation over IP rights is relatively rare: only 5 per cent of UK firms have ever been involved in a legal dispute involving IP rights.

The Words Complained Of

It's an interactive world, with customers being more engaged with a company than before. In addition, social media means there are more 'touch points', and employees have more freedom to talk than they used to have. This means more risk.

This not only relates to statements and behaviour by employees, but also agents, brokers or other business partners whose day to day activities are hard to patrol. Loose talk by them can cause difficulty for the organization.

DEFAMATION, LIBEL AND SLANDER

If you have an active social media presence, the person who tweets or writes posts might say something negative about one of your competitors, who might then sue. The same is true for customer service personnel and those who reply to pre-sales enquiries. Any defamation spoken on the phone or in person would be slander, while written materials are libel.

Employees need to be trained to know what can and cannot be said. This is particularly true about making comparisons with competitors. You might consider having a list of claims and comparisons that staff can or cannot make.

ADVERTISING CLAIMS

False or misleading statements about your product or service will bring you to the attention of a regulator or competitors. Therefore marketing claims should be checked before going live.

Opinions such as 'this is the best telephone on the market' are subjective, and therefore less likely to be attacked; but anything specific such as claims for miles per gallon, battery life or prices must be accurate.

Because review sites are influential, posting fake reviews (shilling) is attractive but wrong; and it can lead the business into trouble. Similarly, defaming a competitor on a review site is wrong and risky.

Contract Risk

Contracts are either simple or complex. A simple contract is for buying a product or a well-established service. You pay money and you get a laptop; or you sign a contract to lease a vehicle. Even an employment contract is relatively straightforward because it is hemmed in by legislation and established practice.

Complex contracts are risky where they have many variables or unknowns. Complex contracts create something new, require decisions to be made, and the methodology or exact outcome is often unknown. Examples could include a supplier that provides a service over a long period of time, the construction or repair of a building, or the creation of software.

This is where contracts can go wrong. If you don't have a clear view of the method of work or the outcomes you expect, you won't know what is likely to go wrong; and therefore you can't build safeguards into the written contract.

Complex contracts particularly affect government organizations that get third-party businesses to deliver services. They also relate to any business that outsources work or buys products from a supplier. Every contract has a lifecycle, whereby:

1. the contract is drawn up;
2. tenders are put out;
3. a winner is chosen;
4. the contractor provides the work;
5. eventually the contract comes to an end.

Thus there are five stages at which risk occurs.

Needless to say, both sides are at risk in a contract. The buyer may get poor value for money, while the contractor risks making a loss.

Most contracts involve uncertainty. The costs may be uncertain, and the difficulties that lie ahead may be unknown. A contract aims to manage risk, but this depends on the relative strength of each party. Where there is one purchaser and many suppliers, the purchaser should have the upper hand. But in many cases the government or business finds there are few organizations with the skills, experience or capital to undertake big contracts.

As we discussed in Chapter 16 on Projects and Chapter 7 on Procurement, partnership sourcing is a way of reducing risk by treating a contractor or supplier as a member of your team. Next we look in brief at each of the five stages of contract risk.

The Five Stages of Contract risk

1. CONTRACT DESIGN

You only discover whether your contract is well designed when you fall out with the other side (Stage 5, below). At that point you ask your lawyer whether you can get out of the contract, or whether you can require the other side to do something.

Failing to draw up a watertight or effective contract leads to problems such as:

1. You get poor value for money, or gain insufficient revenue (a government agency might sell the rights to run trains too cheaply, or pay too much to have its roads swept).
2. You don't include sufficient scope (a software contract might make insufficient allowance for changes).
3. You fail to control IP (so that the IP becomes owned by a contractor who then makes money from it). The author never saw a more horrified look than on the face of a company owner being told that the cartoon characters which symbolized his business were owned by the design agency that had invented them.

You can minimize risk at the design stage by the following:

- have clear procurement and tendering policies and procedures in place;
- define accurately what service you need. Ask yourself the following questions:
  - Who is to do what?
  - When should it happen? In what timeframe?
  - How should it happen?
  - To what standard should it be performed?
  - What is to happen if the other side doesn't deliver?
  - How long should the arrangement last?
  - What happens when it finishes?
  - Is it to be renewed?
- This information needs to be provided by experts. The lawyers can only help you identify the generic risks inherent in a contract. The operations people must have a clear idea of what needs to be done.
- Identify what problems you might encounter when the work is being carried out. How will any necessary changes be made? Good project management (Chapter 16) encompasses 'change control' whereby proposed changes are formally reviewed and agreed in writing.
- Ensure that terms and conditions in the contract limit your risk (though they are not always enforceable).
- Assess the likely risks. Consider different scenarios. What changes might happen in the external environment, such as a recession or the arrival of new technology?
- Do you understand all the words in the contract? If you don't, have a lawyer explain them. Are the words clear? If they are ambiguous, a judge may agree with the other side's opinion as to what they mean. The author once amended a contract in such a way that the payment of

£20,000 was referred to twice. The other side then claimed I owed £40,000. My lawyers said I would probably lose the case if it went to court, so I had to pay.

- Ensure that a dispute goes to arbitration rather than to court, since the latter is usually expensive and slow.
- Beware of urgency compromising thoroughness (the ETTO principle discussed in Chapter 6 on Health and Safety). The phrase 'marry in haste, repent at leisure' is applicable here. If you want to get the contract signed quickly, you may discover errors and omissions later on. In my case, £20,000 worth of payments plus lawyers' fees.
- Consider agile contracting, whereby you phase the work, and fund each successive milestone. The agreement will indicate that you intend to agree, and re-negotiate if necessary, each phase.

2. AND 3. TENDERING AND APPOINTMENT

Contracting with an unsuitable supplier (one whose service is poor or who goes bust), leads to a loss of reputation, both inside the business and – if you're in unlucky – in the media.

You should check that the other side is what they claim to be. Some organizations are merely someone working from their back bedroom, without assets or capability. Check they are financially sound and legally constituted. Do they have the power to do what they say they will do? Will you be damaged by being involved with them?

Beware also of collaboration among suppliers. We examine this in Chapter 7 on Procurement.

When it comes to pricing a job, a 'cost-plus' contract is the safest bet for the contractor. A fixed price involves most risk. For the organization that needs the service, the reverse is true.

4. IMPLEMENTATION

When the service is underway, the supplier might perform badly. Equally, some contracts fail because the client fails to manage it properly. You minimize risks during the implementation phase by the following:

- actively manage the contract, to ensure the contractor does the work properly and avoid any reduction in service quality;
- having good quality information that enables you to see when the contract is not performing to the required standard.

Meanwhile the contractor can reduce its risks by sharing them. The contractor might reduce the capital cost, or get specialist work done by sub-contractors.

5. TERMINATION

If the contract reaches its allotted end without problem, all is well and good. But what if you need to terminate the contract early whether because needs have changed, or because the delivery is poor? As we've seen, it is now you'll discover whether the design of the contract document was robust.

Whoever writes the contract tilts it in their favour. So if you wrote it, there is usually no excuse for a wording that doesn't help you resolve the issue in the event of a problem.

If you are employing a contractor, and there is no get-out clause, no specifics about the service to be delivered, and no penalties for poor performance, it will probably cost you to exit from the contract.

In commercial tenancy leases, the need for early termination is a big issue for the tenant. In these times of change it is hard to forecast what your location, size and business model should be several years hence. In retailing, a successful business model can quickly deteriorate as new trends develop.

Exiting early from a building almost always involves financial penalties, especially in a weak property market where the landlord has no incentive to negotiate a good deal with the tenant. Break clauses are usually the things that facilitate an early escape, though even these rarely seem to occur at the right moment if you want to leave. Giving adequate notice to the landlord usually adds a further period of time. Simply walking away will only lead to legal action, unless you intend to declare insolvency.

RISKS IN THE WORDING

Good lawyers know from experience that even innocent sounding words in a contract can have major repercussions, which is why you should submit any contract to a lawyer's inspection before signing it. The other person can seem amiable, but they have their own interests at heart, not yours. Here are some of the pitfalls to be aware of:

Obligation: some words commit you do doing something, such as 'shall', 'will', 'is to', and 'must'. This is known as 'deterministic language'. Words like 'may' or 'might' give you a get-out. Requiring the other side to do things is, naturally, better.

Partnership: if you agree to a partnership, this has a specific meaning with heavy consequences. You will become liable for their inadequacies. It is best to explicitly state that no partnership is created.

Expression of intent: if you say you will give the other side a contract when the details are agreed, you lose your negotiating power. This creates a contract in itself, and the other side no longer needs to accept obligations to perform their work properly.

Too wide a scope: you should define what you will do as narrowly as possible. Otherwise, you may be agreeing to deliver more than you can deliver or had imagined or had included in your costs. The words 'none', 'every', 'all' and 'complete' do not allow you to deviate in any way.

Rendering the other side harmless: do not accept a clause that says that you will indemnify the other side or hold them harmless. It gives them freedom to behave badly, and could make you liable to pay damages if they affect a third party in some way.

Warranties: words like 'warrant', 'guarantee' and 'insure' can lumber you with problems if things don't go according to plan, even if it isn't your fault.

Vague words: 'sufficient, 'suitable' and periodic' allow the other side to disagree as to the meaning of those words.

Obligation to pay within a timeframe: avoid giving the other side the right to demand payment within 30 days. If you are the contractor, the reverse is true.

Product Liability Litigation

The first big product liability claims were against asbestos. Staff who contracted asbestosis initially got small sums from workers' compensation schemes. Then in 1969, Thelma Borel, the widow of a dead asbestos worker, won $79,000 in a court action against Fibreboard Paper Products, an asbestos firm.

The award triggered claims from thousands of other sick asbestos workers in the US. It meant that other asbestos firms were liable for illness caused by their products. It also meant that people could sue firms directly.

The courts were also appalled by the asbestos companies' duplicity in hiding medical evidence. It led to larger and punitive damages, with James Cavett, a retired boiler, maker receiving $2.3 million in 1982.

Apart from asbestos, there have also been big awards against manufacturers of silicone breast implants, against accountants, against pharmaceutical companies, against tobacco firms and against firms which caused pollution incidents.

Courts and governments seem increasingly keen to see that firms are made to pay heavily for their dangerous failings. The Dalkon Shield case saw A.H. Robins Co. set up a $2.3 billion trust fund in 1989 to pay compensation to the victims of its faulty intrauterine contraceptive devices. They only finished processing claims in late 1999. Even cases such as that pale in comparison to the $255 billion being paid by tobacco companies to the US Government in compensation for smoking-related health costs.

Product liability occurs in unexpected ways, and in the US it can be expensive. A court in Albuquerque, New Mexico ordered McDonald's to pay $2.7 million in damages (later reduced to $480,000) because its coffee was 'too hot'. Stella Liebeck, an 81-year-old grandmother, had tried to prise the lid off her coffee cup by holding the polystyrene cup between her legs. The coffee spilt over her legs and she was scalded.

The jury said the punitive damages were intended to be a 'serious message' to the fast food industry. Two of the jurors had even wanted to hand out even higher compensation. The company said that its coffee was 'substantially cooler than the coffee you would make at home'.

The upshot is to avoid products and services that are likely to end up in court, to fully test new products before launching them, and issue appropriate warnings. After that, you have no more chance of being hit with a headline-grabbing case than any other business.

Sometimes, companies launch or continue to sell products when they might have been expected to have known about problems they caused; and this can give rise to legal risks. The weight loss drug Mediator continued to be sold for years in France after it was withdrawn in Spain and Italy, and never having been licensed in the UK or the US. The drug is believed to have killed hundreds of people from heart valve damage. Louis Servier, the company's chief, was placed under investigation for manslaughter.

CASE STUDY: DISABLED PEOPLE AND LEGISLATION

Bauermeister, a coffee bean roaster, was fined $100,000 in the US after refusing to let an employee who had suffered a manic turn back to work.

In a case brought by the US Equal Opportunity Commission (EEOC, eeoc.gov) the company discharged the employee after learning of his bipolar diagnosis.

The UK's Disability Discrimination Act requires businesses to make 'reasonable adjustments' to their premises, to make them accessible to disabled people. The fines for non-compliance are up to £50,000. Such legislation is also appearing in other countries. In the US, businesses have been fined $1,000 (£560) a day for being in breach of the regulations.

With disabled people ready to assert their rights, businesses need to ensure that their premises, in particular retail outlets, are accessible. Some smaller businesses are reluctant to invest, believing that some of the costs, such as the installation of lifts, are excessive. But some experts say the costs are overstated and that businesses are dragging their feet. The National Register of Access Consultants (nrac.org.uk) can provide advice.

The Legal Risks of Email

In the US, one in five employers (21 per cent) has had employee email and instant messages (IM) subpoenaed in the course of a lawsuit or regulatory investigation. Another 13 per cent have faced workplace lawsuits triggered by employee email. This indicates the risks of email. Legal risks include the following:

> *Legal liability for emails sent by employees.* In the UK, Norwich Union was forced to pay £450,000 in an out-of-court settlement, after an employee sent an email falsely stating that their competitor Western Provident Association was in financial difficulties. Sexually offensive emails may also be the source of litigation from offended employees.
>
> *Court-ordered retrieval of emails:* Email records are increasingly used in lawsuits since they tend to contain important evidence. For instance, you may be required to provide all emails about someone alleging they were bullied. According to epolicy.com, a court ordered a US top 500 company to hand over any email that mentioned the name of a former employee who was suing the company for wrongful dismissal. The company faced the prospect of searching more than 20,000 back-up tapes, containing millions of messages, at a cost of $1,000 per tape. The total potential cost for that electronic search was $20 million.

Other risks include:

> *Confidentiality breaches:* Employees may send confidential information about the business either intentionally or unintentionally to other firms or individuals. With the large number of questions being posed by customers in emails, there is a big risk that someone will send out confidential information.
>
> *Lost productivity:* Companies can lose productivity due to employees spending time on personal email and internet use.
>
> *Viruses from downloads and opening attachments:* Ever more sophisticated fraudulent emails and spyware risk infecting the company's network with viruses.
>
> *Network congestion:* Emailing large picture files or mp3 files can slow down the company's computer system.
>
> *Loss of reputation:* In the US, A Federal Communications Commission (FCC) employee inadvertently sent a rude joke entitled 'Nuns in Heaven' to 6,000 journalists and government officials on the agency's email list. This created bad publicity for the FCC. A badly written email, or one containing unprofessional remarks, can also give a bad impression of the company. At UK law firm Norton Rose two of their employees started the 'Claire Swire' email, a sexually explicit email that was read by over ten million people around the world. Since the company in question was a law firm, this email will have damaged the company's reputation.

SET AN EMAIL POLICY

Email-policy.com recommends setting a comprehensive email policy. Some topics you might include are as follows:

> *Safe use of internet and email:* Employees should be told not to open attachments from unknown emails, to be suspicious of all attachments, and not to download files from the net. They should also be told that visiting some popular websites can cause spyware to be installed on to the PC.

Netiquette: Give employees advice on writing emails, in order to maintain the company's reputation and deliver effective customer service. Any email sent from the company could end up on an internet forum, and so discretion should be maintained at all times.

Personal use: Employees should not make excessive personal use of the company's email system or internet access, nor use company computers for their own business activities. You may want to set an internet Acceptable Use Policy (AUP).

Offensive content: The email system should not be used to create or send offensive or derogatory messages. Emails are highly public messages.

Confidential data: Employees should take extra care not to divulge confidential information or trade secrets.

Reporting abuse. Employees should be given an email contact to refer abuse.

Privacy: Employees should be told that their emails may be monitored.

Other steps to take:

Train staff: Many staff work on email customer service desks nowadays. You should train them to answer emails effectively, rather than assume they can do it. Product knowledge is also essential.

Use standard replies: Programs like Answertool or FocalScope allow companies to paste pre-written answers into emails or other documents. This reduces the risk that the wrong information or inconsistent messages will be sent.

Permanently delete emails: Unless the business has a statutory requirement to keep records, emails should be automatically deleted after a fixed period. This will preclude the business from having to search for, and hand over, old emails. On the other hand, some emails need to be kept, and these should be saved.

Use an email disclaimer: Adding a disclaimer to the bottom of an email may protect the business against legal liability, negligent mis-statement, entering into contracts and liability for spreading a virus.

Monitor emails: Email monitoring programs can check for abusive words or people's names in stored emails. It thus warns the company of abuse, though it doesn't stop them being sent.

Monitor internet usage: If you are concerned about staff wasting time on the internet, you can install software that records the websites visited by staff, and the time taken on each site. To avoid invasion of privacy litigation, you must tell employees that their internet viewing habits will be monitored,

Keep confidential information safe: Keep confidential information off any server that is accessible to the average employee. This includes personnel records, company accounts, new product development information and survey data.

REDUCING THE RISK OF LITIGATION

The best solution is not to get sued. How do you prevent people from taking action against you? Here are six steps to take:

1. *Have good contracts*: Prevent litigation by ensuring that your legal contracts are watertight, and are signed. Ensure that employees and customers understand what they are being offered.
2. *Manage human resources professionally*: Manage staff recruitment, disciplinary meetings and terminations by the book. Make sure everything is documented in writing. And ensure everyone is treated equally. Never act hastily, especially when it comes to firing people. And don't ignore bad workforce practices. Never overlook bullying, or a failure to adopt safe working practices – these things can come back to haunt you.

3. *Don't over-sell or over-claim.* Point out problems or deficiencies in writing before the other side gets committed. People sometimes litigate because their expectations were not met.
4. *Keep solid paperwork.* Make sure that sales people keep records and get the signatures.
5. *Train staff.* Make sure everyone knows what the risks are, and how to manage them.
6. *Act ethically.* Treat all stakeholders with respect. Don't use others' patents or designs without their approval. Companies that regard customers as ignorant are putting the business at risk.

How to Minimize the Impact of Litigation?

You can take steps to minimize the effect of future litigation. Two options are as follows:

Reduce your physical assets by sub-contracting production or offshoring it. This minimizes the company's exposure to predatory litigation.

Protect your assets – by separating ownership of different assets, you can ensure that individuals (notably directors) and assets are protected from being seized in the event of the business being successfully sued. This can include operating from within limited liability status, to having separate ownership of different assets.

Countering Litigation

Once you receive a solicitor's letter, what steps can you take?

- *Avoid going to court if possible.* Court cases are expensive and time consuming, and their outcome is uncertain. There is no certainty that the innocent party will prevail: a court case is a piece of theatre, with juries being swayed by the rhetoric of barristers, and the lies and half-truths of witnesses. In short, try to settle out of court.
- *Get the CEO to talk directly with the litigants,* if possible. This can result in the parties coming to an amicable arrangement. The CEO is the one person in the business with the authority and flexibility to do a deal.
- *Leave your ego behind.* If the litigation becomes a battle of wills, the result could cost you dearly, and may not be in the best interests of the business.
- *Decide how central the litigation is to your business.* If the issue doesn't matter, it may be that you can afford to yield. You may have more important battles to fight. On the other hand, if the litigation goes to the heart of what you do, you will have to face it.
- *Employ legal advisors whose advice you trust.* But take charge of the process. Don't assume that experts know everything. Remember that lawyers make lots more money when cases go to court.
- *Weigh up the strengths of the litigant.* Never fight someone who is bigger or more powerful than you. There are exceptions: Richard Branson has succeeded several times against the much bigger British Airways. But small guys often lose to bigger ones.
- *Some litigants' anger can be assuaged by apologies.* When people go to court, it is sometimes because they have been goaded by the prevarication and elusiveness of the organization. It can be better to make a personal visit, and spend time listening to the complainant. Small amounts of time spent early in the case can prevent years of slow moving and costly litigation.
- *Use an arbitration service.* This will usually be cheaper and less confrontational than going to court.

- *Learn from the experience.* What caused the litigation? Was it a lack of internal control, a loss of ethics or a slippery floor? Whatever the cause, ensure that it cannot happen again; and check whether you could get sued in related areas. For example, a weakness in health and safety could indicate a weakness in environmental issues.

Risk Assessment

By answering the questions below, you can check to see how vulnerable your business is to liability, litigation, and IP problems.

Topic	Question	✓
Risk management	Have the organization failed to conduct a legal risk assessment?	
Litigation	Has the organization been sued in the last 12 months?	
Customers	Is the business at risk of litigation from customers through product or service weakness or defamation?	
	Do you market controversial goods or services?	
Pressure groups	Is there a possibility of litigation by pressure groups?	
Employees	Is it likely that the organization could be sued by employees for wrongful dismissal, unsafe working practices or bullying?	
	Has the company failed to implement best practice for disabled employees and customers?	
Regulation and governance	If public, has the organization failed to implement the relevant governance standards?	
	Is the business at risk of failing to meet regulatory requirements?	
IP	Has the organization failed to actively assess and manage its IP?	
Total points scored:		

Score: 0–3 points: low risk. 4–6 points: moderate risk. 7–10 points: high risk.

The Appendix has a summary of the checklists. By entering the results of this one, you can compare the risk of litigation and IP issues against other categories of risk.

23
Business Continuity: For When Things Go Badly Wrong

You wake up to find a major gale blowing. Many roads are impassable, but with difficulty you make it to work. There you find the phone lines are down, so your telephone and internet connections are both out.

A large tree has fallen on to the spray booths, so your production is stopped. The production director says it'll take some weeks, possibly a month, to repair the damage and install new robots. Until they're back in operation, all production is halted.

'Got any ideas, Charlie?' you say. The production director shakes his head.

When Things Go Badly Wrong

Despite precautions, things can go wrong. A disaster may strike at any time. So the organization needs a contingency plan for each of its main risks. More importantly, it must be able to continue operating. Some risks, like fire and flood, bring an organization to a standstill.

Companies can usually forecast the kind of crisis they will suffer, because each industry has its own problems. To take just two examples:

- Retailers risk armed robbery, staff theft and shoplifting – including wrongful arrest.
- Any large employer which is reducing its staff number may suffer worsening industrial relations. This can lead to an industrial dispute or strike.

However, organizations also suffer from 'Black Swan' events. These are unforeseen incidents which no one had imagined. The volcanic ash cloud that stopped airplanes flying in Europe came as a complete surprise to everyone. It halted supplies of microchips, flowers and mail.

The Biggest Threats

The loss of IT is the single biggest perceived threat to UK organizations, according to a survey by the Chartered Management Institute (managers.org.uk), as shown in Table 23.1.

Table 23.1   **Which disruptions would have a major impact on the organization?**

Loss of IT	67
Loss of access to site	56
Loss of telecommunications	55
Loss of skills	53
Fire	51
Loss of electricity/gas	51
Loss of people	51
Damage to corporate image, brand or reputation	51
Extreme weather, for example, flood or high winds	45
Terrorist damage	43
Negative publicity/coverage	42
Malicious cyber attack	42
Loss of water/sewage	36
Transport disruption	35
Employee health and safety incident	34
Supply chain disruption	34
Customer health/product safety incident	28
Environmental incident	27
Industrial action	27
School/childcare closures	18
Pressure group protest	17

Source: CMI

Sources of Crisis

The survey in the table above reflects what managers perceive as major threats. But the biggest *real* sources of crisis are in white-collar crime, defects, recalls, and mismanagement, according to a survey by the Institute of Crisis Management (ICM, crisisconsultant.com), shown in Table 23.2. This is based on cases reported mainly in the US media.

A UK-based survey (Table 23.3) gives a different picture. When asked, organizations were most tasked by extreme weather, followed by a loss of IT and the loss of key people.

SLOW-BURN VERSUS SUDDEN CATASTROPHE

According to ICM, 61 per cent of all crises are the slow-burn smouldering type, rather than a sudden crisis. They start small, and take days or months before they get out of control. In other words, management knew about the problem long before it hit the media.

Of these, 39 per cent become sudden crises, including industrial accidents, workplace violence, and natural disasters such as storms and floods.

ICM believes that many of the sudden crises were really smouldering ones that were ignored or unrecognized before they blew up. If management had paid attention or acknowledged that a small problem could become a crisis, the problems might have been averted. This includes health and safety problems.

Table 23.2   Crises reported in the media

Crisis Categories (% of total crises each year)	%
White-collar crime	19
Mismanagement	11
Workplace violence	10
Casualty accidents	9
Labour disputes	8
Financial damages	8
Facility damage	8
Class action lawsuits	7
Defects and recalls	5
Consumer activism	5
Discrimination	3
Whistleblowing	3
Sexual harassment	2
Environmental	1
Executive dismissal	1
Hostile takeover	0
Total	100

Source: ICM

THE CAUSES: MANAGEMENT OR THE UNEXPECTED?

ICM also believes that most crises stem from management rather than employees or the environment. Terrorist activities and natural disasters are responsible for few corporate headlines. Corporate scandals, white-collar crime, defects, recalls and other management issues are responsible for half of all crises.

CRISES BY INDUSTRY

The most crisis-prone US industries are shown in Table 23.4. The industries vary year by year.
 Other industries that often suffer crises include accounting and audit services, investment banking, restaurants and food service, and supermarkets.

WHAT DRIVES CONTINUITY MANAGEMENT?

There are many different drivers. And knowing what is pushing you to have a continuity plan in place tells us something about the kind of plan you need.

Customers: The biggest and most immediate impact of a crisis is probably on the organization's customers. In a hospital this would be patients, while in a law firm it would be clients.

Table 23.3   **Disruptions experienced the previous year by UK organizations**

Threats	%
Extreme weather, for example, flood/high winds	49
Loss of IT	39
Loss of people	34
Loss of telecommunications	24
Industrial action	22
School/childcare closures	22
Transport disruption	20
Loss of access to site	20
Loss of key skills	19
Employee health and safety incident	16
Supply chain disruption	15
Loss of electricity/gas	14
Negative publicity/coverage	13
Damage to corporate image/reputation/brand	10
Loss of water/sewerage	8
Pressure group protest	8
Customer health/product safety incident	7
Environmental incident	6
Fire	6
Malicious cyber attack	6
Terrorist damage	2

Source: Chartered Management Institute

Table 23.4   **Top ten most crisis-prone industries (US)**

Most crisis-prone industries (ranked by percentage of database entries)	
Air transport	1
Pharmaceuticals	2
Gas/oil production	3
Banking	4
Securities and commodities dealers	5
Electric utilities	6
Automobiles	7
Telecommunications	8
Shipping	9
Software	10

Source: ICM

Regulation: The UK's Civil Contingencies Act requires local authorities, the health service, police, rail, airports and ports to have an emergency plan in place (see below).

Insurance companies: A business continuity plan will reduce the likelihood of your business needing an insurance payout, so insurance companies are keen on BCM. Many reduce the cost of insurance to businesses that have continuity plans in place, while others regard it as a requirement before agreeing to provide a policy.

Corporate governance: Legislation such as the UK's Corporate Governance Code requires companies listed on a stock exchange to put systems in place to ensure the survival of the business. Failure to provide an emergency plan would put directors in breach of their responsibilities.

Continuity in the public sector: The emergency services attend to catastrophes on a daily basis, and are called out when storms, floods and crashes take place. Therefore continuity planning is even more essential for these organizations.

The UK's Civil Contingencies Act requires Category 1 responders (police, fire, ambulance, health service trusts and others) to plan for emergencies. It also puts legal obligations on Category 2 organizations (mainly transport and utilities firms).

In particular these organizations have to test their emergency planning to make sure they are robust.

Taking Steps to Minimize a Crisis

Having identified, with the help of this book, the most likely area of crisis, you should plan to minimize the impact of such a crisis. Your aim should be to:

- reduce the likelihood of a crisis;
- reduce the effect of the crisis;
- minimize its duration.

The main steps to be taken are listed below:

1. set goals;
2. carry out forecasting and scenario planning;
3. identify the risks;
4. assign roles and responsibilities;
5. develop a written emergency plan for each potential crisis area;
6. Control or mitigate the risks;
7. test the plan: practise emergency procedures.

We examine these in detail in the rest of this chapter.

1. Set Goals

Objectives should ideally be SMART: specific, measurable, attainable, realistic and timed. For example they need to be measurable so you can see if you have succeeded or not, especially when auditing the system.

CASE STUDY: US POSTAL SERVICE BCM OBJECTIVES

- Preparing personnel for potential emergencies by periodically testing and updating all business continuity plans.
- Limiting the number of decisions that must be made following a significant service interruption (those that cause a portion of a facility to be disabled).
- Eliminating the need to develop new procedures during the recovery process.
- Minimizing the recovery time to restore critical applications and core business processes and functions.
- Increasing organizational credibility with customers, business partners and stakeholders by formalizing documentation processes to ensure availability and accuracy of the information for stakeholders.
- Supporting and enhancing compliance with federal directives, standards, and business continuity best practices (for example, NSPD 51/HSPD 20, FCD 1 & 2, FISMA 2002, NIST 800, DRI International (DRII) Professional Practices for Business Continuity Professionals).
- Fostering business relationships with the Postal Service enterprise through better IT organizational understanding of the business.
- Positive marketing of BCM capabilities. (Effective BCM allows the Postal Service to provide high-service levels and thus win business.)

---

2. Introduce Forecasting and Scenario Planning

Once-successful companies fall when they stay with the formula that helped them grow, and don't notice the trends or forecast the changes that will reduce the sales of their products. IBM stuck with mainframe computers for a long time and overlooked the growing importance of desktop computers.

Some management gurus believe that, in this age of uncertainty, forecasting is bound to be wrong, and makes companies over-reliant on faulty assumptions. That is why scenario planning is a better solution. As the name implies, it gets the company to plan for several different future possibilities. It is a 'what if' exercise. It asks questions like:

- What if the price of raw materials doubled?
- What if we lost our biggest two customers?

Scenario planning has traditionally been used to plan markets. For risk management, it can be applied to all potential hazards, and makes the company confront unpleasant possibilities.

Scenario planning helps the company think the unthinkable, and has helped companies overcome threats if they occur. Scenario planning also encourages staff to challenge long-held assumptions, and to act in advance of change. Table 23.5 has several examples.

Some commentators categorize risk by the acronym STEEPLE, standing for Social, Technological, Environmental, Ethical, Political and Economic. While this can be helpful, it doesn't always work for every organization, and sometimes serves to omit important considerations. Sometimes, too, factors don't necessarily fit into seven neat categories; for example if the seaport you depend upon closes, it doesn't matter whether the closure was due to a strike or bad weather, the fact is you can't use the port.

So while the alphabetical listing used here may seem crude, it may encourage you to add more factors of your own.

3. Identify the Risks

You need to identify which parts of the business are critical. These are as follows. You can use the mnemonic TIOPPS, for Technology, Information, Operations, People, Premises and Supplies.

Table 23.5   Planning scenarios

Issue	Examples of event
Competitors	Pacific rim competition introduces an innovative and low-cost alternative to your product.
	The market moves to no-frills, low-cost products. Excess capacity grows in the industry.
Costs	Wages growth or low-cost competitors make the company unprofitable.
Cultural	Buyers' or customers' criteria change.
Customers	Customers start buying a different type of product. For example, consumers switch from frozen to fresh or chilled products.
	Markets splinter into niches of ever-reducing size. The market matures, and sales falter.
Death	Manual workers die from the risks described in the Health and Safety chapter.
	Every year students at universities commit suicide or die during a drunken prank.
Demographic	Mass population movements caused by civil war or hunger.
	Loss of population from the area causes severe skill shortages.
Distribution	New distribution channels emerge, through which new competitors sell their products.
	A leading retailer decides to de-list the company's products.
	Leading retailers decide to offer cheaper, own-label versions of your product.
Economic	Inflation rises substantially.
	A recession occurs.
Energy	Energy prices rise greatly.
Environmental	Lack of water creates food shortages, hunger, regional tensions, perhaps wars.
	Rising sea levels cause a crisis, with cities and coastal areas flooded.
Extortion	An extortionist tampers with the company's products.
Fire	Fire destroys the company's production facilities.
	Fire burns down the head office, together with all records.
Flooding	The organization's main building becomes unusable due to floods.
Health	A new pandemic causes many staff to be off sick and threatens the health of the rest. It hampers supplies and cuts sales.
IT	Data loss due to hard drive failure.
	A virus wipes out the company's data.
	An extortionist launches denial-of-service attack on the company's website, making it inaccessible to customers for several days.
Key worker	A key member of staff dies.
Political	Major depression causes a rise of the extreme right, fascist government comes to power.
	Riots or revolution in a Middle Eastern country that provides raw materials, components or customers.
Pressure groups	Consumer movements attack the business.
Regulation and legal	The government bans an important raw material, or imposes restrictions on a product's use.
Social	Social unrest disrupts the market or halts delivery of components.
Suppliers	Important raw materials become scarce.
Technology	A new material provides better performance. This material cannot be used on the company's existing production plant.
	Investment in high-cost plant is needed to remain competitive.
Transport	Road, rail, airports or sea ports closed.
	Fuel oil unavailable due to strikes.
Weather	Storms can disrupt transport and communications.
	Cold, rain or abnormal weather can affect harvests for agribusiness.
Workforce	A strike halts production for ten weeks.
	Third-party dispute: nurses strike, causing a cancellation of operations and therefore sales.

TECHNOLOGY

This includes computers, telephones, plant and equipment. To minimize the effect of computer loss, you might implement the following:

- use cloud-based services (though you need a plan for the cloud service to fail);
- use computers elsewhere in the business, perhaps at another location;
- use computers provided by a third-party supplier;
- ensure maintenance contracts are in place.

INFORMATION

Most companies of any size depend on a database, especially for processing sales, managing operations and paying staff. When that information becomes corrupted or unavailable, a crisis can ensue.

You can aim to prevent this by ensuring that backups are taken regularly, and are tested properly.

If computers are down, it may be possible to continue operations temporarily with paper versions of computer forms and records, or to get a printout from an offsite backup.

OPERATIONS

You need to know which processes are essential to the business. This includes operations carried out by suppliers, contractors and business partners. You can manage this by the following:

- Have written procedures for mission-critical processes.
- Locate a firm who could perform this function in the event of a crisis.
- Have reciprocal arrangements with a competitor. The author once encountered a firm of printers he'd never used who said they had recently done a job for him. The printer accurately described the nature of the work, involving cardboard sample boxes. It turned out that the original printer who had been given this work had suffered equipment failure. Rather than admit defeat, they had quietly sub-contracted the work to a competitor, and then presented the work as their own. The job was done to a high standard, and if the competitor had not revealed the secret, we would never have known. As it was, we felt it demonstrated that our normal printer was able to deal with problems as they arose.

PEOPLE

Do any members of staff perform unique roles? If they were to leave or die, would mission critical activities stop? You can reduce the risk by:

- training others in doing the work;
- have written procedures for the work;
- plan succession;
- use interim managers;
- use agency staff.

Hierarchical organizations, where people work in 'silos', or where there is inter-departmental wrangling are more at risk because there will be a lack of mutual support and communication will be weaker. This is especially true of BCM where a crisis requires people from different departments to work together as a team.

It's therefore important to try and build good relationships across the organization. Team-building exercises or meetings allied to good communications will help protect the organization if disaster strikes. In a unionized workplace, good relationships with unions will help to ensure that sabotage doesn't happen and carelessness is reduced.

PREMISES

Do you have premises which, if you were denied access to it, would cause the business a big problem? Given the ease of remote working, many admin staff (finance, HR and legal) can work from home or temporary premises, assuming they can get access to corporate data (for example the ERM system). But others need to be at a specific workplace. You can minimize this problem by the following:

- workspace in the same building, perhaps displacing other functions;
- other premises, such as hotels or temporary office rental, or emergency room suppliers;
- working from home.

We discuss various types of crisis centre in the following chapter.

SUPPLIES

If the business can't get its components or raw materials, output comes to a standstill. As we saw in Chapter 5, even services businesses are partly product-based (such as restaurants, which need food, cooking oil and so on, and hospitals, which need dressings, oxygen and so on). The effect of interrupted supplies can be reduced by the following:

- have two or more suppliers for critical goods;
- keep records of alternative suppliers;
- ensure suppliers have risk management and continuity plans in place;
- keep extra supplies of critical goods on hand.

ANALYSING BUSINESS PROCESSES

It is worth mapping the organization's main business processes, as in Figure 23.1. This may reveal some unexpected connections and vulnerabilities.

Each process usually contains many complicated activities. Figure 23.2 is a simplified order process. For the sake of simplicity it omits how stock gets into the warehouse, how stock checks are made, what happens in the case of out of stock items and what happens to damaged stock. Nor does it consider how picking is carried out, and what robots and computers are involved; nor what happens in the case of a breakdown. In other words, you need to have a detailed view of each process, to understand where problems can occur.

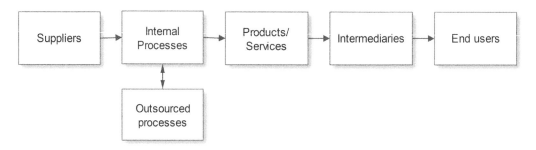

Figure 23.1   Business processes

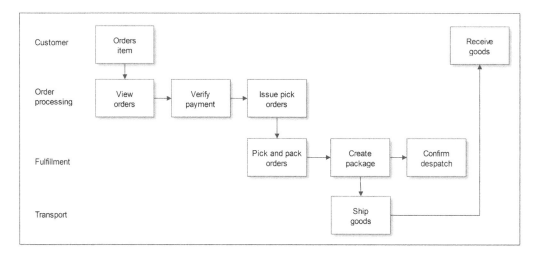

Figure 23.2   Order process

You need to interview each member of staff involved in the process, to verify what steps take place. You also need to know who in the organization is next in the process. Often a process has several branches, as when finance, stores, purchasing or production planning are involved.

SINGLE POINT OF FAILURE

Following a growing disagreement with management, a medium-sized company's book keeper announced she was leaving. At that point the managers realized she was the only person who used the Sage accounts software; and that it might be password protected.

Nor were they sure that she had been taking backups, and therefore if she wiped the data out of spite or locked the system, would the business be able to recover its data? Moreover, there would be no handover because the relationship had broken down badly and the book keeper was in no mood to stay. The managers worked out how to get into the software, and how to take a backup, and the data was thereby protected.

This highlights the fact that finance is one of the areas that the rest of the business is effectively excluded from. So it's worth checking whether people have an isolated role and what would happen if they didn't show up one day. Nor is it sensible to assume that all members of staff will always remain loyal, no matter how mild and honourable they seem to be.

RESOURCES AND ASSETS

Each of the staff involved in the process will use computers, buildings, information and telecommunications. You should log these because some of them will represent a vulnerability.

THIRD PARTIES

The organization never works alone. Each process may involve suppliers, lawyers, estate agents, landlord, temporary staff agents, engineers, web host, telecoms suppliers and outsourced contractors. Each of these may be a critical link.

SEASONALITY

Some crises are seasonal. This can be because bad weather is likely to disrupt transport.

In other cases, the organization may have seasonal activities. Thanksgiving and Christmas or New Year all bring their own special issues. School and universities have specific needs coinciding with the start of the academic year.

HOW LONG HAVE YOU GOT?

Each part of your organization can remain out of action for a different length of time (Table 23.6). Here are a few possible examples:

- lab freezers can only be down for 12–24 hours before tissue samples will be ruined;
- an IT-based ordering system could be run manually, using paper and pencil, for a week before becoming chaotic;
- customers could wait 12 hours to receive conformation about their deliveries, and might not even notice the absence of routine information;
- a purchasing department could order goods by telephone almost indefinitely, albeit with extra people to liaise with production;
- a website that produces 80 per cent of the company's orders cannot be down for more than one day.

Some use the phrase Maximum Acceptable Outage (MAO). Others use the phrase 'Maximum Tolerable Period of Disruption' (MTPD). It is also known as the Recovery Time Objective (RTO), which is the time in which you need a business process to be restored after a disruption.

Each process will have a different RTO, depending on the needs of customers and how long it will take to get the generator working, the spare part to be flown in or an alternative supplier agreed.

Some activities may be categorized as 'fail safe', that is to say, they must not fail. They include a hospital's emergency room power or a bank's processing of financial transactions. This can be achieved by having duplicate or redundant systems.

Other activities – non fail-safe ones – may be classified according to the MAO. You will measure this in hours or days. The plan must allow the activity to resume within the RTO. In some activities there will be a ramping up of activity from a limited start. For example you might achieve 25 per cent of normal client contacts by three days time and 100 per cent output within one week.

Table 23.6   Outage timescales

Outrage timescale	Effect	Impact
Half a day	Only company staff know there is a problem	Imperceptible
1 day	Comments on Twitter	Inconvenient
5 days	First customers complain	Embarrassing
15 days	Comments in the trade. Customers seek new suppliers.	Threatens the organization's survival
30 days	Media comment. Trade credit dries up. Snowball effect starts. Massive defection of customers.	Terminal

A *backlog* often occurs, whether orders to be processed, products to be made or clients to be seen. You should determine where the bottleneck occurs, and plan how this might be overcome. For example, you might take on temporary staff, buy in finished goods or outsource some of the work.

THREE WAYS TO ENSURE THAT YOUR SCENARIO PLANNING WILL FAIL

1. Assume that the future will be an extension of the past. In reality, the future never goes in a straight line.
2. Opt for the scenario that management prefers, the one that suits your business. This is the triumph of optimism over reality. Assuming that customers will stay loyal to the organization has sunk many businesses.
3. Build your business around your chosen scenario. What happens if the future turns out to be different?

The BBC's Director General lost his job and the company was thrown into a crisis following a BBC Radio 4 news report critical of the government. In all the BBC's scenario planning, that possibility had been dismissed as 'simply unbelievable'.

HOW BIG DOES A CRISIS HAVE TO BE? BUSINESS IMPACT ANALYSIS

Crises come in different sizes. The issue is how much of an impact it has on the business. In continuity planning, we're concerned only with critical processes, those whose disruption is unacceptable. For example, while the staff social club might be important to its members, it doesn't matter if it's out of operation for a few weeks.

The same might apply to the cafeteria, to the photocopier on the second floor, or the shed where Estates keeps the lawnmower. Whereas if the order processing system breaks down, the business soon comes to a halt. This is shown in Table 23.7.

Similarly in a local authority, much can go wrong in a Libraries department or Strategic Planning, but no one will get very upset, apart from the librarians and the planners. By contrast, if salaries aren't paid or gritting lorries don't turn out on frosty nights, a big problem will ensue. You can put a figure on the importance of any function by estimating:

- the cost of lost business;

Table 23.7   Business Impact Assessment

	Insignificant	Moderate	Serious	Catastrophic
IT	One-hour outage.	One-day loss of data, restored overnight.	Loss of several major clients plans, requiring clients to provide data again. One-week delay.	Permanent, non-recoverable data loss covering the whole of the ERM. Clients defect. Business future is threatened.
Health and safety	Slip, trip or fall, requiring first ad. No bones broken.	Accident involving hospitalization of one or more people.	Serious injury or death of one individual.	Death of several individuals.
Clients	Loss of a single client in the bottom 5 per cent of our income range within a 12-week period.	Loss of clients amounting to 10 per cent of revenue.	Loss of a major client representing at least 20 per cent of revenue.	Loss of one-third of clients by value.

- the cost of lost customers or contracts;
- fines for failing to meet relevant laws and regulations.

This will allow you to create a spreadsheet (Table 23.8) showing how long you can tolerate the loss of a business process, and your goal for resumption of that process.

Table 23.8   Recovery Time Objective

Issue	Department	Impact (1–5) Maximum Tolerable Period of Disruption (MTPD)	Recovery Time Objective (RTO)
Electricity supply	Admin	Two hours	One day
Telephones	Admin	One hour	One day
Email	IT	Half a day	Half day
IT server	IT	Half a day	Three days
Database	IT	One day	Three days
Supplies from supplier 1	Despatch	One week	One week
Courier offtake of parcels	Despatch	Three days	One week
Packaging supplies	Despatch	Three days	One week

In 'How long have you got', we looked at the MAO. Each type of activity can be tolerated for different lengths of time, and it's the time beyond that that the impact costs money. The scale of the outage or the scale of the impact on the business determines the scale of the crisis.

At this point you can create a table listing each risk, its probability and severity, on a 1–5 scale.

From this you can create a Risk Matrix (Figure 23.3), showing the individual risks, numbered here R1 to R8 – though in practice there would be many more.

Now comes the job of managing those risks. As we saw in Chapter 4, there are four ways of treating risk: you can Accept it, Reduce it, Share it or Avoid it. Figure 23.4 illustrates the four main categories.

Impact				
	High	R1		R4
	Moderate	R6	R3	
	Low	R2 R5		R7 R8
		Rare	Quite probable	Highly probable

Probability

Figure 23.3 Risk Matrix

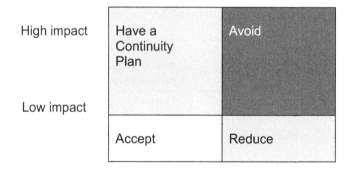

High impact	Have a Continuity Plan	Avoid
Low impact	Accept	Reduce
	Improbable	Very probable

Figure 23.4 Four categories for managing continuity risks

High-impact, low-probability events

This is at the heart of continuity management. It is those 'Black Swan' events that one hopes will never happen. This includes pandemics, the destruction of a vital port or a fire that engulfs the whole plant. Such a plan might involve the emergency services. These are the ones you must create a continuity plan for.

High-impact, high-probability events

The organization cannot afford high-impact events that are very likely to happen. Therefore it must ensure they don't. This could be by exiting from a market. Or if the business was in an area at high risk of flooding, you could move location.

Low-impact events, whether likely or not

Here we are into risk management, as opposed to continuity management. Therefore, such risks need not concern us here. We discuss how to manage these risks, particularly the low impact but high probability in each chapter (for example, health and safety, production and so on).

Organizations tend to view low-impact, low-probability events as something to accept. Low-impact and high-probability events should be reduced, by instigating controls, better training and all the other methods we discuss in this book.

4. Assign Roles and Responsibilities

Someone must take responsibility for continuity. That includes:

* creating the plan in the first place;
* deciding that a crisis has happened, and authorizing staff to implement the emergency plan;
* overseeing the emergency plan.

In large organizations, a director, possibly the finance director, will be responsible. There will also be a continuity manager who in medium-sized organizations may also be the risk manager. Business continuity is also known as resilience management, disaster recovery or contingency planning.

Most businesses avoid creating a specialist continuity department because operations staff need to take ownership of emergencies.

As Figure 23.5 shows, there are four levels of activity:

* working group;
* director with responsibility for continuity;
* continuity or emergency planning manager;
* department heads;
* department champions.

We examine each in the following section.

APPOINT A WORKING GROUP

In a large organization, a working group will be tasked with devising a continuity strategy and allocating resources. This is a scene-setting group. It should have a representative from all relevant departments. Thus it will need people with skills in the law, operations, finance, HR, purchasing and technology. It will also report to senior management on the outcome of audits and exercises.

Figure 23.5   Roles and responsibilities in continuity

NOMINATE A DIRECTOR

You should give a Board director overall responsibility, to ensure that the plan is carried out.

The director or their nominated deputy is the person who decides whether a crisis has occurred and therefore the continuity plan must be invoked.

Research by CMI shows that 44 per cent of managers say the MD or CEO is the BCM sponsor, while 19 per cent said BCM rests with the chief operating officer (COO). There was also a wide range of other titles, including BCM specialists, operations director, quality manager and roles unique to certain sectors.

NOMINATE A BUSINESS CONTINUITY MANAGER

The company should nominate a business continuity manager. This might be the risk manager or a line manager (for example, the production director). A large organization will need several people to be involved, with each risk having its own emergency planning manager. Roles and responsibilities must be defined in writing so that, if a crisis occurs, the chain of command will be clear.

The continuity manager must have a detailed knowledge of all crises for which they will be responsible. They should be involved in the risk assessment and will be responsible for drawing up the emergency plan.

The manager (or someone nominated by them) should be available 24 hour a day, 365 days a year. The system must take into account the emergency planning manager's holidays, sickness and absence on business. Crises happen when people least expect them.

OTHER ROLES AND RESPONSIBILITIES

Departmental heads must be responsible for continuity in their own areas. They need to train staff what do in the case of an emergency. Some staff will have specific jobs in an emergency. Telephonists, as we saw above, may have to handle many more callers than usual. Engineering staff have to switch to crisis mode. The telephone number of the emergency contact should be prominently posted, and staff should know to whom they should report, and what their function would be during a crisis.

Many crises will result in legal action against the company. A lawyer who understands the nature of corporate crisis should be involved in risk assessment and policy setting.

Some organizations create *continuity champions*. These will raise awareness of risk in their departments, and watch out for unnoticed risks.

IS BCM PART OF RISK MANAGEMENT? OR IS IT THE OTHER WAY AROUND?

Some organizations have an advanced BCM culture, while others regard it simply as one element in their overall risk planning.

To a degree it depends on the company's products, markets and history. Some industries lend themselves more to BCM. They include businesses that are dependent on their technology; those with large, well-known brands; and those operating in risky environments.

BCM implies that the organization is liable to suffer a big crisis. And so companies in uneventful markets such as wholesaling, agriculture or plastics manufacture might not find it relevant. By contrast, a utility company would be concerned about leaving a city in darkness.

So the usefulness of BCM depends on the impact of the crisis on the public or customers. But while they might not have the drama of TV cameras and a switchboard jammed with callers, ordinary companies can suffer their own smaller but no less important emergency. So it is worth planning for the worst.

Table 23.9 has a comparison between risk management and business continuity. Not everyone will agree with each comment, and there is considerable overlap between the two. Each shades in to the other, and every organization manages its threats slightly differently. Both can be managed by the same person or team. Nevertheless it may be useful to see where the differences lie.

In some organizations, there is a convergence between risk management and continuity management. In others they continue to be run as separate functions.

Nathaniel Forbes, a continuity expert (NathanielForbes.com) has called the profession 'The Department of Extraordinarily Unlikely Events'. This highlights the specialist nature of the work – planning for things that will probably never happen. As a result, top management is sometimes reluctant to engage with continuity professionals.

Where risk and continuity are managed separately, there tends to be some competition between the two. This is partly because business continuity people see risk management as part of continuity, and risk managers think the opposite.

However, there are savings to be made and efficiencies to be had by amalgamating these functions where they exist separately.

---

5. Prepare an Emergency Plan

The company should have an emergency plan that is divided into sections covering all probable eventualities.

It may be known as the incident plan, continuity plan, contingency plan, emergency plan or disaster recovery plan (depending on corporate culture and the extent of plain speaking).

Whatever you call it, it sits on a shelf, being periodically dusted off and tested, with everyone hoping it will only be used for small scale crises.

Table 23.9   Comparison of risk management versus business continuity

Risk Management	Business Continuity
Manages likely risks	Focuses on unlikely risks
Manages small to medium risks	Focuses on extreme, business-critical risks
Deals with the everyday	Thinks the unthinkable
Applies controls	Builds resilience
Tactical	Strategic
Says, 'Let's control what we're doing'	Asks, 'What needs to be done?'
Bottom-up processes	Top-down process
Has the attention of senior management	Senior management is less concerned

---

CASE STUDY: SOMETIMES IT TAKES ONE NARROW ESCAPE AND ANOTHER MAJOR INCIDENT TO GET STARTED IN BUSINESS CONTINUITY

IP Mirror Pte Ltd (ipmirror.com) is an internet domain name registrar based in Singapore. IP Mirror first encountered continuity issues when the Severe Acute Respiratory Syndrome (SARS) epidemic struck Singapore. Although IP Mirror was not directly affected by the global SARS pandemic, it motivated them to consider the potential impacts of such events.

This spurred Janna Lam, IP Mirror's CEO, to investigate BCM.

Although her original aim of adopting BCM was to address the flu epidemic, Ms Lam soon realized that her company was exposed to other threats.

During a major earthquake in Taiwan, IP Mirror's operations were suspended for a number of hours as the earthquake severely damaged several undersea telecommunications cables, which disrupted 98 per cent of the communications within the Asia Pacific region. IP Mirror is an internet-based company, so having no access to the web was a massive problem.

With no BCM plans in place, IP Mirror was unable to respond, and its operations came to a standstill.

Having suffered this outage, IP Mirror became more determined to build up its resilience.

This was reinforced when Ms Lam discovered that a business partner with BCM certification had recovered and relocated its business operations in just a few hours after encountering the onslaught of massive floods in Queensland Australia.

Ms Lam said, 'Facing so many potential threats nowadays, there needs to be a back-up plan that we can fall back on, and be sure our employees are trained to manage crises.

As the nature of our business is dependent on the internet, it is our priority that our customers get zero or minimal downtime if our business processes were to be compromised.

If you want your customers to trust you to provide effective services, you have to prove to them you can be relied on.

Having BCM certification has allowed us to present ourselves as the trusted internet domain name registrar. It has increased our customers' confidence in our services and given us a competitive advantage over our competitors.'

PLAN SECTIONS

The plan will have separate sections dealing with each type of crisis, including:

- product or service quality failure;
- environmental pollution;
- health and safety accident;

- human resource incident (such as suicide);
- fire and explosion, including loss of buildings, telephones and so on;
- security failure, including extortion, kidnap and ransom;
- fraud;
- financial crisis;
- IT or internet failure;
- industrial relations problem (and the ensuing lack of production);
- problems specific to the business (some examples of which are discussed below);
- other problems affecting the corporate reputation, for example ethical failure.

HEADINGS

The plan should contain the following headings.

> *Objectives:* What is the purpose of this part of the plan?
>
> *Assumptions:* Are you expecting the road to your premises to be open, or specific members of staff to be available? What if they aren't?
>
> *Dependencies:* To restore computer files after a virus has wiped them may depend on a) there being power, b) the backup has being taken, and c) having a member of staff who knows how to restore the backup.

An increase in the likelihood of one risk can make another more likely. The loss of a key supplier may result in a lack of stock, which may produce a lack of sales, which could mean a cash flow crisis.

> *Invoking the plan:* Who has the authority to invoke the plan? Smaller incidents can be dealt with by less senor managers.
>
> *Action:* What needs to be done? Does a command centre need to be set up, and if so where? Who will do each job?
>
> *Contacts:* Names and contact details of relevant staff members, emergency services, and/or appropriate suppliers or business partners.
>
> *Resources:* The plan should have a list of relevant documents, such as a building blueprint. It might also need passwords to access specific files.
>
> *Communications:* Who should be contacted – staff, the emergency services, customers and/or the regulatory authorities? And who should do that?
>
> *Checklist:* A checklist will ensure that all tasks get completed.

ESSENTIAL CHARACTERISTICS OF THE PLAN

> *Versioned:* The plan needs to be versioned. That normally means putting a new number on it each time it is updated. That way, no one will work off an old version, only to discover the contacts list is out of date.
>
> *Relevant format:* Plans used to be kept in A4 folders or ring binders. But they can also be kept in the cloud or on computers or tablets. Sometimes paper versions are best because they don't need technology to be read. On the other hand they are less likely to be updated, and take more effort to circulate.
>
> *Slimline:* The plan should have few pages, and not be a tome, since it will be used in moments of tension.

Accessible: The plan must be available to everyone who needs it. So it must be close at hand. Equally, since it will contain confidential information, it should not be widely circulated. Some copies must be kept away from the organization's main building, in case it becomes unusable.

Communicated: Each section of the plan should be discussed with those members of staff who will be involved.

Tested: The plan should be regularly tested and updated.

SEVERITY

The plan should analyse a range of outcomes for each type of crisis, ranging from minor problems to major catastrophes.

This will encourage a response which matches the severity of the situation. The plan should have an escalation process or flow chart that maps the scale of the problem to the people responsible. The more serious the problem, the higher up the chain of command should go.

As Figure 23.6 shows, by identifying the scale of the problem, you can determine who should take control of it.

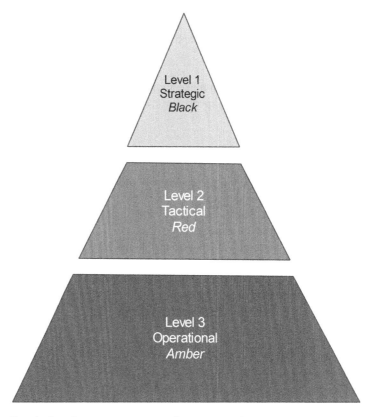

Figure 23.6   Escalating the response according to severity

THE PACE METHOD

All too often, the emergency doesn't happen the way it was expected. That is why it is sensible to have a Plan B. It is even better to have Plans C and D, in case the first one doesn't work out. The military have an expression, known as PACE. It stands for:

- Primary
- Alternate
- Contingency
- Emergency

The expected or Primary meeting will be in the CEO's office. In the event of that not being available, it will be held in an Alternate room, namely the training room. If neither is available, the meeting will take place in the Contingency location: the local pub. If that too is not available, the meeting will use the Emergency format – in the open air.

The PACE method can be used for communications (from landline to sending a handwritten note), and for operations (from using your own facilities to employing a third-ranked supplier).

6. Control or Mitigate the Risks

Having identified the risks, you need to decide how they should be controlled. You can do this in a table format, as shown in Table 23.10.

We have looked at assessing risks earlier in this chapter and also in Chapter 2. And we examined how to control them in Chapter 4.

Table 23.10   Controlling the risks

Department: Estates

Activity or process	Risk	Severity	Probability	Mitigation measures
Car parking	Risk of accident between motorists, cyclists or pedestrians in the car park	M	M	Create marked pedestrian and cycle lanes
	Drivers turning right across traffic on leaving premises Risk of collision	H	L	Put up 'no right hand turn' signs
	Excessive driver speed through tunnels Risk of accidents.	H	M	Install ramps at the entrance and exit to the tunnels to slow drivers down

Figure 23.7 relates the severity and probability of a risk on one axis, and the extent of control measures on the other. It shows five risks, labelled R1 to R5. The chart suggests that:

- even the most serious and probable risks can be simply monitored if you have sufficient controls in place;
- an uncontrolled, high-impact/high-probability risk needs to be managed or improved urgently;
- insignificant and improbable risks that are well controlled can effectively be simply accepted.

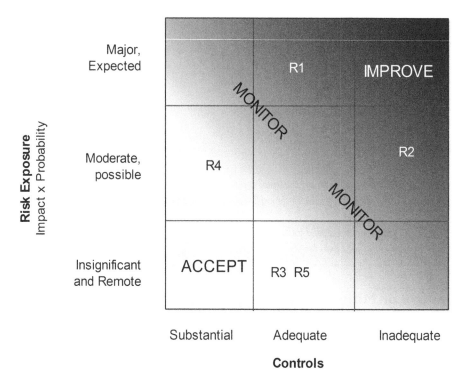

Figure 23.7   Monitoring, accepting and improving risks

CASE STUDY: SHOOTING THE MESSENGER

Olympus sacked its CEO, Michael Woodford after only two weeks in the job, saying that 'major differences had arisen between Mr Woodford and other management regarding the direction and conduct of the company's business'.

It later turned out that Woodford had been asking inconvenient questions about the company's prior acquisitions, including US$2.2 billion spent on acquiring British medical equipment maker Gyrus Group, and US$687 million the company had paid to a middle man during that deal.

The Japanese newspaper *Sankei* subsequently alleged that the company's payments could be linked to the Yakuza, Japanese organized crime.

In an unprecedented turnaround, Tsuyoshi Kikukawa, Olympus' Chairman and President, later apologized and resigned. The company admitted its accounting practices were 'inappropriate' and revealed it had attempted to cover up losses on earlier investments.

The company's share price plummeted by 75 per cent, but subsequently recovered.

Many commentators saw the crisis as a clash of values; and believed that that the company had been trying to rectify earlier fraud and errors without openly admitting them. The arrival of a British-born CEO had brought these matters into the open, and the scandal had thus developed.

---

7. Testing the Plan: Practising Emergency Procedures

An emergency plan is useless if not regularly tested. Companies should carry out regular fire practice, not just the testing of alarm bells. If the drill is not regularly done, deaths could result.

Similarly, emergency procedures should be carried out to test other crises. They should be carried out an inconvenient time, such as Saturday evening, to see whether the plan works. Testing the plan will involve unforeseen problems. Key personnel will be on holiday or simply dining out. No one will know the location of important keys. Or the five-pin plug on the emergency generator won't fit the three-wall pin socket.

The scale of the practice will depend on the potential disaster. Major airlines practise jamming their own switchboard with calls from 'distraught relatives' following a simulated air crash.

You must be able to implement emergency procedures without delay. Recent disasters have demonstrated the importance of being able to respond instantly. Failure to act swiftly increases the scale of the disaster and the level of criticism directed at the company.

You can post emergency procedures on notice boards throughout the site, and have it mentioned at departmental meetings. Staff must know what to do and whom to contact.

When it comes to major incidents, the most commonly used scenarios are:

- you suffer a major loss of staff, perhaps due to illness;
- you are denied the use of your premises, for example due to building collapse;
- you lose your IT systems;
- a major supplier ceases delivering essential materials;
- a key partner stops operating, due possibly to bankruptcy.

Around half of organizations with a continuity plan have exercised their plan in the last year, according to research by CMI. But 17 per cent say their plan has never been tested.

The most common method used when testing a plan is the desktop exercise, followed by IT backup exercises and tests of remote working facilities. Only 22 per cent conduct a full emergency scenario when exercising their plans, doubtless because of the havoc it would cause.

When carrying out an exercise, it is convenient to telescope the timeframe, since it is inconvenient to play it out over days and weeks. You should also log events, so that you can learn from them afterwards.

CASE STUDY: THE LINCOLNSHIRE TRAIN DISASTER

Eight people were killed near Lincoln when a train derailed after slamming into a slow moving agricultural vehicle on the East Coast main line. To make matters worse, a second accident occurred when a lorry, trying to escape the traffic jam caused by the derailment, killed or injured a dozen children and created a chemical spill.

Fire and rescue teams had to deal not only with the screams of the passengers trapped in the wrecked train on a muddy embankment, but also the mayhem of a major incident in a congested town centre.

Except of course that the incident was a test created by Lincolnshire Council to see how its emergency procedures worked in real life. According to *The Guardian* newspaper, it brought home many truths: you can't rescue dying people off a train if the power lines are still live and the emergency workers would get electrocuted. People on the train had to wait for an hour to be treated while rescue workers seemingly stood by.

Apart from paramedics, police, the hospital and the fire service, the test involved helicopters, chaplains, rail company officials cancelling train services, and women from the Royal Voluntary Service handing out tea and biscuits in emergency accommodation.

'Hysterical parents', 'disruptive bystanders' and 'non-English speaking foreigners' put the emergency services to the test. And the cultural differences between the various services added to the tension, with the hierarchical police having to work with the more collaboratively minded council staff.

Meanwhile Karen Spencer, the council's Strategic Communications Officer, pointed out that many people with smartphones would have been tweeting and uploading pictures faster than the council was able to respond. They all agreed: the exercise showed there were lessons to be learnt.

BUILD A LONG-TERM TELATIONSHIP WITH THE MEDIA

A good relationship with the media will ensure that, should a crisis ever occur, the organization can expect a fair hearing. It also ensures that journalists will expect the best rather than the worst, when it comes to news about the company.

If the company builds trust with the media, they are less likely to write damaging articles if a crisis occurs. Journalists will find it difficult (though not impossible) to write accusing articles if they have only ever received useful, honest and accurate information over a long period.

In the next chapter we will examine how the company should deal with the media if a crisis occurs.

Continuity Management and ISO 22301

A management system provides written procedures for staff, so that they know what actions they should take. We've previously looked at management systems in Chapters 5 and 8, on Operations and the Environment.

You should write procedures for all areas of the business which are at risk, and for all contingency plans. The procedures should then be regularly audited, to make sure staff are complying with them. You should set trigger levels, which we discuss in the next chapter, to determine whether to a crisis is taking place, and how severe it is (See 'Severity' above).

ISO 22301 is a management standard for business continuity. It is designed to help you deal with corporate-level risks (loss of power for several days), rather than operational ones (such as a brief internal IT failure). It should outline the company's process for dealing with a crisis, rather than write procedures for each type of catastrophe.

The system has five main clauses (numbered 4–10), which are shown below. To aid comprehension which we have placed them in a chart (Figure 23.8), in a sequence that hopefully makes more sense to the reader.

ISO 22313 provides useful guidance on implementing ISO 22301. Among many other topics it includes a list of items to write in business continuity policy, an explanation of the Business Continuity Management System's (BCMS) roles and responsibilities, and examples of goals for the BCMS. It will save you time if you are implementing a BCMS.

CLAUSE 4: SETTING THE ORGANIZATION IN ITS CONTEXT

You need to identify the organization's main products and services, and its areas of high risk. This will let you to design a suitable BCMS.

CLAUSE 5: MANAGEMENT NEEDS TO LEAD

As with other management standards such as ISO 9001, the organization's management needs to be engaged. It should do the following:

- set a continuity policy and objectives;
- ensure the system reflects the needs of the organization;
- integrate the system into the organization's business processes;

- provide resources and people for the BCMS;
- communicate the importance of BCM;
- ensure that the BCMS achieves its expected outcomes;
- ensure continual improvement.

CLAUSE 6: PLAN FOR CONTINUITY

You should show how the system will manage the risks identified. The system must:

- be consistent with your business continuity policy;
- ensure the minimum level of output that will permit the organization to achieve its objectives. In other words, when a crisis occurs, what needs to carry on working?
- be measurable;
- take into account applicable requirements (for example legislation);
- be monitored and updated.

CLAUSE 7: SUPPORT

The organization must allocate competent staff to operate the system, and ensure they are trained.

The system must specify how it will communicate the system internally and externally. Relevant procedures must be documented.

CLAUSE 8: OPERATION

Having made its plan, the business must now implement it. This requires the following:

- carry out a risk assessment;
- implement a strategy that will allow the business to continue operating after a crisis – the business should write procedures as to how that will happen;
- carry out exercises and tests, to ensure the system is working.

CLAUSE 9: EVALUATING PERFORMANCE

The business needs to:

- carry out internal audits to ensure the system is working;
- evaluate the results of those audits at management reviews.

CLAUSE 10: IMPROVEMENT

Improve the system through corrective and preventive actions. Typically, the management review will require changes to be made to the organization's processes and procedures.

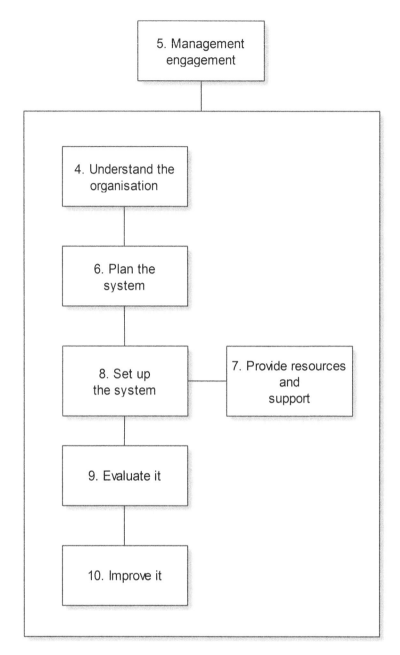

Figure 23.8   The main clauses of ISO 22301

Continuity Software

Many off-the-shelf BCM programs are available. Suppliers include the following:

*   Sunguard, sungard.co.uk
*   Clearview, clearview-continuity.com
*   RSA Archer, emc.com

- Recovery planner, recoveryplanner.com
- Metric Stream, metricstream.com
- Continuity Logic, continuitylogic.com
- Strategic BCP, strategicbcp.com

Crisis management software should include:

- risk assessment;
- business process maps;
- continuity or recovery plans;
- inventory of assets and personnel;
- list of suppliers;
- exercise plans and reports.

The program should include lists of email addresses, interactive checklists, and be activated from anywhere. It should be web-based, and will be independent of the business' internet or email system. It must be accessible to people from different sites, and keep them abreast of what is going on.

The department store group John Lewis developed its own internet-based crisis management program after failing to find a suitable product on the market. Named Crisis Commfile, it provides real-time communication between multiple locations during an emergency, allowing those responsible for resolving the crisis to view current and past dialogue.

Risk Assessment

By answering the questions below, you can assess the company's vulnerability to business continuity. Score one point for every 'Yes' answer.

Question	✓
Has the business conducted scenario planning in the last 12 months?	
Does the business have an emergency planning manager or continuity manager?	
Is there an emergency response plan?	
Has the emergency plan ever been tested?	
Is the business free from any secrets that, if disclosed, would cause major embarrassment?	
Has the industry been out of the media spotlight for the last 12 months?	
The industry is *not* one of the categories shown in Table 23.4	
Do you have a documented system for contacting management outside work hours?	
Has the business the guaranteed use of alternative premises it could occupy in a crisis? Or does it have a mirror data site, with its own standby desks for mission-critical activities?	
Has the business adopted a continuity management system such as ISO 22301?	
Total points scored:	

Score: 0–3 points: high risk. 4–6 points: moderate risk. 7–10 points: low risk.

The Appendix has a summary of the checklists. By entering the results of this one, you can compare the scale of business continuity risk against other categories of risk.

24
How to Survive a Crisis

Only last week the finance director was showing you a picture of his house. He called it a farmhouse, but it's sixteenth century and has a moat round it. You thought you smelled alcohol on his breath and there was a film of sweat on his face.

You didn't think much about it at the time. But he hasn't been at his desk for a couple of days. And the bank has been phoning the CEO, which is unusual. Ken in accounts says the business has an overdraft the size of Alaska. That can't be true because the organization has never been busier, the clients are upmarket and your quotes are never cheap.

The phone interrupts your thoughts. It's the CEO. 'Can you come to my office now?' he says, without the usual pleasantries. You put on your jacket. The staff seem to be watching your face. There's a sick feeling in your stomach.

How a Crisis Develops

There are instant crises and slow burners. Some come with advance warning, and others with none at all.

Most people think a crisis arrives with a bang. But as we saw in the last chapter, 61 per cent of crises have a long smouldering fuse. They take a long time to blow up. This is particularly true of financial crises.

These slow-fuse crises often pass through quite predictable phases, with the signals gradually becoming clearer. Management's response is equally predictable, often beginning with incomprehension and a refusal to face facts.

The future success of the company depends on management's efficiency in dealing with the unfolding crisis. The main features are listed in Figure 24.1. It shows how the stages can be identified, and how the company can manage them.

EXTERNAL CHANGE, ERROR AND CRISIS

A crisis results from an action or error which can usually be foreseen. It could be an internal problem, such as management's failure to innovate. Or it could be an external act, such as new legislation. It may be a combination of both. Whatever the cause, the crisis often has the same disastrous effect.

It is useful to have a management information system which warns the company of an impending crisis. There will be different triggers for each area.

- in finance, certain ratios will turn negative;
- in the environment, meters which monitor effluent levels will exceed a given toxicity;
- in health and safety, the number of minor accidents will reach a specific level.

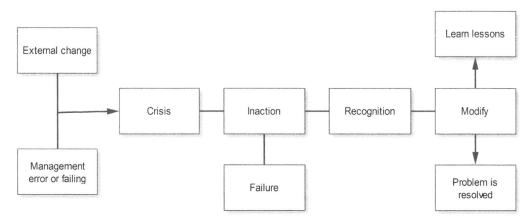

Figure 24.1   Stages of a crisis

Rejection and inaction

Management often fails to take action when the crisis first develops. It often refuses to acknowledge that there is a problem. If managers accept that a problem exists, they often believe that it is not a real crisis. The company hopes that sales will improve or that the bank will change its mind or the situation will rectify itself.

Managers believe that life can go on as before, often with minor changes. This leads to a period of inactivity. If this continues for too long, the company fails. It either goes bust or is bought. The initial delay is often fatal.

Recognition

Eventually (and this may take days or even years) management admits that there is a problem. This can lead to panic. Different views emerge as to how the problem should be resolved.

The views polarize on public admission, with PR people proposing openness and corporate legal advisers cautioning silence. The company's failure to comment at a time of mounting information about the scale of the problem leads to increased press speculation.

Modify or mitigate

The company faces the problem and takes decisive action to rectify it. This may include withdrawing from a project, admitting to the world that if faces huge write-downs or making staff redundant. Now it has a chance of survival.

Good often comes from this process. The organization may emerge fitter, and more aware of the dangers. Management and staff will want to avoid a repetition of the nightmare. The company will be more competitive, and less complacent. Changes that were once unthinkable now take place.

Learn the lessons

The organization needs to understand how the crisis arouse and take steps to ensure that similar problems don't recur. For example, a health and safety incident may be symptomatic of underlying problems that need to be addressed.

Dealing with Disaster: Eight Steps

In the previous chapter we examined how the company should prepare for an emergency. Now is the time to consider what to do if the emergency actually occurs. The activities are shown below.

1. assess whether a crisis exists;
2. activate the crisis team;
3. set up a crisis centre;
4. gather the information;
5. assess the scale of the crisis;
6. manage the crisis;
7. communicate with stakeholders;
8. re-start the business, or return to normal operations.

Next we look at each of these activities in more detail.

1. ASSESS WHETHER A CRISIS EXISTS

When the company is facing major problems, it may be difficult to determine whether a crisis really exists. Therefore the plan should define a crisis. Apart from the trigger levels discussed above, a crisis might entail:

* a threat to the long-term reputation of the business;
* a potential cost equal to a proportion of annual profit (say, 25 per cent), whether in sales decline, the costs of a recall or in clean-up costs;
* the risk of widespread litigation;
* an interruption of business of a given period of time (say, a week).

2. ACTIVATE THE CRISIS TEAM

You need a cross-disciplinary team, including subject experts and senior management. Specialists might include HR, IT, PR, operations, estates or counsellors.

Senior management must break out of the routine of standard meetings and reviews, switching into a crisis response mode instead. Regular work is predictable and therefore comforting, but wrong at this stage.

At the same time, one or more senior managers must be released to carry on the business. This will ensure that ongoing work is not neglected.

Team members should be trained on how to respond in the event of them being called upon to participate in the crisis team.

3. SET UP A CRISIS CENTRE

The company should set up a crisis centre, which should be insulated from the day-to-day operation of the business. The crisis centre should be equipped with all relevant information, including contact names and phone numbers. It needs administrative support, with photocopying and other equipment. Remember that in the case of flood, fire or bombing, the head office may not be available.

The company may need to publicize an emergency number. Like everything else in crisis management, this should have been discussed with the telephone company in advance. The emergency number will then have to be manned with staff who, in turn, will need to be trained and briefed.

A recovery work area or a dedicated mirroring operation?

A bomb, explosion, flood, storm or other disaster may render your work area unusable. In such cases, you will need alternative premises to work from.

For many organizations, their call centre, order entry operation or trading floor is critical to their continued success. If they fail, the BCM plan needs to get them running again quickly.

Business can rent work areas from third-party suppliers, or use alternative buildings of their own. Third-party suppliers can bring in fully equipped IT and telephone equipment and desks that will re-create the business. Options for business recovery include:

- *Dedicated*, where the offices are permanently available for the business.
- *Syndicated*, available from within eight to 24 hours from the call for critical processes, and 24–72 hours for other services. These are sometimes rented on a 25:1 ratio, that is, each desk is rented out to 25 clients. You should ascertain the ratio, and find out who else is renting the desks. And the question must be asked: what happens if the site is occupied by another firm when you 'invoke'; that is, when you require the service?
- *Mobile*, where the recovery firm brings mobile recovery trailers to you. If so, there must be parking facilities, and these must be accessible. Police have been known to move such vehicles on.

The third-party organization should be involved in any rehearsals or testing that you carry out.

There is a risk that the recovery operation may not work when the time comes. There are many factors that can hamper recovery, such as adverse weather, differences in IT and telephony systems, or the use of older or newer software. Your staff may be unfamiliar with the recovery company's switchboard.

It may take several hours for technicians to iron out the glitches in the system. At the very least, there will be delays and some facilities will not be available.

Given the always-on nature of today's business, speed of recovery is essential. Businesses that use real-time data processing, such as banks or e-businesses, require 'never-fail' or non-stop dependability.

A recovery that takes 24 hours is simply unacceptable for these services, and so you may need a data-mirroring operation in an off-site location, along with standby computing and telecommunications services. These cost considerably more than conventional recovery services.

The BCM plan we discussed in the previous chapter should assess the impact of lost services, and determine what delays are acceptable. This is turn will dictate what work area recovery plan is chosen.

Requirements for the crisis centre

The crisis centre needs to be equipped with the following:

- telephones;
- computers;
- internet access;
- access to the organization's main database or ERP system;

- video conferencing equipment;
- printer and photocopier;
- TV with satellite connection;
- projector and screen;
- whiteboards;
- clocks showing time in different time zones, if the organization has multiple locations;
- food supplies;
- toilets;
- rest areas;
- records, emergency plans, contact lists, building plans;
- sign in and out board, so the team knows who is where;
- marker pens, writing pads and flip charts.

4. GATHER THE INFORMATION

In a crisis, news is often muddled. It is important to get accurate information. To achieve this, the organization needs a form that will gather the information for each kind of crisis. A typical form is shown in Table 24.1.

Table 24.1   Sample emergency form

Environmental incident form	
Division/department	
Describe the incident	
Injuries or deaths	
Damage caused	
Spillages etc.	
Witnesses	
Reported to police or other authority	Police report no. (time and date)
Response of police	
Corporate legal department informed (name, time, date)	
Corporate legal department's opinion and recommendation	
Reported to insurance company (date/time)	
News media's interest	
External affairs opinion and recommendation	
Nominated spokesperson	Home phone number
Work phone number	
Strategy to be adopted	
Report completed by	Date

5. ASSESS THE SCALE OF THE CRISIS

Having established that a crisis exists, the business should evaluate its scale. It can do this by undertaking a Business Impact Analysis, as shown in Table 24.2. In the example shown, the company is undergoing a huge increase in demand which, while attractive to the finance director, is likely to pose major future problems with customers.

The timescale of the problem is short, and helps to concentrate the directors' minds. The company has little control over the disaster, but a degree of financial freedom to determine its own future.

There are several options, for both the short- and medium-term, with the most important issue of crucial issue of getting more finished stock into the warehouse.

Table 24.2   Impact analysis

Problem	Competitor has gone into liquidation, resulting in demand for our product exceeding capacity by 30 per cent.
Scale of the problem	Important customers may get increasingly irate.
	Customers may start to buy from overseas competitors.
	It is difficult to absorb the new customers.
Timescale	The problem is urgent. Some patterns are now out of stock, with Production Planning quoting three months for delivery. Other patterns will run out in five to 30 days.
	Installation of new plant would take six months.
Control	Future solutions are under our control.
Choices	*Short term*
	1. buy product from overseas competitors;
	2. ration supply;
	3. add extra shift;
	4. increase productivity.
	Medium term
	Extra plant can be leased or bought.
Action agreed	Initially, ration supplies, giving priority to best customers.
	Boost productivity – assign engineers to maximizing output, possibly by allocating more fork-lifts and so on to speed materials handling.
	Assess costs of extra shift.
	Review problem in one week.

6. MANAGE THE CRISIS

Later in this chapter we examine some specific types of disaster, because each requires a different type of response. A financial scandal is different from a gas explosion.

But the overall strategy remains the same. The organization has to halt the problem, admit the scale of the disaster, make changes so that it won't happen again and communicate with stakeholders.

As Figure 24.2 shows, there are three main stages: responding to the problem, resolving it and recovering from it.

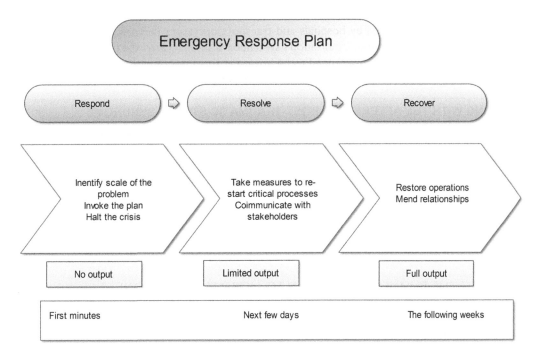

Figure 24.2   Timing of the response to a crisis

7. COMMUNICATE WITH STAKEHOLDERS

In a crisis, stakeholders will include the workforce, customers and the media, and may also involve the local community and the public.

An emergency notification system can telephone people very quickly. It can send a pre-determined message to hundreds or thousands of people almost instantly. This automation means that there is one less job to do; and it lets you test your staff's preparedness. Suppliers include the following:

- Mir3 (mir3.com)
- RapidReach (rapidreach.co.uk)
- Criticall (criticall.co.uk)
- AlertCascade (alertcascade.co.uk)
- Vocal (vocal.co.uk)
- RapidReach (rapidreach.co.uk)

It is easy to overlook *customers* if the crisis does not directly affect them (for example, in a health and safety problem). But customers will see the newspaper or television reports, and will be concerned. Uppermost in the mind of wholesalers, retailers or users will be the organization's ability to continue meeting their needs.

Sales people should therefore be taken off their daily routine. They should speak to all major customers, reassuring them that supplies are not threatened. In some cases the work could be given to a telemarketing agency, because it can contact customers rapidly.

In the case of crises faced by hospitals and transport operators (airlines, ferries, trains and buses), the public may deluge the organization's switchboard. In such cases, the phones need to be re-routed to a call centre, which will need to be briefed at short notice.

Later in this chapter we consider how best to deal with the media.

DEALING WITH THE AFTERMATH: POST-TRAUMATIC STRESS SYNDROME

After an incident involving explosion, fire or flood, the survivors may need medical attention, food, clothing and warmth. Then they may require help to reach a destination, or financial compensation.

Post-traumatic stress is now recognized as a serious problem. The business may provide the victims of major incidents and their families with trained counsellors who help them come to terms with their experience. Survivors feel a mixture of emotions, including depression, fear and guilt at having survived while their colleagues or friends did not. Counsellors may encourage the survivor to relive his experience a sufficient number of times to reduce its intensity.

The organization must be able to locate counsellors in a hurry: this needs to be planned in advance.

---

8. RESTART THE BUSINESS OR RETURN TO NORMAL OPERATIONS

After the crisis, the company should learn from the situation, and aim to prevent it from reoccurring. It should investigate how the problem occurred, and take steps. This might involve staff training or the adoption of new procedures.

The company also has to rebuild confidence, and to ensure that sales are maintained. It may need to relaunch itself, to tell the world that it is back. Activity that attracts headlines can be useful. After its benzene scare, Perrier relaunched the product as new Perrier, and organized a photo opportunity, showing the old bottles being crushed and recycled.

Those affected may also be threatening litigation. In some cases, litigation is inevitable, but where possible it is better to settle out of court. Legal action is slow and therefore expensive. Juries in some countries can also hand out swingeing fines. The company should demonstrate contrition to those affected, and should speak to them in person. One building firm sends flowers to customers who complain, and then sends a sales manager to visit the person. The customer is usually more conciliatory after receiving the flowers.

TAKING ADVANTAGE OF A CRISIS

In 'Strategic Risk Taking', Aswath Damodaran suggests five ways you can take advantage of bad times to gain an advantage over your competitors.

1. Have better and more timely information about events as they occur and their likely consequences. You need good feedback from finance, sales people and market research. This lets you respond better to the situation.

2. Act faster than your competitors in response to changed circumstances. Looking back on her arrival as CEO at troubled Thomas Cook, Harriet Green said most employees were still in post, despite annual losses of £485m. She got rid of 2,500 staff, closed 195 travel agency branches, and disposed of businesses.

3. Use your experience of similar crises in the past. This helps you respond more effectively than organisations which have forgotten or are unfamiliar with what happened in the past. Knowing what you did in a previous downturn can give you a good road map. If you lack that experience, you can tap into the knowledge that's inside the heads of older managers.

4.   Use your resources to help you to ride out the rough periods that follow a crisis better than the rest of the sector. If you have reserves, you can buy distressed assets or businesses cheaply. Or you may be able to hire staff with experience of new growth markets.

5.   Be more flexible in the way you use your finances and operations. For example, be ready to change the way you bill clients or package your products.

These advantages, says Damodaran, can let you emerge from a crisis more competitive than you were prior to the event.

How to Handle the Media

Throughout the crisis, the business has to tell people what it is doing. Customers are often the most important group, but many other groups are often reached via the media.

Successful handling of the media has many benefits. It lets the company present its side of the story. It gets the public, customers or shareholders on the organization's side. It can even help to create a tide of opinion against a perceived aggressor (whether a company which plans a takeover, or a regulatory authority intent on taking the company to court). The first stage is to create an emergency PR plan.

DESIGNING A PR EMERGENCY PLAN

In a crisis, the PR person will be responsible for managing the media. One important task is to nominate a spokesperson, which we consider next.

NOMINATE A SPOKESPERSON

It is best to adopt an open attitude towards the media. If the company fails to answer their enquiries or is 'unavailable for comment', journalists will write their version of the story. And it will lack any mitigating facts that the company might provide.

The company should therefore ensure that a company spokesperson is nominated to handle press enquiries. The representative may not be the PR person. The words often sound better coming from the chief executive.

The spokesperson must be authorized to talk to the press. That authorization must allow him or her to make statements without having to clear them in advance. Otherwise, the company will spend too long debating the wording of an emergency press statement. Equally, the spokesperson must have access to the chief executive for briefings and debate.

INVEST IN MEDIA TRAINING

The company should institute a media training programme. This will involve a training day (or half day), in which a media specialist plays the role of an investigating journalist in a crisis. The training should include television training, using a video camera to record the executive's answers and body posture.

Practice in answering questions will help the executive respond effectively to questions concerning admission of liability, compensation and remedial activity.

MANAGING THE MEDIA INTERVIEW

The rules for managing an interview (especially on TV) are as follows:

- Lean forward in your chair. Sitting back makes you look disinterested.
- Look at the interviewer, not the camera. Don't look at the ground: you will seem shifty. Keep your hands still.
- Be primed with the answers. Be ready to counter false statements and rumours. If the interviewer says, 'People say your plant has a history of accidents,' say 'That's not true. Until yesterday, we had no accidents for a full year.'
- Keep talking – it stops the interviewer from asking another difficult question.
- Keep promoting the generalized good things about your company. Say, 'Safety is always uppermost on our minds, and we have a safety committee which is meeting as we speak…'
- Concentrate on the steps you're taking. 'We have drafted in a team of 20 people to clean up the mess … we are recalling all the faulty products … I've spoken personally to the family… '

KEEPING THE PRESS INFORMED

The company should keep giving the media information. Otherwise, the media will resort to inventing its own news, because a journalist's job is to file stories. When an oil tanker went aground off the Shetlands, and nothing newsworthy happened for several days, journalists started writing apocalyptic stories about dead otters. In fact, the only otter to die had been run over by a TV crew. Providing the media with information sets the agenda and keeps the company in control.

SUMMARY: STEPS TO TAKE IN CRISIS PR

- Address the problem. Don't try to conceal the issue and claim there is no problem. That will simply portray the company as a liar or at best not failing to recognize the true situation.
- Stop the problem. This could mean ceasing production of the product or turning off a valve.
- Instigate emergency procedures quickly. Activate the emergency plan. Collect faulty stock from customers, or start mop-up operations.
- Liaise with media and customers. Set up a hotline if appropriate. Be honest with people.

Managing Ten Types of Crisis

You need to identify the most likely scenarios, and have a plan for each. The crisis team needs to be able to open the relevant plan and see what they need to do.

Below we look at the ten typical crises: faulty products, pollution, health and safety incidents, fire and explosion, security, fraud, financial crisis, computer failure, industrial dispute and pandemic. We also consider how to deal with the problem for which no plan exists.

The actual crisis will rarely mirror the scenario exactly as planned. Therefore your plans should be modular. That is, you should have generic modules for specific activities, such as:

- power outage;
- fire;
- evacuating buildings;
- isolating buildings: stopping people from entering them;
- medical containment: your response to an epidemic;

- managing grief;
- managing death or injury;
- loss of buildings, facilities or IT;
- communication with internal stakeholders (staff, students);
- external communication with media and other stakeholders.

I. PRODUCT QUALITY FAILURES

A product quality failure is relatively simple if contained within the factory gates. Once it is in the distribution chain, the problem becomes more complicated. The first task is to get the faulty product back to the warehouse and destroyed. For the makers of branded goods, this becomes a major task, but one which has reduced in scale with the growing concentration of retail power. If you sell products, you should have a recall plan in place.

The subsequent actions – making amends and talking to the trade, media and public – may be strategically even more important: In the US, Tylenol survived and even prospered as a brand despite a contamination threat, due to management's media handling and prompt action. Other brands have lost substantial brand share when faced with similar problems.

2. POLLUTION

Pollution is likely to be caused by human error or by bad housekeeping. The first job is to stop the pollution from continuing to occur. This may involve shutting off a tap, or even stopping an entire factory.

The company has to inform the regulatory authorities immediately. Then it has to define the extent of the problem. How serious is it? What is the impact on wildlife or the environment? Then the company should identify the possible solutions. This may involve a clean-up operation. If you have the potential to pollute, you should have emergency pollution control plans carefully prepared, and those plans should be regularly practised.

3. HEALTH AND SAFETY INCIDENTS

Health and safety incidents are often traumatic for the company, its employees and the local community if they involve the injury or the death of a worker.

Many incidents need never happen. Workers get careless, and management does not insist on adhering to standards.

In health and safety incidents, the firm should adopt a contrite posture, apologizing for the incident (without necessarily implying liability), and offering sympathy and support to the victim and their family.

The company should stop all work that involves the process or machine that caused the accident. A report should be written, new procedures adopted and lessons learnt. Prepare a statement for the media explaining the steps that you have taken.

4. FIRE AND EXPLOSION

Fire is the one threat that most companies are prepared for. But planning often stops at the point where staff are safely outside the building at the assembly point.

Companies need to plan for the loss of the building, stock, telephones and computers. Specialist firms provide business recovery services, including:

- waste management firms to clear debris;
- power supplies, perhaps involving portable generators;
- temporary accommodation.

Fire often destroys plant and equipment, work in progress, finished stock and records. This prevents the company from delivering goods, invoicing and chasing money.

After a fire, the task is to start production, possibly in temporary premises or through sub-contractors.

Sometimes the fire is minor but the subsequent water damage can be extensive.

It is also imperative to talk to all customers, especially major ones, and explain how and when their orders will be delivered. If customers are aware of the problem, they are much more likely to be tolerant of delays in delivery.

5. SECURITY FAILURE, INCLUDING EXTORTION, KIDNAP AND RANSOM

Extortion, kidnap and ransom are serious threats for companies which trade in the world's dangerous zones. It is therefore dangerous to send staff abroad without reviewing the arrangements for their safety.

The world's danger zones shift as politics change, so it is important to stay abreast of changes.

The company should have a travel policy that limits the number of executives travelling together. Those who drive executives or key people should be trained in defensive driving techniques.

6. FRAUD

When fraud is uncovered, companies find it difficult to believe that a trusted company servant has behaved so dishonourably. Yet, all too often, the fraud could have been prevented by taking the steps outlined in this book.

The company should have a clear contingency plan for fraud. This involves preserving the evidence, and informing the authorities. Levels of fraud vary in severity, and some sensitivity should be adopted when dealing with it.

In one court case, a woman about to give birth was jailed for stealing a small sum of money, despite the fact that she had replaced it before the fraud was uncovered. Though the sentence was outside the company's power, it brought the firm some hostile headlines.

7. FINANCIAL CRISIS

No management ever wants to plan for a financial crisis. Nevertheless, the company should know what cuts could be made in the company's costs if revenue falls below a given point.

Likewise, if you're threatened with a takeover bid, don't wait until a hostile bid materializes before deciding what to do. There are many strategies you can adopt and which would make the business less attractive to a buyer. Likewise, you should have your PR machine ready to run, should the bid occur.

8. COMPUTER FAILURE

IT failure occurs, as we have seen, for all kinds of reasons, ranging from a flooded basement to a hacker. Organizations that are especially at risk should define the nature of their computing risks and take appropriate action.

You should assume for planning purposes that your IT is unusable, and then decide how you would survive without them. Your action might include getting loan servers and duplicate software from the computer supplier or a computer recovery firm, or moving into third-party business continuity offices.

Cloud computing reduces the risk of IT loss, though it isn't without its own risks.

9. INDUSTRIAL DISPUTE

In the case of an industrial dispute, the business might lessen the impact of the crisis by having built buffer stocks or having more than one production plant. A more positive approach would be to actively manage industrial relations or outsourcing certain activities so that disputes do not arise.

The crisis phase of the plan should cover areas such as contacting customers, and reassuring them about continuity of service. It should also cover communications with staff about the company's policies and plans. The question of maintaining production must also be addressed.

10. PANDEMIC

Pandemics are a 'Black Swan' event, coming out of nowhere, spread either by tourists returning from abroad, brought in by migrating birds or developed from a virus in farm animals.

Pandemics grow in intensity, with the number of cases increasing daily, reaching a crisis point, and then subsiding. This could give you enough time to set in place some of the procedures listed below.

You have a responsibility to provide a safe place of work, and with a contagious virus being spread, your workplace may no longer be safe. Staff should be educated, using emails and posters on how viruses are spread, and how to avoid infection. This includes enhanced personal hygiene and thorough handwashing. The organization should implement extra cleaning of the building, including handrails and door knobs. It may also need to co-operate with the health service in a vaccination programme

Face-to-face meetings should be discouraged. Wherever possible, staff should be able to work from home. This will require IT staff training people to access work computers remotely. When using remote access, staff will have more problems than usual, so you may need more IT helpdesk staff.

This increased access will also increase the vulnerability of corporate IT systems from viruses, so measures need to be taken to protect the system.

You might also want to use video conferencing. Staff will need training to use this.

You will find that with staff being absent through illness or home working, it is hard to maintain output. Therefore you will need either to release staff from other jobs or hire temporary staff.

Eventually the virus will lose its potency, and staff will start returning to work. Catching up with delayed work will be a priority: you may need temporary staff.

Unforeseen Problems

Murphy's Law will ensure that the problem that affects the organization will be none of the ones for which you have contingency plans.

If this happens, the business should have a general contingency plan for handling PR problems (for example, an ethical failure). This can be adapted to meet the specific requirements of differing scenarios.

The crisis team can also take the plan that most closely resembles the problem, and work from that.

Risk Assessment – Crisis Management

By answering the questions below, you can assess the company's vulnerability to business continuity. Score one point for every 'Yes' answer.

Question	✓
Do you have a management information system that will warn the company of an impending crisis?	
Do you have a crisis plan that identifies cross-disciplinary team members?	
Do you have a stated location for use as a crisis centre?	
Could your call centre, manufacturing sites or operations area continue to operate with minimal delay if they lost access to their building?	
Have you nominated a senior company spokesperson to handle press enquiries?	
Do you have an emergency notification system?	
Can your phones be re-routed if necessary to a call centre?	
Is there provision for a telemarketing agency to make contact with customers and stakeholders?	
Have senior executives received media training?	
Do you have the ability to rapidly provide counselling to victims?	
Total points scored:	

Score: 0–3 points: high risk. 4–6 points: moderate risk. 7–10 points: low risk.

The Appendix has a summary of the checklists. By entering the results of this one, you can compare the scale of business continuity risk against other categories of risk.

25
Risk Management Software: 11 Brands to Consider

Roger the FD is enraged. Apparently the Risk guys have bought some shiny new risk management software. But it won't talk to Roger's SAP, which cost a fortune and is supposed to be the future of the business. Meanwhile Anna, who isn't a team player, says she can't understand the risk software: it's too complex for her team to use. And Miguel prefers his Excel spreadsheets.

It's like a nursery school playground, you think. You can feel a headache coming on.

Why You Need Software

A cohesive approach to risk has become an essential part of an organization's management.

Risk management is no longer focused on isolated risks (for example, financial or reputational ones), but should be a company-wide process that covers all types of risk and all parts of the organization.

It needs to systematically identify and manage risks that affect the success of your operations.

This allows you to make decisions based on a better understanding of its risks. And that in turn avoids under- or over-investment and resourcing.

The organization therefore needs an easy to use, flexible and scalable tool that allows it to efficiently and transparently monitor its risks across different areas and levels of their organizations.

The system should deliver a structured, systematic and integrated approach to risk management.

This means having an automated tool that guides your risk management activities. It should use a common methodology, based on the organization's risk criteria.

A good system will provide the organization with a 'helicopter view' of all risks and their status, in contrast to the disjointed information usually held on departmental spreadsheets. The spreadsheet approach is widespread in organizations, and that itself creates a big risk, as we saw in Chapter 4.

A company-wide system will help to eliminate surprises, and will ensure that you focus not only on your high-probability/high-impact risks but also on the less 'high-profile' risks that are often under the radar but can escalate in probability and impact over time.

Typical Functionality Provided by Risk Management Software

At its core, ERM software should deliver a technological solution to meet the organization's need to manage its governance, risk and compliance (GRC) process.

It must also help the organization to manage its day-to-day work and change management process.

Ideally, the software should simply be an automation of a business' existing GRC policies and procedures. The functionality of risk management software varies from product. Some ERM products cover the entire risk management, governance and compliance sphere, while others are limited to specific areas of risk management.

Overall, some of the functionality available on ERM software includes:

- modelling business processes;
- identifying and assessing risks;
- testing and monitoring of controls;
- reporting and escalating risks;
- analysing risks;
- monitoring policy;
- managing audits;
- assessing compliance;
- managing incidents;
- managing project risk;
- regulatory reporting;
- document management;
- BCM.

Benefits of Using Risk Management Software

Some of the main advantages of using risk management software are as follows.

- enterprise-wide risk visibility (through interactive reports and risk dashboards that incorporate charts, graphs and heat maps);
- real-time warning systems that expose any looming 'surprises' in good time;
- organized approach to risk management;
- risk aggregation – one event on its own may not pose much risk to the smooth flow of business processes, however, a set of distinct events affecting different areas of the organization may create a substantial risk to the business as a whole;
- foster risk accountability – every risk has one or more managers assigned to it;
- ensure consistent application of standards, policies and procedures;
- identify opportunities for process improvement and risk reduction;
- develop organization-wide awareness of risk management;
- compliance with statutory and regulatory requirements;
- having readily available data for reporting to shareholders, rating agencies and other stakeholders.

Drawbacks of Using Risk Management Software

ERM software is vital. However, there are drawbacks that one must take into consideration before, during, and after ERM software implementation.

- *Complex to build* – For large organizations, the number, depth and diversity of potential risks is vast. Full implementation may take a while and considering the evolving nature of the modern organization, making changes to the parameters can be an arduous task.
- *Illusion of safety* – Automation of ERM can give a false sense of comfort and lead to a casualness that does not augur well for the organization overall. The organization may be ambushed by emerging risks that may not have been present or captured during initial ERM software implementation.
- *Incompatibility with the organization's existing processes and structure* – While small organizations with a fairly simple organizational structure may not find much difficulty alighting their structure and processes with ERM software, large organizations do not have any such luxury. It is improbable that a large organization's existing structure will be a perfect match for any ERM software. There may be need to restructure operations to ensure near perfect conformity.
- *Incompatibility with other business systems* – For ERM software to work as required, it must draw upon data from other business systems. The integration process may demand the development of interfaces to manage the transition of data from business systems to the ERM software.
- *Cost* – The most expensive ERM software is not necessarily the best. However, there is a correlation between functionality and price. The best ERM software does not come cheap and can significantly dent IT budgets of even large organizations. There are open source risk management software solutions but these are usually too simplistic to handle the demands of a major business.

Choosing the Right Software

Picking the right software is a complex process. The choice must be driven by stakeholders and risk management, and not the IT department.

Sometimes an organization chooses a complex piece of software when it actually needs something simpler. Hence the software should be scalable. This will allow the business to start risk management in a small way, and grow as it becomes more mature and has a better understanding of risk management.

You may find that using a Tier 2 supplier (one which is less well known and whose products are more tailored or more narrowly focused) may provide a better solution than Tier 1 suppliers, even though the latter may seen more reliable or have a better sales team.

It may be wise to use the KISS approach (Keep It Simple Stupid), because it is hard to implement complex ERM software in a large organization.

In making your choice you will need to instigate effective project management. This will ensure that the decision-making process is objective and controlled, rather than ad-hoc.

For the remainder of this chapter, we review 11 leading brands of risk software. Needless to say, software changes rapidly; so some of this information may be superseded by the time you read it.

Reviews of 11 Enterprise Risk Management (ERM) Software Brands

BWISE

URL – http://www.bwise.com/
Introduction – Founded in 1994, BWise is a major global player in the ERM, governance and compliance software space with particular strengths in business process management. With

hundreds of customers spanning over 80 countries, BWise Risk Management solution allows companies to comply with not just best practice risk management standards such as ISO 31000 but also industry regulations like Basel, SOX and Solvency.

Open Source or Proprietary – Proprietary.

Cost – Available on request.

Scope – GRC.

Industry – Manufacturing, energy/utilities, financial services, and pharmaceuticals.

Ease of Use, Implementation and Integration – BWise is one of the most mature risk management platforms available in the market today, a fact that has seen it consistently rated among the best by respected market research firms such as Forrester Research. It has a comprehensive risk library that enables clients to implement it quickly.

On-Premise or SaaS (Software-as-a-Service) – Available as both On-Premise and SaaS.

Case Study – France Telecom is one of the world's largest telecommunication and wide band operators in the world. The company was keen to replace an internally developed control system with an integrated enterprise-wide risk management solution. The goal was to get a more flexible, better performing easy to use system for managing internal controls. BWise' functionality met France Telecom's pre-defined criteria and was implemented within time and budget.

OPENPAGES (IBM)

URL – http://www-01.ibm.com/software/analytics/openpages/

Introduction – Open Pages was acquired by IBM, and is now managed by IBM software's business analytics division. It is strong on risk analytics and builds on the capabilities of IBM's Business Analytics offering, Cognos. Open Pages has strength in risk management, policy management and audit management. While it is one of the leading products in the ERM software space, there have been concerns on the quality of its reporting integration.

Open Source or Proprietary – Proprietary.

Cost – Available on request.

Scope – GRC

Industry – Banking, insurance, energy/utilities, healthcare/pharmaceuticals, travel/transport.

Ease of Use, Implementation and Integration – Not known.

On-Premise or SaaS – Available as both SaaS and On-Premise.

Case Study – Carnival Corporation, one of the world's largest vacation and cruise companies was keen on ensuring the operational, financial and compliance procedures and controls met management expectations. The cruise company was looking for a solution that provided a consistent, comprehensive and near-real-time view of risks across its 13 separate subsidiaries. There are also the new reporting requirements precipitated by SOX.

Carnival Corporation settled on IBM's Open Pages and was able to map its business processes in each subsidiary, identify the risks, monitor risks and create a consistent appraisal of risks across the enterprise. The result was improved staff productivity and convenient compliance with SOX.

Carnival's Costa Concordia cruise liner sank off Italy's Tuscany coast, allegedly caused by the captain deviating from the ship's computer-controlled route, and sailing too close to the shore while entertaining a female dancer on the bridge. It serves to remind us that the best plans and all the software in the world can't stop employees from doing stupid things. Thirty-two people died as a result.

THOMSON REUTERS

URL – http://accelus.thomsonreuters.com/solutions/risk-management/risk-management
Introduction – Thomson Reuters' Accelus Enterprise GRC suite provides a framework for identifying and managing enterprise risks. Its risk management module optimizes overall performance by firming up internal controls, streamlining business processes and proactively managing risk.
Open Source or Proprietary – Proprietary.
Cost – Around $50,000 a year for the SaaS version
Scope – GRC.
Industry – Banking, insurance and heavily-regulated industries.

Ease of Use – Thomson Reuters has a long reputation as a global content provider in both the general news and the financial information. This and its experience of serving sophisticated end users has led to the creation of this user friendly Enterprise GRC product. It includes regulatory compliance tracking and process workflow.
Ease of Implementation – It is easy to implement, particularly the SaaS model.
Ease of Integration – Thomson Reuters GRC platform was originally built for the financial services industry. Its functionality is therefore most suited to banks and insurance companies. That said, Thomson Reuters has made strides in attracting clients from other sectors.
On-Premise or SaaS – Available as either On-Premise and SaaS.
Case Study – Large US-based confectioner.
One of America's largest confectioners was looking for an enterprise GRC solution that would facilitate compliance with SOX. The use of disparate Microsoft Office documents for scheduling, planning and reporting was no longer practical and sustainable. After evaluating a number of solutions, the confectioner settled for Thomson Reuters Enterprise GRC. One of the main factors in their decision was the ability of Enterprise GRC to accommodate both internal audit and SOX data in a single, central and secure location. Also, the availability of a SaaS model meant implementation would be quick and upgrades would not require specialised IT skills.

A project manager from Thomson Reuters was available to help in the preparation, formatting and upload of data from the disparate legacy systems to the new one. Key processes, accounts, controls and strategies for SOX compliance were identified and applied.

SAS

URL – http://www.sas.com/software/risk-management/index.html
Introduction – SAS is known for enterprise-wide risk analytics and operational risk management. Other than its strength in risk management, the software supports audit, compliance and fraud monitoring functionality as well as social media tools for keeping tabs on risks to the business' reputation.
Open Source or Proprietary – Proprietary.
Cost – Available on request.
Scope – GRC.
Industry – Banking, insurance, energy, utilities, government
Ease of Use, Implementation and Integration – SAS is available in at least 15 languages. It is able to visualize complex relationships between processes, policies, risks and controls. While SAS's

strengths in risk management are difficult to dispute, there are areas of potential improvement particularly in audit management for large corporations.

On-Premise or SaaS– Available as both On-Premise and SaaS.

Case Study – Vattenfall is the fifth largest electricity generating company in Europe operating in Sweden, Germany, Finland, Denmark, Poland, Netherlands and the UK. It is involved in all stages of the value chain including distribution, transmission, generation and sales. Vattenfall chose SAS Enterprise GRC Solutions to manage the slew of risks that faced its complex business. The risks include operational, financial, legal, market, political and environmental risks. SAS Enterprise GRC monitors close to 1,000 risk metrics drawn from roughly 200 end users.

The implementation has allowed Vattenfall to streamline not just its risk management process but also have a workflow that delivers reports every quarter – something the company had been unable to do previously. Vattenfall has managed to assign assignments and tasks to specific users, groups or roles in the system. Overall, the main benefits the company derived from SAS Enterprise GRC include large data processing, risk event monitoring, detailed event assessment and risk incident reporting.

SAP BUSINESSOBJECTS RISK MANAGEMENT

URL – http://www.sap.com/solutions/analytics/governance-risk-compliance/risk-management/index.epx

Introduction – SAP's BusinessObjects Risk Management can identify risk events, track risk indicators and make available the kind of information that assist responsible and prudent management decisions. A business can map business goals on to their respective value drivers. SAP BusinessObjects Risk Management allows organizations to understand system risk and its impact on processes, value and performance.

Open Source or Proprietary – Proprietary.

Cost – Available on request.

Scope – Risk management. However, SAP has a larger set of integrated programs that cover the greater GRC domains.

Industry – Almost every major industry including government, manufacturing, pharmaceuticals, financial services and transport,

Ease of Use, Implementation and Integration – As one of the most established vendors of enterprise software in the world, SAP has a large network of vendors including partnerships with such major players as PwC, Deloitte, Novell, Oversight Systems, Computer Associates and SenSage. However, these partnerships also add an extra layer of cost.

Though vendor consultancy fees are not limited to SAP Risk Management and GRC products, the use of consultants is pervasive in SAP compared with other packaged risk management solutions. Customers may have little option but to purchase separate licences for the different SAP GRC products if they want to enjoy the impressive advanced capabilities of the system as outlined in SAP's product brochures.

On-Premise or SaaS – Available as both On-Premise and SaaS.

Case Study – Baker Hughes Inc. is an oil and gas services company with its headquarters in Houston, TX. The company serves oil and gas operators by engineering technologies, products and services that enable better expense management, risk reduction and increased productivity. Despite its successes, Baker Hughes was struggling to integrate its project risk management. Several manual processes and techniques had to be used and these were time consuming and prone to error.

Baker Hughes settled for SAP BusinessObjects Risk Management. The implementation ensured the creation of a central risk register that all relevant members of staff could access them. It has allowed every division within the organization to have a consistent view of risks.

ORACLE FUSION ENTERPRISE GRC MANAGER

URL – http://www.oracle.com/us/solutions/corporate-governance/risk-financial-governance/enterprise-GRC-manager/overview/index.html

Introduction – Oracle Fusion Enterprise GRC Manager is part of the Oracle Fusion suite of applications. It provides a comprehensive platform for managing enterprise risk, governance and compliance. It is a complete platform for the unified management of enterprise management, governance and compliance activities across the organization.

Open Source or Proprietary – Proprietary.

Cost – Available on request. However, prospective clients can use an onsite calculator http://www.oracle.com/us/logical-apps-calculator-152846.html to determine their potential savings from using Oracle Fusion GRC.

Scope – GRC.

Industry – Given the global dominance of Oracle in the enterprise software space particularly in the management of Big Data, it is perhaps not surprising that Enterprise GRC Manager is found in almost every industry. The reputation of Oracle's Big Data management solutions has made cross-selling easy. Industries where Enterprise GRC Manager has found application include financial services, transport, energy and utilities, manufacturing, healthcare, pharmaceuticals, real estate, telecommunication and government. Oracle has a massive vendor network worldwide that customers can benefit from.

Ease of Use, Implementation and Integration – Enterprise GRC Manager is highly adaptive to different industries and allows for a staggered approach to implementation. Customers can start with a basic rollout that focuses on key processes such as financial management before gradually rolling it out to eventually cover all functions within the bank. Like SAP, the dependence on consultants and vendors for implementation and second line support can significantly contribute to overall costs.

On-Premise or SaaS – Available as both On-Premise and SaaS.

Case Study – CSX Corporation provides rail services mainly for heavy cargo including gravel, sand, crushed stone, metal, coal, machinery and chemicals. It ships cargo to and from steel manufacturers, electricity power plants, sea ports and industrial plants. The company's network spans much of the eastern US including access to the Great Lakes, Mississippi River, Saint Lawrence Seaway, the Gulf Coast and ports on the Eastern seaboard. CSX relied on Enterprise GRC Manager to automate labour-intensive risk management and compliance processes. The system rapidly resolved existing end user access conflicts within the company's existing Oracle Financials deployment.

METRICSTREAM

URL – http://www.metricstream.com/solutions/risk_management.htm

Introduction – MetricStream is a significant player in the enterprise GRC market with a focus on large multinational corporations. MetricStream's ERM solution is built to ensure compliance with regulations, corporate policies, best practice and quality.

Open Source or Proprietary – Proprietary.

Cost – Small business deployment: US$75,000 to US $150,000; medium-sized business deployment: US$250,000 to US$500,000; large corporation deployment: US$750,000 to US$1 million. There are estimates only and will vary from one company to another.

Scope – Risk management.

Industry – Pharmaceuticals, healthcare, manufacturing, food and financial services.

Ease of Use, Implementation and Integration – MetricStream enjoys one of the highest levels of customer satisfaction. This is in part due to the company's ability to quickly respond to new demands and develop functionalities that address emerging challenges. However, one of its major drawbacks is the inability to support any other RDBMS (relational database) other than Oracle.

On-Premise or SaaS – Available as both On-Premise and SaaS.

Case Study – A global Fortune 500 corporation with its headquarters in the US was grappling with the challenges of regulatory requirements in multiple jurisdictions and the need to capture risks across numerous processes. A review of the existing risk management framework unearthed major vulnerabilities, accountability gaps and control inconsistencies. The company settled for MetricStream's Risk Assessment and Analysis module to tackle these challenges.

Following the implementation of MetricStream, the company enjoyed more consistent reporting, integrated management of risks, identification of key risks, strengthening of internal controls and enhanced efficiency.

SYMBIANT RISK SUITE

URL – http://www.symbiant.co.uk/Risk-Suite/

Introduction – Symbiant Risk Suite is a web-based risk management solution built around key best practice standards such as ISO 31000. With more than 20,000 clients worldwide but mainly in the UK and Australia, the Symbiant Risk Suite has found application everywhere from government agencies and multinational corporations, to small businesses and not for profit organizations. It allows businesses to identify, analyse, manage and monitor risks and controls.

Open Source or Proprietary – Proprietary.

Cost – Starts from £2 per user per month.

Scope – Risk management.

Industry – Financial services, construction, industrial, telecommunications, IT, retail, travel, government, not for profit.

Ease of Use, Implementation and Integration – Compared with the products discussed earlier, the Symbiant Risk Suite is perhaps the easiest to use and least difficult to implement. This has a lot to do with the fact that unlike other products in the market (particularly those used by large enterprises), Symbiant does not have the complex GRC functionality present in more sophisticated products.

On-Premise or SaaS – Available as both On-Premise and SaaS.

RSA ARCHER RISK MANAGEMENT

URL – http://www.emc.com/security/rsa-archer/rsa-archer-risk-management.htm

Software – RSA Archer Risk Management (RSA ARM) is a tool from EMC that allows organizations to actively identify risks to the business operations, reputation, technology infrastructure and finances. RSA ARM is part of EMC's broader enterprise GRC portfolio that includes risk, compliance, audit, policy, incident, threat, business continuity and vendor management.

Open Source or Proprietary – Proprietary.
Cost – A medium-sized enterprise deployment starts from US$55,000. The availability of a per-module license model with unlimited use/users allows for simpler, direct understanding of the cost and savings.
Scope – Risk management.
Industry – Government, retail, healthcare, manufacturing, financial services, energy, telecommunication and IT.
Ease of Use, Implementation and Integration – RSA ARM started out as a tool for IT GRC but has since expanded to accommodate other processes in the business. It comes prebuilt with a content library of over 10,000 assessment questions that allow customers to quickly get started.
On-Premise or SaaS – Available as both On-Premise and SaaS.

CURA ENTERPRISE RISK MANAGEMENT

URL – http://www.curarisk.com/index.php/solutions/enterprise-risk-management.html
Software – Cura provides a powerful and flexible framework for managing risk, allowing organizations to identify, analyse, evaluate and treat both risks and opportunities. While Cura's clients have predominantly been in Southern Africa, Oceania and Asia, the company is making strides in the European and North American markets.
Open Source or Proprietary – Proprietary.
Cost – Available on request. Pricing is either per-user or per-module.
Scope – ERM.
Industry – Mainly mining, financial services and pharmaceuticals. However, Cura ERM has also found application in healthcare, construction, IT, telecommunications, transport and utilities/energy
Ease of Use, Implementation and Integration – Cura ERM has excellent quantitative and qualitative analytical capabilities and is well suited for not only ERM in general but also more specific areas of risk management such as operational risk and project risk management. Some customers have raised concerns with its reporting ability especially for large multi-subsidiary multi-division organizations.
On-Premise or SaaS – Available as both On-Premise and SaaS.
Case Study – Anglogold Ashanti is one of the world's largest gold mining companies in the world with operations in 21 countries. Listed on the Australian, Johannesburg, London and New York Stock Exchanges, it has more than 60,000 members of staff. Anglogold Ashanti has relied on CuraRisk's ability to create consistent, assured, and simplified risk management and compliance processes. This has led to better and faster visibility of known and emerging risks.

ACTIVE RISK MANAGER (ARM)

URL – http://www.activerisk.com/products/active-risk-manager-arm/
Software – Active Risk Manager (ARM) claims to be the world's leading ERM software package. The company says that unlike traditional, compliance-focused 'GRC' solutions, ARM delivers more value and capability to its users. From managing project and programme risk through to strategic business planning, ARM helps organizations identify, analyse, control, monitor, mitigate and report on risk across the enterprise.
Open source or proprietary – Proprietary.
Cost – Available on request.
Scope – ERM.

Industry – ARM is used at London Underground, Crossrail, Lockheed Martin, EADS, US Department of Homeland Security, UK MOD, Saudi Aramco, Rio Tinto, Bechtel and Skanska.

Ease of Use, Implementation and Integration – Based on Microsoft .NET Framework that links to the client's choice of Oracle or Microsoft SQL database.

On-Premise or SaaS – On Premise.

Case study – '[Crossrail] realized that we needed a consistent approach to managing risk across the program. A key part of getting that consistent approach was to implement a common RMS. We went through a competitive procurement process and we selected ARM. We have now implemented ARM within our own organization and mandated its use across our supply chain to our major tier one contracting organizations … ARM has been able to drive consistency across this enormous program.' Rob Halstead, Head of Risk Management, Crossrail Ltd.

26
Do You Need a Risk Management System? ISO 31000

Your plant in Romania has suffered a big IT failure. 'Does anyone know what systems those guys were using?' you ask. 'They're a law unto themselves,' your assistant says, with a shrug.

'And why are their reports so vague, so lacking in numbers?' you say. But your colleagues just look blank.

Maybe it's time you got some better controls in place, with everyone using one system. At least you'd know what was going on.

The Need for Standardization

If you work for an organization of any size, you'll have divisions engaged in different activities, in various countries, each with their own patterns of behaviour. While a certain level of localism helps to foster entrepreneurialism and respond to market needs, it can also hamper the organization's ability to manage its risks. So while McDonalds' menu varies slightly in Oklahoma, Odessa and Osaka, the company needs to maintain common standards across the world to ensure a consistent service. Using standardized methods and getting standardized reports reduces risk and improves the company's knowledge. For this to happen requires several things:

* a common method of assessing, controlling and reporting risks;
* written procedures covering the main risks;
* a common vocabulary for risk;
* a process for checking that everything is working properly.

All this can be wrapped inside a risk management system (RMS). It lets management be reasonably sure that business risks are being tackled professionally.

Various RMS have developed, but ISO 31000 is the most recent and advanced. In this chapter we examine the main elements of the standard, and see how organizations have set about implementing it. Firstly though, a quick look at the concept of a management system.

What is a Management System?

A management system is a set of structures and procedures that ensure the organization can carry out the tasks required to achieve its objectives. Some people call it a 'framework'.

There are management standards for all kinds of activity, including quality, environmental impact and IT security, many of them hailing from the International Organization for Standardization (ISO, iso.org). Your organization is probably already using one or more of these standards, so introducing a risk management standard needs to be thought through, to prevent a proliferation and duplication of paperwork.

The most well-known management system is ISO 9000, the quality standard. Contrary to its title and popular belief, ISO 9000 is not really about quality. It is actually concerned with *consistency*. ISO 9000 audits ensure that jobs are done they way they're supposed to, which is a way of reducing risk. If a bakery uses ISO 9000, its cakes will be more alike in taste, consistency and appearance. ISO 9000 reduces the possibility of them being under-cooked or burnt.

Similarly, a RMS seeks to identify the major risks, and manage them systematically. Like ISO 9000, a RMS should mean 'no surprises'. The principles involved in a RMS are shown in Table 26.1.

Table 26.1   Principles of an RMS

Task	Effect
Understand your risks	Discover what risks the organization might suffer, and assess their likely impact
Set a risk policy	Decide on the business' attitude to risk
Decide how to deal with risks. Write procedures to manage them	This ensures that there is an agreed method for managing the company's risks
Assign roles and responsibilities	Determine who will take responsibility for managing risks
Train staff. Communicate effectively	Make sure that people are engaged in managing risk
Keep records	This allows you to see when or if important areas are moving outside the area of safety
Conduct regular internal audits	Audits ensure that every risk is regularly checked
Regularly review audit findings	This makes sure that management examines the audit findings, and takes corrective action
Institute a contingency plan	The plan will guide the business through a period of crisis
Engage external assessors	The system will be examined by independent outsiders, who are free from internal politics or culture

THE DEBT TO DR DEMING

Today's management systems are largely based on the work of W Edwards Deming, an American statistician and the man who popularized the principles of Plan, Do, Check, Adjust,[1] as shown in Figure 26.1.

---

1   Deming originally used the phrase 'Plan, Do, Check, Act'. But since 'Act' could cause confusion with the 'Do' phase, the term 'Adjust' is more widely used these days.

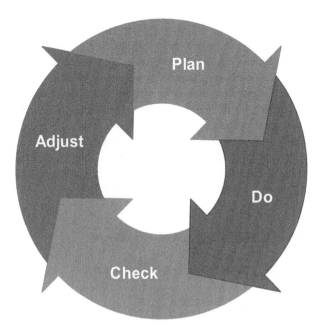

Figure 26.1   The Deming cycle

Translated into risk management, this goes as follows:

1. *Plan how you're going to manage risk*: Decide what risks need to be controlled. Specify how you will control them. Have written procedures for every major risk, so that staff know what is expected of them.
2. *Do what you say*: Staff carry out the work in the prescribed manner. Then show that you've done it – keep records showing that you managed your risks in the way you planned.

    Reports will show whether the work is within tolerances; and audits will show whether staff are doing their work in the required manner – for example whether computer backups are being taken. Any problems should be reported to management.
3. *Audit the business* to check that people are doing what they are supposed to.
4. *Make any necessary changes*. The final phase is when staff or management takes action to rectify any mistakes shown up in the audit or reports. It may involve retraining, changing a procedure or adjusting a machine. This stage also involves continuous improvement. You need to ask yourself: how could we reduce the risk still further?

After the Adjust stage, the cycle starts again. Are your plans still fit for purpose?

As we will see later in this chapter, different organizations have come up with different systems. The ISO model differs from the US COSO model, which we review later in this chapter. But they all work on the same principle, as shown above. To be effective, a RMS must detail all the company's major risks, ranging from environmental to corporate governance.

Benefits of a Risk Management System

As we saw in Chapter 1, it is better to tackle risk in an integrated and systematic way rather than work on individual elements. Working on isolated problems means that you may overlook some risks, and fail to control other ones.

A RMS may also be a good defence in law. If you can show that the company had a system in place for managing risk, and it was therefore taking reasonable steps to prevent a crisis, the judge or regulator is likely to look more favourably on the business in the aftermath of a crisis.

Companies can write fine sounding policies that are routinely ignored by the workforce. To be credible, therefore, the RMS should be verified by a respected independent authority.

Furthermore, the system must conform to a recognized standard. This yardstick tells stakeholders that the system is set at a meaningful level.

However, ISO 31000 is not a system that can be certified. In Section 1, Scope, the standard says it is 'not intended for the purpose of certification'. Thus the standard serves only to help an organization identify and manage its risks. This is a recognition that it could clash with or duplicate activity in an existing system such as ISO 9000 or OHSAS 18000.

Disadvantages of a Risk Management System

Opponents of management systems usually have two complaints:

1. The system is unwieldy and bureaucratic.
2. The company only installs the management system because a major customer or regulator requires it.

These two problems are not the fault of the management system, but of the management that implements it. Nevertheless, it is important to guard against such problems. Implementing a RMS should be done with a light touch, and done in a way that everyone learns from it.

TAILORING THE SYSTEM TO YOUR ORGANIZATION

The tail shouldn't wag the dog, and the RMS should fit your organization. Paul Hopkin of AIRMIC (Association of Insurance and Risk Managers, airmic.com) has suggested the acronym PACED. It stands for:

- Proportionate.
- Aligned with the company's activities.
- Comprehensive and structured.
- Embedded in the organization's processes.
- Dynamic. It changes as the organization does.

Meanwhile the ISO has said that risk management should do the following:

- create value – resources expended on mitigating risk should be less than the consequence of inaction, or the gain should exceed the pain;
- be an integral part of organizational processes;
- be part of the decision-making process;
- explicitly address uncertainty and assumptions;
- be systematic and structured;
- be based on the best available information;
- be tailorable;
- take human factors into account;
- be transparent and inclusive;
- be dynamic, iterative and responsive to change;
- be capable of continual improvement and enhancement;
- be continually or periodically reassessed.

A third danger is that the RMS will become routine and ossified – that it will fail to anticipate new dangers. Management can be lulled into a false sense of security, never dreaming that a problem could arise from an entirely unexpected direction. So you have to stay alert to new risks over time.

ISO 31000

As we have seen, ISO 3100 is the world's major RMS (allowing for the importance of COSO in the USA). In the sections that follow, we examine its structure.

HOW ISO 31000 WORKS

As with any management system, there is a lot of jargon. But once you've become familiar with that, it becomes straightforward. Figure 26.2 overleaf shows the process involved in the standard.

Figure 26.2 is not the conventional way of showing the standard. The ISO itself (iso.org) uses a more complex figure with no less than 17 arrows, many of them double-headed.

ESTABLISH THE CONTEXT (CLAUSE 4.3.1)

To establish the context you need to identify the wider issues that could put your organization at risk, or bring opportunity. It's not enough to consider your traditional risks, such as health and safety: ISO 31000 wants you to look at the broader picture. The standard recognizes two 'contexts' – the external and internal.

THE EXTERNAL CONTEXT (CLAUSE 5.3.2)

For the external context, you need to know who your external stakeholders are. They could be local, national or international, and include people who live in the shadow of your plants, local councillors, national politicians, regulators or civil servants. They can also include suppliers and business partners. Have a look at Figures 2.5 and 2.6 in Chapter 2.

Then there are the key drivers and trends that will affect your business.

- Stakeholder values, perceptions and relationships: What do stakeholders expect of you, in terms of ethical behaviour, support for the community, or levels of employment? The higher your profile, the more visible you will be; and therefore the risk is higher.
- Social, cultural, political and legal issues: What social trends are taking place, for example the formation of households, house buying or attitudes towards business? What is going on in the political world that might affect you?
- Regulation: How are you regulated? What are the risks of failing an inspection or being the subject of an investigation?
- Technology: What trends are taking place? What opportunities or threats might they bring?
- Finance and the economy: Is the country in a period of growth or recession? Is the mood optimistic or pessimistic? What risks does the business cycle bring?
- Nature: Are you affected by climate change? Is adverse weather likely to be a problem?
- Competition: Who are your competitors? How strong are they?

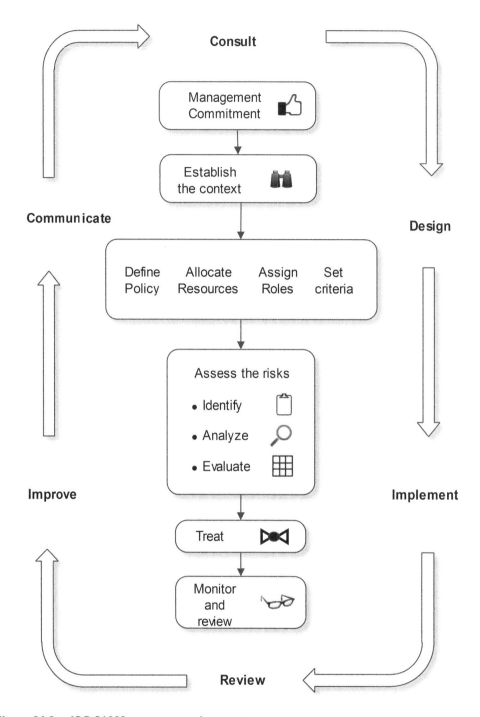

Figure 26.2   ISO 31000 structure and process

We discussed these factors in Chapter 2. Have a look at the PESTLE chart and the SWOT analysis in Figures 2.1 and 2.2.

THE INTERNAL CONTEXT (CLAUSE 5.3.3)

The internal context relates to your own organization. It includes the following.

- *Who are your internal stakeholders*, such as employees, trades unions, business partners, and shareholders? What are their views? How could they affect you?
- *The organization's governance.* What are the organization's goals? How is the business managed and structured, and is there a risk structure in place? Is it a publicly quoted, publicly owned or a privately held business? Entrepreneurial or mature? Risk hungry or risk averse? Based in many countries or just one? What policies does it have, for example on ethical issues or governance?
- *Your contractual relationships.* Are you engaged in any JVs or partnerships? Do you outsource work? These relationships are a source of both opportunity and harm.
- *Capabilities.* How much money is in the bank, and therefore how much can you put at risk? How profitable is the business? How much debt is it carrying? What technological strengths or weaknesses does it have? How able are its people?
- *Culture.* Do staff understand the nature of risk, or will you need to educate them? Do you employ mainly professionals or blue-collar workers?
- *Standards.* Do you have ISO 9000 or any of the other management standards?

DEFINE A RISK POLICY (CLAUSE 4.3.2)

A risk policy statement defines your organization's commitment to managing risk, and outlines its general approach.

The policy might explain why the business has developed a risk policy, what your appetite for risk is, what the scope of the system is (what will be covered), who is responsible for managing risk and how risk will be audited.

You also have to communicate the policy, so that staff are aware of it, and so they know that risk is part of the organization's goals. The CEO needs to be seen to embrace risk management. This means the CEO should promote the need for good risk management.

NOTTINGHAMSHIRE RISK POLICY (EXTRACT)

Nottinghamshire Police will employ a formal, structured process for the identification, evaluation and response to corporate risks. The Force will seek to identify threats, opportunities and vulnerabilities at the earliest opportunity and then measure their likely effect on the achievement of its business priorities. Wherever practicable, the Force will endeavour to apply a proportionate level of resources to control known risks in order to preserve the quality of its service provision, whilst maintaining value for money.

The Chief Officer Team will seek to obtain regular assurance that the controls put in place to mitigate risk exposure throughout the organization are effective and proportionate. This will be enabled through the maintenance of risk registers that are reviewed and updated in line with a formal procedure, and the production of an annual report on the efficiency and effectiveness of corporate risk management throughout the organization (as part of the Annual Governance Statement).

The Force will maintain a Strategic Risk Register, and separate risk registers for every division and department, recording them on the Orchid Risk Management System.

RISK POLICIES FOR SPECIFIC AREAS

The organization needs a policy for each area of risk. Most large companies have environmental policies and health and safety policies, while fewer have a defined policy on fraud or computer failure.

The policies should be as brief and simple as possible to ensure that staff read and understand them.

Defining a policy is not simply a paper exercise. The policy should be translated into strategy statements which are implemented at the lowest suitable level. For example, a quality policy should be implemented in each manufacturing plant.

ALLOCATE RESOURCES (CLAUSE 4.3.5)

To make the system work you need people and money, and this means allocating resources to it. You need to give staff responsibilities, as we see below. And people need training, so this needs to be included here.

The standard only suggests you 'consider' setting up procedures (clause 4.3.5). This means it lacks the definitiveness of ISO 9001 which requires them. Without written procedures, people can't be held accountable, performance measured or systems audited.

ASSIGN ROLES AND RESPONSIBILITIES (CLAUSES 4.3.2 AND 5.3.4)

Most organizations of any size will have someone who performs the role of 'chief risk officer'. There may also be 'risk champions' in each department who look out for possible risks.

In addition, every risk needs an 'owner', someone who will take responsibility for it. Hence the need to assign responsibilities. We considered this in Chapter 3, Figure 3.3.

Members of staff should be responsible for managing risks in their own area. This can form part of their job description, and be a criterion for measuring performance in annual appraisals. This will encourage them to think about risk. You can also empower people to modify their work environment, within the scope of their authority.

Be aware, however, that the more junior the staff, the more risk averse they are, and that can lead to delay and inefficiency. Junior members of staff often want to cross every T and dot every i, without realizing that the risks can be low. For example, you can add many controls to the recruitment process, and make sure every step is carried out exactly. But the risk of being sued for minor infringements of the recruitment process is quite small, and you can end up making the system too burdensome.

Managers take risks only when they are comfortably above their target, according to Zur Shapira of New York University. As a corollary, they won't take risks when they are at a survival level.

SET RISK CRITERIA (CLAUSE 5.3.5)

You need to know in advance how the organization will respond to a particular threat or opportunity. For that reason you need to set out risk criteria. These reflect the company's appetite for risk, in terms of its probability and severity. They will help a risk owner know whether a specified level risk is acceptable or not. This is often set in a heat map. One is shown in Figure 2.15.

ASSESS YOUR RISKS (CLAUSE 5.4)

We looked at risk assessment in Chapter 2. ISO 31000 breaks this down into three distinct phases, as shown in Figure 26.2. These are: Identify, Analyse and Evaluate.

IDENTIFY THE RISKS

In Chapter 2 we looked at mechanisms for identifying your risks. They included questionnaires, brain-storming sessions and historical data. You can seek advice from experts, both inside and outside the organization. In uncovering risks, you will need to identify their possible causes and also the potential consequences.

In establishing the context (4.3.1, 5.3.2, and 5.3.3) earlier, you will have begun to see where risks were lurking, and probably started adding them to the list or 'risk register' (see Chapter 2 and Table 2.8).

Businesses sometimes come up with over 100 risks, which are later honed down to 10–20 serious ones.

ANALYSE YOUR RISKS (5.4.3)

Analysing the risks is only a small step beyond Identifying them. Here you will decide which risks are the serious ones. You do that by considering the likelihood of an event happening (probability), and the seriousness of its consequences (the severity).

EVALUATE YOUR RISKS (5.4.4)

In ISO 31000 parlance, when you evaluate your risks you compare them against your criteria (is the risk acceptable to us?).

At this stage you will decide what to do with each risk, whether to Accept, Avoid, Share or Control it (see Figure 4.1).

In practice, identifying analysing and evaluating risks are likely to happen at the same time: once you identify a risk you will ask yourself how serious it is and how you should manage it.

TREATING YOUR RISKS (CLAUSE 5.5)

You will need a treatment plan for each. This has to be documented (that is, written down). Many organizations keep a risk register. The plan is likely to be contained in a Manual, whose contents we outlined in Chapter 4.

Risk should be a component of all procedures, rather than being part of a stand-alone RMS. Moving risk outside everyone's everyday work life makes it less likely that risks will be spotted or controlled.

All new projects, for example, should have a process to guide them; and risk assessment and treatment should be part of that.

MONITOR AND REVIEW YOUR PROCESS (CLAUSE 5.6) AND YOUR FRAMEWORK (4.5)

Monitoring and reviewing allow you to:

• evaluate how effective your risk treatment is;
• watch for changes in the inside and outside environment, and identify emerging risks.

This makes sense, for if a risk isn't monitored, it won't be managed.

You will need to define who is to be responsible for monitoring and reviewing the system.

The information you collect should be helpful to management. It should support the business in reducing cost, identifying problems and making the workplace safer and better.

The monitoring should be planned, and be part of the organization's regular performance measurement system. In other words it should fit in with the quality systems or the company's audit programme. And the results should be reported to management so they can review it and take corrective action. They in turn should be ready to change the plan as new circumstances arise.

KEEP A RECORD OF RISK MANAGEMENT ACTIVITIES (CLAUSE 5.7)

As we've seen, you need to monitor your risks; and that entails keeping records.

You can require staff to record incidents, near misses and insurance loss events. Many of these records will already be in use. Most organizations have a book or database to record health and safety accidents.

Your records will demonstrate that you're managing your risks. You can also use those records to review and improve your risk management.

The data should be kept to a minimum and be put on an intranet. Given the breadth of the system, from corporate governance to slippery warehouse floors, the system could easily turn into an unresponsive, unwieldy beast. The documentation could include the following:

• organization charts;
• descriptions of roles and responsibilities for risk;
• risk register, describing major risks;
• policies;
• operating procedures, that is, a description of how you control the most important risks;
• process flowcharts;
• key performance indicators;
• results of audits;
• results of reviews.

COMMUNICATE AND CONSULT WITH YOUR STAKEHOLDERS (CLAUSE 5.2)

The standard emphasizes the value of two-way communication. This implies that your communication should include listening and consultation, rather than top-down statements.

Risk management communication tends to be one-way, telling people what to do. But the business should also engage in a dialogue with stakeholders. This in itself can be a risk: you can receive unwelcome answers or criticism. But unless you start a dialogue with the workforce, local community and advisors, you will be reliant on just your own knowledge.

The risk manager should liaise closely with the PR and marketing departments, so that the risk management message is regularly explained.

Some of the best ways to communicate with stakeholders are as follows:

Team briefings: Used to trickle policy and priorities down through an organization.

Intranet: Good for communicating with employees. Some companies provide incentives, such as free or discounted tickets to events, on the intranet. This is used as a means to get people to visit the intranet, and in so doing see the company's policies and news. The intranet could provide a risk portal, containing lists of the company's known risks, categorized by region, plant, process or market. The risk portal could encourage staff to suggest new risks, perhaps using Sharepoint or a content management system that allows users to post material to the site.

Internet: A good way to communicate with shareholders, pressure groups, young people, the local community and potential employees – in fact just about everyone.

Annual report: As we saw previously, analysts are unimpressed with companies' reporting of their risks in financial statements, with 53 per cent rating them 'fair' to 'poor'.

Email: Most people are overwhelmed with email, so it is a less effective method of communication than it once was. Nevertheless, it gets to every desk; so at the very least you can use it defensively to demonstrate that you have, for example, disseminated corporate policies to staff.

Notice boards: Another good way of communicating news to employees.

Newsletters: A solution often used to communicate with employees in an informal way.

Mail: Since shareholders receive financial news by post, this can be a good time to explain how the company tackles risk. Mail is another option for firms who communicate routinely with customers by post, such as utilities.

On pack: A good way to communicate with end users, and warn them of the risks involved in your product. This applies especially to businesses that sell pharmaceuticals, tools, plant, equipment or machinery.

The needs of different stakeholders

Different stakeholders need different communications. These can be summarized as follows.

- *Board and senior management:* Information to help them make policy decisions, set priorities, and allocate resources.
- *Workforce:* Information to help them avoid suffering physical harm.
- *Management:* Help in managing the company's risks.
- *Local residents:* Reassurance about CSR issues, such as environmental matters.
- *Investors and stockholders:* Information about the company's ability to survive the uncertainties of the market.
- *Customers:* Reassurance about the company's and brands' integrity. Information about the products' safety or environmental credentials.
- *Partners and suppliers:* You should raise their awareness of risk, and create a dialogue about shared risks and how they might be controlled.

BENEFITS AND DISADVANTAGES OF ISO 31000

There are benefits and disadvantages in implementing ISO 31000. Below we look at the main ones.

ARGUMENTS IN FAVOUR OF ISO 31000

Every organization needs to manage its risks in a controlled way. And ISO 31000 provides a good structure for that.

Over 40 countries have now recognized ISO 31000 as their national risk management standard. This includes Russia (GOST R ISO 31000), Canada (CAN/CSA-ISO 31000) and Argentina (IRAM-ISO 31000).

ARGUMENTS AGAINST ISO 31000

Opponents question why they should install multiple standards with all the duplication it entails, while good management practice points towards integrated systems.

Clause 4 of the standard tells us that ISO 31000 does not set out to describe a management system, and that it is up to individual businesses to integrate risk management within its overall management system. This will avoid a proliferation of systems. Nevertheless, the very existence of ISO 31000 produces more handbooks, manuals or procedures.

ISO 31000: ONE COMPANY'S FINDINGS

Using the principles of ISO 31000, the enterprise risk management programme at Dubai Aluminium (Dubal, www.dubal.ae) has achieved the following results:

- Improved standardization. It has streamlined and harmonized risk management processes across the enterprise.
- Increased collaboration and accountability. It has broken down organizational silos and improve decision-making.
- Increased risk protection. It enables decision makers to determine the potential impact of risk and develop an action plan.
- Provided an integrated, real-time view of risks. There are advanced risk heat maps, charts, dashboards, and trending analyses that strengthen transparency into risk and control management.
- Introduced a mature approach to risk management. It offers a structured and systematic approach to risk management.
- Delivered greater efficiency and reduced costs. It has automated manual processes such as risk-control assessment, reporting, remediation and audit trails.

Other Risk Management Systems

Apart from ISO 31000, there are two other major risk management standards, the IRM standard and COSO.

IRM'S RISK MANAGEMENT STANDARD

The IRM developed a risk management standard (Figure 26.3), with help from the Association of Insurance and Risk Managers (AIRMIC) and ALARM, the National Forum for Risk Management in the Public Sector (alarm-uk.org).

With the arrival of ISO 31000, IRM decided to maintain its support for the IRM standard because it was well established and provides a practical means of managing risk.

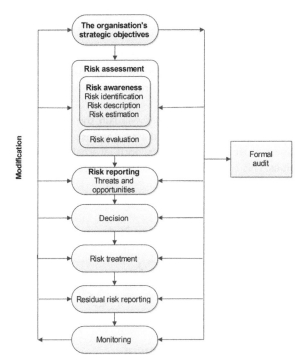

Figure 26 3   IRM risk management standard

Source: IRM

In this book we won't go through the IRM process because ISO 31000 uses the same broad approach.

You can get more details from the IRM (theirm.org).

THE COSO FRAMEWORK

'COSO' refers to the 'Committee of Sponsoring Organizations of the Treadway Commission', a private sector initiative set up to investigate the causes of fraudulent financial reporting. The Chairman was James C. Treadway, hence the committee's name.

COSO is important because the US SEC, the Federal regulator, requires public companies to maintain a system of internal control under SOX. It has named only one, COSO, as suitable for that purpose.

COSO is more limited than ISO 31000, because it focuses on corporate governance and internal controls. It doesn't look at the broader issue of risk management.

It seeks to ensure that companies are run honestly, and aims to achieve that by having transparent and accurate financial reporting. It concentrates on compliance – that is, obeying the rules set down by legislation or a regulator. It is aimed primarily at US publicly quoted companies.

As with ISO 31000, the Framework sets out guidelines rather than instructions, which means that no organization can get audited against it. In the following section, we'll examine its main tenets. But to avoid duplication we won't go into detail unless it deviates significantly from the methodologies discussed above.

COSO structure

The COSO document 'Enterprise Risk Management – Integrated Framework' explains how businesses can integrate risk management into their day-to-day controls. In the Framework, COSO identifies three areas where the business needs objectives:

- *Operations.* The company must define how it will run the business in a way that meets its operational and financial goals, and safeguard assets against loss.
- *Reporting.* The business must provide reports, both internal and external, on its financial and non-financial activities. This can include reliability, timeliness, transparency or other measures required by regulators or the business itself.
- *Compliance.* The business must adhere to legislation.

These run along the top of the box shown in Figure 26.4.

Figure 26.4   COSO Framework: objectives and components depicted in a three-dimensional matrix

Source: COSO

They are supposed to run like the words in a stick of rock through the 17 principles defined by COSO. We look at the 17 components below, but to clarify matters, COSO wraps the principles inside five components:

1. Control environment
2. Risk assessment
3. Control activities
4. Information and communication
5. Monitoring.

The five components are shown by horizontal rows, and the organization's units by the third dimension.

The 17 principles are as follows. They will be recognizable to any reader familiar with the UK's Corporate Governance Code.

Control environment

1. The organization demonstrates a commitment to integrity and ethical values.
2. The Board of directors demonstrates independence from management and exercises oversight of the development and performance of internal control.
3. Management establishes, with Board oversight, structures, reporting lines, and appropriate authorities and responsibilities in the pursuit of objectives.
4. The organization demonstrates a commitment to attract, develop and retain competent individuals in alignment with objectives.
5. The organization holds individuals accountable for their internal control responsibilities in the pursuit of objectives.

Risk assessment

6. The organization specifies objectives with sufficient clarity to enable the identification and assessment of risks relating to objectives.
7. The organization identifies risks to the achievement of its objectives across the entity and analyses risks as a basis for determining how the risks should be managed.
8. The organization considers the potential for fraud in assessing risks to the achievement of objectives.
9. The organization identifies and assesses changes that could significantly impact the system of internal control.

Control activities

10. The organization selects and develops control activities that contribute to the mitigation of risks to the achievement of objectives to acceptable levels.
11. The organization selects and develops general control activities over technology to support the achievement of objectives.
12. The organization deploys control activities through policies that establish what is expected and procedures that put policies into action.

Information and communication

13. The organization obtains or generates and uses relevant, quality information to support the functioning of internal control.
14. The organization internally communicates information, including objectives and responsibilities for internal control, necessary to support the functioning of internal control.

15. The organization communicates with external parties regarding matters affecting the functioning of internal control.

Monitoring activities

16. The organization selects, develops and performs ongoing and/or separate evaluations to ascertain whether the components of internal control are present and functioning.
17. The organization evaluates and communicates internal control deficiencies in a timely manner to those parties responsible for taking corrective action, including senior management and the Board of directors, as appropriate.

VIEW FROM A REGISTRAR

'We ask companies to identify their biggest risks and opportunities,' says one certification body (also known as a Registrar). 'Then we focus on those areas in our audit. As with ISO 9000, we look at their processes. The audit is tailor-made to each customer, and helps them control their risks.

'Risk management is still evolving,' he says. 'ISO 9000 took many years to develop, and so we don't necessarily want an international standard before everyone agrees to what is needed.'

'With risk management, different firms face different risks, and each industry is very different. By contrast, ISO 9000 is very broad and meets the needs of most industries. But even with ISO 9000, some industries, such as automotive, are developing their own variants of the standard.

'It's also worth remembering,' he said, 'that a risk-based assessment can take five times longer than a straightforward ISO 9000 one. We want to be of use to companies as their needs change, rather than imposing undue burdens.

'Another major issue we get involved with is "triple bottom-line reporting": often it's not financial problems that bring companies down but social or ethical issues.'

Risk Assessment

By answering the questions below, you can assess your vulnerability from not having a RMS in place. Tick all applicable boxes.

Topic	✓
Do you have a written policy on risk?	
Have you analysed both internal and external risks?	
Is there a register of risks which is updated annually?	
Do major risks have controls in place?	
Does each major risk have an owner?	
Are the major risks audited?	
Are audit results reviewed by senior management?	
Are continuity plans in place?	
Have continuity plans been tested?	
Is there a process for continuous improvement in the management of risks?	
Total points scored:	

Score: 0–3 points: high risk. 4–6 points: moderate risk. 7–10 points: low risk.

The Appendix has a summary of the checklists. By entering the results of this one, you can compare your need for a RMS against other categories of risk.

27
What Risks Will the Future Bring?

What the Future Will Look Like

The future is the biggest risk of all. It's hard to know what the risks are, because they haven't arrived yet. But ignoring the future is like going out on a winter's day without taking a look at the sky. Just as you'll know whether to take a raincoat, many of tomorrow's risks can be seen on the horizon.

In many ways the future will look like today. For a time traveller who arrives in town today from the 1970s, most things would look reasonably familiar. Supermarkets may be bigger and there is more fresh food, but the products are broadly similar. Petrol stations look familiar. Shoes are still shoes. Men's suits haven't changed much. For many markets, change is slow. We will always need food, housing and clothes.

Other markets, however, look completely different. Mobile phones and the internet were simply unimaginable 40 years ago (though laptops have been around since 1981).

So the question is, which things will change in the future, and how will that affect your organization?

Some changes are catastrophic, such as civil war, nuclear war or a sudden new ice age. These apocalyptic changes are virtually impossible to plan for, for they require us to stock up on candles and baked beans, and live in the woods.

For the purposes of this book, we have to assume a reasonably consistent future. We have to presuppose that political change in the West will be consensual, and that no 1930s-style financial meltdown will occur. It is equally probable that some countries (for example in parts of Africa) will suffer famine and civil war.

The rest of the world will continue to develop, albeit with disruption in many markets. Therefore the question is: what changes are we likely to see?

There are two major types of change: cyclical and linear. Cyclical change keeps swinging back and forth, such as boom and bust. This often creates change over the short term – through the business cycle – but comes back to the same point each time.

Linear change is the small or big changes that move us to a new permanent point. They can be step-wise (the arrival of the mobile phone) or incremental (a gradual increase in living standards).

We look at cyclical and linear change next, followed by a review of the four mega changes – in technology, resources, social change, and climate change.

Cyclical Change: The Swing of the Pendulum

If people like cars with sharper outlines today, in five years' time they will want cars with softer lines. This thirst for change applies not only to hemlines and other fashion items, but also to many other areas of life including politics (Figure 27.1). People's votes swing between parties that support the community and those that promote the individual.

Thus politicians decide introduce greater control over businesses (for example, more health and safety legislation); and when these controls begin to mount, the population votes for politicians who want to relax those constraints.

In economics we're subject to cycles, from growth to boom and bust and back again. There are cycles in exchange rates, housing prices and investors' behaviour (where they hop from equities to property and then to government bonds).

A business that doesn't know where it is in the cycle risks falling sales, or may be unable to manage its peaks and troughs. When the number of house sales reduce towards the end of the business cycle, estate agencies with multiple offices find they have too many offices and too many staff. Being aware of the cycle will help you plan – it is better to restrict investment and recruitment at the peak of the market, because sales might slump thereafter.

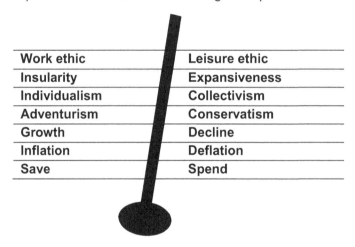

Work ethic	Leisure ethic
Insularity	Expansiveness
Individualism	Collectivism
Adventurism	Conservatism
Growth	Decline
Inflation	Deflation
Save	Spend

Figure 27.1   The swing of the pendulum: cyclical change

Linear Change

Some change is linear. That is, it goes in a straight line. For example, product quality gradually improves, and year by year the population is getting taller.

Maslow's 'hierarchy of human needs' (Figure 27.2) shows how linear change works. Maslow believed that people's most basic need is for shelter and food – this was the caveman's main agenda for each day. When these needs are met, people need safety, followed by affection and the esteem of others.

Finally, people opt for self-actualization. They look for ways to express themselves, whether by writing a novel or becoming assertive.

The theory is unproven. But it can be seen in the holiday market where, as affluence rises, people tire of popular holiday resorts and begin to seek more exclusive destinations. Or else they choose holidays where they learn new skills.

Similarly, when people have all the electrical goods they need, such as a fridge or a radio (Maslow's physiological area), they seek experiences instead (esteem and self-actualization).

But products will continue to be important. People aspire to 'statement products', ones that communicate certain values or characteristics ('I'm rich', 'I'm careful with money', or 'I'm fit and healthy'). Many brands carry strong associations, from Lidl to LVMH (Louis Vuitton, Moët, Hennessy, lvmh.com). These brands signal the owner's values or aspirations.

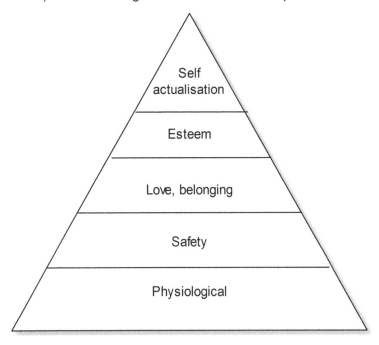

Figure 27.2   Maslow's hierarchy of human needs

As long as living standards continue to rise in the long run, people will seek better quality products, experiences and services (Figure 27.3). As food takes a smaller share of people's budget, consumers will experiment with new, more exotic foods. And they will seek higher-quality food rather than the cheapest brand. Some supermarket own-label brands, once the cheapest, are sometimes more expensive than the conventional brand leader.

Meanwhile, people living in the developing world will still aspire to products that they associate with the rich West. So the quality of products will vary throughout the world, while also increasing.

Some Western markets are unlikely to see much change. Gardeners, for example, will still want watering cans and seeds (albeit ones offering more value added, such as drought-resistant seed). Even in markets like automotive, change is usually incremental rather than revolutionary. Until or unless governments decide to outlaw the petrol engine, there will be simply a gradual improvement in safety and fuel economy, allied to styling changes dictated by fashion. Even electric cars, sometimes seen as revolutionary, have been around since 1888.

But it would be wrong to imagine there will be no pain, or that all markets will prosper.

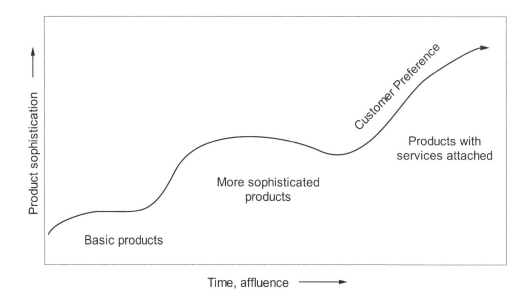

Figure 27.3   Linear change

The Four Big Changes

Major changes are coming. The big changes are in technology, social change, climate change, and resources change. These are the PESTLE-type factors we saw in Chapter 2.

> *Changes in technology*: Previous editions of this book mused philosophically about the coming together of voice and data communications, and the growth of computing power. But this edition sees something altogether more far reaching, even malevolent, namely the potential widespread loss of jobs due to automation. We look at this in more detail below, and see what changes your business might need.
>
> *Political, social and economic change*: Political and social change is much harder to guess. But the changes will undoubtedly reflect certain patterns, specifically the linear change and cyclical change we referred to earlier. Knowing what changes might happen will allow you to modify your products and services.
>
> *Climate change*: It's easy to think that climate change will simply make us grumble that autumn isn't as crisp as it used to be. But the problem is much deeper than that. It's about the loss of productive land, leading to harvest failures and starvation in many parts of the world, plus the creeping inundation of cities and shorelines, setting millions of people on the move. This may not be a problem for small businesses that trade in their local area. But most readers of this book have extended supply chains and customers in distant lands. We consider how businesses can adapt to this threat.
>
> *Shortage of resources*: There is a perfect storm headed our way, with climate change, a growing world population, a thirst for raw materials and finished goods, an increasing demand for water, and rising food prices. Later in this chapter we see how specific industries might adapt to this problem.

Let's look in more detail at these four major areas of change: technology, social change, climate change, and resources change.

Change in Technology

The internet has decimated the high street, with one in six UK shops empty, and with some prophesying that the number could rise to 40 per cent. Especially hard hit were book and record stores, whose customers went to Amazon.

Newspapers and magazines lost their readers to Google, while their recruitment advertising went to sites like Monster.com.

The improved communication afforded by the internet and email has also contributed to the growth of offshoring and outsourcing, with thousands of jobs going to China and India.

Disintermediation (customers buying direct rather than going through a broker) has hit insurance brokers and travel agents (Figure 27.4).

Other industries are vulnerable, including education, law firms and real estate agents.

Some industries however, remained intact. Heavy goods like pet food and furniture remained in retail, while those where the consumer needed to try on items, such as clothing and footwear, have largely remained in the high street, despite the growth of Zappos in the USA.

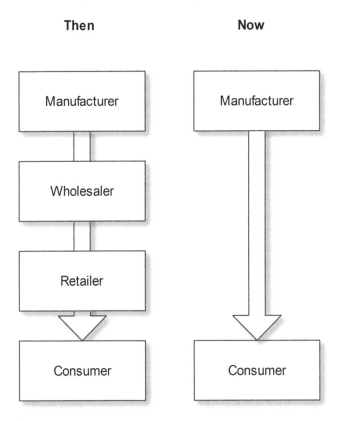

Figure 27.4   Disintermediation

Arguably, the low-hanging fruit has been plucked, and the consumer markets will stabilize in terms of online market share. In addition, the split between online and bricks and mortar will become less clear as traditional stores add stronger online offerings, and online businesses get a physical presence.

The high street will shrink, and become more focused on personal service and indulgence shopping; while landlords will need to develop flexible premises and models. Retailers will need to adapt to consumers' needs, and identify what they can't buy online.

GROWTH IN COMPUTING POWER AND TELECOMMUNICATIONS

We have grown accustomed to the changes brought about by the internet, but there are other changes on the horizon. Many of these changes are being facilitated by the growth in computing power and its convergence with telecommunications (the internet), something we turn to next.

Near Field Communication (NFC) such as London's Oyster Card and Barclays Pingit will facilitate the growth of new services, along with cashless payments such as Square. Identification methods such as fingerprint, iris and voice recognition will also have an impact. Much of this will involve the mobile phone whose potential hasn't yet been reached.

The changes in technology will be driven by Moore's Law, which states that computer power will double every 18 months. This means that computing power continues to become cheaper and more powerful, and that it will have hitherto unexpected uses.

Formulated by Gordon Moore of Intel in 1965, this law has held true for over 30 years, and shows no signs of tailing off. The associated technologies such as lithography (for printed boards) are likely to give way to others, which will enable computing power to continue its growth. Some believe that the law is self-fulfilling prophecy – in that companies like Intel put huge efforts into making the law come true. Whether or not this is true, the outcome is a continuing race to develop ever more powerful processing power, which in turn makes technological advances possible.

New storage devices, such as DNA storage, and the processing power of a quantum computer may overcome current IT limitations.

Although computing power has grown in a linear manner, the changes created by new technology will be neither gradual nor linear. History suggests that technical change is discontinuous. Things will seem to be steady or moving forward gradually, and then take off in a new direction. Such changes are hard to predict. For example, no one realized that mobile phones would be heavily used for a primitive form of communication – texting.

Computing power will also become embedded in everyday objects, giving us feedback and control that isn't possible today. According to Paul Saffo, the Director of the Institute for the Future (iftf.org), there is more computing power in the average electronic watch than existed in the entire world before 1960.

THE IMPACT ON JOBS

Perhaps the biggest impact of technology will be automation, followed by job losses. Robotics and automation are set to cut swathes of jobs. This has already happened in repetitive manual jobs, such as factories, and is set to happen in personal service such as waiting staff, baristas and reception staff.

Most jobs are mundane ones, and these are most at risk from automation. Table 27.1 shows the top ten occupations in the USA. You can probably see how automation will render many of these people unnecessary. Most of these jobs require education 'less than high school', to quote the US Department of Labor.

Already automated supermarket checkouts have replaced cashiers. They can accept vouchers, dispense change and ensure you pay for everything in your trolley.

Table 27.1   Most populous jobs, by category, USA

Rank	Occupation	Millions
1	Retail salespersons	4.5
2	Cashiers	3.3
3	Office clerks	3.0
4	Combined food preparation and serving workers, including fast food	3.0
5	Registered nurses	2.7
6	Customer service representatives	2.4
7	Waiters and waitresses	2.5
8	Secretaries and administrative assistants	2.3
9	Janitors and cleaners,	2.3
10	Laborers and freight, stock, and material movers, by hand	2.1

Source: US Department of Labor

Meanwhile robot security guards patrol warehouses. Unlike humans, they don't get tired, take sickies or skive off work.

And then there are the robots. Unlike older robots that had to be programmed to do one simple job, such as welding or painting a specific car, newer robots are small, flexible, have vision, can learn, and can make decisions. And they cost less than a human's annual salary.

Automation and robotics will remove the need for much physical labour, as we have seen above. Already there are self-guiding vacuum cleaners and lawn mowers. According to Hans Moravec of the Robotics Institute at Carnegie Mellon University (ri.cmu.edu), robots will, aided by the growth in computing power, become more intelligent than human beings later this century.

Robots find some tasks hard, such as going upstairs. But technologists are already making this happen.

As a business you will be able to replace many members of staff with computers and robots. From the robot that delivers food to a café table to the online automated assistant, you will be able to automate many jobs at low cost. Some businesses will retain people to provide personal service. But many customers prefer the speed and privacy that automation brings. When ATMs were first introduced, many customers said they wouldn't want to use them, preferring instead to speak with bank clerks. But they quickly changed their minds and stopped going into branches. So for many markets only the elderly will require human interaction.

Other businesses will boast about retaining the human touch, to disguise their lack of technical competence. In failing to automate, they will die out.

THE LOSS OF ROAD-BASED JOBS

The on-demand driverless car is set to cause huge unemployment, among taxi-, bus-, van- and lorry drivers. And driverless cars have fewer accidents than human drivers. So, according to Chunka Mui, writing in Forbes, there will be a big reduction in car accidents and therefore A&E visits, insurance claims and body shops, along with cuts in staff working in A&E, insurance and car repair.

There will be a big reduction in the need for car parks, since the cars will always be on the road. Similarly, car sales will fall, once the switch to driverless cars has been made. Government revenue from speeding tickets will disappear. Petrol stations and car washes will vanish. Car sales will plummet as people will summon them by mobile phone when needed. Even planes will have no pilots.

OTHER TECHNOLOGIES

Nanotechnology is the technology of very small things, which involves building products by moving atoms around. This makes it possible to manufacture anything at low cost. It includes:

- very small objects, such as nanorobots, performing surgery, killing cancer cells, or removing pollutants from water;
- conventional objects made from nanotechnology, including food made from atoms, and self-healing materials.

According to Foresight Technology (foresightgroup.eu), nanotechnology will bring about a revolution in manufacturing. This technology may allow things to be made for not much more than the cost of their raw materials. It is also likely to converge with other sciences and technologies, just as computers and telecoms did in the 1990s. NBIC (nanoscience, biotechnology, information technology and cognitive science) is a heady cocktail for new products.

- *Robotics* will remove the need for much physical labour. This can already be seen in self-guiding vacuum cleaners and lawn mowers. According to Hans Moravec of the Robotics Institute at Carnegie Mellon University (ri.cmu.edu), robots will become more intelligent than human beings later this century, aided by the growth in computing power. However, some experts believe that it will take many years before robots will be walking around the home. John Petersen, founder of the Arlington Institute (arlingtoninstitute.org), says that programming computers to behave like a human is difficult.
- *Molecular engineering* and biotechnology will extend peoples' lives and improve their quality. We are moving into an era where people can choose their babies' gender and other attributes. Stem cell technology will create new treatments from curing cancer to ending baldness. We will have a greater understanding of the role of individual genes (or groups of genes) particularly in triggering disease. We will see the development of tailored drugs that will manage cancers and other diseases more effectively. Using drugs and subcutaneous sensors and motors, we should be able to conquer diseases like Parkinsons.

  Synthetic biology uses 'biobricks', standardized DNA sequences that we can pop into living cells. Each of these new life forms will each have different characteristics or tools. They can be nanobots, equipped with Boolean logic gates (yes/no options) that will seek out cells which have been previously tagged, and communicate with them. This will include reprogramming damaged cells.
- *Carbon nanotubes* – the likely successor to silicon in computing, monitors, batteries and other products.
- *Fuel cell cars* – the end of the internal combustion engine, as well as a means of powering other objects such as laptop computers.
- *Flexible displays and memory devices* that can be integrated into clothing, or rolled up and stuffed into a pocket (OLEDS and plastic transistors).

- *Wearable computers* will monitor the individual's health, and allow more portable music or satellite navigation devices.
- *Voice recognition* may become more prevalent ('Open the pod bay doors, HAL' says the astronaut Dave Bowman in the film 2001 'A Space Odyssey').
- *3D printing* will allow individuals to make or order personalized products in unexpected ways. Among other things, it will allow people to make their own guns or bombs, or mobile phones. The principle of 'Rip, Mod and Fab' allows anyone to capture a real object with a scanner, model it on a computer and fabricate it with a 3D printer. It then becomes as cheap to make one product as to make 10,000. This means a shoe manufacturer can receive someone's footprint and their design preferences, and make a tailor-made shoe for them. Whether people will download a design and make the product in their home, thus bypassing the manufacturer, remains to be seen.
- *Moving from the inanimate to biology*, we can create living tissue using bioprinters that deposit cells on to gel, allowing the cells to fuse and grow.
- *Unmanned aerial vehicles* (UAVs or drones) will allow the transportation of objects, surveying and crime prevention.
- *Big Data* will permit greater accuracy in provide better information for business and research, such as drug development. It also has many other uses: BT has used the data from its track sensing and fault chasing systems to discover where people were stealing its underground copper cable. But Big Data also encourages an invasion of privacy by government or corporations.
- *Decentralized digital currencies*, such as Bitcoin, will facilitate the exchange of goods and services. They will challenge national or supranational currencies (the euro), just as Paypal fostered online payments when traditional banks failed to offer such a service.
- *Low-power embedded and connected sensors* embedded in everything from street lamps, appliances, cows, shoes and medical devices like asthma inhalers. This, the 'internet of things', will give rise to even more Big Data, with devices constantly talking to each other. It's the scenario where your home lets you know that your partner's train has been delayed, and that the plants need watering.
- *Renewable energies*, especially wind and solar, will grow in importance as they fall in price. Germany will soon get 30% of its energy from renewables, according to the New York Times, and its utility companies' profits have collapsed. Some US states are pushing for 20 or 30 per cent renewable energy. Consumers are installing their own panels, feeding energy into the grid. Experts say that energy generation is in a period of turmoil as big as anything in its 130-year history, and as great as the changes that have affected the airline, music and phone business. China could solve its pollution problems and become a leader in renewables.

NON-DISRUPTIVE TECHNOLOGIES

Many technologies around the corner merely enhance or supersede existing products. These include the fridge that warns you when food is past its sell-by date or biosensors that monitor your health.

However, while companies need to adopt these new technologies to stay in business, they won't cause a step-wise change in the market. Thus new forms of cryptography or computer grid networking will make life easier and more flexible but won't change the way the consumer does things. And the factory that produces silicon-based memory will merely switch to making plastic ones. Such technology includes radio frequency ID tags (RFID).

And while molecular machines will allow doctors to perform nanosurgery, repairing arteries and removing clots, this will still require a surgeon with the knowledge of the body, prescribing and controlling the nanobots in the patient's body. So even this most radical of technologies will in some cases merely give additional tools to existing technicians.

Some experts believe that the keyboard will still be everywhere, since it is the perfect device for inputting large amounts of text very fast. Voice-activated systems will also be common, but in the office they may not be very practical.

THE RISKS OF NEW TECHNOLOGIES

Cellular telephones and the internet have given people unprecedented ability to communicate. But there is also an opposing trend. New technology will allow greater control over the individual by the state or the corporation over the individual. For example, tracking devices can be used by the police to restrict an offender's movements, and keep them in a geographic area. The same technology can let an abusive spouse track their partner, and identify where they go and for how long.

In time, therefore, there will be calls for democratic controls on the technology, as we have seen with genetically modified organisms (GMOs) and stem cells.

There is also a risk that the new technologies – for example, powerful new diseases or new military technologies – could be stolen and misused by malevolent or mad individuals to wreak havoc on civil society. Even self-replicating machines pose dangers. UAVs could be used for snooping, crime and remote killing. This means there may have to be more stringent controls.

Biological technologies are evolving so fast that many scientists are unaware of the consequences, in the same way that nuclear scientists in the 1950s could not foresee the dangers of nuclear proliferation, miniaturization of nuclear hardware and the theft of nuclear devices. As Bill Joy, co-founder of Sun Microsystems, points out, we are moving from 'from wet to dry', no longer working with test tubes but instead clipping lengths of DNA to alter life forms. 'Matter programming' may lead to unexpected activities by these altered life forms, or their abuse by criminals or terrorists.

Similarly, the arguments over GM (genetically modified) food demonstrated that the scientists had not thought sufficiently about the risks of GM crops. These include the possibility that crops would become resistant to pests or herbicides, or wipe out or harm wildlife. It could also mean biotechnology companies taking an increasing control of the world's agriculture.

Socio-Economic Change

Many factors are at work in the economy and in people's behaviour. Here are the main ones:

> *Rising wealth.* As society moves forward, household wealth rises year by year. There is a possibility that the growth in affluence will be interrupted by the implosion of China or even the US, rising oil prices, stock market decline, inflation, a loss of consumer confidence, bank failures and job losses. But while some of them may occur, we have to assume that the world will continue to turn.
>
> Increased affluence will lead people to seek better quality goods and services. As *C.K. Prahalad* indicated in *The Fortune at the Bottom of the Pyramid*, millions of people in the East will gain unprecedented incomes in the coming decades, and will therefore want the same kinds of material goods as they see in the West.
>
> Within the Western population, there will be wide variations – not everyone wants to engage in self-actualization, and people's purchasing behaviour will continue to be affected by their age

and social class. In addition, some of the population will be affected by redundancy, a decline in their wages or other factors that reduce their spending power.

These people, along with the squeezed middle classes, will consume at the bottom end of the market, buying budget goods in budget stores, while a big proportion of the market will be content to buy no-frills services such as Ryanair flights. In *The Inventor's Dilemma*, Clayton Christensen describes how companies overshoot their market, providing products that are unnecessarily good. His examples included the shift from mainframe computers to desktops, and customers who abandoned department stores to shop at low-price drugstores.

Continued globalization. The loss of jobs in the West will be accompanied by the rise of China as the biggest economy in the world. As we have seen earlier in this book, offshoring of jobs will affect manufacturing and clerical work, plus some professional and management work. Personal services will be less affected by offshoring.

Increased proportion of older people. This will boost demand in some markets (for example, healthcare), and have negative effects elsewhere, such as making recruitment harder and increasing government's costs.

Innovative new products, leading to changed behaviour. As we see elsewhere in this chapter, there will be new opportunities in nanotechnology and molecular engineering that will lead to new products in healthcare, personal services and other markets. This will change people's behaviour, as mobile phones and the internet has done.

Massive unemployment. If predictions about job losses due to automation come to pass, and if there are insufficient new jobs to take their place, huge unemployment will occur. The majority of today's jobs are still the low wage, routine manual or clerical work they always have been, and these are especially at risk.

Unlike the past, today's technology will not give rise to jobs. Computer and robotic based technologies are doing away with the need for human staff, and we could have 50%–75% unemployment in the future. That in turn could lead to social unrest and major upheavals.

Many occupations will decline, largely due to technology. They include:

Loan or mortgage interviewers, switchboard operators; and other office and administrative support occupations.

Many middle class jobs. Computers can write newspaper articles, decide whether to buy or sell shares, do accounts and write reports.

Many production workers' jobs will disappear due to advances in technology, such as faster machines and automated processes. Electronic business will automate other jobs such as waiting staff, wholesale and retail buyers; postal office mail sorters, processing machine operators; railway staff, and meter readers.

Drivers. The on-demand driverless vehicle could cost the jobs of 232,000 taxi and limousine drivers in the US, along with 647,000 bus drivers and 125,000 truck drivers. In fact there are said to be 3 million such jobs in the USA, and 70 million worldwide, and all could disappear. Then take out related jobs such as traffic police and court staff. There could be 90% fewer cars, leading to a collapse of the car industry.

Medical. Medical diagnostic devices will give you a diagnosis, tell you what drugs to take, and then dispense them to you, without the need for doctors and nurses. Here's what Emmy, a former lab worker said:

"At the laboratory, we used to use glass pipettes to make dilutions of one sample of blood at a time, and then look at it in a counting chamber and count the cells. Then we would do some calculations – on paper. Now you put a drop of blood in a machine and it does the rest and gives you a graphic representation of the calculations. And that's what put me out of work."

There is a strong commercial imperative to replace humans with computers and robots, and employers will do this as soon as they can. With prices falling and intelligence rising, it is only a matter of time. 'Lights out manufacturing', where there are no humans, is upon us. People will look back on the era when jobs were only offshored as a time of good, steady jobs.

In the past technology helped to create jobs, because it simply replaced human muscle. But now computers are able to replace human minds.

Others are less pessimistic. Writing in the Observer newspaper, Will Hutton believes that humans will always have needs that must be filled. He believes there are four major growth areas:

1. *Micro production.* Hutton foresees a big growth in micro brewers, craft bakers, small film makers, micro energy producers, personalised tailors, and small software houses. These will use the internet and micro production techniques to produce inexpensive goods that are customised for the individual customer.
2. *Human wellbeing.* There will be growth in advising, coaching, caring, mentoring, doctoring, nursing, and teaching roles. Medical provision will grow, with replacement organs, skin and limbs opening up new specialisms and industries.
3. *Jobs relating to the problems of water and food shortages.* There will be new forms of mining on land, and farms on the oceans, along with space exploration to find new minerals and energy sources.
4. *Powerful and inexpensive data will give rise to new forms of communication*, relating to software, learning, information, and security.

The reader will have to make up their own mind whether these roles will replace the millions of office workers and delivery drivers who get made redundant. How much demand can there be for micro breweries and film makers? If there is growth it may be in the following fields, even allowing for the fact that many jobs will be displaced by automation:

- Teachers and child care (a big employment group).
- Health (another big employment group, but with increased employment due to the ageing population).
- Computer occupations.
- Food preparation and serving workers, including fast food and waiters and waitresses. Eating out and food preparation will grow.
- General and operations managers.
- Janitors and cleaners, for people who are time-poor and money-rich.

This paints a picture of a society taking on:

- Manual labour – to wait at tables, prepare food and guard our homes.
- Computer people, as IT continues to grow.
- Teachers, because it's a big profession.
- Health workers, to support an ageing population and a medical industry that has ever more ways to heal people.
- Managers to keep the whole thing going.

SOCIAL CHANGE

There has been a longstanding trend towards *individualism* and social freedom. We have seen a loss of deference, and reduced satisfaction with church and monarchy, allied to a decline in church going in the West. In the US, a third of adults under 30 are now without a religious affiliation, according to the Pew Research Center (pewresearch.org). There has been a similar fall in membership of established political parties, and a disinclination to vote.

But people still have their *spiritual and emotional needs*. People are keen to belong – to identify themselves as being part of a movement, a spiritual organization or community. This poses a problem for established churches, and opportunities for new political parties and religions.

The same loss of deference can be seen in the loss of trust in established medicine, and a reaching out to complementary and alternative medicine, and different religions and beliefs.

Consumers want to be treated as an individual when they buy goods and services in both the private and public sector, and to have their needs met by a personalized service.

Many people are still rooted in their local communities, but less bound by the conventions of their upbringing. They are also more needy and less private in terms of their emotions.

Another factor is people's *dissatisfaction at work*, and their need for a better *work–life balance* and an enhanced private life. This can take the form of having less commitment to their employer, working fewer hours or down-shifting.

If outsourcing and offshoring continue, companies will have a smaller pool of full-time employees, bolstered by contract workers, freelancers, outworkers, part-timers and seasonal workers, and with the majority of non-customer facing labour taking place in the developing world. But a gulf between growing corporate profits and reduced labour compensation and security could lead to unrest, higher taxes, protectionism or other government legislation.

Population change is something we have alluded to. This takes the form of the post-1945 baby bulge moving gradually through the population, getting gradually older. This results in a large affluent and adventurous group of people in later life. This is an attractive market for many businesses. It also means reductions in the size of the younger workforce, which will create problems for some companies. Equally, the Gen X, Gen Y and Millennial employees may be more compliant than the baby boomers who preceded them.

Obesity will be a worldwide problem. This together with smoking is a class-based phenomenon, with fewer professionals and managers overweight. Obesity (and its ugly daughters, diabetes and heart disease) is likely to be a growing problem for healthcare providers. It also requires new solutions for airlines and restaurants (wider seats), and new opportunities for other industries, such as gyms and fitness trainers.

Political Change

Political changes are both linear and cyclical. Over time businesses become subject to ever more stringent rules. On the other hand, voters oscillate between laissez-faire and dirigiste governments.

Thus it is difficult to predict whether governments will choose to effect social change, such as the following:

Leftist	Rightist
Reduce the gap between rich and poor by increasing taxes	Extend the gap between rich and poor by cutting welfare
Support employees	Support employers
Cut 'red tape'	Increase legislation on businesses

There are four main types of economic model:

- *The Anglo-Saxon*: a frontier-spirit type of economy that protects the rights of the individual and hostile to big government. It provides fewer safeguards for the sick, weak and the old. This has led the US through decades of growth, but whether the population will continue to want this remains to be seen.

- *The Nordic or European*: an egalitarian, interventionist model that provides more support for its citizens, possibly to the detriment of growth and entrepreneurialism. Some chafe against their loss of individual freedom and choice.
- *Authoritarian Eastern model*: a protectionist, centralized economy that pursues economic development while limiting political freedom. This is seen in China, where some in the population now demand political rights.
- *Caliphate*: A model based on religious teaching, seen in the Middle East.

It is unclear which route nations will choose in the future, and is it possible that populations will vacillate between the four types, or versions of them.

Meanwhile supranational groupings will continue to strengthen, such as ASEAN and the EU. At the same these groups will be vulnerable to implosion and to nationalist or separatist breakaways.

It is likely, however, that rising living standards will lead to higher standards of living and more regulation being imposed on business; but these are unlikely to affect the best-run businesses. Indeed, they can wipe out marginal, sweat-shop type businesses, to the advantage of larger or better-run organizations.

Climate and Weather Change

Global warming creates changeable weather patterns, with an increase in mean temperatures in some areas, and violent weather in others. Rising sea levels are another development. These new weather patterns can alter markets. For example:

A rise in average temperature could:

- increase the demand for one type of product and hamper another. For example, we could see increased sales of soft drinks, salads and light clothes, while reducing sales of heaters. Alternatively the reverse could be true if rising CO_2 levels stop the action of the Gulf Stream and make the UK cold;
- change crop production, with British farmers discovering they can grow Mediterranean crops.

More unpredictable or hostile weather could have the following effects:

- make sales harder to predict, leading to swings in demand in weather-related industries;
- make logistics less reliable, with more hold-ups occurring;
- stop normal sales activity, as in the case of tropical storms;
- damage or destroy property and other assets.

Organizations can adapt to changing weather patterns by:

- altering their products to suit changed markets;
- reducing dependency on suppliers and customers in geographical areas that are at risk from bad weather;
- learning to predict and adapt to a new seasonality, for example the annual arrival of a stormy season;
- making their buildings and other assets more storm-proof;

- ensuring that staff are able to cope with weather emergencies, for example by practising emergency drills.

Meanwhile some low-lying coastal areas will get covered by the sea as the years progress, leading to the loss of property and tourist businesses. Conflicts will grow between residents desperate to save their livelihoods and governments reluctant to fight a losing battle with the weather.

Since weather affects many markets, directly or indirectly, each business will have to assess the likely impact of weather on their products, markets and suppliers; and take practical steps. The Swiss are building new resorts in higher alpine slopes to counteract the loss of snow in traditional resorts. This meanwhile has led to outcries by environmentalists.

Finally, changing weather patterns can bring big rewards to companies that see risk as an opportunity.

Changes in Resources

Though population growth has reduced, there are an extra 8,000 people on the planet every hour. This means there is a growing demand for water, food and energy. As Figure 27.5 shows, they are linked. To grow more food we need more water. To get more water, we need more energy to cool power plants. To get more energy we have been using biofuels which creates a food shortage. Each makes demands on the other.

Despite mankind's ingenuity, we risk a shortage of energy, water or food. The answers lie in energy efficiency, in its broadest terms. This means finding new ways to reduce the demand for water. For example, a growing demand for meat around the world is increasing the need for grain, which in turn needs more water. An alternative would be lab-produced meat.

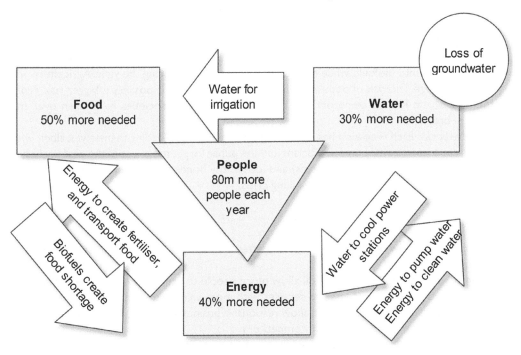

Figure 27.5   The water, food, energy nexus

These are not just issues for government. If energy costs rise, raw food costs rise or the availability of water declines – this has an impact on most businesses and large organizations. We may have to create a world in which we get by with less energy or different energy, with different foods and with less water usage.

Threats and Opportunities in the Future

Businesses must learn to live with the threats and seize the opportunities. The risks include the following:

- *Continued terrorist activity.* The anger felt in many Islamic societies towards Western governments will either abate or continue to manifest itself in acts of terrorism, depending in part on whether the West intervenes in Islamic states. Conflict between Shia and Sunni will continue, resulting in unstable societies. Other terrorism groups, possibly based on leftist ideology, may emerge.
- *Risk of chemical, biological, or nuclear disaster* (the so-called 'bioterror and bioerror').
- *Risk of nanotechnology error* which we discussed earlier in this chapter.
- *Breakdown of law and order* due to income disparities and global instability.
- *New forms of infectious diseases* in the ways that AIDS, Sars and Ebola have affected many populations.
- *Risk of environmental disaster,* associated with climate change.
- *Military adventures* by either the West, Islamic countries, Russia or China, leading to a clash of cultures.
- *Rising food prices.* Rising population numbers means the planet will need to feed an extra 70 million people each year. The developing world has been getting a taste for Western-style higher protein diet, notably beef and pork. It takes 7kg of grain to create 1kg of beef, so there is growing competition for the world's arable crops. At the same time, some of the crop has been diverted into biofuels, while climate change has been reducing the yield. Agriculture will need to change. The risk of population revolt over food prices is possibly a bigger risk. From the days of the Roman emperors and the French revolution, societies have fallen over the price of bread.
- *Unstable banks.* Each recession has been caused by banks over-reaching themselves, albeit with a different scenario every time. Politicians and the public forget the crises of the past (including the devastating 1933 US banking crisis) and loosen the bonds that previous governments set, thus setting the scene for the next banking crisis.

And the opportunities include the following:

- *Technology* will allow businesses to provide better and cheaper products and services, in part through automation.
- *Rising affluence* around the world will allow more people to acquire the things they want, and let organisations offer products and services with more value added and higher margins.
- *Changes in fashion and technology* will allow responsive businesses to gain competitive advantage by changing more rapidly than their competitors.
- *New ideas* will be adopted more rapidly, thanks to improved communication and greater acceptance of new ideas.
- *The many challenges* that face society will offer businesses equal opportunities for solving them.

Appendix:
The Final Risk Checklist

This checklist summarizes the checklists located at the end of chapters. To complete the assessment, mark the level of risk you recorded for each. This will identify your greatest areas of risk.

Chapter	Low risk	Medium risk	High risk
3. Who does what? People and their roles			
4. Treating risk. Avoid, accept, control or share?			
5. Product and service problems: operations and production risk			
6. Managing health and safety			
7. Procurement problems			
8. Preventing environmental damage			
9. From drought to flooding: risk opportunities from adverse weather			
10. Protecting against fire			
11. On your guard: how to maintain security			
12. Extremists, terrorism and international risks			
13. Pre-empting fraud and theft			
14. Staying financially healthy			
15. Avoiding IT disaster			
16. Minimizing project risk			
17. Corporate governance: how to comply			
18. Risk is all about people			
19. The ethical dilemma for organizations			
20. How to succeed in marketing by being aware of risk			

Chapter		Low risk	Medium risk	High risk
21.	Diversification, acquisition, divestment and partnership			
22.	Legal risks, intellectual property and minimizing the chances of getting sued			
23.	Continuity: for when things go badly wrong			
24.	How to survive a crisis			
26.	Do you need a risk management system? ISO 31000			

Index

Printed and bound by CPI Group (UK) Ltd, Croydon, CR0 4YY

18/10/2024

01776204-0011